Topics in Fourier Analysis and Function Spaces

Topics in Fourier Analysis and Function Spaces

Hans-Jürgen Schmeisser
Hans Triebel

Friedrich-Schiller-Universität, Jena, GDR

A Wiley Interscience Publication

JOHN WILEY & SONS

Chichester · New York · Brisbane · Toronto · Singapore

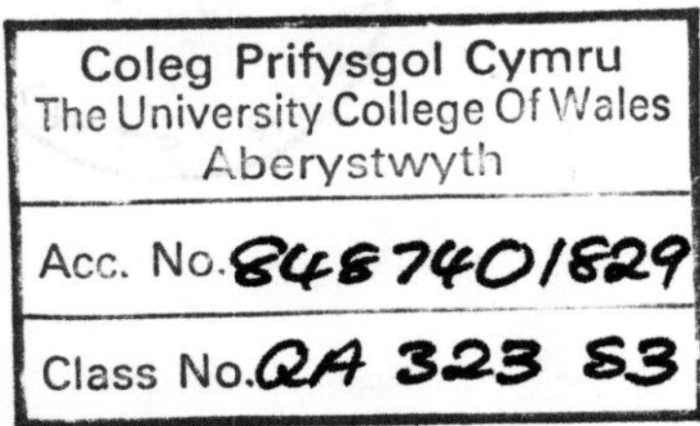

British Library Cataloguing in Publication Data:

Schmeisser, Hans-Jürgen
Topics in Fourier analysis and function spaces. –
1. Function spaces
I. Title II. Triebel, Hans
515.7 QA323
ISBN 0 471 90895 9

Library of Congress Cataloging-in-Publication Data:

Schmeisser, Hans-Jürgen.
Topics in Fourier analysis and function spaces.
Includes index.
1. Fourier analysis. 2. Function spaces.
I. Triebel, Hans. II. Title.
QA403.5.S36 1987 515′.2433 85-22706
ISBN 0 471 90895 9

Printed in the German Democratic Republic.

Preface

Several classes of function spaces attracted much attention in recent times. For example:

(i) Isotropic (non-homogeneous or homogeneous) spaces of Besov-Hardy-Sobolev type on the Euclidean n-space R_n. These spaces include the Sobolev, Besov-Lipschitz, Bessel-potential spaces as well as the Hardy spaces and BMO.

(ii) The anisotropic generalization of the spaces from (i).

(iii) The periodic spaces of Besov-Hardy-Sobolev type on the n-torus T_n, which are closely connected with multiple trigonometric series.

(iv) Spaces of the above type with dominating mixed smoothness properties, which are useful in order to study mapping properties of integral operators.

(v) Weighted spaces of Besov-Hardy-Sobolev type on R_n.

(vi) Abstract spaces of Besov type which are based on the spectral theory of self-adjoint operators in Hilbert spaces.

Of course, there are many other classes of function spaces which are of interest, e.g. spaces of the above type on (smooth and non-smooth) domains in R_n, on manifolds and Lie groups etc.

The two monographs [I] and [T][1]) dealt mainly with isotropic (non-homogeneous) spaces of Besov-Hardy-Sobolev type on R_n and on smooth domains, including applications to boundary value problems for elliptic differential equations. The present book deals with some other classes of function spaces listed above. In Chapter 2 we give a systematic study of the spaces of Besov-Hardy-Sobolev type with dominating mixed smoothness properties. The corresponding theory for periodic spaces is developed in Chapter 3, including applications to the theory of multiple trigonometric series, in particular to the problem of strong summability. Chapter 4 deals with anisotropic spaces of Besov-Sobolev type. However somewhat in contrast to the two preceding chapters we are not interested in a systematic theory. On the contrary, we consider rather special questions: traces on curves and connected problems, including boundary value problems for semi-elliptic differential equations. Chapter 5 has two rather different sections. In Section 5.1. weighted spaces of the described type on R_n in the framework of ultra-distributions are considered. This section depends on the relevant sections of Chapter 1, which deal with weighted spaces of entire analytic functions on R_n. The latter spaces are in some sense the "building blocks" in the Fourier-analytical approach to function spaces of the above type. "Dyadic decompositions" reduce the above spaces of Besov-Hardy-Sobolev type to these "building blocks". The Chapters 2 and 3 are based on these ideas. Section 5.2. is an exception. Here we study what happens if one replaces the dyadic decomposition by a congruent decomposition (modulation spaces). In Chapter 6 we examine these decomposition methods from an abstract point of view. It turns out that these decompositions coincide with some spectral decompositions of appropriate self-adjoint positive-definite operators in Hilbert (and Banach) spaces. We introduce corresponding abstract spaces and sketch several applications, e.g. to classical orthogonal expansions and to Bessel transforms.

The Chapters 1 and 6 may be considered as updated versions of the corresponding chapters in [F]. The material of the remaining Chapters 2–5 is presented here for the first time in a book.

---

[1]) [F], [I], [S], [T] refer to the corresponding books of the second named author, cf. the list of references.

The six chapters are essentially self-contained. There are few exceptions. Chapter 2 and Section 5.1. depend on the relevant parts of Chapter 1. Furthermore, the reader of Chapter 3 should be prepared to consult occasionally results from Chapter 1.

It is expected that the reader has a working knowledge of the basic facts of Functional and Fourier Analysis and of the nowadays classical parts of the theory of function spaces.

The book is organized by the decimal system. "Definition 6.2.2/2" refers to Definition 2 in Subsection 6.2.2., "Theorem 1.5.3" means the theorem in Subsection 1.5.3., etc. All unimportant positive constants will be denoted by c (with additional marks if there are several c's in the same estimate). Remarks which contain historical comments or additional references are characterized by a star*.

Jena, Spring 1985

H.-J. Schmeisser
H. Triebel

Contents

1. Spaces of Entire Analytic Functions 12

1.1. Introduction 12

1.2. Preliminaries I: Ultra-Distributions 14

1.2.1. Definitions 14
1.2.2. Basic Properties 16
1.2.3. Paley-Wiener-Schwartz Theorems for Ultra-Distributions 18
1.2.4. Distributions 19

1.3. Preliminaries II: Spaces of L_p-Type and Maximal Inequalities 20

1.3.1. Basic Spaces and Weighted Spaces 20
1.3.2. Mixed Spaces 21
1.3.3. Maximal Inequalities 23

1.4. Weighted Inequalities of Plancherel-Polya-Nikol'skij Type for S_ω-Functions 24

1.4.1. Admissible Weights and Measures 24
1.4.2. A Maximal Inequality 26
1.4.3. Inequalities for Weights of Type $R(\omega)$ 28
1.4.4. Inequalities for Atomic Measures 29
1.4.5. Inequalities for Measures of Type M^h and Weights of Type $K_h(\varrho, \mu)$ 30
1.4.6. Complements (A Discussion about Admissible Weights and Measures) 32

1.5. Weighted L_p-Spaces of Entire Analytic Functions 35

1.5.1. Definition 35
1.5.2. Inequalities 36
1.5.3. Basic Properties 37
1.5.4. Further Properties 38
1.5.5. Shannon's Sampling Theorem, Isomorphism Properties, and Schauder Bases 41

1.6. Mixed $L_{\bar{p}}$-Spaces of Entire Analytic Functions 44

1.6.1. Definition 44
1.6.2. Inequalities of Nikol'skij Type 44
1.6.3. Properties 47
1.6.4. A Maximal Inequality 47

1.7. Fourier Multipliers for Weighted L_p-Spaces of Entire Analytic Functions 49

1.7.1. Introduction and Definitions 49
1.7.2. Basic Multiplier Theorem 51
1.7.3. Convolution Algebras 51
1.7.4. Some Prerequisites: Besov-Sobolev Spaces 52
1.7.5. Special Multiplier Theorem 54
1.7.6. Discrete Multiplier Theorem 56
1.7.7. Comments 58

1.8. Fourier Multipliers for Mixed $L_{\bar{p}}$-Spaces of Entire Analytic Functions 59

1.8.1. Definition and Basic Multiplier Theorem 59
1.8.2. Convolution Algebras 60
1.8.3. Main Multiplier Theorem 60

1.9. Vector-Valued Weighted L_p-Spaces 63

1.9.1. The Spaces $L_p^\Omega(R_n, \varrho(x), l_q)$ 63
1.9.2. The Spaces $L_p(R_n, \varrho(x), l_q)$ 65
1.9.3. The Spaces $L_p(R_n, \varrho(x), l_2)$ 67
1.9.4. The Spaces $L_p(R_n, \varrho(x))$ 71

1.10. Vector-Valued Mixed $L_{\bar{p}}$-Spaces of Entire Analytic Functions 74

1.10.1. Definitions 74
1.10.2. Maximal Inequalities 75
1.10.3. Fourier Multipliers 77

2. Spaces with Dominating Mixed Smoothness Properties 80

2.1. Introduction 80

2.2. Definitions and Basic Properties 81

2.2.1. Definitions 81
2.2.2. Other Types of Spaces 83
2.2.3. Equivalent Quasi-Norms and Elementary Embeddings 87
2.2.4. Basic Properties 92
2.2.5. Fourier Multipliers 95
2.2.6. Lifting Property and Related Equivalent Quasi-Norms 98
2.2.7. Maximal Inequalities and Equivalent Quasi-Norms 100

2.3. Representations and Special Cases 102

2.3.1. Bessel-Potential and Sobolev Spaces with a Dominating Mixed Derivative 102
2.3.2. Representations of Nikol'skij-Type 105
2.3.3. Characterizations of $S_{\bar{p},\bar{q}}^{\bar{r}}F$ by Differences 108
2.3.4. Characterizations of $SB_{\bar{p},\bar{q}}^{\bar{r}}$ and $S_{\bar{p},\bar{q}}^{\bar{r}}B$ by Differences 118

2.4. Embedding Theorems 127

2.4.1. Embedding Theorems for Different Metrics 127
2.4.2. Traces 133

2.5. Notes and Comments 136

2.5.1. Mapping Properties of Integral Operators 136
2.5.2. Approximation 139
2.5.3. Vector-Valued Besov Spaces 140
2.5.4. Interpolation 140
2.5.5. Further Properties 141
2.5.6. Further Spaces 141

3. Periodic Spaces 142

3.1. Introduction 142

3.2. Preliminaries 142

3.2.1. Distributions on T_n 142
3.2.2. Fourier Coefficients and Fourier Series 143

3.2.3. Periodic Distributions on R_n 144
3.2.4. Maximal Inequalities 145

3.3. Basic Inequalities and Multipliers for Trigonometric Polynomials 146

3.3.1. Trigonometric Polynomials 146
3.3.2. Nikol'skij's Inequality 147
3.3.3. Convolution Inequalities 149
3.3.4. Fourier Multipliers 150
3.3.5. Maximal Inequalities 152

3.4. Vector-Valued L_p-Spaces of Periodic Functions 153

3.4.1. Maximal Inequalities and Fourier Multipliers 153
3.4.2. Fourier Multipliers for $L_p(T_n, l_q)$, $1 < p < \infty$, $1 < q < \infty$ 155
3.4.3. Special Multipliers for $L_p(T_n, l_q)$, $1 < p < \infty$, $1 < q < \infty$ 157
3.4.4. The Spaces $L_p(T_n)$, $1 < p < \infty$ 160

3.5. The Spaces $B^s_{p,q}(T_n)$ and $F^s_{p,q}(T_n)$ 162

3.5.1. Definitions and Basic Properties 162
3.5.2. The Spaces $F^s_{\infty,q}(T_n)$ 165
3.5.3. Lizorkin Representations 166
3.5.4. Periodic Besov and Sobolev-Lebesgue Spaces 167
3.5.5. Embedding Theorems for Different Metrics 169
3.5.6. Dual Spaces 171
3.5.7. Remarks on Further Properties 172

3.6. Complex Interpolation and Fourier Multipliers 172

3.6.1. Real and Complex Interpolation 172
3.6.2. Complex Interpolation for the Spaces $B^s_{p,q}(T_n)$ and $F^s_{p,q}(T_n)$ 174
3.6.3. Fourier Multipliers for $B^s_{p,q}(T_n)$ and $F^s_{p,q}(T_n)$ 176
3.6.4. Fourier Multipliers for $L^A_p(T_n, l_q)$ 177

3.7. Approximation and Strong Summability 178

3.7.1. Characterizations of $B^s_{p,q}(T_n)$ and $F^s_{p,q}(T_n)$ by Approximation 178
3.7.2. Approximation by Partial Sums and Smooth Means 181
3.7.3. Strong Summability of Partial Sums 185
3.7.4. Strong Summability of Smooth Means 190

3.8. Remarks 195

3.8.1. Periodic Spaces and Weighted Spaces 195
3.8.2. Periodic Spaces and Abstract Spaces 195

4. Anisotropic Spaces 196

4.1. Introduction 196

4.2. Definition and Preliminaries 196

4.2.1. Definitions 196
4.2.2. Equivalent Norms 198
4.2.3. Embeddings and Traces 199
4.2.4. Interpolation and Duality 201

4.3. Anisotropic Inequalities of Hardy Type 202

4.3.1. Inequalities in One Dimension 202
4.3.2. Inequalities in Two Dimensions 204

4.4. The Spaces $\dot{B}^{\bar{s}}_p(R_2)$ and $\dot{W}^{\bar{s}}_p(R_2)$ 208

4.4.1. The System $\Psi^{\bar{a}}(R_2)$, an Inequality 208
4.4.2. The Spaces $\dot{B}^{\bar{s}}_p(R_2)$ and $\dot{W}^{\bar{s}}_p(R_2)$ with $s > 0$ and $\bar{s}$ Non-Critical 210
4.4.3. The Spaces $\dot{B}^{\bar{s}}_p(R_2)$ and $\dot{W}^{\bar{s}}_p(R_2)$, the General Case 211
4.4.4. Properties of the Spaces $\dot{B}^{\bar{s}}_p(R_2)$ 214

4.5. Traces of $\dot{B}^{\bar{s}}_p(R_2)$ and $\dot{W}^{\bar{s}}_p(R_2)$ on Curves 217

4.5.1. Introduction 217
4.5.2. Weighted Besov Spaces on R_1^+ 218
4.5.3. Weighted Besov Spaces on Curves 220
4.5.4. Traces of the Spaces $\dot{B}^{\bar{s}}_p(R_2)$ and $\dot{W}^{\bar{s}}_p(R_2)$: The Main Theorem 222
4.5.5. Traces of the Spaces $\dot{B}^{\bar{s}}_p(R_2)$ and $\dot{W}^{\bar{s}}_p(R_2)$: Complements 225
4.5.6. Traces of the Spaces $B^{\bar{s}}_p(R_2)$ and $W^{\bar{s}}_p(R_2)$ 226

4.6. Anisotropic Spaces on the Unit Circle 228

4.6.1. Definitions 228
4.6.2. Equivalent Norms 230
4.6.3. Traces 232

4.7. Weighted Anisotropic Sobolev Spaces on Domains 233

4.7.1. Definitions and Problems 233
4.7.2. Extensions and Traces 236

4.8. Boundary Value Problems for Semi-Elliptic Differential Equations 237

4.8.1. Introduction 237
4.8.2. A Priori Estimates 238
4.8.3. L_2-C^∞-Theory 239
4.8.4. The Main Theorem 239
4.8.5. Complements 240

5. Further Types of Function Spaces 241

5.1. Weighted Spaces of Besov-Hardy-Sobolev Type 241

5.1.1. Introduction and Definitions 241
5.1.2. Basic Properties 242
5.1.3. Mapping Properties 245
5.1.4. Equivalent Quasi-Norms 249
5.1.5. Complements 252

5.2. Modulation Spaces 253

5.2.1. Introduction 253
5.2.2. Definitions 254
5.2.3. Traces 255
5.2.4. Further Properties 256

6. Abstract Spaces 258

6.1. Introduction 258

6.2. Abstract Besov Spaces 258

6.2.1. Multipliers for Banach Spaces 258
6.2.2. Definitions 260
6.2.3. Basic Theorem 262

6.3. Abstract Potential Spaces 263

6.3.1. The Operators Λ^α 263
6.3.2. Interpolation 265
6.3.3. Multipliers 268

6.4. Fourier Series 269

6.4.1. Introduction 269
6.4.2. Lacunary Series 270
6.4.3. Fourier Coefficients 272

6.5. Examples and Comments 275

6.5.1. Spaces on R_n 275
6.5.2. Periodic Spaces, Multiple Trigonometric Series 277
6.5.3. Generalization of the Abstract Approach 279
6.5.4. Classical Orthogonal Expansions 281
6.5.5. Bessel-Transform 283

References 285

Index 298

1. Spaces of Entire Analytic Functions

1.1. Introduction

Let Ff be the Fourier transform of a tempered distribution $f \in S'(R_n)$, defined on the Euclidean n-space R_n. If Ff has a compact support then $f = f(x)$ can be identified with an analytic function of exponential type on R_n. This follows from the Paley-Wiener-Schwartz theorem.

$$\|f \mid L_p(R_n)\| = \left(\int_{R_n} |f(x)|^p \, dx \right)^{1/p} \tag{1}$$

has the standard meaning, $0 < p \leqq \infty$ (usual modification if $p = \infty$), where dx stands for the Lebesgue measure. If Ω is a compact subset of R_n, then

$$L_p^\Omega(R_n) = \{f \mid f \in S'(R_n), \operatorname{supp} Ff \subset \Omega, \|f \mid L_p(R_n)\| < \infty\} \tag{2}$$

makes sense. These quasi-Banach spaces (Banach spaces if $1 \leqq p \leqq \infty$) play a crucial role in the theory of function spaces. Some assertions connected with (2) have a long history. We mention three of them:

(i) If $\{x^l\}_{l=1}^\infty$ is an appropriate set of points in R_n (e.g. lattice points, where the length of the mesh is sufficiently small) then

$$\left(\sum_{l=1}^\infty |f(x^l)|^p \right)^{1/p} \sim \|f \mid L_p(R_n)\|, \quad 0 < p \leqq \infty, \tag{3}$$

in the sense of equivalent quasi-norms (modification if $p = \infty$).

(ii) If $0 < p \leqq q \leqq \infty$ and if α is an arbitrary multi-index then

$$\|D^\alpha f \mid L_q(R_n)\| \leqq c \|f \mid L_p(R_n)\|. \tag{4}$$

(iii) If $\Omega = \{y \mid |y_j| \leqq b;\, j = 1, \ldots, n\}$ is a cube where $b > 0$ is given, and if $1 < p < \infty$, then

$$f(x) = \sum_{k \in Z_n} f\left(\frac{\pi}{b} k\right) \prod_{j=1}^n \frac{\sin (bx_j - k_j\pi)}{bx_j - k_j\pi}, \tag{5}$$

with $Z_n = \{k \mid k = (k_1, \ldots, k_n) \in R_n,\ k_j \text{ integer}\}$ (convergence in $L_p(R_n)$).

The admissible sets $\{x^l\}_{l=1}^\infty$ in (3) and the constant c in (4) depend on Ω (and p, q, α) but not on $f \in L_p^\Omega(R_n)$. Inequalities of type (3) and (4) and representation formulas of type (5) are well-known. (3) is essentially due to M. Plancherel, G. Polya [1], cf. also R. P. Boas [1, pp. 197–199]. (4) has been extensively used by S. M. Nikol'skij [1, 2] (at least if $p \geqq 1$). The one-dimensional case of (5) is essentially the well-known Shannon sampling theorem which plays a basic role in communication, control theory and data processing, cf. C. E. Shannon [1] and in particular P. L. Butzer [1] for a discussion of this aspect and further references. A fairly comprehensive treatment of the spaces $L_p^\Omega(R_n)$ has been given in the first chapter of [T], cf. also J. Peetre [4, Chapter 11]. There exist many modifications and generalizations of spaces of type (2) and of the above assertions, in particular of (4). First of all (and earlier from a historical point of view) we have to mention the periodic counterpart of the above theory dealing with trigonometric polynomials. We return to this subject in Section 3.1. Furthermore inequalities of type (4) have also been considered for the classical orthogonal expansions (Jacobi, Hermite, Laguerre polynomials). A unified treatment (from an abstract point of view) and many references may be found in R. J. Nessel, G. Wilmes [2].

The aim of this chapter is threefold. We consider two generalizations of the above sketched theory. First we replace the L_p-spaces quasi-normed via (1) by mixed $L_{\bar{p}}$-spaces quasi-normed via

$$\|f \mid L_{\bar{p}}(R_2)\| = \left(\int\limits_{R_1}\left(\int\limits_{R_1} |f(x_1, x_2)|^{p_1}\, \mathrm{d}x_1\right)^{p_2/p_1} \mathrm{d}x_2\right)^{1/p_2} \tag{6}$$

where $\bar{p} = (p_1, p_2)$ with $0 < p_1 \leqq \infty$ and $0 < p_2 \leqq \infty$ (usual modifications if $p_1 = \infty$ and/or $p_2 = \infty$). Then $L^{\Omega}_{\bar{p}}(R_2)$ is the obvious counterpart of (2). Our second task is somewhat more ambitious. If μ is an arbitrary Borel measure on R_n, then

$$\|f \mid L_{p,\mu}(R_n)\| = \left(\int\limits_{R_n} |f(x)|^p\, \mathrm{d}\mu\right)^{1/p} \tag{7}$$

is the generalization of (1), $0 < p \leqq \infty$ (usual modification if $p = \infty$). We define $L^{\Omega}_{p,\mu}(R_n)$ in the same way as in (2). We shall study spaces of this type. A typical problem is for which Borel measures ν and λ the Plancherel-Polya-Nikol'skij inequality (4) can be generalized to

$$\|D^{\alpha} f \mid L_{q,\nu}(R_n)\| \leqq c\|f \mid L_{p,\lambda}(R_n)\|, \; f \in L^{\Omega}_{p,\lambda}(R_n), \quad 0 < p \leqq q \leqq \infty. \tag{8}$$

If one looks for natural conditions for ν and λ such that (8) holds, then one is faced from the very beginning with a serious problem. Let, for example, $\mathrm{d}\nu = \mathrm{d}\lambda = \varrho(x)\, \mathrm{d}x$, where $\varrho(x)$ is a positive weight-function. If the weight $\varrho(x)$ and its inverse $\varrho^{-1}(x)$ are of at most polynomial growth, e.g. $\varrho(x) = (1 + |x|)^{\varkappa}$ where $\varkappa$ is an arbitrary real number, then it is rather easy to extend the theory for the unweighted spaces $L^{\Omega}_p(R_n)$ to these special weighted spaces $L^{\Omega}_{p,\nu}(R_n)$ within the framework of the usual tempered distributions $S'(R_n)$, inclusively inequalities of type (4). On the other hand, if $\varrho(x)$ or $\varrho^{-1}(x)$ grow very rapidly, e.g. $\varrho(x) = \mathrm{e}^{|x|}$ or $\varrho(x) = \mathrm{e}^{-|x|}$, the inequalities of type (8) and the underlying spaces $L^{\Omega}_{p,\nu}(R_n)$ do not make sense. But what about weights $\varrho(x)$ in between, e.g. $\varrho(x) = \mathrm{e}^{|x|^{\beta}}$ or $\varrho(x) = \mathrm{e}^{-|x|^{\beta}}$ with $0 < \beta < 1$? It comes out that one has for those weights a rather satisfactory theory which is even richer than the corresponding theory for the unweighted spaces. But one has to pay a price: The theory of the usual tempered distributions $S'(R_n)$ is inadequate for such general weights, one needs the more general and in many respects also more natural, but unfortunately also more complicated, theory of Beurling's ultradistributions. Then it will be not a surprise that the true limit of the admissible growth of $\varrho(x)$ is closely connected with the Denjoy-Carleman non-quasi-analyticity condition (for a formulation of the latter cf. e.g. L. Hörmander [2, I, p. 23]). In this chapter we develop the theory of the (unweighted) mixed spaces $L^{\Omega}_{\bar{p}}(R_2)$ as well as the theory of the (unmixed) weighted spaces $L^{\Omega}_{p,\mu}(R_n)$ with a preference in favour of the latter. There is no difficulty to combine these two theories, cf. B. Stöckert [1, 2]. However it is not our aim to present most general cases. On the contrary, we wish to describe the typical techniques as simple as possible. This explains also our restriction to $n = 2$ in the case of the mixed spaces, cf. (6). We develop the theory of the spaces $L^{\Omega}_{p,\mu}(R_n)$ and $L^{\Omega}_{\bar{p}}(R_2)$ for their own sake. On the other hand, these spaces are the building blocks of the corresponding Besov spaces. In other words: Chapter 1 is also the basis of the corresponding considerations in the Chapters 2 and 3 as far as the corresponding Besov spaces are concerned. Beside (mixed and weighted) Besov spaces we study in these chapters also corresponding spaces of Hardy-Sobolev type. For these spaces one needs the vector-valued counterpart of the above spaces $L^{\Omega}_{p,\mu}(R_n)$ and $L^{\Omega}_{\bar{p}}(R_2)$, i.e. the spaces $L^{\Omega}_{p,\mu}(R_n, l_q)$ and $L^{\Omega}_{\bar{p}}(R_2, l_{\bar{q}})$ and some modifications, where l_q (and $l_{\bar{q}}$) stand for the usual (mixed) sequence spaces. The third aim of this chapter is to develop the theory of these spaces in an extent which is sufficient for our later purposes.

The chapter is organized as follows. The necessary preliminaries are collected in the Sections 1.2. and 1.3. without proofs: Ultra-distributions (Section 1.2.) and spaces of L_p-type, including maximal inequalities (Section 1.3.). Afterwards the weighted spaces are treated in the Sections 1.4., 1.5., 1.7., and 1.9. (Plancherel-Polya-Nikol'skij inequalites, Fourier multipliers, l_q-valued spaces). The corresponding theory for the mixed spaces may be found in the Sections 1.6., 1.8., and 1.10. A reader who is not interested in the weighted case but in the mixed case need not read the theory of ultra-distributions. It is assumed that he/she is familiar with the usual theory of tempered distributions on R_n, but for his/her convenience we collect few facts in 1.2.4. (mostly in order to fix our notations).

1.2. Preliminaries I: Ultra-Distributions

1.2.1. Definitions

Let R_n be the real Euclidean n-space. The general point is denoted by $x = (x_1, \ldots, x_n)$. As usual $|x|$ stands for the distance of x from the origin.

Definition 1. *Let $\mathfrak{M}$ be the collection of all real-valued functions $\omega(x)$ on R_n which can be represented as $\omega(x) = \sigma(|x|)$, where $\sigma(t)$ is an increasing continuous concave function on $[0, \infty)$ with the following properties:*

(i) $\sigma(0) = 0,$

(ii)
$$\int_0^\infty \frac{\sigma(t)}{1+t^2}\,\mathrm{d}t < \infty, \tag{1}$$

(iii) *there exists a real number c and a positive number d such that*
$$\sigma(t) \geqq c + d\log(1+t), \quad t \geqq 0, \tag{2}$$
holds.

Remark 1. It is easy to see that every function $\omega(x) \in \mathfrak{M}$ is subadditive:
$$\omega(x+y) \leqq \omega(x) + \omega(y) \quad \text{with} \quad x \in R_n \quad \text{and} \quad y \in R_n. \tag{3}$$

Examples. The two most prominent examples of functions $\omega(x) \in \mathfrak{M}$ are given by
$$\omega(x) = \log(1+|x|)^d \quad \text{with} \quad d > 0 \tag{4}$$
and
$$\omega(x) = |x|^\beta \quad \text{with} \quad 0 < \beta < 1. \tag{5}$$

Next we recall the definition of the Fourier transform F and its inverse F^{-1}. Let $L_1(R_n)$ be the usual space of all complex-valued Lebesgue-integrable functions on R_n. Let $x\xi = \sum_{j=1}^n x_j\xi_j$ be the scalar product in R_n, where $x \in R_n$ and $\xi \in R_n$. If $\varphi(x) \in L_1(R_n)$ then
$$(F\varphi)(\xi) = (2\pi)^{-n/2} \int_{R_n} \mathrm{e}^{-\mathrm{i}x\xi}\,\varphi(x)\,\mathrm{d}x, \quad \xi \in R_n. \tag{6}$$

If one replaces $-\mathrm{i}$ in (6) by i, then one obtains the inverse Fourier transform $(F^{-1}\varphi)(\xi)$.

Definition 2. *Let $\omega(x) \in \mathfrak{M}$.*
(i) D_ω is the collection of all fuctions $\varphi(x) \in L_1(R_n)$ with compact support, such that

$$\|\varphi\|_{\omega,\lambda} = \int_{R_n} |F\varphi(x)| \, e^{\lambda\omega(x)} \, dx < \infty \quad \text{for all} \quad \lambda > 0 \tag{7}$$

holds.
(ii) S_ω is the collection of all functions $\varphi(x) \in L_1(R_n)$ such that both $\varphi(x)$ and $(F\varphi)(x)$ are infinitely differentiable functions on R_n with

$$p_{\alpha,\lambda}(\varphi) = \sup_{x \in R_n} e^{\lambda\omega(x)} \, |D^\alpha\varphi(x)| < \infty \tag{8}$$

and

$$\pi_{\alpha,\lambda}(\varphi) = \sup_{x \in R_n} e^{\lambda\omega(x)} \, |D^\alpha(F\varphi)(x)| < \infty \tag{9}$$

for all multi-indices α and all positive numbers λ. The locally convex topology of S_ω is defined via the semi-norms $p_{\alpha,\lambda}$ and $\pi_{\alpha,\lambda}$.

Remark 2. In (8) and (9) we used the standard abbreviation $D^\alpha = \dfrac{\partial^{|\alpha|}}{\partial x_1^{\alpha_1} \ldots \partial x_n^{\alpha_n}}$ where $\alpha = (\alpha_1, \ldots, \alpha_n)$ is a multi-index (i.e. the α_j's are non-negative integers) and $|\alpha| = \sum_{j=1}^{n} \alpha_j$ is its length. The above two definitions coincide essentially with corresponding definitions in G. Björck [1]. As fas as it is convenient for us we follow Björck's representation of this subject in the sequel and use his notations.

Remark 3. If $\omega(x)$ is given by (4) then S_ω coincides with the usual Schwartz space $S = S(R_n)$ of all complex-valued rapidly decreasing infinitely differentiable functions on R_n and $D_\omega = D(R_n)$ is the usual space of test functions. If $\omega \in \mathfrak{M}$ then we have

$$S_\omega \subset S \quad \text{(topological embedding)}. \tag{10}$$

This follows from (2). Furthermore, as it will be proved in Remark 1.2.2/3 the space S_ω is dense in S.

Definition 3. *Let $\omega(x) \in \mathfrak{M}$. The topological dual of S_ω (equipped with the strong topology) is denoted by S'_ω.*

Remark 4. We recall that the topological dual of S_ω is the collection of all linear continuous functionals on S_ω. By (10) and the density of S_ω in S we have (in the sense of the usual interpretation)

$$S' \subset S'_\omega, \tag{11}$$

where $S' = S'(R_n)$ has the usual meaning, i.e. the collection of all tempered distributions on R_n. The elements of S'_ω are denoted as ultra-distributions or ω-tempered distributions. They generalize the usual tempered distributions.

By Definition 2(ii) the Fourier transform and its inverse are one-to-one mappings from S_ω onto itself. Hence, the following definition makes sense.

Definition 4. *Let $\omega(x) \in \mathfrak{M}$ and $f \in S'_\omega$. The Fourier transform Ff and the inverse Fourier transform $F^{-1}f$ of f are given by*

$$(Ff)(\varphi) = f(F\varphi) \quad \text{and} \quad (F^{-1}f)(\varphi) = f(F^{-1}\varphi), \tag{12}$$

respectively, for all $\varphi \in S_\omega$.

Remark 5. Obviously, $Ff \in S'_\omega$ and $F^{-1}f \in S'_\omega$. Furthermore, F and F^{-1} are one-to-one mappings of S'_ω onto itself (F^{-1} is also in this sense the inverse of F and vice versa). By (11) and (12) Definition 4 generalizes the usual Fourier transform and its inverse on S'.

Remark* 6. It is not the aim of this section to give a systematic introduction to the theory of ultra-distributions. We are mainly interested in the ultra-distributions S'_ω and their properties. However we wish to mention that one can introduce ultra-distributions D'_ω as the topological dual of D_ω (after furnishing D_ω with an appropriate locally convex topology which generalizes the topology of D), cf. G. Björck [1, p. 358]. Then one obtains Beurling's ultra-distributions, cf. A. Beurling [1], which have also been extensively considered in G. Björck [1]. Other systematic treatments of the theory of ultra-distributions are given by C. Roumieu [1, 2], I. M. Gel'fand, G. E. Šilov [1], and H. Komatsu [1]. Further references and historical remarks may be found in I. M. Gel'fand, G. E. Šilov [1] and J. L. Lions, E. Magenes [1, III, pp. 22–24].

Remark 7. We discuss the assumptions from Definition 1. The restriction to radial functions $\omega(x)$ is convenient for us but not necessary. One can consider more general subadditive functions $\omega(x)$ in the sense of (3), cf. G. Björck [1]. Condition (2) is also mostly for convenience. It ensures that D_ω and S_ω are subspaces of D and S, respectively. The heart of Definition 1 is condition (1). One can prove the following assertion: *Let $\omega(x) = \sigma(|x|)$ where $\sigma(t)$ is an increasing continuous concave function on $[0, \infty)$ with $\sigma(0) = 0$. Let D_ω be defined as above. Then D_ω is non-trivial if and only if* (1) *holds.* (Non-trivial means that there exist non-identically vanishing functions $\varphi(x) \in D_\omega$.) This crucial characterization of (1) is due to A. Beurling [1], cf. also G. Björck [1, p. 359]. (We described only a partial case of Beurling's result.) This shows that Beurling's condition (1) is closely connected with the Denjoy-Carleman non-quasi-analyticity condition, cf. G. Björck [1] and L. Hörmander [2, I, p. 23]. We add a warning. The reader who followed the last discussion may have the impression that any notion of "generalized functions" beyond the ultra-distributions (with all their possible modifications) is senseless. However this is not the case. Beyond the ultra-distributions begins the realm of Sato's hyperfunctions. We refer to L. Hörmander [2, I, Chapter 9] for a comparatively elementary presentation of this subject.

1.2.2. Basic Properties

We collect some basic properties. We omit those proofs which can be found in G. Björck [1]. We recall that $x\tau$ with $x \in R_n$ and $\tau \in R_n$ stands for the scalar product in R_n. Furthermore, if $\alpha = (\alpha_1, \ldots, \alpha_n)$ is a multi-index and $x = (x_1, \ldots, x_n) \in R_n$ then we write $x^\alpha = x_1^{\alpha_1} \ldots x_n^{\alpha_n}$.

Proposition 1. *Let $\omega(x) \in \mathfrak{M}$.*
(i) *D_ω is dense in S_ω.*
(ii) *(Resolution of unity.) Let Ω be a compact subset of R_n, covered by a finite number of open balls $K_1, \ldots, K_N$. Then there exist real-valued functions $\varphi_j(x) \in D_\omega$ $(j = 1, \ldots, N)$ with the following properties:*

$$\operatorname{supp} \varphi_j \subset K_j; \qquad 0 \leqq \varphi_j(x) \leqq 1 \quad \text{if} \quad x \in R_n; \tag{1}$$

$$\sum_{j=1}^{N} \varphi_j(x) = 1 \quad \text{if} \quad x \in \Omega. \tag{2}$$

(iii) *S_ω is a topological algebra under pointwise multiplication and under convolution, where the latter is given by*

$$\varphi * \psi = F^{-1}(F\varphi F\psi) \quad \text{if} \quad \varphi \in S_\omega \quad \text{and} \quad \psi \in S_\omega.$$

(iv) *Multiplication by x^α and differentiation generate continuous operators in S_ω.*
(v) *Translations $\varphi(x) \to \varphi(x - \tau)$ with $\tau \in R_n$, dilations $\varphi(x) \to \varphi(ax)$ with $a \neq 0$ real, and multiplications $\varphi(x) \to e^{ix\tau} \varphi(x)$ with $\tau \in R_n$, generate continuous operators in S_ω.*

Remark 1. These are counterparts of well-known properties for the classical spaces D and S (which are included in our approach if one chooses $\omega(x) = \log(1 + |x|)$). In the same way as in the case of the usual distributions some of the above properties can be carried over from S_ω to S'_ω: Multiplications by $\varphi \in S_\omega$ or by x^α or by $e^{ix\tau}$, translations, dilations and differentiations generate continuous operators in S'_ω (the definitions are exactly the same as for S'). Furthermore, in the same way as for the usual tempered distributions one defines the support of $f \in S'_\omega$. Finally one has the following assertion: *D_ω is the collection of all functions $\varphi \in S_\omega$ with compact support* (this is not obvious by the above definitions).

Proposition 2. *Let $\omega(x) \in \mathfrak{M}$.*

(i) *S_ω and S'_ω are complete locally convex spaces. The Fourier transform F and its inverse F^{-1} are one-to-one mappings from S_ω onto itself and from S'_ω onto itself.*

(ii) *Multiplications by $\varphi \in S_\omega$ or by x^α or by $e^{ix\tau}$, translations, dilations and differentiations generate continuous operators in S'_ω.*

(iii) (*Approximation properties*). *Let $\varphi \in S_\omega$ with $\varphi(0) = 1$ and* supp $F\varphi \subset \{y \mid |y| \leqq 1\}$. *Then we have*

$$\varphi(\varepsilon x)\,\psi(x) \to \psi(x) \quad \text{in} \quad S_\omega \quad \text{if} \quad \varepsilon \downarrow 0 \quad \text{and} \quad \psi \in S_\omega \tag{3}$$

and

$$\varphi(\varepsilon\cdot)\,f \to f \text{ in } S'_\omega \quad \text{if} \quad \varepsilon \downarrow 0 \quad \text{and} \quad f \in S'_\omega. \tag{4}$$

If $Ff \in S'_\omega$ has a compact support then $\varphi(\varepsilon\cdot)\,f \in S_\omega$.

Proof. S_ω is a Fréchet space. Consequently, its strong dual is complete. Remark 1.2.1/5 and Remark 1 prove the remaining assertions from (i) and (ii). We prove (iii). For any multi-index α, any $\lambda > 0$ and any $\psi \in S_\omega$ we have

$$e^{\lambda\omega(x)}\, D^\alpha\psi(x) \to 0 \quad \text{if} \quad |x| \to \infty. \tag{5}$$

It follows that

$$p_{\alpha,\lambda}(\varphi(\varepsilon x)\,\psi(x) - \psi(x)) \to 0 \quad \text{if} \quad \varepsilon \downarrow 0 \tag{6}$$

holds. On the other hand we have

$$\begin{aligned}[D^\alpha F\varphi(\varepsilon\cdot)\,\psi(\cdot)]\,(x) &= (2\pi)^{-n/2} \int_{R_n} \varepsilon^{-n}(F\varphi)\left(\frac{y}{\varepsilon}\right)(D^\alpha F\psi)\,(x-y)\,dy \\ &= (2\pi)^{-n/2} \int_{R_n} (F\varphi)\,(y)\,(D^\alpha F\psi)\,(x - \varepsilon y)\,dy.\end{aligned}$$

We use (1.2.1/3) and

$$(2\pi)^{-n/2} \int_{R_n} (F\varphi)\,(y)\,dy = \varphi(0) = 1.$$

If $\lambda > 0$ then it follows that

$$\begin{aligned}&e^{\lambda\omega(x)}\,|(D^\alpha F\varphi(\varepsilon\cdot)\,\psi(\cdot) - D^\alpha F\psi)\,(x)| \\ &= e^{\lambda\omega(x)}\,(2\pi)^{-n/2} \left| \int_{R_n} (F\varphi)\,(y)\,[(D^\alpha F\psi)\,(x-\varepsilon y) - (D^\alpha F\psi)\,(x)]\,dy \right| \\ &\leqq (2\pi)^{-n/2} \int_{|y|\leqq 1} e^{\lambda\omega(y)}\,|(F\varphi)\,(y)|\,e^{\lambda\omega(x-y)}\,|(D^\alpha F\psi)\,(x-\varepsilon y) - (D^\alpha F\psi)\,(x)|\,dy\end{aligned}$$

holds. $\omega(x) = \sigma(|x|)$ is uniformly continuous on R_n, because it is generated by a continuous increasing concave function. Then the last estimate yields

$$\pi_{\alpha,\lambda}(\varphi(\varepsilon x)\,\psi(x) - \psi(x)) \leqq c\varepsilon \sum_{j=1}^{n} \pi_{\alpha,\lambda}\left(\frac{\partial}{\partial x_j}\, D^\alpha\psi(x)\right), \tag{7}$$

where c is independent of ε. Now, (6) and (7) prove (3). Let $f \in S'_\omega$. Let B be a bounded set in S_ω. Then we have

$$\sup_{\psi \in B} |(\varphi(\varepsilon\cdot) f - f)(\psi)| \leqq \sum_{|\alpha| \leqq k} \sup_{\psi \in B} (p_{\alpha,\lambda} + \pi_{\alpha,\lambda}) (\varphi(\varepsilon x) \psi(x) - \psi(x)),$$

where λ and k are some numbers. (7) and a corresponding assertion for the semi-norms $p_{\alpha,\lambda}$ show that the right-hand side of the last estimate tends to zero if $\varepsilon \downarrow 0$. This proves (4). Finally, the last assertion in (iii) follows from the Paley-Wiener-Schwartz theorem for ultra-distributions which will be formulated in the next sub-section.

Remark 2. The mapping properties of F and F^{-1} show that (3) and (4) can be reformulated as follows:

$$[F^{-1}\varphi(\varepsilon\cdot) F\psi](x) \to \psi(x) \quad \text{in} \quad S_\omega \quad \text{if} \quad \varepsilon \downarrow 0 \quad \text{and} \quad \psi \in S_\omega \tag{8}$$

and

$$F^{-1}\varphi(\varepsilon\cdot) Ff \to f \quad \text{in} \quad S'_\omega \quad \text{if} \quad \varepsilon \downarrow 0 \quad \text{and} \quad f \in S'_\omega. \tag{9}$$

Remark 3. We give the postponed proof that S_ω is dense in S, cf. Remark 1.2.1/3: First we recall that the functions $\psi \in S$ whose Fourier transform $F\psi$ has compact support are dense in S. Let ψ be such a function. Then it follows from Proposition 2(iii) that $\psi(x) \in S \subset S'_\omega$ can be approximated in S by $\varphi(\varepsilon\cdot)\psi \in S_\omega$.

Remark 4. Let μ be a Borel measure in R_n such that

$$\varphi \to \int_{R_n} \varphi(x)\, d\mu$$

generates an element of S'_ω for some $\omega(x) \in \mathfrak{M}$. We wish to identify this element with μ. In order to justify such an identification we must prove the following: If

$$\int_{R_n} \varphi(x)\, d\mu = 0 \quad \text{for all} \quad \varphi \in S_\omega \tag{10}$$

then μ is the 0-measure. We approximate a given function $\psi \in D$ by (8). The supports of $F^{-1}\varphi(\varepsilon\cdot) F\psi$ are uniformly bounded if $\varepsilon \leqq 1$. Furthermore we have $F^{-1}\varphi(\varepsilon\cdot) F\psi \in D_\omega$, cf. (1.2.1/7). Now it follows that (10) can be extended to $\varphi \in D$. But this situation is well-known and it follows that μ is the 0-measure. In particular, *any locally integrable function on R_n (with respect to the Lebesgue measure) which generates an element of S'_ω for some $\omega(x) \in \mathfrak{M}$ in the above sense can be identified with this element.* (Of course, two functions are identified if they coincide a.e.). In this sense we have

$$e^{\lambda\omega(x)} \in S'_\omega \quad \text{for any real number } \lambda. \tag{11}$$

1.2.3. Paley-Wiener-Schwartz Theorems for Ultra-Distributions

An analytic function f of n complex variables will be denoted by $f(z)$ with $z = x + iy$, $x \in R_n$ and $y \in R_n$. If $f(x)$ with $x \in R_n$ can be extended to an analytic function of n complex variables, then this extension will also be denoted by $f(z)$.

Theorem 1. *Let $\omega(x) \in \mathfrak{M}$. The following two assertions are equivalent:*

(i) $\varphi(x) \in S_\omega$ *and* $\operatorname{supp} F\varphi \subset \{y \mid |y| \leqq b\}$,

(ii) *$\varphi(z)$ is an entire analytic function of n complex variables, for any $\lambda > 0$ and any $\varepsilon > 0$ there exists a constant $c_{\lambda,\varepsilon}$ such that*

$$|\varphi(z)| \leqq c_{\lambda,\varepsilon}\, e^{(b+\varepsilon)|y| - \lambda\omega(x)} \tag{1}$$

holds for all $z = x + iy$ with $x \in R_n$ and $y \in R_n$.

Theorem 2. *Let $\omega(x) \in \mathfrak{M}$. The following two assertions are equivalent:*

(i) $f \in S'_\omega$ *and* $\operatorname{supp} Ff \subset \{y \mid |y| \leqq b\}$,

(ii) *$f(z)$ is an entire analytic function of n complex variables, there exists a real number λ such that for any $\varepsilon > 0$*

$$|f(z)| \leqq c_\varepsilon \, e^{(b+\varepsilon)|y|+\lambda\omega(x)} \tag{2}$$

holds for all $z = x + iy$ with $x \in R_n$ and $y \in R_n$, where c_ε depends on ε, but not on z.

Remark 1. It is clear how to understand these two theorems: If (i) of Theorem 2 holds, then $f \in S'_\omega$ is a regular distribution $f(x)$, which can be extended to an analytic function $f(z)$ which satisfies (2). Vice versa, if (ii) of Theorem 2 holds, then the restriction of $f(z)$ to R_n satisfies (i). Similarly for Theorem 1.

Remark 2. These theorems are due to G. Björck [1, pp. 365 and 379]. If one compares our formulation with Björck's formulation then one should take into consideration that D_ω coincides with those functions from S_ω which have compact support, and that E'_ω (in Björck's notation) coincides with those ultra-distributions from S'_ω which have a compact support.

Remark* 3. Both theorems extend the famous Paley-Wiener-Schwartz theorem for rapidly decreasing functions and tempered distributions, cf. e.g. L. Hörmander [2, I, p. 181]. Extensions of this classical theorem to several types of ultra-distributions are known: Cf. L. Hörmander [2, II, p. 137], C. Romieu [2, pp. 158 and 184] and H. Komatsu [1, I, § 9 and 1, II, Theorem 1.1]. In H. Triebel [4, II, Appendix] we proved a corresponding assertion for the Gel'fand-Šilov ultra-distributions.

1.2.4. Distributions

For the benefit of the reader who skipped the Subsections 1.2.1.–1.2.3. (because he/she is not interested in weighted spaces, but in mixed spaces, or for other reasons) we give a brief description of the (usual) distributions, mostly in order to fix our notations. Extended descriptions of this theory may be found in L. Schwartz [1], K. Yosida [1] or L. Hörmander [2, I].

The Euclidean n-space is always denoted by R_n with the general point $x=(x_1,\ldots,x_n)$ and the usual norm $|x|$ (distance of $x \in R_n$ from the origin). Let $S = S(R_n)$ be the Schwartz space of all complex-valued rapidly decreasing infinitely differentiable functions on R_n, equipped with the usual locally convex topology. Let $S' = S'(R_n)$ be the collection of all tempered distributions on R_n, i.e. the topological dual of S, equipped with the strong topology. If $x \in R_n$ and $\xi \in R_n$ then $x\xi = \sum_{j=1}^{n} x_j \xi_j$ stands for the usual scalar product in R_n. If $\varphi \in S$ then the Fourier transform $F\varphi$ of φ is defined by

$$(F\varphi)(x) = (2\pi)^{-n/2} \int_{R_n} e^{-ix\xi} \varphi(\xi) \, d\xi, \quad x \in R_n. \tag{1}$$

The inverse Fourier transform $F^{-1}\varphi$ of φ is given by (1) where one must replace $-i$ by i. One extends F and F^{-1} in the usual way from S to S'. Recall that F yields an isomorphic mapping from S onto itself and from S' onto itself. Similarly for F^{-1}. Finally we recall the famous Paley-Wiener-Schwartz theorem. An analytic function f of n complex variables will be denoted by $f(z)$ with $z = x + iy$, $x \in R_n$ and $y \in R_n$. If $f(x)$ with $x \in R_n$ can be extended to an analytic function of n complex variables, then this extension will also be denoted by $f(z)$. Then the Paley-Wiener-Schwartz theorems read as follows:

1. *The following two assertions are equivalent:*

(i) $\varphi(x) \in S$ *and* $\operatorname{supp} F\varphi \subset \{y \mid |y| \leqq b\}$,

(ii) $\varphi(z)$ *is an entire analytic function of n complex variables, for any* $\lambda > 0$ *and any* $\varepsilon > 0$ *there exists a constant* $c_{\lambda,\varepsilon}$ *such that*

$$|\varphi(z)| \leqq c_{\lambda,\varepsilon}(1 + |x|)^{-\lambda} \, \mathrm{e}^{(b+\varepsilon)|y|} \tag{2}$$

holds for all $z = x + \mathrm{i}y$ *with* $x \in R_n$ *and* $y \in R_n$.

2. *The following two assertions are equivalent:*

(i) $f \in S'$ *and* $\operatorname{supp} Ff \subset \{y \mid |y| \leqq b\}$,

(ii) $f(z)$ *is an entire analytic function of n complex variables, there exists a real number* λ *such that for any* $\varepsilon > 0$

$$|f(z)| \leqq c_\varepsilon (1 + |x|)^{\lambda} \, \mathrm{e}^{(b+\varepsilon)|y|} \tag{3}$$

holds for all $z = x + \mathrm{i}y$ *with* $x \in R_n$ *and* $y \in R_n$, *where* c_ε *depends on* ε, *but not on* z.

Proofs may be found in L. Schwartz [1, p. 272], K. Yosida [1, VI.4] and L. Hörmander [2, I, 7.3.]. Of course these formulations are the direct counterparts of the two theorems in 1.2.3.

1.3. Preliminaries II: Spaces of L_p-Type and Maximal Inequalities

1.3.1. Basic Spaces and Weighted Spaces

We fix few notations which we use in the sequel. Let μ be a Borel measure on R_n. Then

$$\|f \mid L_{p,\mu}\| = \left(\int_{R_n} |f(x)|^p \, \mathrm{d}\mu\right)^{1/p}, \quad 0 < p < \infty, \tag{1}$$

and

$$\|f \mid L_{\infty,\mu}\| = \mu - \operatorname*{ess\,sup}_{x \in R_n} |f(x)| \tag{2}$$

are standard notations. If $1 \leqq p \leqq \infty$ then (1) and (2) are norms. If $0 < p < 1$, then (1) is a quasi-norm.

We recall that $\|\cdot \mid A\|$ in a complex linear space A is called a quasi-norm if the following properties hold:

(i) $\|a \mid A\| > 0$ *if* $a \neq 0$,

(ii) $\|\lambda a \mid A\| = |\lambda| \, \|a \mid A\|$ *if* $a \in A$ *and* λ *complex*,

(iii) *there exists a constant c such that*

$$\|a_1 + a_2 \mid A\| \leqq c(\|a_1 \mid A\| + \|a_2 \mid A\|) \tag{3}$$

for all $a_1 \in A$ *and* $a_2 \in A$ *holds*. Of course, if $c = 1$ is an admissible number in (3) then $\|\cdot \mid A\|$ is a norm. A quasi-normed space is called a quasi-Banach space if it is complete (i.e. any fundamental sequence in A with respect to $\|\cdot \mid A\|$ converges).

If $0 < p \leqq \infty$ then $L_{p,\mu} = L_{p,\mu}(R_n)$ denotes the set of all Borel measurable complex-valued functions on R_n such that $\|f \mid L_{p,\mu}\| < \infty$. It is well-known that $L_{p,\mu}$ is a quasi-Banach space (Banach space if $p \geqq 1$). We write $L_{p,\mu}$ instead of $L_{p,\mu}(R_n)$ if there is no danger of confusion. If μ is the Lebesgue measure, then we put $L_p = L_{p,\mu}$ and $\|f \mid L_p\| = \|f \mid L_{p,\mu}\|$. If $\mathrm{d}\mu = \varrho^p(x)\,\mathrm{d}x$, where $\varrho(x)$ is a non-negative weight function in R_n, then we prefer later on the notation

$$L_p(R_n, \varrho(x)) = L_{p,\mu}(R_n) = L_{p,\mu}, \quad 0 < p \leqq \infty, \tag{4}$$

(it is clear what this means if $p = \infty$). Obviously, this space is quasi-normed (normed if $p \geqq 1$) via $\|\varrho f \mid L_p\|$. If μ is an atomic measure in R_n then $L_{p,\mu}$ is essentially the usual sequence space l_p. In particular if μ is concentrated in the lattice-points

$$Z_n = \{k \mid k = (k_1, \ldots, k_n) \in R_n,\ k_j \text{ integer}\} \tag{5}$$

then

$$\|a_k \mid l_p\| = \Big(\sum_{k \in Z_n} |a_k|^p\Big)^{1/p} \quad \text{if} \quad 0 < p < \infty \tag{6}$$

and

$$\|a_k \mid l_\infty\| = \sup_{k \in Z_n} |a_k|. \tag{7}$$

Of course this is a severe abuse of notations. More correct versions of $\|a_k \mid l_p\|$ would be $\|\{a_k\}_{k \in Z_n} \mid l_p\|$ or at least $\|\{a_k\} \mid l_p\|$. However we have to handle so many expressions of this type (sometimes part of more complicated constructions) that we are eager to save parentheses (for sake of clarity!). Finally we need later on the following l_q-valued version of (4): Let again $\varrho(x)$ be a non-negative weight-function in R_n and let $0 < p < \infty$. Then we put

$$\|f_k \mid L_p(R_n, \varrho(x), l_q)\| = \Big(\int_{R_n} \varrho^p(x) \Big(\sum_{k=0}^{\infty} |f_k(x)|^q\Big)^{p/q} \mathrm{d}x\Big)^{1/p} \tag{8}$$

if $0 < q < \infty$ and

$$\|f_k \mid L_p(R_n, \varrho(x), l_\infty)\| = \Big(\int_{R_n} \varrho^p(x) \sup_{k \in Z_1} |f_k(x)|^p \,\mathrm{d}x\Big)^{1/p} \tag{9}$$

(cf. the above lamentations about the abuse of notations). Of course $L_p(R_n, \varrho(x), l_q)$ is the corresponding quasi-Banach space with respect to the underlying measure $\mathrm{d}\mu = \varrho^p(x)\,\mathrm{d}x$ (Banach space if $p \geqq 1$ and $q \geqq 1$). If $\varrho(x) = 1$ then we write $\|f_k \mid L_p(R_n, l_q)\|$ or $\|f_k \mid L_p(l_q)\|$ instead of (8), (9), and $L_p(R_n, l_q)$ or $L_p(l_q)$ instead of $L_p(R_n, \varrho(x), l_q)$.

1.3.2. Mixed Spaces

Let μ_j with $j = 1, \ldots, n$ be Borel measures on R_1. Let $\mu = \prod_{j=1}^{n} \mu_j$ be the corresponding product measure on R_n. Let $\bar{p} = (p_1, \ldots, p_n)$ with $0 < p_j \leqq \infty$. If $f = f(x) = f(x_1, \ldots, x_n)$ is a complex-valued Borel-measurable function on R_n, then we put

$$\|f \mid L_{\bar{p},\mu}\| = \Big(\int_{R_1} \Big(\cdots \Big(\int_{R_1} \Big(\int_{R_1} |f(x_1, x_2, \ldots, x_n)|^{p_1} \mathrm{d}\mu_1\Big)^{p_2/p_1} \mathrm{d}\mu_2\Big)^{p_3/p_2} \cdots\Big)^{p_n/p_{n-1}} \mathrm{d}\mu_n\Big)^{1/p_n} \tag{1}$$

with the usual modifications if some of the p_j's equal infinity, cf. (1.3.1/2). Of course, x_1 in (1) corresponds to μ_1 etc. and we obtain the left-hand side of (1) via a successive application of the L_{p_1,μ_1}-quasi-norm, L_{p_2,μ_2}-quasi-norm etc. to f. Let $L_{\bar{p},\mu} = L_{\bar{p},\mu}(R_n)$ be the collection of all complex-valued Borel-measurable functions $f(x)$ in R_n with $\|f \mid L_{\bar{p},\mu}\| < \infty$. For sake of brevity we simplify (1) as follows,

$$\|f \mid L_{\bar{p},\mu}\| = \|f(x_1, \ldots, x_n) \mid L_{p_1,\mu_1} \mid \ldots \mid L_{p_n,\mu_n}\|. \tag{2}$$

If $\mu_1, \ldots, \mu_n$ are the usual Lebesgue measure on R_1 (and consequently, μ is the Lebesgue measure on R_n) then we write $\|f \mid L_{\bar{p}}\|$ and $L_{\bar{p}}$ instead of $\|f \mid L_{\bar{p},\mu}\|$ and $L_{\bar{p},\mu}$, respectively. $L_{\bar{p},\mu}$ is a quasi-Banach space $\Big($Banach space if $\min_{j=1,\ldots,n} p_j \geqq 1\Big)$. Spaces of this type (mixed L_p-spaces) have been studied extensively in A. Benedek, R. Panzone [1]. A survey may also be found in O. V. Besov, V. P. Il'jin, S. M. Nikol'skij [1, Chapter 1]. We recall that $L_{\bar{p},\mu}$ can also be defined via vector-valued quasi-Banach spaces. If $n = 2$, then we have

$$L_{\bar{p},\mu} = L_{p_2,\mu_2}(R_1, L_{p_1,\mu_1}(R_1)) \tag{3}$$

with the usual interpretations. It is clear how to extend (3) to arbitrary n's.

We are mostly interested in spaces $L_{\bar{p},\mu}$ where the μ_j's are Lebesgue measures or atomic measures. Let $n = 2$ (this is sufficient for our later purposes). Let $\bar{p} = (p_1, p_2)$ and $\bar{q} = (q_1, q_2)$ with $0 < p_j \leqq \infty$ and $0 < q_j \leqq \infty$, ($j = 1, 2$). We put

$$\|a_{j,k} \mid l_{\bar{q}}\| = \left(\sum_{k=0}^{\infty}\left(\sum_{j=0}^{\infty}|a_{j,k}|^{q_1}\right)^{q_2/q_1}\right)^{1/q_2} \tag{4}$$

and (as a special case of our above notations)

$$\|f(x_1, x_2) \mid L_{\bar{p}}\| = \left(\int_{R_1}\left(\int_{R_1}|f(x_1, x_2)|^{p_1}\,\mathrm{d}x_1\right)^{p_2/p_1}\mathrm{d}x_2\right)^{1/p_2}, \tag{5}$$

(usual modifications if one of the q_j's or p_j's equals infinity). As for the abuse of notations in (4) we refer to the preceding subsection. In the same way we must interpret the following definitions. Let $\{f_{j,k}(x_1, x_2)\}_{j,k=0}^{\infty} \subset L_{\bar{p}} = L_{\bar{p}}(R_2)$. Then we put

$$\|f_{j,k}(x_1, x_2) \mid Sl_{\bar{q}}(L_{\bar{p}})\|$$
$$= \left(\sum_{k=0}^{\infty}\left(\int_{R_1}\left(\sum_{j=0}^{\infty}\left(\int_{R_1}|f_{j,k}(x_1, x_2)|^{p_1}\,\mathrm{d}x_1\right)^{q_1/p_1}\right)^{p_2/q_1}\mathrm{d}x_2\right)^{q_2/p_2}\right)^{1/q_2} \tag{6}$$

and

$$\|f_{j,k}(x_1, x_2) \mid L_{\bar{p}}(l_{\bar{q}})\|$$
$$= \left(\int_{R_1}\left(\int_{R_1}\left(\sum_{k=0}^{\infty}\left(\sum_{j=0}^{\infty}|f_{j,k}(x_1, x_2)|^{q_1}\right)^{q_2/q_1}\right)^{p_1/q_2}\mathrm{d}x_1\right)^{p_2/p_1}\mathrm{d}x_2\right)^{1/p_2} \tag{7}$$

(usual modifications if one of the q_j's or p_j's equals infinity). Of course, there exist other combinations than that ones in (6) and (7). However we concentrate our studies on spaces which are related to these two typical and most interesting cases. Similar as in (3) we denote the quasi-Banach spaces which are generated by the quasi-norms (4)–(7) by

$$l_{\bar{q}} = l_{q_2}(l_{q_1}), \tag{4'}$$

$$L_{\bar{p}} = L_{p_2}(R_1, L_{p_1}(R_1)), \tag{5'}$$

$$Sl_{\bar{q}}(L_{\bar{p}}) = l_{q_2}(L_{p_2}(R_1, l_{q_1}(L_{p_1}(R_1)))), \tag{6'}$$

and

$$L_{\bar{p}}(l_{\bar{q}}) = L_{p_2}(R_1, L_{p_1}(R_1, l_{q_2}(l_{q_1}))), \tag{7'}$$

respectively. Of course, definitions of this type can be extended from $n = 2$ to arbitrary n's. Obviously, the spaces in (4')–(7') are Banach spaces if the involved parameters are larger than or equal to 1.

Often we handle simultaneously several integration variables and summation indices. In those cases it will be helpful to indicate the chosen integration variable and summation index: In this sense we write occasionally $L_{p_i|x_i}$ instead of L_{p_i} in order to indicate that the L_{p_i}-quasi-norm is taken with respect to x_i or $l_{q_i|j}$ instead of l_{q_i} in order to indicate that the summation in the l_{q_i}- quasi-norm is taken with respect to j. Because we treat the weighted spaces and the mixed spaces separately in this book, there is no danger of confusion. However to avoid any chance of misunderstandings we distinguished between L_{p_i,x_i} (weighted space) and $L_{p_i|x_i}$ (unweighted space).

1.3.3. Maximal Inequalities

Let $f(x)$ be a complex-valued locally Lebesgue-integrable function on R_n. Then

$$(Mf)(x) = \sup |B|^{-1} \int_B |f(y)| \, \mathrm{d}y \tag{1}$$

is the Hardy-Littlewood maximal function, where the supremum is taken over all balls B centered at x. We have

$$(Mf)(x) \geqq |f(x)| \quad \text{for almost every } x \in R_n. \tag{2}$$

Let $1 < p \leqq \infty$. Then exists a constant c such that

$$\|Mf \mid L_p\| \leqq c\|f \mid L_p\| \tag{3}$$

holds for all $f \in L_p = L_p(R_n)$. This is the famous Hardy-Littlewood maximal inequality. (3) is trivial if $p = \infty$, (3) cannot be extended to $p = 1$. If B in (1) stands for arbitrary balls with $x \in B$, or for cubes with $x \in B$, or for cubes centered at x, then the corresponding maximal functions can be mutually estimated from above and from below with the help of constants which are independent of $f(x)$. In particular, (3) holds for any of these maximal functions. Further assertions and historical remarks can be found in E. M. Stein [1, § 1], cf. also Remark 4.

There exist many modifications and extensions of the classical inequality (3). We formulate a theorem which is due to R. J. Bagby [1]. Let μ_j with $j = 1, \ldots, m$ be σ-finite Borel measures on R_1 and let $\mu = \prod_{j=1}^{m} \mu_j$ be the corresponding product measure on R_m. Let ν be the Lebesgue measure on R_1 and let $\mu\nu$ be the product measure on R_{m+1}. Let $f(t, y)$ with $t = (t_1, \ldots, t_m) \in R_m$ and $y \in R_1$ be a locally integrable (with respect to $\mu\nu$) complex-valued function on R_{m+1}. Then we put

$$(Mf(t, \cdot))(y) = \sup \frac{1}{2r} \int_{|z-y|<r} |f(t, z)| \, \mathrm{d}z, \tag{4}$$

where the supremum is taken over all $r > 0$. (As usual $\mathrm{d}z$ stands in (4) for the Lebesgue measure.) This is the one-dimensional parameter-valued version of (1).

Theorem 1. *Let $1 < p < \infty$ and $\bar{q} = (q_1, \ldots, q_m)$ with $1 < q_j \leqq \infty$ if $j = 1, \ldots, m$. Let μ be the above measure on R_m. Then there exists a constant c such that*

$$\int_{R_1} \|(Mf(t, \cdot))(y) \mid L_{\bar{q},\mu}\|^p \, \mathrm{d}y \leqq c \int_{R_1} \|f(t, y) \mid L_{\bar{q},\mu}\|^p \, \mathrm{d}y \tag{5}$$

holds for all $f(t, y) \in L_p(R_1, L_{\bar{q},\mu})$.

Remark 1. Similar as in (1.3.2/2) we write (5) as follows,

$$\|(Mf(t, \cdot))(y) \mid L_{\bar{q},\mu} \mid L_p\| \leqq c\|f(t, y) \mid L_{\bar{q},\mu} \mid L_p\|. \tag{6}$$

Remark 2. Of special interest in Theorem 1 are atomic measures μ. Then $L_{\bar{q},\mu} = l_{\bar{q}}$ is a mixed sequence space. This generalizes the one-dimensional case of the fundamental theorem by C. Fefferman, E. M. Stein [1] to the $l_{\bar{q}}$-valued version of (3). We formulate the n-dimensional version of this theorem.

The notation $L_p(l_q)$ has been introduced at the end of 1.3.1.

Theorem 2. *Let $1 < p < \infty$ and $1 < q \leqq \infty$. There exists a constant c such that*

$$\|Mf_k \mid L_p(l_q)\| \leqq c\|f_k \mid L_p(l_q)\| \tag{7}$$

holds for all sequences $\{f_k(x)\}_{k=0}^{\infty}$ of complex-valued locally Lebesgue-integrable functions on R_n.

Remark 3. As we said, Theorem 2 is due to C. Fefferman, E. M. Stein [1] and Theorem 1 is due to R. J. Bagby [1]. They are the basis for some assertions for the weighted spaces (Theorem 2) and the (unweighted) mixed spaces (Theorem 1) treated in this book.

Remark* 4. Techniques based on maximal inequalities, maximal functions and maximal operators are very fashionable in Fourier analysis in order to study spaces, Fourier multipliers, fractional and singular integrals etc. In the last 10 or 15 years appeared several hundred papers on this subject. The historical roots can be found in E. M. Stein [1]. More recent results on some aspects may be found in the book by M. de Guzmán [1], the survey by B. Muckenhoupt [2] covers inequalities of different type, cf. also E. T. Sawyer [1]. We restrict ourselves to those papers which are closely connected with inequalities of type (3), (5), and (7). First we recall Muckenhoupt's A_p-condition, which has been introduced in B. Muckenhoupt [1]: *Let $1 < p < \infty$ and let $\omega(x) > 0$ a.e. be a locally integrable function on R_n such that $\omega^{-1/p-1}(x)$ is also locally integrable on R_n. Then $\omega(x)$ satisfies the A_p-condition if and only if*

$$\sup \left(\frac{1}{|Q|} \int_Q \omega(x)\, dx \right) \left(\frac{1}{|Q|} \int_Q \omega^{-\frac{1}{p-1}}(x)\, dx \right)^{p-1} < \infty, \tag{8}$$

where the supremum is taken over all cubes Q in R_n. Let $d\mu = \omega(x)\, dx$ and let $L_{p,\mu} = L_{p,\mu}(R_n)$ be the corresponding space from 1.3.1. Muckenhoupt's surprising result reads as follows: *Let $1 < p < \infty$. There exists a positive constant c such that*

$$\|Mf \mid L_{p,\mu}\| \leqq c \|f \mid L_{p,\mu}\| \tag{9}$$

holds for all $f \in L_{p,\mu}$ if and only if $\omega(x)$ satisfies the A_p-condition. The original proof by B. Muckenhoupt [1] has been simplified by R. R. Coifman, C. Fefferman [1] and recently by M. Christ, R. Fefferman [1]. Further references can be found in E. T. Sawyer [1], J. L. Rubio de Francia [1], M. A. Leckband, C. J. Neugebauer [1] and R. A. Kerman, A. Torchinsky [1] (in particular about problems of type (9) but with different weight functions on the right-hand side and on the left-hand side, and extensions of (9) to Orlicz spaces). Another interesting problem is the combination of (7) and (9). This can be done on the basis of the papers by C. Fefferman, E. M. Stein [1] and R. R. Coifman, C. Fefferman [1]: *Let $1 < p < \infty$ and $1 < q \leqq \infty$. Then there exists a positive constant c such that*

$$\|Mf_k \mid L_{p,\mu}(l_q)\| \leqq c \|f_k \mid L_{p,\mu}(l_q)\| \tag{10}$$

holds for all sequences $\{f_k(x)\}_{k=0}^{\infty} \in L_{p,\mu}(l_q)$ if and only if $\omega(x)$ satisfies the A_p-condition. As in (9) we have $d\mu = \omega(x)\, dx$ and $L_{p,\mu} = L_{p,\mu}(R_n)$ are the spaces from 1.3.1. This result can be found in K. F. Andersen, R. T. John [1] and H. Triebel [6, p. 102], cf. also H. P. Heinig [1]. The anisotropic and weighted extension of Bagby's result may be found in V. M. Kokilašvili, J. Rakosník [1]. The interesting one-to-one connections beween (scalar) weighted maximal inequalities and (unweighted) vector maximal inequalities has been emphasized by J. L. Rubio de Francia [2], cf. also A. P. Calderón [1] for some aspects. We shall not use (9) or (10) in our considerations. However it is almost obvious that a combination of our techniques below (which are based on (3) and (7) as far as the weighted unmixed spaces are concerned) with (9) and (10) allows to extend the class of admissible weight functions. But we shall not follow this path.

1.4. Weighted Inequalities of Plancherel-Polya-Nikol'skij Type for S_ω-Functions

1.4.1. Admissible Weights and Measures

We use the previous notations: $\mathfrak{M}$ has the meaning of Definition 1.2.1/1 and Z_n is the lattice from (1.3.1/5). If $k = (k_1, \ldots, k_n) \in Z_n$ and $h > 0$ then

$$Q_k^h = \{x \mid x \in R_n,\ hk_j \leqq x_j < h(k_j + 1) \quad \text{with} \quad j = 1, \ldots, n\} \tag{1}$$

yields a decomposition of R_n. Of course, $x = (x_1, \ldots, x_n)$ in (1).

Definition. (i) *Let $\omega(x) \in \mathfrak{M}$. Then $R(\omega)$ denotes the collection of all Borel-measurable real functions $\varrho(x)$ on R_n such that there exists a positive constant c with*

$$0 < \varrho(x) \leqq c\varrho(y)\, e^{\omega(x-y)} \tag{2}$$

for every $x \in R_n$ and every $y \in R_n$.

(ii) *Let $h > 0$. Then M^h denotes the collection of all Borel measures μ in R_n with*

$$\mu(Q_k^h) = 1 \tag{3}$$

for every cube Q_k^h, where $k \in Z_n$.

(iii) *Let $\omega(x) \in \mathfrak{M}$, $\varrho(x) \in R(\omega)$ and $\mu \in M^h$. Then $K_h(\varrho, \mu)$ denotes the collection of all Borel-measurable real functions $\varkappa(x)$ on R_n with the following two properties:* (a) *there exists a positive number c with*

$$0 \leqq \varkappa(x) \leqq c\varrho(x) \tag{4}$$

for every $x \in R_n$, and (b) *there exist two positive numbers δ and c', and a Borel-measurable subset G of R_n with $\mu(G \cap Q_k^h) \geqq \delta$ for every cube Q_k^h and*

$$\varkappa(x) \geqq c'\varrho(x) \quad \text{for every} \quad x \in G. \tag{5}$$

Remark 1. We discuss part (i) of the definition. Let $\varrho(x) \in R(\omega)$. From (2) with $y = 0$ and Remark 1.2.2/4, in particular (1.2.2/11) follows

$$\varrho(x) \in S'_\omega. \tag{6}$$

Furthermore, (2) yields

$$c\, e^{-\omega(x)} \leqq \varrho(x) \leqq c'\, e^{\omega(x)}, \quad x \in R_n, \tag{7}$$

where c and c' are appropriate positive numbers. This shows that the growth of $\varrho(x)$ is not only restricted from above but also from below. If $\omega_1(x) \in \mathfrak{M}$, $\omega_2(x) \in \mathfrak{M}$ and $\lambda > 0$, then $\lambda\omega_1(x) \in \mathfrak{M}$ and $\omega_1(x) + \omega_2(x) \in \mathfrak{M}$, cf. Definition 1.2.1/1. Furthermore if $\varrho_1(x) \in R(\omega_1)$ and $\varrho_2(x) \in R(\omega_2)$ then

$$\left.\begin{aligned} &\lambda\varrho_1(x) \in R(\omega_1), && \varrho_1^\lambda(x) \in R(\lambda\omega_1),\\ &\varrho_1(x) + \varrho_2(x) \in R(\omega_1 + \omega_2), && \varrho_1^{-1}(x) \in R(\omega_1),\\ &\varrho_1(x)\,\varrho_2(x) \in R(\omega_1 + \omega_2), && \frac{\varrho_1(x)}{\varrho_2(x)} \in R(\omega_1 + \omega_2). \end{aligned}\right\} \tag{8}$$

Remark 2. We describe few examples. If $\omega(x) \in \mathfrak{M}$, then

$$\varrho(x) = e^{\omega(x)} \in R(\omega). \tag{9}$$

This follows from the subadditivity (1.2.1/3) of $\omega(x)$. In particular,

$$\varrho(x) = (1 + |x|)^d \in R(\log(1 + |x|)^d), \quad d > 0, \tag{10}$$

and

$$\varrho(x) = e^{|x|^\beta} \in R(|x|^\beta), \quad 0 < \beta < 1, \tag{11}$$

cf. (1.2.1/4) and (1.2.1/5). These are the most interesting examples. The first one, i.e. $\varrho(x) = (1 + |x|)^d$ with $d > 0$ can be treated in the framework of the usual distributions, but not the second one, i.e. $\varrho(x) = e^{|x|^\beta}$ with $0 < \beta < 1$. It leads to the so-called Gevrey ultra-distributions.

Remark 3. We discuss part (ii) of the definition. The measures from M^h have essentially a lattice structure. Of special interest are the modified Lebesgue measure $d\mu = h^{-n}\, dx$ and atomic measures $\mu(Q_k^h) = \mu(\{x^k\}) = 1$, where x^k is a point in Q_k^h.

Remark 4. We discuss part (iii) of the definition. Our goal is to prove inequalities of type (1.1/8) where the measures ν and λ have the structure from part (iii) of the above definition, i.e. $\nu = \varkappa(x)\,\mu$ with $\mu \in M^h$ and $\varkappa(x) \in K_h(\varrho, \mu)$, and similarly for λ. In rough terms: The general admissible measures ν and λ are divided in a measure μ which has essentially lattice structure and which controls the local behaviour, and a weight function $\varkappa(x)$ which controls the global growth (but which may be also zero on a set of infinite Lebesgue measure).

Remark 5. Finally we mention few interesting examples. Let $h > 0$ and let μ with $\mathrm{d}\mu = h^{-n}\,\mathrm{d}x$ be the modified Lebesgue measure. Then we have

$$\prod_{j=1}^{n} |x_j|^{d_j} \in K_h\left(\prod_{j=1}^{n}(1+|x_j|)^{d_j}, \mu\right) \quad \text{with} \quad d_j \geqq 0 \tag{12}$$

and

$$|x|^d \in K_h((1+|x|)^d, \mu) \quad \text{with} \quad d \geqq 0. \tag{13}$$

1.4.2. A Maximal Inequality

We use the notations S_ω, $\mathfrak{M}$ and $R(\omega)$ from 1.2.1. and 1.4.1., respectively. If Ω is a compact subset of R_n and $\omega(x) \in \mathfrak{M}$, then we put

$$S_\omega^\Omega = \{\varphi \mid \varphi \in S_\omega, \operatorname{supp} F\varphi \subset \Omega\}. \tag{1}$$

Furthermore, $\nabla\varphi$ stands for the gradient of the function φ, and Mf is the maximal function from (1.3.3/1).

Theorem. *Let $\omega(x) \in \mathfrak{M}$, $\varrho(x) \in R(\omega)$, and $0 < r < \infty$. Let Ω be a compact subset of R_n. Then there exist two positive numbers c_1 and c_2 such that*

$$\sup_{z \in R_n} \varrho(x-z)\frac{|\nabla\varphi(x-z)|}{1+|z|^{n/r}} \leqq c_1 \sup_{z \in R_n} \varrho(x-z)\frac{|\varphi(x-z)|}{1+|z|^{n/r}} \leqq c_2[(M|\varrho\varphi|^r)(x)]^{1/r} \tag{2}$$

holds for all $\varphi \in S_\omega^\Omega$ and all $x \in R_n$.

Proof. Step 1. We choose a function $\psi \in S_\omega$ with $(F\psi)(x) = 1$ if $x \in \Omega$, cf. Proposition 1.2.2/1(ii). If $\varphi \in S_\omega^\Omega$ then we have $F\varphi = F\varphi \cdot F\psi$ and $\varphi = F^{-1}(F\varphi \cdot F\psi)$. By well-known properties of the Fourier transform it follows that

$$\varphi(x) = c \int_{R_n} \varphi(x)\,\psi(x-y)\,\mathrm{d}y, \quad x \in R_n, \tag{3}$$

where c is an appropriate constant. We replace x in (3) by $x - z$, apply $\partial/\partial x_1$, and use (1.4.1/2). Then we have

$$\varrho(x-z)\left|\frac{\partial\varphi}{\partial x_1}(x-z)\right| \leqq c \int_{R_n} \varrho(y)\,|\varphi(y)|\,\mathrm{e}^{\omega(x-z-y)}\left|\frac{\partial\psi}{\partial x_1}(x-z-y)\right| \mathrm{d}y \leqq c' \int_{R_n} \varrho(y)\,|\varphi(y)|\,\mathrm{e}^{-\lambda\omega(x-z-y)}\,\mathrm{d}y, \tag{4}$$

where λ is an arbitrary positive number. The latter estimate comes from (1.2.1/8), the constant c' depends on λ. We divide both sides of (4) by $1 + |z|^{n/r}$ and use the inequality

$$\frac{1+|x-y|^{n/r}}{1+|z|^{n/r}} \leqq c(1+|x-y-z|^{n/r}), \quad x \in R_n, y \in R_n, z \in R_n. \tag{5}$$

Then it follows that

$$\sup_{z\in R_n} \varrho(x-z)\frac{|\nabla\varphi(x-z)|}{1+|z|^{n/r}} \leqq c \sup_{z\in R_n} \int_{R_n} \varrho(y)\frac{|\varphi(y)|}{1+|x-y|^{n/r}} \mathrm{e}^{-\lambda'\omega(x-z-y)}\,\mathrm{d}y$$
$$\leqq c' \sup_{w\in R_n} \varrho(x-w)\frac{|\varphi(x-w)|}{1+|w|^{n/r}}, \tag{6}$$

where λ' is an appropriate positive number with $\lambda' < \lambda$. We used (1.2.1/2). We remark that the right-hand side of (6) is finite, cf. (1.4.1/7) and (1.2.1/8).

Step 2. We need an auxiliary assertion. Let $g(z)$ be a complex-valued continuously differentiable function in the closed unit ball $B = \{y \mid |y| \leqq 1\}$ in R_n. By the mean value theorem we have

$$|g(z)| \leqq c \min_{w\in B} |g(w)| + c \sup_{w\in B} |\nabla g(w)|$$
$$\leqq c\left(\int_B |g(w)|^r\,\mathrm{d}w\right)^{1/r} + c\sup_{w\in B}|\nabla g(w)|, \quad z\in B. \tag{7}$$

Let B_δ be an arbitrary closed ball of radius $\delta > 0$ in R_n. Then there exists a constant c such that for all $\delta > 0$ and all functions $g(z)$, which are continuously differentiable in B_δ,

$$|g(z)| \leqq c\delta \sup_{w\in B_\delta}|\nabla g(w)| + c\delta^{-n/r}\left(\int_{B_\delta}|g(w)|^r\,\mathrm{d}w\right)^{1/r}, \quad z\in B_\delta, \tag{8}$$

holds. We may assume that $B_\delta = \{y \mid |y| \leqq \delta\}$. If one replaces $g(z)$ in (7) by $g(\delta z)$, then (8) follows from (7).

Step 3. We apply (8) with $\delta \leqq 1$ to $\varphi(x-w)$. We use

$$c_1\varrho(u) \leqq \varrho(v) \leqq c_2\varrho(u) \quad \text{if} \quad |u-v| \leqq 1, \tag{9}$$

where the positive constants c_1 and c_2 are independent of $u \in R_n$ and $v \in R_n$. Then we have

$$\varrho(x-w)\,|\varphi(x-w)| \leqq c\delta \sup_{|y|\leqq\delta} \varrho(x-w-y)\,|\nabla\varphi(x-w-y)|$$
$$+ c\delta^{-n/r}\left(\int_{|y|\leqq\delta} \varrho^r(x-w-y)\,|\varphi(x-w-y)|^r\,\mathrm{d}y\right)^{1/r}. \tag{10}$$

The last integral can be estimated from above by

$$\left(\int_{|u|\leqq|w|+1} \varrho^r(x-u)\,|\varphi(x-u)|^r\,\mathrm{d}u\right)^{1/r} \leqq c(1+|w|^{n/r})\,[(M|\varrho\varphi|^r)(x)]^{1/r}.$$

We put this estimate in (10). Then we divide both sides by $(1+|w|^{n/r})$ and take afterwards the supremum with respect to $w \in R_n$. The result reads as follows,

$$\sup_{w\in R_n} \varrho(x-w)\frac{|\varphi(x-w)|}{1+|w|^{n/r}} \leqq c\delta \sup_{w\in R_n} \varrho(x-w)\frac{|\nabla\varphi(x-w)|}{1+|w|^{n/r}}$$
$$+ c\delta^{-n/r}[(M|\varrho\varphi|^r)(x)]^{1/r}, \tag{11}$$

where c is independent of $\delta \leqq 1$. By (6), the first term on the right hand side of (11) can be replaced by the left-hand side of (11) multiplied with $c\delta$, where c is independent of δ. If we choose δ sufficiently small then we obtain the last part of (2) (we recall that all terms in (11) are finite). The first part of (2) follows from (6).

Remark* 1. The proof coincides essentially with the corresponding proof in [F, 1.3.2]. However we wish to emphasize that some key ideas in the above proof are due to J. Peetre [3]. Cf. also C. Fefferman, E. M. Stein [2], where the technique of maximal functions of the above type has been developed.

Remark 2. *Let* $\omega(x) \in \mathfrak{M}$, $0 < r < \infty$ and Ω be given. Then the constants c_1 and c_2 in (2) depend only on the constant c in (1.4.1/2) (and not on the special choice of $\varrho(x)$).

1.4.3. Inequalities for Weights of Type $R(\omega)$

One of our main goals in this chapter is the proof of inequalities of type (1.1/8). A first step will be done in this subsection where we specialize $\mathrm{d}\nu = \mathrm{d}\lambda = \varrho(x)\,\mathrm{d}x$ with $\varrho(x) \in R(\omega)$ and $\omega \in \mathfrak{M}$, cf. 1.4.1. and 1.2.1., respectively. Of course $\mathrm{d}x$ stands for the Lebesgue measure. S_ω^Ω has been defined in (1.4.2/1), Ω again denotes a compact subset of R_n. Finally $\|\cdot \mid L_p\|$ has the same meaning as in 1.3.1 and D^α has been explained in Remark 1.2.1/2.

Proposition. *Let* $\omega(x) \in \mathfrak{M}$, $\varrho(x) \in R(\omega)$ *and* $0 < p \leqq q \leqq \infty$. *Let* Ω *be a compact subset of* R_n. *If* α *is a multi-index, then there exists a positive constant c such that*

$$\|\varrho D^\alpha \varphi \mid L_q\| \leqq c \|\varrho\varphi \mid L_p\| \tag{1}$$

holds for all $\varphi \in S_\omega^\Omega$.

Proof. Step 1. First we prove that

$$\|\varrho\varphi \mid L_q\| \leqq c\|\varrho\varphi \mid L_p\|; \quad 0 < p \leqq q \leqq \infty. \tag{2}$$

We use (1.4.2/3) and obtain in the same way as in (1.4.2/4) that

$$\varrho(x)\,|\varphi(x)| \leqq c \int_{R_n} \varrho(y)\,|\varphi(y)|\, \mathrm{e}^{\omega(x-y)}\, |\psi(x-y)|\, \mathrm{d}y \tag{3}$$

holds for all $x \in R_n$. If $1 \leqq p < \infty$ and $q = \infty$ then (2) follows from Hölder's inequality (and (1.2.1/8)). If $0 < p < 1$ then we have

$$\varrho(x)\,|\varphi(x)| \leqq c\Big(\sup_{y \in R_n} \varrho(y)\,|\varphi(y)|\Big)^{1-p} \int_{R_n} \varrho^p(y)\,|\varphi(y)|^p\, \mathrm{d}y. \tag{4}$$

On the left-hand side of (4) we take the supremum with respect to $x \in R_n$. Because $\|\varrho\varphi \mid L_\infty\| < \infty$ we obtain (2) with $q = \infty$ and $0 < p < 1$. Finally if $0 < p < q < \infty$ then (2) follows from the just proven facts and

$$\|\varrho\varphi \mid L_q\| \leqq \|\varrho\varphi \mid L_\infty\|^{1-\frac{p}{q}}\, \|\varrho\varphi \mid L_p\|^{\frac{p}{q}}.$$

Step 2. If $\varphi \in S_\omega^\Omega$ then we have $D^\alpha\varphi \in S_\omega^\Omega$. Then iteration of (1.4.2/2) yields

$$\varrho(x)\,|D^\alpha\varphi(x)| \leqq \sup_{z \in R_n} \varrho(x-z)\,\frac{|D^\alpha\varphi(x-z)|}{1+|z|^{n/r}} \leqq c[(M|\varrho\varphi|^r)(x)]^{1/r}$$

where $0 < r < \infty$. If $0 < p \leqq \infty$, then we choose $0 < r < p$ (with $r = 1$ if $p = \infty$). It follows that

$$\|\varrho D^\alpha\varphi \mid L_p\| \leqq c\|[M|\varrho\varphi|^r]^{1/r} \mid L_p\| = c\|M|\varrho\varphi|^r \mid L_{p/r}\|^{1/r}.$$

By the Hardy-Littlewood maximal inequality (1.3.3/3) we have

$$\|\varrho D^\alpha\varphi \mid L_p\| \leqq c\||\varrho\varphi|^r \mid L_{p/r}\|^{1/r} = c\|\varrho\varphi \mid L_p\|. \tag{5}$$

This proves (1) if $p = q$. The general case follows from (5) and (2).

Remark*. Section 1.4. of this book is a modified version of [F, 1.3] which, in turn, is based on H. Triebel [4, II] (where in the latter paper we used the Gel'fand-Šilov ultra-distributions instead of the more suitable Beurling-Björck ultra-distributions). If $\varrho(x) = 1$ and $1 \leqq p \leqq q \leqq \infty$ then (1) is the famous Nikol'skij inequality, which has been extensively used in the theory of function spaces by

S. M. Nikol'skij and the Russian school, cf. S. M. Nikol'skij [2] and O. V. Besov, V. P. Il'jin, S. M. Nikol'skij [1]. The extension of (1) with $\varrho(x) = 1$ to $0 < p \leqq q \leqq \infty$ was the basis for the study of spaces of Besov type in [T], cf. also [S] and J. Peetre [4, in particular Chapter 11]. (In some sense [T, 1.3] is the unweighted counterpart of Section 1.4. of this book, cf. also [T, Chapter 6] for a brief description of the subject treated in this section.) Beside the extension to $0 < p \leqq q \leqq \infty$ of the classical (unweighted) Nikol'skij inequality there exist many other modifications and generalizations: Nikol'skij inequalities for trigonometric polynomials and for orthogonal polynomials, extensions to general measures and weights (some of them will be treated in the sequel), abstract generalizations etc. References and detailed descriptions may be found in S. M. Nikol'skij [2] and R. J. Nessel, G. Wilmes [1, 2]. Weighted inequalities of type (1) (but with other weights) can also be obtained if one uses (1.3.3/9) in the above proof instead of (1.3.3/3). But this is not in our program. We refer in this context to B. Jawerth [2]. Unweighted inequalities of type (1) for mixed $L_{\bar{p}}$-spaces will be discussed in 1.6.2., their weighted versions (in the sense of this section) may be found in B. Stöckert [1, 2].

1.4.4. Inequalities for Atomic Measures

In this subsection $\|a_k \mid l_p\|$ is given by (1.3.1/6) or (1.3.1/7), $0 < p \leqq \infty$. Furthermore, $\mathfrak{M}$, $R(\omega)$, S_ω^Ω and $\|\cdot \mid L_p\|$ have the same meaning as in the preceding subsection. The cubes Q_k^h with $k \in Z_n$ have been defined in 1.4.1.

Proposition. *Let $\omega(x) \in \mathfrak{M}$, $\varrho(x) \in R(\omega)$ and $0 < p \leqq \infty$. Let Ω be a compact subset of R_n. There exist three positive numbers h_0, c_1 and c_2 such that*

$$c_1 \|\varrho(x^k)\, \varphi(x^k) \mid l_p\| \leqq h^{-n/p} \|\varrho\varphi \mid L_p\| \leqq c_2 \|\varrho(x^k)\, \varphi(x^k) \mid l_p\| \tag{1}$$

holds for all h with $0 < h \leqq h_0$, all sets $\{x^k\}_{k \in Z_n}$ with $x^k \in Q_k^h$ and all $\varphi \in S_\omega^\Omega$.

Proof. Step 1. We prove the right-hand side of (1). Without restriction of generality we may assume that $h \leqq 1$. Let $\varphi \in S_\omega^\Omega$ and $x \in Q_k^h$. Then we have

$$|\varphi(x)| \leqq |\varphi(x^k)| + ch \sup_{z \in Q_k^h} |\nabla\varphi(z)| \leqq |\varphi(x^k)| + ch \sup_{|x-z| \leqq c'} |\nabla\varphi(z)|, \tag{2}$$

where c and c' are independent of h and φ. By (1.4.2/9) we have

$$\varrho(x)\, |\varphi(x)| \leqq c\varrho(x^k)\, |\varphi(x^k)| + ch \sup_{|x-z| \leqq c'} \varrho(z)\, |\nabla\varphi(z)|. \tag{3}$$

If $p < \infty$ then (3) yields

$$\begin{aligned} &\int_{R_n} \varrho^p(x)\, |\varphi(x)|^p \, dx \\ &\leqq ch^n \sum_{k \in Z_n} \varrho^p(x^k)\, |\varphi(x^k)|^p + ch^p \int_{R_n} \Big(\sup_{|x-z| \leqq c'} \varrho(z)\, |\nabla\varphi(z)| \Big)^p (x)\, dx, \end{aligned} \tag{4}$$

where c and c' are independent of h and φ. Let $0 < r < p$. Then (1.4.2/2) and the Hardy-Littlewood maximal inequality (1.3.3/3) show that the second term on the right-hand side of (4) can be estimated from above by

$$ch^p \int_{R_n} [(M|\varrho\varphi|^r)(x)]^{\frac{p}{r}} \, dx \leqq c' h^p \int_{R_n} |\varrho\varphi|^{r\frac{p}{r}} (x)\, dx = c' h^p \|\varrho\varphi \mid L_p\|^p, \tag{5}$$

where c and c' are independent of h and φ. We put this estimate in (4). Then we obtain the right-hand side of (1), provided that $0 < h \leqq h_0$ and h_0 is sufficiently small. If $p = \infty$ then the right-hand side of (1) follows from (3) and (1.4.3/1) with $p = q = \infty$.

Step 2. We prove the left-hand side of (1). The case $p = \infty$ is trivial. Let $p < \infty$. Instead of (3) we begin with

$$\varrho(x^k)\,|\varphi(x^k)| \leqq \varrho(x)\,|\varphi(x)| + ch \sup_{|x-z|\leqq c'} \varrho(z)\,|\nabla\varphi(z)|\,, \quad x \in Q_k^h\,. \tag{6}$$

We integrate over R_n (with respect to x) and use (5). Then we obatin the left-hand side of (1).

Remark. What about the number h_0 in the proposition? It depends on $\omega(x)$, $\varrho(x)$, p, and Ω, but not on the chosen systems $\{x^k\}_{k\in Z_n}$ with $x^k \in Q_k^h$. Estimates for the admissible values of h_0 would be interesting. As an easy consequence of Proposition 1.4.6/1 one has the following assertion: If Ω contains a cube with the side-length $2b$, then we have $h_0 \leqq \pi/b$. We return to problems of this type in 1.4.6.

1.4.5. Inequalities for Measures of Type M^h and Weights of Type $K_h(\varrho, \mu)$

Our aim is twofold. First we wish to extend Proposition 1.4.4 to arbitrary measures $\mu \in M^h$ and weights $\varkappa(x) \in K_h(\varrho, \mu)$ in the sense of Definition 1.4.1. Our second aim is to prove Plancherel-Polya-Nikol'skij inequalities (1.1/8), where ν and λ of the type $\varkappa(x)\,\mathrm{d}\mu$ with $\mu \in M^h$ and $\varkappa(x) \in K_h(\varrho, \mu)$. We recall that we use the notations from Subsection 1.4.1., furthermore, S_ω^Ω has been defined in (1.4.2/1) and $L_{p,\mu}$ has the same meaning as in 1.3.1.

Theorem 1. *Let $\omega(x) \in \mathfrak{M}$, $\varrho(x) \in R(\omega)$ and $0 < p \leqq \infty$. Let Ω be a compact subset of R_n. Let $\mu \in M^h$ and $\varkappa(x) \in K_h(\varrho, \mu)$ with $0 < h < h_0$, where h_0 has the same meaning as in Proposition* 1.4.4. *There exist two positive numbers c_1 and c_2 such that*

$$c_1\|\varrho\mu \mid L_p\| \leqq \|\varkappa\varphi \mid L_{p,\mu}\| \leqq c_2\|\varrho\varphi \mid L_p\| \tag{1}$$

holds for all $\varphi \in S_\omega^\Omega$.

Proof. Step 1. Let h with $0 < h < h_0$ be fixed. Let $p < \infty$. Because (1.4.4/1) holds for every sequence $\{x^k\}_{k\in Z_n}$ with $x^k \in Q_k^h$ we have

$$\begin{aligned}\int_{R_n} \varrho^p(x)\,|\varphi(x)|^p\,\mathrm{d}\mu &\leqq \sum_{k\in Z_n} \sup_{z\in Q_k^h} \varrho^p(z)\,|\varphi(z)|^p \\ &\leqq c \sum_{k\in Z_n} \inf_{z\in Q_k^h} \varrho^p(z)\,|\varphi(z)|^p \leqq ch^{-n} \int_{R_n} \varrho^p(x)\,|\varphi(x)|^p\,\mathrm{d}x\,.\end{aligned} \tag{2}$$

One has a similar estimate if $p = \infty$. The roles of $\mathrm{d}\mu$ and $\mathrm{d}x$ in (2) can be changed. This proves (1) with $\varkappa = \varrho$.

Step 2. Next we prove that

$$\|\varrho\varphi \mid L_{p,\mu}\| \leqq c\|\varkappa\varphi \mid L_{p,\mu}\| \tag{3}$$

holds. Let G be the set from Definition 1.4.1(iii). Then we introduce the measure ν, defined by

$$\nu(B) = \sum_{k\in Z_n} \frac{\mu(B \cap G \cap Q_k^h)}{\mu(G \cap Q_k^h)}\,, \tag{4}$$

where B is an arbitrary Borel set in R_n. We have $\nu \in M^h$. Let δ be the number from Definition 1.4.1(iii) and let χ_G be the characteristic function of G. By Step 1 we know that $\|\varrho\varphi \mid L_{p,\mu}\|$ and $\|\varrho\varphi \mid L_{p,\nu}\|$ are equivalent. Consequently we have

$$\|\varrho\varphi \mid L_{p,\mu}\| \leqq c\|\varrho\varphi \mid L_{p,\nu}\| \leqq c\delta^{-1/p}\|\varrho\chi_G\varphi \mid L_{p,\mu}\|\,. \tag{5}$$

By (1.4.1/5) the right-hand side of (5) can be estimated from above by $c'\|\varkappa\varphi \mid L_{p,\mu}\|$. This proves (3). By (1.4.1/4) the reverse inequalitiy is trivial. Hence, the both sides of (3) are equivalent to each other. This assertion and (1) with $\varkappa = \varrho$ prove (1) in the general case.

Theorem 2. *Let* $\omega(x) \in \mathfrak{M}$, $\varrho(x) \in R(\omega)$ and $0 < p \leqq q \leqq \infty$. *Let* Ω *be a compact subset of* R_n. *Let* $\mu_1 \in M^h$ *and* $\varkappa_1(x) \in K_h(\varrho, \mu_1)$, *let* $\mu_2 \in M^h$ *and* $\varkappa_2(x) \in K_h(\varrho, \mu_2)$. *There exists a positive number* h' *with the following property: If* $0 < h < h'$ *and if* α *is a multi-index then exists a constant* c *such that*

$$\|\varkappa_1 D^\alpha \varphi \mid L_{q,\mu_1}\| \leqq c\|\varkappa_2 \varphi \mid L_{p,\mu_2}\| \tag{6}$$

holds for all $\varphi \in S_\omega^\Omega$.

Proof. We recall that $D^\alpha\varphi \in S_\omega^\Omega$ if $\varphi \in S_\omega^\Omega$. Then the proof is an immediate consequence of Theorem 1 and Proposition 1.4.3.

Remark. This is the Plancherel-Polya-Nikol'skij inequality (1.1/8) for rather general measures $d\nu = \varkappa_1(x)\, d\mu_1$ and $d\lambda = \varkappa_2(x)\, d\mu_2$. The extension from S_ω^Ω to $L_{p,\lambda}^\Omega(R_n)$ is a technical matter and will be done later on, cf. 1.5.2.

Examples. *Let* $0 < p \leqq q \leqq \infty$ *and* $d > -\frac{n}{q}$. *Let* α *be a multi-index. There exists a positive constant* c *such that*

$$\||x|^d D^\alpha \varphi \mid L_q\| \leqq c b^{|\alpha| + n\left(\frac{1}{p} - \frac{1}{q}\right)} \||x|^d \varphi \mid L_p\| \tag{7}$$

holds for all $b > 0$ *and all*

$$\varphi \in S, \quad \operatorname{supp} F\varphi \subset \{y \mid |y| \leqq b\}. \tag{8}$$

We prove this assertion. Let $b = 1$ and let Ω be the unit cube. If $d \geqq 0$, then (7) follows from (6) and (1.4.1/13). If $0 > d > -\frac{n}{q} \geqq -\frac{n}{p}$ then we choose measures μ_1 and μ_2 given by

$$d\mu_1 = \max(1, |x|^{dq})\, dx \quad \text{and} \quad d\mu_2 = \max(1, |x|^{dp})\, dx. \tag{9}$$

If $h > 0$ is given, then small modifications (dilations on the cubes Q_k^h) yield measures which belong to M^h. Furthermore, if $\varkappa_1(x) = \varkappa_2(x) = \varrho(x) = (1 + |x|)^d$ then (7) with $b = 1$ follows from (6). We recall the well-known formula

$$[F\varphi(a\cdot)](\xi) = a^{-n}(F\varphi)(a^{-1}\xi), \quad a > 0. \tag{10}$$

In particular, if φ satisfies (8) then we can apply (7) with $b = 1$ to $\varphi(b^{-1}\cdot)$. However this coincides with (7) for general b. The proof is complete. Inequalities of type (7) are well-known (at least if $1 \leqq p \leqq q \leqq \infty$ and $d = 0$). They are the basis in order to study (weighted and unweighted) Besov spaces. The explicit dependence of the involved constant on b is of interest in this context. Unfortunately if the weights are more complicated then one has no such simple arguments as above. On the other hand, the above arguments can be applied to the weight $\prod_{j=1}^n |x_j|^{d_j}$ with $d_j \geqq 0$, cf. (1.4.1/12): *If* α *is a multi-index and if* $0 < p \leqq q \leqq \infty$ *then exists a positive constant* c *such that*

$$\left\|\prod_{j=1}^n |x_j|^{d_j} D^\alpha \varphi \mid L_q\right\| \leqq c b^{|\alpha| + n\left(\frac{1}{p} - \frac{1}{q}\right)} \left\|\prod_{j=1}^n |x_j|^{d_j} \varphi \mid L_p\right\| \tag{11}$$

holds for all $b > 0$ *and all* φ *from* (8). One can extend (11) to $d_j > -\frac{1}{q} \geqq -\frac{1}{p}$. This follows in the same way as above, cf. (9). Inequalities of type (11) are of interest also for mixed spaces, where L_p and L_q are replaced by $L_{\bar{p}}$ and $L_{\bar{q}}$, respectively.

1.4.6. Complements (A Discussion about Admissible Weights and Measures)

In 1.4.1. we defined measures $\mu \in M^h$ and weights $\varkappa(x) \in K_h(\varrho, \mu)$ where $\varrho(x) \in R(\omega)$ and $\omega(x) \in \mathfrak{M}$ are given functions. In the two theorems of the preceding subsection we proved equivalence assertions, cf. (1.4.5/1), and Plancherel-Polya-Nikol'skij inequalities, cf. (1.4.5/6), for measures ν of the type $d\nu = \varkappa(x)\, d\mu$. In rough terms: We break up the general admissible measure ν into a measure μ which has a lattice structure and which regulates the local behaviour and a weight $\varkappa(x)$ which controls the admissible growth (or decay). The crucial ingredients in the two theorems in 1.4.5. are the lattice-constant h_0 in Theorem 1.4.5/1 and the growth restrictions for $\varrho(x) \in R(\omega)$, given by (1.4.1/2) and the Denjoy-Carleman quasi-analyticity condition (1.2.1/1). As far as the growth is concerned,

$$\varrho(x) = e^{\omega(x)} \quad \text{and} \quad \varrho(x) = e^{-\omega(x)} \quad \text{with} \quad \omega(x) \in \mathfrak{M} \tag{1}$$

are extreme cases. The following questions arise: Is the lattice structure natural and what about h_0? Are the growth-decay restrictions from (1) natural? The following two propositions and the discussions in the subsequent remarks shed some light on these questions. It turns out that our approach characterized by the lattice structure and the growth restrictions is fairly natural.

All notations have the same meaning as in 1.4.5. (and 1.4.1.), in particular, h_0 is the lattice-constant from Theorem 1.4.5/1 and Proposition 1.4.4. Furthermore, we specialize the compact subset Ω by the cube

$$\Omega = Q_b = \{y \mid y = (y_1, \ldots, y_n) \in R_n, |y_j| \leqq b\}, \tag{2}$$

where b is a positive number.

Proposition 1. *Let $\omega(x) \in \mathfrak{M}$, $\varrho(x) \in R(\omega)$ and $0 < p \leqq \infty$. Let $\Omega = Q_b$ with $b > 0$. Then $h_0 \leqq \frac{\pi}{b}$.*

Proof. We assume that Proposition 1.4.4 with $\Omega = Q_b$ holds for some number h_0 with $h_0 > \frac{\pi}{b}$. Let $0 < a < b$ such that $h_0 < \frac{\pi}{a}$. We recall the well-known formula

$$F\left(\prod_{j=1}^{n} \frac{\sin a\xi_j}{\xi_j}\right)(x) = c\chi_a(x), \quad x \in R_n, \tag{3}$$

where $\chi_a(x)$ is the characteristic function of Q_a. Let $\psi(x) \in S_\omega$ be a non-trivial function with

$$\operatorname{supp} F\psi \subset B_\delta = \{y \mid |y| \leqq \delta\}, \tag{4}$$

where $\delta > 0$ is a given number, cf. Proposition 1.2.2/1(ii). By Theorem 1.2.3/1 we have

$$\varphi(x) = \psi(x) \prod_{j=1}^{n} \frac{\sin ax_j}{x_j} \in S_\omega \tag{5}$$

and

$$\operatorname{supp} F\varphi = \operatorname{supp}\left(F\psi * F\left(\prod_{j=1}^{n} \frac{\sin ax_j}{x_j}\right)\right) \subset Q_a + B_\delta \subset Q_b$$

if we choose δ sufficiently small. Hence, $\varphi \in S_\omega^{Q_b}$. Because $h_0 > \frac{\pi}{a}$ we can choose points $x^k \in Q_k^{h_0}$ from Proposition 1.4.4 such that $\varphi(x^k) = 0$, cf. (5). But this contradicts (1.4.4/1).

Remark 1. Of peculiar interest is the question whether $\left\{x^k = \frac{\pi}{b} k\right\}_{k \in Z_n}$ is an admissible set in the sense of Proposition 1.4.4 with $\Omega = Q_b$. We discuss this problem in 1.5.5., where we give affirmative answers for special weights $\varrho(x)$ (in particular for $\varrho(x) = 1$).

Proposition 2. *Let* $\omega(x) = \sigma(|x|)$, *where* $\sigma(t)$ *is an increasing continuous concave function on* $[0, \infty)$ *with* $\sigma(0) = 0$ *and*

$$\int_0^\infty \frac{\sigma(t)}{1+t^2} \,\mathrm{d}t = \infty. \tag{6}$$

Let $0 < p \leqq \infty$. *Let* $\varphi(x) \in S$ *such that* $F\varphi$ *has a compact support and*

$$\|\mathrm{e}^{\omega(x)} \varphi(x) \mid L_p\| < \infty. \tag{7}$$

Then we have $\varphi(x) = 0$.

Proof. Step 1. First we prove that there exists an increasing continuous concave function $\tilde{\sigma}(t)$ on $[0, \infty)$ such that

$$\tilde{\sigma}(0) = 0, \qquad \int_0^\infty \frac{\tilde{\sigma}(t)}{1+t^2} \,\mathrm{d}t = \infty \tag{8}$$

and

$$\frac{\tilde{\sigma}(t)}{\sigma(t)} \to 0 \quad \text{if} \quad t \to \infty. \tag{9}$$

If $\sigma(2^j) = c_j \cdot 2^j$, then $\{c_j\}_{j=1}^\infty$ is a monotonically decreasing sequence of positive numbers. Furthermore

$$\int_0^\infty \frac{\sigma(t)}{1+t^2} \,\mathrm{d}t = \infty \quad \text{if and only if} \quad \sum_{j=1}^\infty c_j = \infty. \tag{10}$$

In particular, we find a sequence of integers L_k with $0 = L_1 < L_2 < \dots$ such that

$$\sum_{j=L_k+1}^{L_{k+1}} c_j \geqq 1 \quad \text{with} \quad k = 1, 2, 3, \dots \tag{11}$$

Let $a_j = \frac{c_j}{k}$ if $L_k + 1 \leqq j \leqq L_{k+1}$. Then $\{a_j\}_{j=1}^\infty$ is also a monotonically decreasing sequence of positive numbers and we have

$$\sum_{j=1}^\infty a_j = \infty. \tag{12}$$

Let $\tilde{\sigma}(t)$ be the polygonal line in the plane with the corner-points $(0, 0)$ and $(2^j, a_j \cdot 2^j)$, where $j = 1, 2, 3, \dots$ We may assume that $\tilde{\sigma}(t)$ is concave. Otherwise we construct the smallest concave majorant of $\tilde{\sigma}(t)$ which is also a polygonal line and the set of its corner-points is an infinite subset of the corner-points $(0, 0)$ and $\{(2^j, a_j \cdot 2^j)\}_{j=1}^\infty$ (and this is sufficient for our purposes). If $2^j \geqq N(\varepsilon)$ then we have $\tilde{\sigma}(2^j) \leqq \varepsilon\sigma(2^j)$. It follows that $\tilde{\sigma}(t) \leqq \varepsilon\sigma(t)$ holds for all $t \geqq N(\varepsilon)$. This proves (9). Finally, (8) follows from (12) and (10) (with $\tilde{\sigma}$ instead of σ and a_j instead of c_j).

Step 2. After this preparation, the proof of the proposition is rather easy. Let $\varphi(x)$ be the above function. By Step 1 we have

$$\|\mathrm{e}^{\lambda\tilde{\omega}(x)} \varphi(x) \mid L_p\| < \infty \quad \text{for all} \quad \lambda > 0 \tag{13}$$

with $\tilde{\omega}(x) = \tilde{\sigma}(|x|)$. Let $0 < \varepsilon < 1$ and $1 < q < \infty$. Then it follows from $\varphi \in S$, (13) and Hölder's inequality that

$$\int_{R_n} e^{\lambda\varpi(x)} |\varphi(x)| \, dx \leq c \left(\int_{R_n} e^{\lambda q \varpi(x)} |\varphi(x)|^{\varepsilon q} \, dx \right)^{1/q} < \infty$$

holds if $\varepsilon q = p$. Now we can apply Remark 1.2.1/7 to $(F\varphi)(x)$, cf. also Definition 1.2.1/2. We obtain $(F\varphi)(x) = 0$. This proves the proposition.

Remark 2. The above proposition shows that our theory becomes meaningless if the weight $\varrho(x) = e^{\omega(x)}$ grows too rapidly. The limit is given by (1.2.1/1) and $\omega(x) = \sigma(|x|)$. In particular, the above proposition can be applied to $\sigma(t) = t$, i.e. to $e^{\omega(x)} = e^{|x|}$. But in this case one has a simpler argument at hand: Let e.g. $e^{|x|} \varphi(x) \in L_1(R_n)$, $\varphi \in S$, and supp $F\varphi$ compact. If we fix $x_2, \ldots, x_n$, then $(F\varphi)(z, x_2, \ldots, x_n)$ is an analytic function in the strip $|\operatorname{Im} z| < 1$ in the complex plane. Because supp $F\varphi$ is compact (in R_n), we obtain $(F\varphi)(x) = 0$.

Remark 3. How to understand that our theory breaks down if the underlying weights decay too rapidly, e.g. like $e^{-|x|^\beta}$ with $\beta \geq 1$. The simplest case is $\beta = 2$. Let Q_b be given by (2). Let us assume that we find a constant c such that

$$\|e^{-|x|^2} \varphi(x) \mid L_\infty\| \leq c \|e^{-|x|^2} \varphi(x) \mid L_2\| \tag{14}$$

holds for $\varphi \in S^{Q_b}$. We use the approximation procedure from Proposition 1.2.2/2, cf. also the proof of Proposition 1. Then it follows from (1.2.4/3) and Lebesgue's bounded convergence theorem that (14) holds for all $\varphi \in S'$ with supp $F\varphi \subset Q_a$ and $0 < a < b$. In particular, (14) holds for polynomials $\varphi(x)$. But the set

$$\{e^{-|x|^2} \varphi(x),\ \varphi(x) \text{ polynomial}\}$$

is dense in $L_2(R_n)$ (these are linear combinations of the n-dimensional Hermite functions). Then it follows from (14) by completion that every L_2-function is bounded. This is a contradiction.

Remark 4. The above considerations can be extended to other weights. We restrict ourselves to the one-dimensional case $n = 1$, mostly in order to have immediate references. Let us assume that we find a constant c such that

$$\|e^{-|x|} \varphi(x) \mid L_\infty\| \leq c \|e^{-|x|} \varphi(x) \mid L_2\| \tag{15}$$

holds for all $\varphi \in S$ with supp $F\varphi \subset [-b, b]$, where b is a positive number. Then it follows in the same way as in the last remark that (15) holds for arbitrary polynomials $\varphi(x)$. However

$$\{e^{-|x|} \varphi(x),\ \varphi(x) \text{ polynomial}\}$$

is dense in $L_2(R_1)$, cf. G. Freud [1, pp. 79 and 84]. Then one has again a contradiction.

Remark 5. What remains is the gap between $e^{-|x|}$ and $e^{-\omega(x)}$ with $\omega(x) \in \mathfrak{M}$. This gap is filled by weights $e^{-\omega(x)}$ where $\omega(x) = \sigma(|x|)$ satisfies the hypotheses of Proposition 2, in particular (6). We restrict ourselves again to the one-dimensional case $n = 1$. Let $\omega(x)$ be an arbitrary positive function on R_1. Then $e^{\omega(x)}$ is called a Bernstein weight if the polynomials on R_1 are dense in

$$C_{e^\omega} = \{f(x) \mid f(x) \text{ continuous on } R_1,\ e^{-\omega(x)} f(x) \to 0 \text{ if } |x| \to \infty\},$$

equipped in the usual way with the norm

$$\|f \mid C_{e^\omega}\| = \sup_{x \in R_1} e^{-\omega(x)} |f(x)| < \infty .$$

Necessary and sufficient conditions for Bernstein weights may be found in N. J. Achiezer [1] and S. N. Mergeljan [1]. We formulate a result which fits in our context, cf. S. N. Mergeljan [1, p. 126]: *Let $\omega(x) = \omega(-x)$ and let $\omega(e^x)$ be a convex function on $[0, \infty)$. Then $e^{\omega(x)}$ is a Bernstein weight if and only if*

$$\int_0^\infty \frac{\omega(x)}{1 + x^2} \, dx = \infty . \tag{16}$$

This fits very well in our context, cf. (6). We return to our original task. Let $e^{\omega(x)}$ be a Bernstein weight (e.g. in the sense of Mergeljan's theorem). It is no serious restriction if we assume that the approximation procedures from the last two remarks with respect to $\|e^{-|x|^2}\varphi(x) \mid L_\infty\|$ and $\|e^{-|x|}\varphi(x) \mid L_\infty\|$ work also for $\|e^{-\omega(x)}\varphi(x) \mid L_\infty\|$. For those weights $e^{-\omega(x)}$ Plancherel-Polya-Nikol'skij inequalities cannot be valid: Let us assume that we find a constant c such that

$$\left\|e^{-\omega(x)}\frac{d\varphi}{dx} \mid L_\infty\right\| \leqq c\|e^{-\omega(x)}\varphi(x) \mid L_\infty\| \tag{17}$$

holds for all $\varphi \in S$ with supp $F\varphi \subset [-b, b]$ and $b > 0$. Then one obtains (17) successively for polynomials $\varphi(x)$ and arbitrary elements $\varphi(x) \in C_{e^\omega}$. But the latter is a contradiction.

1.5. Weighted L_p-Spaces of Entire Analytic Functions

1.5.1. Definition

Let $\omega(x) \in \mathfrak{M}$, $\varrho(x) \in R(\omega)$ and $0 < p \leqq \infty$. Let Ω be a compact subset of R_n. Then we introduced in Proposition 1.4.4 (and Theorem 1.4.5/1) an lattice-constant $h_0 = h_0(\omega, \varrho, p, \Omega)$. In order to extend Theorem 1.4.2, Proposition 1.4.4 and the two theorems in 1.4.5. from S_ω^Ω to some ultra-distributions (and to introduce corresponding L_p-spaces) we must care about this lattice-constant h_0. If Ω_1 and Ω_2 are two compact subsets of R_n with $\Omega_1 \subset \Omega_2$, then it follows from Proposition 1.4.4 (and its proof) that any admissible lattice-constant h_0 with respect to Ω_2 is also an admissible lattice-constant with respect to Ω_1: The smaller sets have the larger lattice-constants. This shows that the following construction is quite reasonable: Let

$$H = H(\omega, \varrho, p, \Omega) = \sup h_0(\omega, \varrho, p, \bar{\Lambda})$$

where the supremum is taken over all open sets Λ with $\Omega \subset \Lambda$ and all admissible lattice-constants h_0 (we recall that $\bar{\Lambda}$ stands for the closure of Λ). In general we have $H < \infty$: If, e.g., Ω contains a cube with the side-length $2b$, then we have $H \leqq \frac{\pi}{b}$, cf. Remark 1.4.4 (and Proposition 1.4.6/1).

Definition. *Let $\omega(x) \in \mathfrak{M}$, $\varrho(x) \in R(\omega)$ and $0 < p \leqq \infty$. Let Ω be a compact subset of R_n. Let H be the above number. Let $0 < h < H$, $\mu \in M^h$, and $\varkappa(x) \in K_h(\varrho, \mu)$. Then $L_p^\Omega(\varkappa, \mu)$ is given by*

$$L_p^\Omega(\varkappa, \mu) = \{f \mid f \in S'_\omega, \text{ supp } Ff \subset \Omega, \|\varkappa f \mid L_{p,\mu}\| < \infty\}. \tag{1}$$

Remark. These are the spaces $L_{p,\nu}^\Omega(R_n)$ which we announced in 1.1., cf. (1.1/8). As we have seen in the preceding section it is not only convenient but also natural to split the measure ν in a measure $\mu \in M^h$ with lattice-structure and a weight $\varkappa(x) \in K_h(\varrho, \mu)$, i.e. $d\nu = \varkappa^p(x)\, d\mu$ (with $\varkappa^\infty(x) = \varkappa(x)$). We shall see that $L_p^\Omega(\varkappa, \mu)$ is a quasi-Banach space. Let p with $0 < p \leqq \infty$ and the compact set Ω be fixed. We discuss the interdependence of the measures and weights in the spaces $L_p^\Omega(\varkappa, \mu)$. First we assume that $\omega(x) \in \mathfrak{M}$, $\varrho(x) \in R(\omega)$, $0 < h_j < H$, $\mu_j \in M^{h_j}$ and $\varkappa_j(x) \in K_{h_j}(\varrho, \mu_j)$ with $j = 1,2$. We recall that $H = H(\omega, \varrho, p, \Omega)$ is independent of the μ's and $\varkappa$'s. Then we prove in Remark 1.5.3/1 that the spaces $L_p^\Omega(\varkappa_1, \mu_1)$ and $L_p^\Omega(\varkappa_2, \mu_2)$ coincide. The next interesting problem is the following: Let the Borel measure ν in R_n be given. Then we define $L_{p,\nu}^\Omega = L_p^\Omega(\varkappa, \mu)$ via a representation $d\nu = \varkappa^p(x)\, d\mu$ (with $\varkappa^\infty(x) = \varkappa(x)$), $\mu \in M^h$, $\varkappa(x) \in K_h(\varrho, \mu)$ with $\varrho(x) \in R(\omega)$, $\omega(x) \in \mathfrak{M}$ and $0 < h < H$, where $H = H(\omega, \varrho, p, \Omega)$ is the above lattice-constant (of course, we assume that representations of this type exist). In Remark 1.5.3/2 we prove the following assertion: If one has two representations, in particular,

$$0 < h_1 < H(\omega_1, \varrho_1, p, \Omega) \quad \text{and} \quad 0 < h_2 < H(\omega_2, \varrho_2, p, \Omega) \tag{2}$$

holds, then the two spaces $L^{\Omega}_{p,\nu}$ (based on the two representations) coincide. As a special case we mention the following consequence: If $\mu \in M^h$ and $\varkappa(x) \geqq 0$ in the above definition are given, then $L^{\Omega}_p(\varkappa, \mu)$ is independent of ϱ and ω (of course, again we assume that appropriate ϱ's and ω's with the necessary properties exist). In this sense $L^{\Omega}_p(\varkappa, \mu)$ from (1) is independent of ϱ and ω. This justifies our notation.

Examples. We refer to the concrete measures and weights in 1.4.1.

1.5.2. Inequalities

Our next aim is to extend the maximal inequality from Theorem 1.4.2, and the two theorems in 1.4.5. from S^{Ω}_{ω} to the above spaces $L^{\Omega}_p(\varkappa, \mu)$. All notations have the above meaning. Furthermore, μ_L stands for the Lebesgue measure.

Theorem. *Let $\omega(x) \in \mathfrak{M}$, $\varrho(x) \in R(\omega)$ and $0 < p \leqq \infty$. Let Ω be a compact subset of R_n.*

(i) (*Maximal inequality*). *Let $0 < r < p$. Then there exists a positive number c such that*

$$\left\| \sup_{z \in R_n} \varrho(x-z) \frac{|f(x-z)|}{1+|z|^{n/r}} \mid L_p \right\| \leqq c \|\varrho f \mid L_p\| \tag{1}$$

holds for all $f \in L^{\Omega}_p(\varrho, \mu_L)$.

(ii) (*Equivalence*). *Let $\mu \in M^h$ and $\varkappa(x) \in K_h(\varrho, \mu)$ with $0 < h < H$, where H is the lattice constant from* 1.5.1. *There exist two positive numbers c_1 and c_2 such that*

$$c_1 \|\varrho f \mid L_p\| \leqq \|\varkappa f \mid L_{p,\mu}\| \leqq c_2 \|\varrho f \mid L_p\| \tag{2}$$

holds for all $f \in L^{\Omega}_p(\varrho, \mu_L)$.

(iii) (*Plancherel-Polya-Nikol'skij inequality*). *Let $p \leqq q \leqq \infty$. Let $\mu_1 \in M^h$ and $\varkappa_1(x) \in K_h(\varrho, \mu_1)$, let $\mu_2 \in M^h$ and $\varkappa_2(x) \in K_h(\varrho, \mu_2)$. There exists a positive number H' with the following property: If $0 < h < H'$ and if α is a multi-index then exists a constant c such that*

$$\|\varkappa_1 D^{\alpha} f \mid L_{q,\mu_1}\| \leqq c \|\varkappa_2 f \mid L_{p,\mu_2}\| \tag{3}$$

holds for all $f \in L^{\Omega}_p(\varkappa_2, \mu_2)$.

Proof. Step 1. We prove (i). If $p = \infty$ then (1) is trivial. Let $p < \infty$. Let $\psi(x) \in S^{\Omega}_{\omega}$. Then (1.4.2/2) and the maximal inequality (1.3.3/3) $\left(\text{with } \frac{p}{r} \text{ instead of } p\right)$ yield

$$\left\| \sup_{z \in R_n} \varrho(x-z) \frac{|\psi(x-z)|}{1+|z|^{n/r}} \mid L_p \right\| \leqq c \| M|\varrho\psi|^r \mid L_{p/r}\|^{1/r} \leqq c' \| |\varrho\psi|^r \mid L_{p/r}\|^{1/r}$$
$$= c' \|\varrho\psi \mid L_p\| . \tag{4}$$

This proves (1) with $\psi \in S^{\Omega}_{\omega}$ instead of f. Let $f \in L^{\Omega}_p(\varrho, \mu_L)$. Let $\varphi(x)$ be the function from Proposition 1.2.2/2(iii). We have $\psi_{\varepsilon}(x) = \varphi(\varepsilon x) f(x) \in S_{\omega}$, where we may assume that $\varepsilon \leqq 1$. Furthermore,

$$\operatorname{supp} F(\psi_{\varepsilon}) \subset \operatorname{supp} F[\varphi(\varepsilon \cdot)] + \operatorname{supp} Ff \subset \Omega',$$

where Ω' is the compact set of all points in R_n whose distance to Ω is not larger than 1. Of course, (4) also holds if one replaces Ω by Ω'. We use (4) with ψ_{ε} instead of ψ. Then (1) follows from $\varepsilon \downarrow 0$, (1.2.2/4), and Fatou's lemma.

Step 2. We prove (ii). By (1.4.5/1) and the definition of H we have

$$c_1\|\varrho\psi_\varepsilon \mid L_p\| \leqq \|\varkappa\psi_\varepsilon \mid L_{p,\mu}\| \leqq c_2\|\varrho\psi_\varepsilon \mid L_p\| \tag{5}$$

if ε is sufficiently small. Of course, ψ_ε has the same meaning as in the first step. Then (2) follows again from Fatou's lemma.

Step 3. We prove (iii). We use (1.4.5/6) with ψ_ε instead of φ. If $\alpha = (0, \ldots, 0)$, then $\varepsilon \downarrow 0$ and Fatou's lemma yield the desired result, i.e. $f \in L_q^\Omega(\varkappa_1, \mu_1)$. By the same argument ($\varepsilon \downarrow 0$ and Fatou's lemma) follows now $\frac{\partial}{\partial x_1} f \in L_q^\Omega(\varkappa_1, \mu_1)$ and a corresponding estimate. Iteration proves (iii).

Remark 1. The extension of Proposition 1.4.4 reads as follows: *Let $0 < H_1 < H$. There exist two positive numbers c_1 and c_2 such that*

$$c_1\|\varrho(x^k)\, f(x^k) \mid l_p\| \leqq h^{-n/p}\|\varrho f \mid L_p\| \leqq c_2\|\varrho(x^k) f(x^k) \mid l_p\| \tag{6}$$

holds for all h with $0 < h \leqq H_1$, all sets $\{x^k\}_{k \in Z_n}$ with $x^k \in Q_k^h$ and all $f \in L_p^\Omega(\varrho, \mu_L)$. This follows from the above considerations. All notations have the same meaning as in 1.4.4.

Remark 2. If $q = p$, then we can choose $H' = H$ in (iii). This follows from the above proof.

1.5.3. Basic Properties

We study the spaces $L_p^\Omega(\varkappa, \mu)$ from Definition 1.5.1. All notations have the same meaning as in the preceding subsections. In particular, H is the lattice-constant from 1.5.1. If $f \in L_p^\Omega(\varkappa, \mu)$ with $\|\varkappa f \mid L_{p,\mu}\| = 0$ then it follows from (1.5.2/2) that $\|\varrho f \mid L_p\| = 0$ holds, i.e. $f(x) = 0$. This shows that $L_p^\Omega(\varkappa, \mu)$ is a quasi-normed (normed if $p \geqq 1$) linear space.

Theorem. *Let $\omega(x) \in \mathfrak{M}$, $\varrho(x) \in R(\omega)$ and $0 < p \leqq \infty$. Let Ω be a compact subset of R_n and let $0 < h < H$.*

(i) *Let $\mu \in M^h$ and $\varkappa(x) \in K_h(\varrho, \mu)$. Then $L_p^\Omega(\varkappa, \mu)$ is a quasi-Banach space (Banach space if $1 \leqq p \leqq \infty$). Furthermore,*

$$S_\omega^\Omega \subset L_p^\Omega(\varkappa, \mu) \subset S'_\omega \tag{1}$$

holds (topological embedding).

(ii) *Let $\mu_1 \in M^h$ and $\varkappa_1(x) \in K_h(\varrho, \mu_1)$, let $\mu_2 \in M^h$ and $\varkappa_2(x) \in K_h(\varrho, \mu_2)$. Then*

$$L_p^\Omega(\varkappa_1, \mu_1) = L_p^\Omega(\varkappa_2, \mu_2) \tag{2}$$

holds (equivalent quasi-norms).

Proof. Step 1. Part (ii) is an immediate consequence of Theorem 1.5.2(iii) with $p = q$ and Remark 1.5.2/2.

Step 2. We prove (1). Let $\varphi(x) \in S_\omega^\Omega$. Then it follows from (2), (1.4.1/7), (1.2.1/2) and (1.2.1/8) that

$$\begin{aligned} \|\varkappa\varphi \mid L_{p,\mu}\| &\leqq c\|\varrho\varphi \mid L_p\| \leqq c'\|e^{\omega}\varphi \mid L_p\| \\ &\leqq c''\|e^{\lambda\omega}\varphi \mid L_\infty\| = c'' p_{0,\lambda}(\varphi) \end{aligned} \tag{3}$$

holds, where λ is a sufficiently large number. This proves the left-hand side of (1). In order to prove the right-hand side of (1) we assume that B is a bounded set in

S_ω. Let $f \in L_p^\Omega(\varkappa, \mu)$. Then it follows from (1.4.1/7) and (1.5.2/3) that

$$\begin{aligned}\sup_{\varphi \in B} |f(\varphi)| &= \sup_{\varphi \in B} \left| \int_{R_n} f(x)\, \varphi(x)\, \mathrm{d}x \right| \\ &\leqq \|\mathrm{e}^{-\omega} f \mid L_\infty\| \sup_{\varphi \in B} \|\mathrm{e}^{\omega} \varphi \mid L_1\| \leqq c \|\varrho f \mid L_\infty\| \sup_{\varphi \in B} \|\mathrm{e}^{\lambda\omega} \varphi \mid L_\infty\| \\ &\leqq c' \|\varrho f \mid L_p\| \sup_{\varphi \in B} p_{0,\lambda}(\varphi) \leqq c'' \|\varrho f \mid L_p\| \leqq c''' \|\varkappa f \mid L_{p,\mu}\| \end{aligned} \quad (4)$$

where again λ is a sufficiently large number. The last estimate follows from (2). This proves the right-hand side of (1).

Step 3. We prove that $L_p^\Omega(\varkappa, \mu)$ is complete. Let $\{f_j\}_{j=1}^\infty$ be a fundamental sequence in $L_p^\Omega(\varkappa, \mu) = L_p^\Omega(\varrho, \mu_L)$, where μ_L stands again for the Lebesgue measure. $\{\varrho f_j\}_{j=1}^\infty$ is a fundamental sequence in L_∞, cf. (1.5.2/3). Let ϱf be the limit element. By the middle part of (4) (where we used only $\varrho f \in L_\infty$) we have $f_j \to f$ in S'_ω. Consequently, $Ff_j \to Ff$ in S'_ω and hence supp $Ff \subset \Omega$. If $p = \infty$ then f is the limit element of $\{f_j\}_{j=1}^\infty$ in $L_\infty^\Omega(\varrho, \mu_L)$. If $p < \infty$ then it follows by standard arguments that

$$\|\varrho f_j - \varrho f \mid L_p\| \to 0 \quad \text{when} \quad j \to \infty\,.$$

This proves $f \in L_p^\Omega(\varrho, \mu_L)$ and $f_j \to f$ in $L_p^\Omega(\varrho, \mu_L)$.

Remark 1. We return to the first discussion in Remark 1.5.1. First we assume that $\omega(x) \in \mathfrak{M}$, $\varrho(x) \in R(\omega)$, $0 < h_j < H$, $\mu_j \in M^{h_j}$ and $\varkappa_j(x) \in K_{h_j}(\varrho, \mu_j)$ with $j = 1,2$. Then (2) yields

$$L_p^\Omega(\varkappa_1, \mu_1) = L_p^\Omega(\varrho, \mu_L) = L_p^\Omega(\varkappa_2, \mu_2).$$

Remark 2. We return to the second discussion in Remark 1.5.1. Let ν be a given Borel measure in R_n. Let $0 < p < \infty$ and let Ω be a compact subset of R_n. We assume that ν can be represented as $\mathrm{d}\nu = \varkappa^p(x)\, \mathrm{d}\mu$ with $\mu \in M^h$, $\varkappa(x) \in K_h(\varrho, \mu)$, where $\varrho(x) \in R(\omega)$ and $\omega(x) \in \mathfrak{M}$ are appropriate weights, and $0 < h < H$, where H is the lattice-constant from 1.5.1. Let $L_{p,\nu}^\Omega = L_p^\Omega(\varkappa, \mu)$, where the latter space has the meaning of (1.5.1/1). Now we assume that we have two such realizations of $L_{p,\nu}^\Omega$ via ω_k, ϱ_k, $\varkappa_k$, μ_k and h_k with $k = 1,2$. In particular, (1.5.1/2) holds. Let $f \in L_{p,\nu}^\Omega$ in the sense of the first representation. Then we have $f \in L_p^{\bar{\Lambda}}(\varkappa_1, \mu_1)$ where Λ is an open set with $\Omega \subset \Lambda$ and $h_1 < H(\omega_1, \varrho_1, p, \bar{\Lambda})$. We use the functions $\psi_\varepsilon(x)$ from the proof of Theorem 1.5.2. If $0 < \varepsilon < \varepsilon_0$ then we may assume $\psi_\varepsilon(x) \in S_{\omega_1+\omega_2}^{\bar{\Lambda}}$ and $\psi_\varepsilon(x) \to f$ in $L_p^{\bar{\Lambda}}(\varkappa_1, \mu_1)$ if $\varepsilon \downarrow 0$. If we choose Λ in an appropriate way, then $\{\psi_\varepsilon(x)\}_{0<\varepsilon<\varepsilon_0}$ is also a fundamental sequence (fundamental set) in $L_p^{\bar{\Lambda}}(\varkappa_2, \mu_2)$. Then we have $f \in L_p^\Omega(\varkappa_2, \mu_2)$ which proves the desired independence of the representation. In particular, if $\varkappa$ and μ in Definition 1.5.1 are given then $L_p^\Omega(\varkappa, \mu)$ (with $p < \infty$ at this moment) is independent of the underlying weights $\varrho(x) \in R(\omega)$ and $\omega(x) \in \mathfrak{M}$. Let $p = \infty$. We assume that $\mu \in M^h$ and $\varkappa(x) \geqq 0$ are given and that we have two realizations of $L_\infty^\Omega(\varkappa, \mu)$ in the sense of Definition 1.5.1, in particular, (1.5.1/2) with $h_1 = h_2 = h$ holds. Let $f \in L_\infty^\Omega(\varkappa, \mu)$ (based on ω_1 and ϱ_1, in particular $f \in S'_{\omega_1}$). We use the above approximation, in particular $\psi_\varepsilon(x) \to f(x)$ and $|\psi_\varepsilon(x)| \uparrow |f(x)|$ for every $x \in R_n$. By (1.5.2/3) we have

$$\|\varrho_2 f \mid L_\infty\| = \lim_{\varepsilon \downarrow 0} \|\varrho_2 \psi_\varepsilon \mid L_\infty\| \leqq c \lim_{\varepsilon \downarrow 0} \|\varkappa \psi_\varepsilon \mid L_{\infty,\mu}\| \leqq c' \|\varrho_1 f \mid L_\infty\| < \infty.$$

By (1.4.1/7) and (1.2.2/11) it follows that

$$|f(x)| \leqq \frac{c}{\varrho_2(x)} \leqq c'\, \mathrm{e}^{\omega_2(x)} \quad \text{and} \quad f \in S'_{\omega_2}$$

holds. Hence, we have $f \in L_\infty^\Omega(\varkappa, \nu)$ (based on ω_2 and ϱ_2). This proves the desired independence.

1.5.4. Further Properties

In this subsection we summarize some further properties of the spaces $L_p^\Omega(\varkappa, \mu)$ from Definition 1.5.1. We omit proofs and refer the interested reader to [F, pp. 39–46].

Approximation and Density. Let $L_p^\Omega(\varkappa, \mu)$ be the space from Definition 1.5.1 and let Λ be an open bounded set with $\Omega \subset \Lambda$ such that also $L_p^{\bar{\Lambda}}(\varkappa, \mu)$ makes sense. Let $f \in L_p^\Omega(\varkappa, \mu)$ and let $\psi_\varepsilon(x)$ be the functions from the proof of Theorem 1.5.2. If $p < \infty$ then $\psi_\varepsilon(x) \in S_\omega^{\bar{\Lambda}}$ with $0 < \varepsilon < \varepsilon_0$ approximates f in $L_p^{\bar{\Lambda}}(\varkappa, \mu)$. If $p = \infty$ then we have $\psi_\varepsilon(x) \to f(x)$ pointwise and $\|\varkappa f \mid L_{\infty, \mu}\| = \lim_{\varepsilon \downarrow 0} \|\varkappa \psi_\varepsilon \mid L_{\infty, \mu}\|$. We proved these assertions in Remark 1.5.3/2. What about approximations of $f \in L_p^\Omega(\varkappa, \mu)$ by functions from S_ω^Ω? (I.e., is S_ω^Ω dense in $L_p^\Omega(\varkappa, \mu)$?) By (1.5.3/1) this question makes sense. However, in general, one cannot expect affirmative answers. Let e.g. $\Omega = \{0\}$ and $\varrho(x) = e^{-|x|^\beta}$ with $0 < \beta < 1$. Then any polynomial is an element of $L_p^\Omega(e^{-|x|^\beta}, \mu_L)$, where μ_L stands for the Lebesgue measure. On the other hand, S_ω^Ω with $\omega(x) = e^{|x|^\beta}$ contains no non-trivial elements. In order to describe an affirmative results we assume $\Omega = \bar{\Lambda}$, where Λ is an open bounded set in R_n which has the segment property. The latter means the following: There exist N open balls K_j and N vectors $y^j \in R_n$ such that

$$\Omega \subset \bigcup_{j=1}^{N} K_j \quad \text{and} \quad (\Omega \cap K_j) + t y^j \subset \Lambda \tag{1}$$

if $0 < t \leqq 1$ and $j = 1, \ldots, N$. In [F, p. 40] we proved the following assertion: *Let Λ be a bounded open set in R_n with the segment property. Let $\Omega = \bar{\Lambda}$ and $0 < p < \infty$. Then S_ω^Ω (and even S_ω^Λ) are dense in $L_p^\Omega(\varkappa, \mu)$.* Of course, S_ω^Λ is the collection of all $\varphi \in S_\omega$ with $\operatorname{supp} F\varphi \subset \Lambda$.

Duality I. Let $\varphi(x) \in S_\omega^\Omega$ and $g \in S'_\omega$ with $\omega(x) \in \mathfrak{M}$ and the compact set Ω. We have $g(\varphi) = (F^{-1}g)(F\varphi)$. In other words: Only the restriction of $F^{-1}g$ on Ω, i.e. the restriction of Fg to $-\Omega = \{x \mid -x \in \Omega\}$, is of interest. We wish to study the topological dual of $L_p^\Omega(\varkappa, \mu)$. We may assume that $\mu = \mu_L$ is the Lebesgue measure and that $\varkappa(x) = \varrho(x) \in R(\omega)$. Let $1 < p < \infty$ and $\frac{1}{p} + \frac{1}{p'} = 1$. By the above argument it seems to be reasonable to expect that $L_{p'}^{-\Omega}(\varrho^{-1}, \mu_L)$ is the dual of $L_p^\Omega(\varrho, \mu_L)$ under the natural pairing. However this is a very sophisticated problem and we studied it rather carefully in [F, pp. 41–46]. We mention few results and refer for proofs to [F]. Let χ_Ω be the characteristic function of Ω. If again $\varphi \in S_\omega^\Omega$ and $g \in S'_\omega$ then we have (somewhat heuristic)

$$g(\varphi) = (F^{-1}g)(F\varphi) = (\chi_{-\Omega}F^{-1}g)(F\varphi) = (F\chi_{-\Omega}F^{-1}g)(\varphi). \tag{2}$$

This gives at least the feeling that $g \to F\chi_{-\Omega}F^{-1}g$ should be a bounded operator in $L_{p'}(R_n, \varrho^{-1}(x))$, or by duality, that $f \to F^{-1}\chi_\Omega Ff$ should be a bounded operator in $L_p(R_n, \varrho(x))$ (as far as notations are concerned we refer to (1.3.1/4)). In other words, it seems to be of interest for the above problem whether χ_Ω is a Fourier multiplier in $L_p(R_n, \varrho(x))$. We postpone our further discussion about duality for a moment and mention few relevant results about Fourier multipliers.

Fourier Multipliers for $L_p(R_n, \varrho(x))$. Let $\varrho(x) \in R(\omega)$, $\omega(x) \in \mathfrak{M}$ and $1 < p < \infty$. Let Ω be a rectangle, say, $\Omega = Q_1 = Q$ is the unit cube, cf. (1.4.6/2). We ask for conditions which ensure that χ_Q is a Fourier multiplier in $L_p(R_n, \varrho(x))$, i.e.

$$\|F^{-1}\chi_Q Ff \mid L_p(R_n, \varrho(x))\| \leqq c\|f \mid L_p(R_n, \varrho(x))\| \tag{3}$$

holds for all $f \in L_p(R_n, \varrho(x))$, where c is an appropriate constant (which is independent of f). Obviously, $\varrho F^{-1}\chi_Q \in L_p$ is a necessary condition. By (1.4.6/3) and the local behaviour of $\varrho(x)$, cf. (1.4.1/2), we obtain

$$\frac{\varrho(x)}{\prod_{j=1}^{n}(1+|x_j|)} \in L_p \tag{4}$$

as a necessary condition. Let $\varrho(x) = (1+|x|)^d$, where d is a real number. Then $dp < p - 1$, i.e. $d < \frac{1}{p'}$ with $\frac{1}{p} + \frac{1}{p'} = 1$ is a necessary condition. By duality we have $-\frac{1}{p} < d < \frac{1}{p'}$ as a necessary condition. But this is also sufficient. In other words: *Let $1 < p < \infty$ and $\varrho(x) = (1+|x|)^d$. Then the characteristic function of a rectangle in R_n is a Fourier multiplier for $L_p(R_n, \varrho(x))$ if and only if $-\frac{1}{p} < d < \frac{1}{p'}$.* One can extend this result to more generel weights $\varrho(x)$. Let $n = 1$ (this is not only for sake of simplicity): *Let $\varrho(x) \in R(\omega)$ for some $\omega(x) \in \mathfrak{M}$ and let $1 < p < \infty$. Then the characteristic function of an interval in R_1 is a Fourier multiplier for $L_p(R_1, \varrho(x))$ if and only if $\varrho^p(x)$ satisfies Muckenhoupt's A_p-condition, i.e. (1.3.3/8) with ϱ^p instead of ω holds, where the supremum is taken over all intervals $Q = I$ in R_1.* Of course, if $\varrho(x) = (1+|x|)^d$ then the last condition reduces to the above one, i.e. $-\frac{1}{p} < d < \frac{1}{p'}$. These assertions follow from R. A. Hunt, B. Muckenhoupt, R. L. Wheeden [1] and R. R. Coifman, C. Fefferman [1] (cf. [F, p. 42] for details). Finally we mention the famous assertion by C. Fefferman [1] that *the characteristic function of a ball is a Fourier multiplier in L_p with $1 < p < \infty$ if and only if $p = 2$.* Cf. also B. S. Mitjagin [1] for generalizations.

Duality II. On the basis of the results about Fourier multipliers on $L_p(R_n, \varrho(x))$ and the above heuristical arguments, cf. (2), one can prove the following result. *Let $1 < p < \infty$, $\frac{1}{p} + \frac{1}{p'} = 1$ and $\varrho(x) = (1+|x|)^d$ with $-\frac{1}{p} < d < \frac{1}{p'}$. Let Ω be a closed rectangle in R_n, then we have*

$$(L_p^{\Omega}(\varrho, \mu_L))' = L_{p'}^{-\Omega}(\varrho^{-1}, \mu_L). \tag{5}$$

Of course, (5) should be understood via the dual pairing

$$f(g) = \int_{R_n} f(x)\, g(x)\, \mathrm{d}x, \tag{6}$$

and the norm of f as a functional is equivalent to $\|\varrho^{-1} f \mid L_{p'}\|$. Cf. [F, Theorem 1.4.5 and Remark 1.4.5/2]. However this assertion is not completely covered by the just given references. We must add few additional remarks. In the next subsection we prove that the above spaces $L_p^{\Omega}(\varrho, \mu_L)$ are isomorphic to l_p, in particular they are reflexive. By the arguments from [F, Remark 1.4.5/2] one obtains the above assertions if $p \geqq 2$ and $d \leqq 0$. By duality (reflexivity) one extends this assertion to $p \leqq 2$ and $d \geqq 0$. If one uses [F, Theorem 1.4.5] and its proof one obtains a corresponding assertion for the case $p \leqq 2$ and $d \leqq 0$ and finally, again by duality, for the remaining case $p \geqq 2$ and $d \geqq 0$. In other words, at least if one specializes $\varrho(x) = (1+|x|)^d$ with $-\frac{1}{p} < d < \frac{1}{p'}$, then the awkward additional assumptions in [F, Theorem 1.4.5] which ensure (5) can be omitted. Finally, based on Fefferman's result (and the above duality-reflexivity argument) one can prove the following assertion: Let $1 < p < \infty$

and $p \neq 2$. Let $\varrho(x) = 1$. *Let Ω be a closed ball in R_n (with $n \geqq 2$). Then $L_{p'}^{-\Omega}(\varrho^{-1}, \mu_L)$ is not the dual of $L_p^{\Omega}(\varrho, \mu_L)$ in the sense of the interpretation* (6). Cf. [F, Proposition 1.4.5].

1.5.5. Shannon's Sampling Theorem, Isomorphism Properties, and Schauder Bases

The assertions about Fourier multipliers for $L_p(R_n, \varrho(x))$ formulated in the preceding subsection can be used in order to shed more light upon the structure of the spaces $L_p^{\Omega}(\varrho, \mu_L)$, provided that Ω is a rectangle and $1 < p < \infty$. For sake of simplicity we restrict ourselves to cubes

$$\Omega = Q_b = \{y \mid y = (y_1, \ldots, y_n) \in R_n, |y_j| \leqq b\} \tag{1}$$

with $b > 0$ and the weights $\varrho(x) = (1 + |x|)^d$ with $-\frac{1}{p} < d < \frac{1}{p'}$, where again $1 < p < \infty$ and $\frac{1}{p} + \frac{1}{p'} = 1$. Of course, in this case we may choose $\omega(x) = (|d|+1) \log(1 + |x|)$, cf. (1.2.1/4) and (1.4.1/10). In other words: The usual distributions are completely sufficient in this subsection. Furthermore in this subsection we denote the characteristic function of the cube Q_b by $\chi_b(x)$, somewhat in contrast to our previous notation $\chi_\Omega(x) = \chi_{Q_b}(x)$. First we recall an assertion from [T, p. 21] which covers essentially the n-dimensional version of Shannon's sampling theorem where the latter was mentioned in 1.1., cf. (1.1/5). In order to make this subsection self-contained we include a full proof although we take it over from [T, p. 21]. We recall that

$$S^\Omega = \{\varphi \mid \varphi \in S, \operatorname{supp} F\varphi \subset \Omega\}, \tag{2}$$

cf. (1.4.2/1), and that Z_n stands for the lattice of all integer-valued points in R_n, cf. (1.3.1/5).

Proposition. *Let $\Omega = Q_b$ be given by* (1). *Then any $\varphi(x) \in S^\Omega$ can be represented by*

$$\begin{aligned}\varphi(x) &= \left(\frac{\pi}{2}\right)^{n/2} b^{-n} \sum_{k \in Z_n} \varphi\left(\frac{\pi}{b} k\right) (F^{-1}\chi_b)\left(x - \frac{\pi}{b} k\right) \\ &= \sum_{k \in Z_n} \varphi\left(\frac{\pi}{b} k\right) \prod_{j=1}^{n} \frac{\sin(bx_j - k_j\pi)}{bx_j - k_j\pi}, \quad x \in R_n,\end{aligned} \tag{3}$$

(*absolutely convergent series*).

Proof. As usual, $xk = \sum_{j=1}^{n} x_j k_j$ stands for the scalar product of $x \in R_n$ and $k \in Z_n$. We develop $(F\varphi)(x)$ in Q_b in an absolutely convergent multiple trigonometric series

$$(F\varphi)(x) = \chi_b(x) \sum_{k \in Z_n} a_k \, \mathrm{e}^{-\mathrm{i}\frac{\pi}{b} xk}, \quad x \in R_n, \tag{4}$$

with the Fourier coefficients

$$\begin{aligned}a_k &= (2b)^{-n} \int_{Q_b} (F\varphi)(x) \, \mathrm{e}^{\mathrm{i}\frac{\pi}{b} kx} \, \mathrm{d}x \\ &= (2b)^{-n} (2\pi)^{n/2} (F^{-1}F\varphi)\left(\frac{\pi}{b} k\right) = \left(\frac{\pi}{2}\right)^{n/2} b^{-n} \varphi\left(\frac{\pi}{b} k\right).\end{aligned} \tag{5}$$

The inverse Fourier transform of (4),

$$F^{-1}\left(\chi_b(x) \, \mathrm{e}^{-\mathrm{i}\frac{\pi}{b} xk}\right)(y) = (F^{-1}\chi_b)\left(y - \frac{\pi}{b} k\right), \tag{6}$$

and (5) prove the first representation in (3). In order to prove the second representation in (3) we recall that in the one-dimensional case, i.e. $Q_b = [-b, b]$, we have

$$(F^{-1}\chi_b)(x) = (2\pi)^{-1/2} \int_{-b}^{b} e^{ixy}\, dy = \left(\frac{2}{\pi}\right)^{1/2} b \frac{\sin bx}{bx}, \quad x \in R_1. \tag{7}$$

In the n-dimensional case it follows by multiplication that

$$(F^{-1}\chi_b)\left(x - \frac{\pi}{b}k\right) = \left(\frac{2}{\pi}\right)^{n/2} b^n \prod_{j=1}^{n} \frac{\sin(bx_j - k_j\pi)}{bx_j - k_j\pi}, \quad x \in R_n. \tag{8}$$

Now one obtains the second representation in (3) from the first one and (8). The claimed absolute convergence of the series in (3) follows from $\varphi(x) \in S$.

Remark 1. This is the n-dimensional version of Shannon's sampling theorem for functions from S, cf. (1.1/5). We wish to extend this assertion to functions from $L_p^\Omega(\varrho, \mu_L)$, where $\Omega = Q_b$ is the cube given by (1), $1 < p < \infty$ and $\varrho(x) = (1 + |x|)^d$ with $-\frac{1}{p} < d < \frac{1}{p'}$. Of course, μ_L always stands for the Lebesgue measure. l_p has the usual meaning normed by (1.3.1/6).

Theorem. *Let* $1 < p < \infty$, $\frac{1}{p} + \frac{1}{p'} = 1$ *and* $\varrho(x) = (1 + |x|)^d$ *with* $-\frac{1}{p} < d < \frac{1}{p'}$. *Let* $\Omega = Q_b$ *be given by* (1).

(i) $f \in S'$ *is an element of* $L_p^\Omega(\varrho, \mu_L)$ *if and only if it can be represented as*

$$f(x) = \left(\frac{\pi}{2}\right)^{n/2} b^{-n} \sum_{k \in Z_n} a_k (F^{-1}\chi_b)\left(x - \frac{\pi}{b}k\right) = \sum_{k \in Z_n} a_k \prod_{j=1}^{n} \frac{\sin(bx_j - k_j\pi)}{bx_j - k_j\pi} \tag{9}$$

(*convergence in* S') *with* $\|(1 + |k|)^d a_k \mid l_p\| < \infty$.

(ii) *Any element* $f \in L_p^\Omega(\varrho, \mu_L)$ *can be represented by* (9) *with* $a_k = f\left(\frac{\pi}{b}k\right)$, $k \in Z_n$. *Furthermore,* $L_p^\Omega(\varrho, \mu_L)$ *is isomorphic to* l_p *and the sequence*

$$\left\{\prod_{j=1}^{n} \frac{\sin(bx_j - k_j\pi)}{bx_j - k_j\pi}\right\}_{k \in Z_n} \tag{10}$$

of entire analytic functions in R_n *is an unconditional Schauder basis for* $L_p^\Omega(\varrho, \mu_L)$. *There exist two positive numbers* c_1 *and* c_2 *such that*

$$c_1 \|(1 + |x|)^d f \mid L_p\| \leq \left\|(1 + |k|)^d f\left(\frac{\pi}{b}k\right) \mid l_p\right\| \leq c_2 \|(1 + |x|)^d f \mid L_p\| \tag{11}$$

for all $f \in L_p^\Omega(\varrho, \mu_L)$ *holds (equivalent norms).*

Proof. Step 1. By (6) and (8) we have $(F^{-1}\chi_b)\left(x - \frac{\pi}{b}k\right) \in L_p^\Omega(\varrho, \mu_L)$, cf. also (1.5.4/4). In particular, (10) is a subset of $L_p^\Omega(\varrho, \mu_L)$. By the density assertion from 1.5.4. it follows that S^Ω is dense in $L_p^\Omega(\varrho, \mu_L)$. If $\varphi(x) \in S^\Omega$ then (3) converges in $L_p^\Omega(\varrho, \mu_L)$. Hence, (10) is a dense subset in $L_p^\Omega(\varrho, \mu_L)$.

Step 2. We prove (11). The right-hand side follows from (1.5.2/6) because $\left\{\frac{\pi}{b}k\right\}_{k \in Z_n}$ can always be interpreted as a sub-lattice of an appropriate set $\{x^k\}_{k \in Z_n}$ in the sense of Remark 1.5.2/1. In order to prove the reverse inequality we assume that f is a finite linear combination of the elements from (10), i.e.

$$f(x) = \left(\frac{\pi}{2}\right)^{n/2} b^{-n} \sum_{|k| \leq N} a_k (F^{-1}\chi_b)\left(x - \frac{\pi}{b}k\right). \tag{12}$$

By the above density assertion and (1.5.4/5) and (1.5.4/6) it follows that $\|\varrho f \mid L_p\|$ can be estimated from above by sup $|f(g)|$, where the supremum is taken over all g with $\|\varrho^{-1} g \mid L_{p'}\| \leqq 1$ which can be represented in the same way as in (12). Hence, if

$$g(x) = \sum_{|k| \leqq N'} c_k (F^{-1}\chi_b)\left(x - \frac{\pi}{b} k\right)$$

is an appropriate function, say with $N' \geqq N$, then we have

$$\begin{aligned} \|\varrho f \mid L_p\| &\leqq c \int_{R_n} f(x)\, \overline{g(x)}\, \mathrm{d}x = c \int_{R_n} (Ff)(x)\, \overline{(Fg)(x)}\, \mathrm{d}x \\ &= c \sum_{\substack{l,k \in Z_n \\ |l|,|k| \leqq N}} a_k \bar{c}_l \int_{Q_b} \mathrm{e}^{-\mathrm{i}\frac{\pi}{b}xk}\, \mathrm{e}^{\mathrm{i}\frac{\pi}{b}xl}\, \mathrm{d}x = c \sum_{|k| \leqq N} (2b)^n a_k \bar{c}_k \\ &\leqq c' \left\| \varrho\left(\frac{\pi}{b}k\right) a_k \mid l_p \right\| \left\| \varrho^{-1}\left(\frac{\pi}{b}k\right) c_k \mid l_{p'} \right\|. \end{aligned} \tag{13}$$

In the middle estimate we used (6). From the above considerations follows that the second factor on the right-hand side of (13) can be estimated from above by $c\|\varrho^{-1} g \mid L_{p'}\|$ and hence by a constant which is independent of f. Because $a_k = f\left(\frac{\pi}{b}k\right)$ this proves the left-hand side of (11). By completion, cf. Step 1, we have (11) for all $f \in L_p^\Omega(\varrho, \mu_L)$.

Step 3. We prove the remaining part of (ii). Let $f \in L_p^\Omega(\varrho, \mu_L)$. Let f_N be given by the right-hand side of (9) with $a_k = f\left(\frac{\pi}{b}k\right)$ and $\sum_{|k| \leqq N}$ instead of $\sum_{k \in Z_n}$. By (11), $\{f_N\}_{N=1}^\infty$ is a fundamental sequence in $L_p^\Omega(\varrho, \mu_L)$ and hence in S'. Its limit is f. Consequently, (9) with $a_k = f\left(\frac{\pi}{b}k\right)$ is a representation of f. The other assertions from (ii) are clear. Finally, (i) follows by the same arguments.

Remark 2. In other words, we proved that any $f \in L_p^\Omega(\varrho, \mu_L)$ can be represented as

$$f(x) = \sum_{k \in Z_n} f\left(\frac{\pi}{b}k\right) \prod_{j=1}^{n} \frac{\sin(bx_j - k_j\pi)}{bx_j - k_j\pi}, \quad x \in R_n,$$

where the rate of convergence is given by (11). This is an essential extension of the above proposition and, consequently, of Shannon's sampling theorem.

Remark 3. In Proposition 1.4.6/1 we proved that all admissible lattice-constants h_0 for $L_p^\Omega(\varrho, \mu_L)$ with $\Omega = Q_b$ can be estimated from above by $h_0 \leqq \frac{\pi}{b}$. Let $1 < p < \infty$ and $\varrho(x) = (1 + |x|)^d$ with $-\frac{1}{p} < d < \frac{1}{p'}$. Then it follows from the above theorem that the limiting number $h_0 = \frac{\pi}{b}$ can be realized if we choose the lattice $\left\{\frac{\pi}{b}k\right\}_{k \in Z_n}$.

Remark 4. The theorem can be extended to other weights $\varrho(x)$ for which the affirmative assertions about the Fourier multipliers for $L_p(R_n, \varrho(x))$ from 1.5.4. hold, cf. also [F, 1.4.6].

Remark* 5. The above theorem is a modified version of [F, 1.4.6]. The unweighted case has been treated (essentially) in [I, pp. 239/240], which was based on H. Triebel [1]. In this paper we used some ideas from the unpublished note by J. Peetre [2], cf. also J. Peetre [4, pp. 180 and 190].

1.6. Mixed $L_{\bar{p}}$-Spaces of Entire Analytic Functions

1.6.1. Definition

The main aim of Section 1.6. is to study the mixed spaces $L_{\bar{p}}^{\Omega}(R_2)$, where we replaced $\|f \mid L_p(R_2)\|$ in (1.1/2) by $\|f \mid L_{\bar{p}}(R_2)\|$ from (1.1/6). We restrict ourselves to $n = 2$, however the extension to arbitrary natural numbers $n > 1$ is obvious. Furthermore, we deal only with unweighted mixed spaces. This has the advantage that we do not need the ultra-distributions: The usual distributions are sufficient. We refer to 1.2.4., where we fixed our notations and where we recalled the famous Paley-Wiener-Schwartz theorems. In particular, $S = S(R_2)$ and $S' = S'(R_2)$ have the usual meaning, F and F^{-1} stand for the Fourier transform and its inverse, respectively. Furthermore, we use the notations from 1.3.2., in particular $\|f \mid L_{\bar{p}}\| = \|f(x_1, x_2) \mid L_{\bar{p}}\|$ and $L_{\bar{p}}$ from (1.3.2/5) and (1.3.2/5′), respectively, where $\bar{p} = (p_1, p_2)$ with $0 < p_1 \leqq \infty$ and $0 < p_2 \leqq \infty$.

Definition. *Let $\bar{p} = (p_1, p_2)$ with $0 < p_1 \leqq \infty$ and $0 < p_2 \leqq \infty$. Let Ω be a compact subset of R_2. Then $L_{\bar{p}}^{\Omega}$ is given by*

$$L_{\bar{p}}^{\Omega} = \{f \mid f \in S', \operatorname{supp} Ff \subset \Omega, \|f \mid L_{\bar{p}}\| < \infty\}. \tag{1}$$

Remark. By the Paley-Wiener-Schwartz theorem from 1.2.4. the space $L_{\bar{p}}^{\Omega}$ consists of entire analytic functions of exponential type. In particular, (1) makes sense. If $p_1 = p_2 = p$ then we have the spaces L_p^{Ω} from [T, Definition 1.4.1], cf. also [T, Theorem 1.4.2]. Of course, L_p^{Ω} is also a special case of the spaces $L_p^{\Omega}(\varkappa, \mu)$ from Definition 1.5.1 if μ is the Lebesgue measure and $\varkappa(x) = 1$. The question is whether one can combine $L_{\bar{p}}^{\Omega}$ with the weighted spaces from Definition 1.5.1: Weighted mixed spaces $L_{\bar{p}}^{\Omega}(\varkappa, \mu)$ which should be treated in the framework of ultra-distributions. This can be done, we refer to B. Stöckert [1, 2]. However our aim is different: We restrict our attention to the above spaces $L_{\bar{p}}^{\Omega}$. Furthermore, in contrast to our fairly systematic study of the weighted spaces $L_p^{\Omega}(\varkappa, \mu)$ we deal only with those properties of the mixed spaces $L_{\bar{p}}^{\Omega}$ which we need in Chapter 2. There is no doubt that many other assertions which we derived for weighted spaces have more or less direct counterparts for the mixed spaces.

1.6.2. Inequalities of Nikol'skij Type

We are looking for the unweighted mixed counterpart of (1.4.5/7), cf. also [T, (1.3.2/5)]. We denote inequalities of this type as Nikol'skij inequalities in contrast to the more general Plancherel-Polya-Nikol'skij inequalities which include atomic measures etc., cf. the discussion in 1.1. Let

$$\Omega = Q_{\bar{b}} = \{(x_1, x_2) \mid |x_1| \leqq b_1, |x_2| \leqq b_2\} \tag{1}$$

be a rectangle in R_2 with sides parallel to the axes. Of course, $b_1 > 0$ and $b_2 > 0$. We recall that $L_{p_j \mid y_j}$ stands for the space $L_{p_j}(R_1)$ where the underlying integration has to be taken with respect to the variable y_j, cf. the end of 1.3.2.

Theorem. *Let $\bar{p} = (p_1, p_2)$ and $\bar{u} = (u_1, u_2)$ with $0 < p_1 \leqq u_1 \leqq \infty$ and $0 < p_2 \leqq u_2 \leqq \infty$. Let $\alpha = (\alpha_1, \alpha_2)$ be a multi-index. Let $\Omega = Q_{\bar{b}}$ be given by (1). Then there exists a positive constant c, which is independent of b_1 and b_2, such that*

$$\|D^{\alpha} f \mid L_{\bar{u}}\| \leqq c b_1^{\alpha_1 + \frac{1}{p_1} - \frac{1}{u_1}} b_2^{\alpha_2 + \frac{1}{p_2} - \frac{1}{u_2}} \|f \mid L_{\bar{p}}\| \tag{2}$$

holds for all $f \in L_{\bar{p}}^{\Omega}$.

Proof. Step 1. We begin with some preparations. Let $\psi = \psi(x_1, x_2) \in S$ such that supp $F\psi$ is compact and

$$(F\psi)(x) = 1 \quad \text{if} \quad x = (x_1, x_2) \in \Omega .$$

If $f \in L_{\bar{p}}^{\Omega}$ then we have for $x \in R_2$

$$f(x) = F^{-1}(Ff)(x) = F^{-1}(F\psi Ff)(x) = c \int_{R_2} f(y)\, \psi(x-y)\, dy, \tag{3}$$

where the latter equality comes from the fact that the entire analytic function $f(y)$ is of at most polynomial growth. If $x \in R_2$ is fixed then $[f(\cdot)\, \psi(x - \cdot)](y)$ belongs to S, cf. e.g. (1.2.4/2) and (1.2.4/3), and hence to any L_r with $0 < r \leqq \infty$. Furthermore, we have

$$F[f(\cdot)\, \psi(x - \cdot)](\xi) = c[(Ff)(\eta) * e^{-ix\eta}(F^{-1}\psi)(\eta)](\xi)$$

where $*$ stands for the convolution. Then it follows that

$$\operatorname{supp} F[f(\cdot)\, \psi(x - \cdot)] \subset \operatorname{supp} Ff - \operatorname{supp} F\psi$$

is compact. Let $0 < r \leqq 1$. By the classical Nikol'skij inequality, cf. e.g. [T, p. 22] (or (1.5.2/3)), we have

$$|f(x)|^r \leqq c\| f(\cdot)\, \psi(x - \cdot) \mid L_1\|^r \leqq c' \int_{R_2} |f(y)|^r\, |\psi(x-y)|^r\, dy, \tag{4}$$

where c and c' are independent of $x \in R_2$.

Step 2. We prove that

$$\| f \mid L_{(p_1, u_2)}\| \leqq c \| f \mid L_{\bar{p}}\| \tag{5}$$

holds, $f \in L_{\bar{p}}^{\Omega}$. Let $u_2 = \infty$. We choose $r = \min(1, p_1, p_2)$ in (4). Furthermore, for sake of convenience we assume that $\psi(x) = \psi_1(x_1)\, \psi_2(x_2)$. Then (4) and Hölder's inequality yield

$$\begin{aligned} &\Big(\int_{R_1} |f(x_1, x_2)|^{p_1}\, dx_1\Big)^{1/p_1} \\ &\leqq c \Big\| \int_{R_2} |f(x_1 - y_1, y_2)|^r\, |\psi_1(y_1)\, \psi_2(x_2 - y_2)|^r\, dy_1\, dy_2 \mid L_{(p_1/r)|x_1}\Big\|^{1/r} \\ &\leqq c' \Big(\int_{R_1} \| f(x_1, y_2) \mid L_{p_1|x_1}\|^r\, |\psi_2(x_2 - y_2)|^r\, dy_2\Big)^{1/r} \\ &\leqq c'' \| f \mid L_{\bar{p}}\|, \quad x_2 \in R_1, \end{aligned} \tag{6}$$

(usual modification if $p_1 = \infty$). We take the supremum with respect to $x_2 \in R_1$. Then we have

$$\| f \mid L_{(p_1, \infty)}\| \leqq c \| f \mid L_{\bar{p}}\| . \tag{7}$$

Let $p_2 < u_2 < \infty$. We use (7) and obtain

$$\begin{aligned} \| f \mid L_{(p_1, u_2)}\| &= \Big(\int_{R_1} \| f(x_1, x_2) \mid L_{p_1|x_1}\|^{u_2}\, dx_2\Big)^{1/u_2} \\ &\leqq \sup_{x_2 \in R_1} \| f(x_1, x_2) \mid L_{p_1|x_1}\|^{1 - \frac{p_2}{u_2}}\, \| f \mid L_{\bar{p}}\|^{\frac{p_2}{u_2}} \\ &\leqq c \| f \mid L_{\bar{p}}\| . \end{aligned} \tag{8}$$

Now, (7) and (8) prove (5).

Step 3. We prove that

$$\|f \mid L_{\bar{u}}\| \leqq c\|f \mid L_{(p_1,u_2)}\| \tag{9}$$

holds, $f \in L^{\Omega}_{(p_1,u_2)}$. Let $u_1 = \infty$. We choose $r = \min(1, p_1, u_2)$ in (4). Let again $\psi(x) = \psi_1(x_1)\,\psi_2(x_2)$. Then (4) and Hölder's inequality yield

$$\begin{aligned} |f(x_1, x_2)| &\leqq c\Big(\int_{R_2} |f(x_1 - y_1, x_2 - y_2)|^r\, |\psi_1(y_1)\,\psi_2(y_2)|^r\, dy_1\, dy_2\Big)^{1/r} \\ &\leqq c'\Big(\int_{R_1} \|f(y_1, x_2 - y_2) \mid L_{p_1 \mid y_1}\|^r\, |\psi_2(y_2)|^r\, dy_2\Big)^{1/r}. \end{aligned} \tag{10}$$

In (10) we take the supremum with respect to x_1 and the quasi-norm $\|\cdot \mid L_{u_2}\|$ with respect to x_2. Then we have

$$\begin{aligned} &\|f \mid L_{(\infty,u_2)}\| \\ &\leqq c\Big\|\int_{R_1} \|f(y_1, x_2 - y_2) \mid L_{p_1 \mid y_1}\|^r\, |\psi_2(y_2)|^r\, dy_2 \mid L_{u_2/r \mid x_2}\Big\|^{1/r} \\ &\leqq c\|f \mid L_{(p_1,u_2)}\|\, \|\psi_2 \mid L_r\| \leqq c'\|f \mid L_{(p_1,u_2)}\|. \end{aligned} \tag{11}$$

This proves (9) if $u_1 = \infty$. Let $p_1 < u_1 < \infty$. Then we have

$$\begin{aligned} &\|f(x_1, x_2) \mid L_{\bar{u}}\| \\ &= \Big(\int_{R_1}\Big(\int_{R_1} |f(x_1, x_2)|^{u_1}\, dx_1\Big)^{u_2/u_1} dx_2\Big)^{1/u_2} \\ &\leqq \Big[\int_{R_1} \sup_{y_1 \in R_1} |f(y_1, x_2)|^{u_2\left(1 - \frac{p_1}{u_1}\right)} \Big(\int_{R_1} |f(x_1, x_2)|^{p_1}\, dx_1\Big)^{\frac{u_2}{u_1}} dx_2\Big]^{\frac{1}{u_2}}. \end{aligned} \tag{12}$$

We apply Hölder's inequality with $\dfrac{u_1}{p_1}$ and $\dfrac{u_1}{u_1 - p_1}$ to the right-hand side of (12). We use (11) and obtain that

$$\begin{aligned} \|f \mid L_{\bar{u}}\| &\leqq \Big(\int_{R_1} \|f(y_1, x_2) \mid L_{\infty \mid y_1}\|^{u_2}\, dx_2\Big)^{\frac{1}{u_2}\left(1 - \frac{p_1}{u_1}\right)} \|f \mid L_{(p_1,u_2)}\|^{\frac{p_1}{u_1}} \\ &\leqq c\|f \mid L_{(p_1,u_2)}\|^{1 - \frac{p_1}{u_1}}\, \|f \mid L_{(p_1,u_2)}\|^{\frac{p_1}{u_1}} = c\|f \mid L_{(p_1,u_2)}\| \end{aligned} \tag{13}$$

holds. This proves (9).

Step 4. Let $\alpha = 0$ and $b_1 = b_2 = 1$. Then (2) follows from (9) and (5). If α is an arbitrary multi-index then we apply D^α to both sides of (3). If $0 < r \leqq 1$ then the counterpart of (4) reads as follows,

$$|(D^\alpha f)(x)|^r \leqq c \int_{R_2} |f(x - y)|^r\, |D^\alpha \psi(y)|^r\, dy, \quad x \in R_2. \tag{14}$$

Let $r = \min(1, u_1, u_2)$. We apply the norm $\Big\|\cdot \mid L_{\left(\frac{u_1}{r}, \frac{u_2}{r}\right)}\Big\|$ to (14) and obtain that

$$\|D^\alpha f \mid L_{\bar{u}}\| \leqq c\|f \mid L_{\bar{u}}\| \tag{15}$$

holds, $f \in L^{\Omega}_{\bar{u}}$. Now (15) and (2) with $\alpha = 0$ and $b_1 = b_2 = 1$ prove (2) for any α and $b_1 = b_2 = 1$. Let $f \in L^{\Omega}_{\bar{p}}$, where Ω is given by (1) with arbitrary $b_1 > 0$ and $b_2 > 0$. Then $f\left(\dfrac{x_1}{b_1}, \dfrac{x_2}{b_2}\right)$ belongs to $L^{Q_{\bar{1}}}_{\bar{p}}$, where $Q_{\bar{1}}$ is given by (1) with $\bar{1} = (1, 1)$. Then one obtains (2) by a homogeneity argument.

Remark*. The theorem can be found in B. Stöckert [1]. This paper contains also more general assertions of this type. The Banach space case $1 \leqq p_1 \leqq \infty$ and $1 \leqq p_2 \leqq \infty$ has been proved much more earlier by A. P. Uninskij [1, 2].

1.6.3. Properties

In contrast to the theory of the weighted spaces $L_p^\Omega(\varkappa, \mu)$ we develop the theory of the mixed spaces $L_{\bar{p}}^\Omega$ only in such an extent as it is necessary for our later purposes in Chapter 2. However there is no doubt that many properties which we derived in 1.4. and 1.5. for the weighted spaces $L_p^\Omega(\varkappa, \mu)$ or in [T, 1.3, 1.4] for the spaces $L_p^\Omega = L_{\bar{p}}^\Omega$ with $\bar{p} = (p, p)$, have more or less immediate counterparts for the mixed spaces $L_{\bar{p}}^\Omega$: Maximal inequalities, characterizations via atomic measures, representation theorems, density assertions, duality, isomorphism properties, Schauder bases etc. We refer to B. Stöckert [1]. In this subsection we restrict ourselves to the formulation of some basic facts. In the next subsection we prove a maximal inequality which we need later on. If Ω is a compact subset of R_2 then

$$S^\Omega = \{\varphi \mid \varphi \in S, \operatorname{supp} F\varphi \subset \Omega\} \tag{1}$$

has the previous meaning.

Theorem. *Let Ω be a compact subset of R_2 and let $\bar{p} = (p_1, p_2)$ with $0 < p_1 \leqq \infty$ and $0 < p_2 \leqq \infty$. Then $L_{\bar{p}}^\Omega$ is a quasi-Banach space (Banach space if $1 \leqq p_1 \leqq \infty$ and $1 \leqq p_2 \leqq \infty$). Furthermore,*

$$S^\Omega \subset L_{\bar{p}}^\Omega \subset S' \tag{2}$$

holds (topological embedding).

Proof. By Theorem 1.6.2 we have

$$L_{\min(p_1,p_2)}^\Omega \subset L_{\bar{p}}^\Omega \subset L_{\max(p_1,p_2)}^\Omega. \tag{3}$$

Then (2) follows from (3) and [T, (1.4.2/2)] (or (1.5.3/1)). Obviously, $L_{\bar{p}}^\Omega$ is a quasi-normed linear space. The completeness follows in the same way as in [T, p. 24] (or in Step 3 of the proof of Theorem 1.5.3), where we used (1.6.2/2) with $\alpha = 0$ and $\bar{u} = (\infty, \infty)$.

Remark. Corresponding assertions for $L_p^\Omega = L_{\bar{p}}^\Omega$ with $\bar{p} = (p, p)$ may be found in [T, 1.4.2]. The weighted counterpart has been proved in 1.5.3.

1.6.4. A Maximal Inequality

As has been said in Remark 1.6.1 it is not our intention to develop the theory of the mixed spaces $L_{\bar{p}}^\Omega$ in the same extent as for the unmixed spaces L_p^Ω in [T, Chapter 1] or the (unmixed) weighted spaces $L_p^\Omega(\varkappa, \mu)$ in the preceding respective sections. However we need the estimate (1) below in 1.10.2., and we take the opportunity to complement Theorem 1.6.2 by the maximal inequality (2) below. Then we have (together with Theorem 1.6.2) a full counterpart of [T, Theorem 1.4.1] (unmixed case) and of Theorem 1.5.2 (weighted case) at least as far as Lebesgue measures are concerned. We use the maximal function $(M_2 f)(x) = (Mf(x_1, \cdot))(x_2)$ from (1.3.3/4) with $x = (x_1, x_2)$ and its obvious counterpart $(M_1 f)(x) = (Mf(\cdot, x_2))(x_1)$ (where x_1 and x_2 change their roles).

Theorem. *Let Ω be a compact subset of R_2 and let $\bar{p} = (p_1, p_2)$* with $0 < p_1 \leqq \infty$ and $0 < p_2 < \infty$.

(i) *Let* $0 < r_1 < \infty$ *and* $0 < r_2 < \infty$. *Let* $\alpha = (\alpha_1, \alpha_2)$ *be a multi-index. Then there exist two positive numbers* c_1 *and* c_2 *such that*

$$\sup_{z\in R_2} \frac{|D^\alpha f(x-z)|}{(1+|z_1|^{1/r_1})(1+|z_2|^{1/r_2})} \leqq c_1 \sup_{z\in R_2} \frac{|f(x-z)|}{(1+|z_1|^{1/r_1})(1+|z_2|^{1/r_2})}$$
$$\leqq c_2 (M_2(M_1|f|^{r_1})^{r_2/r_1})^{1/r_2}(x) \tag{1}$$

holds for all $f \in L^\Omega_{\bar p}$ *and all* $x \in R_2$.

(ii) *Let* $0 < r_1 < p_1$ *and* $0 < r_2 < \min(p_1, p_2)$. *Then there exists a positive number* c *such that*

$$\left\| \sup_{z\in R_2} \frac{|f(\cdot - z)|}{(1+|z_1|^{1/r_1})(1+|z_2|^{1/r_2})} \,|\, L_{\bar p} \right\| \leqq c \|f \,|\, L_{\bar p}\| \tag{2}$$

holds for all $f \in L^\Omega_{\bar p}$.

Proof. Step 1. In order to prove the left-hand side of (1) it is sufficient to deal with ∇ instead of D^α, the rest is a matter of mathematical induction, cf. (1.6.2/2). However this modified inequality follows from Step 1 of the proof of Theorem 1.4.2 (obvious modifications, in particular $\varrho(x) = 1$ and $\omega(x) = 0$).

Step 2. Let $Q_\delta = [-\delta, \delta] \times [-\delta, \delta]$ with $0 < \delta < 1$. The obvious counterpart of (1.4.2/8) reads as follows,

$$|f(z)| \leqq c\delta \sup_{w\in Q_\delta} |(\nabla f)(w)| + c\delta^{-\frac{1}{r_1}-\frac{1}{r_2}} \left(\int_{-\delta}^{\delta} \left(\int_{-\delta}^{\delta} |f(w)|^{r_1}\, dw_1 \right)^{\frac{r_2}{r_1}} dw_2 \right)^{\frac{1}{r_2}}, \tag{3}$$

where c is independent of δ with $0 < \delta < 1$, $z \in Q_\delta$ and $f \in L^\Omega_{\bar p}$. In particular, if $x \in R_2$ and $y \in R_2$, then we have

$$|f(x-y)| \leqq c\delta \sup_{\omega\in Q_\delta} |(\nabla f)(x-y-w)|$$
$$+ c\delta^{-\frac{1}{r_1}-\frac{1}{r_2}} \left(\int_{-\delta}^{\delta} \left(\int_{-\delta}^{\delta} |f(x-y-w)|^{r_1}\, dw_1 \right)^{\frac{r_2}{r_1}} dw_2 \right)^{\frac{1}{r_2}}. \tag{4}$$

The last integral in (4) can be estimated from above by

$$\left(\int_{-(1+|y_2|)}^{1+|y_2|} \left(\int_{-(1+|y_1|)}^{1+|y_1|} |f(x-u)|^{r_1}\, du_1 \right)^{r_2/r_1} du_2 \right)^{1/r_2}$$
$$\leqq c(1+|y_1|^{1/r_1})(1+|y_2|^{1/r_2})(M_2(M_1|f|^{r_1})^{r_2/r_1})^{1/r_2}(x). \tag{5}$$

We put (5) in (4) and obtain that

$$\sup_{y\in R_2} \frac{|f(x-y)|}{(1+|y_1|^{1/r_1})(1+|y_2|^{1/r_2})}$$
$$\leqq c\delta \sup_{y\in R_2} \frac{|(\nabla f)(x-y)|}{(1+|y_1|^{1/r_1})(1+|y_2|^{1/r_2})} + c\delta^{-\frac{1}{r_1}-\frac{1}{r_2}} \left(M_2(M_1|f|^{r_1})^{\frac{r_2}{r_1}} \right)^{\frac{1}{r_2}}(x), \tag{6}$$

holds, where c is independent of δ. If we choose δ sufficiently small, then the second inequality in (1) follows from (6) and the first inequality in (1) (with ∇ instead of D^α) which we proved in Step 1. (By (1.6.2/2) all terms in (6) and (1) are finite). The proof of (1) is complete.

Step 3. We prove (2). For sake of clarity we use again the notation $L_{q|x_i}$ as it had been explained at the end of 1.3.2. By (1) and (1.3.3/5) we have

$$\left\| \sup_{z \in R_2} \frac{|f(\cdot - z)|}{(1 + |z_1|^{1/r_1})(1 + |z_2|^{1/r_2})} \mid L_{\bar{p}} \right\|$$
$$\leqq c_1 \| \, \| M_2 (M_1 |f|^{r_1})^{r_2/r_1} \mid L_{p_1/r_2 | x_1} \| \, L_{p_2/r_2 | x_2} \|^{1/r_2}$$
$$\leqq c_2 \| \, \| (M_1 |f|^{r_1})^{r_2/r_1} \mid L_{p_1/r_2 | x_1} \| \, L_{p_2/r_2 | x_2} \|^{1/r_2}$$
$$= c_2 \| \, \| M_1 |f|^{r_1} \mid L_{p_1/r_1 | x_1} \|^{1/r_1} \mid L_{p_2 | x_2} \| \, . \tag{7}$$

We apply the usual maximal inequality (1.3.3/3) to the inner expression on the right-hand side. It follows that (7) can be estimated from above by

$$c \| \, \| |f|^{r_1} \mid L_{p_1/r_1 | x_1} \|^{1/r_1} \mid L_{p_2 | x_2} \| = c \| f \mid L_{\bar{p}} \| \, .$$

This proves (2).

Remark 1. This is the mixed counterpart of Theorem 1.4.2. Cf. also Remark 1.4.2/1 as far as references are concerned. The unmixed counterpart may be found in [T, Theorem 1.3.1].

Remark 2. We used $p_2 < \infty$ in (7), but not in the proof of (1). In other words: (1) holds for all couples $\bar{p} = (p_1, p_2)$ with $0 < p_1 \leqq \infty$ and $0 < p_2 \leqq \infty$.

1.7. Fourier Multipliers for Weighted L_p-Spaces of Entire Analytic Functions

1.7.1. Introduction and Definitions

Let $L_p = L_p(R_n)$ be the usual spaces of complex-valued p-integrable functions on R_n, cf. 1.3.1. Let $1 \leqq p \leqq \infty$ and $1 \leqq q \leqq \infty$. Then $M \in S' = S'(R_n)$ is said to be a Fourier multiplier of type M_p^q if there exists a positive number c such that

$$\| F^{-1} M F f \mid L_q \| \leqq c \| f \mid L_p \| \tag{1}$$

holds for all $f \in S$. We write $F^{-1}MFf$ instead of $F^{-1}[M(Ff)]$. If $p < \infty$ then the extension by continuity yields a linear and bounded operator from L_p in L_q. The case $p = q$ is of special interest. Problems of this type have been studied extensively for more than 40 years. Milestone theorems in this field are connected with the names of J. Marcinkiewicz, S. G. Michlin, L. Hörmander, E. M. Stein and others. Problems of this type have been studied for their own sake, but they are also of crucial interest in the theory of singular integrals, partial differential equations etc. We shall not go into detail, references may be found in E. M. Stein [1], E. M. Stein, G. Weiss [1], [T, 2.2.4] and (as far as more recent developments are concerned) R. E. Edwards [1, in particular Chapter 16]. Fourier multipliers have also been studied for other spaces, in particular for (unweighted and unmixed) spaces of Besov-Hardy-Sobolev type. The interested reader should consult [T, in particular 2.6], where we gave many references, cf. also J. Peetre [4]. Sections 1.7. and 1.8. of the present book deal with Fourier multiplier problems in the weighted spaces $L_p^\Omega(\varkappa, \mu)$ and the mixed spaces $L_{\bar{p}}^\Omega$, respectively. Vector-valued counterparts are treated in 1.9. and 1.10., respectively. Also in the following chapters we return several times to Fourier multiplier problems.

We recall that $L_p^\Omega(\varkappa, \mu)$ always has the meaning of Definition 1.5.1 (inclusively $\omega(x) \in \mathfrak{M}$ and $\varrho(x) \in R(\omega)$, cf. 1.2.1. and 1.4.1.): Weighted spaces of entire analytic functions on R_n. By Theorem 1.5.3 we may assume that $L_p^\Omega(\varkappa, \mu) = L_p^\Omega(\varrho, \mu_L)$, where

μ_L stands for the Lebesgue measure in R_n and $\varrho(x)$ is the weight from Definition 1.4.1(i). Finally $\|\cdot \mid L_p\|$ has the same meaning as in 1.3.1. (with respect to the Lebesgue measure on R_n). We give two different definitions of Fourier multipliers in $L_p^\Omega(\varkappa, \mu)$ and discuss their interrelations in due course. As usual we omit the suffix R_n in our notations.

Definition 1. *Let $\omega(x) \in \mathfrak{M}$ and $\varrho(x) \in R(\omega)$. Let Ω be a compact subset of R_n and let $0 < p \leqq \infty$. Let $M \in S'_\omega$ and $e^{\omega(x)}(F^{-1}M)(x) \in L_1$. Then M is said to be a Fourier multiplier for $L_p^\Omega(\varrho, \mu_L)$ if there exists a constant c such that*

$$\|\varrho F^{-1}MFf \mid L_p\| \leqq c\|\varrho f \mid L_p\| \tag{2}$$

holds for all $f \in L_p^\Omega(\varrho, \mu_L)$.

Remark 1. We recall that $F^{-1}MFf$ stands for $F^{-1}[M(Ff)]$. This definition is the direct counterpart of [T, Definition 1.5.1]. The a priori assumptions for M ensure that $F^{-1}MFf$ makes sense for any $f \in L_p^\Omega(\varrho, \mu_L)$: If $f \in L_p^\Omega(\varrho, \mu_L)$ then $\varrho f \in L_\infty$, cf. Theorem 1.5.2 (iii), and by (1.4.1/2) we have

$$\varrho(x)\,|(F^{-1}MFf)(x)| \leqq c \int_{R_n} e^{\omega(x-y)}\,|(F^{-1}M)(x-y)|\,\varrho(y)\,|f(y)|\,dy, \quad x \in R_n. \tag{3}$$

At the first glance, the a priori assumptions for M seem to be rather restrictive. E.g., a corresponding assumption $F^{-1}M \in L_1$ in order to study Fourier multipliers for L_p with $1 \leqq p \leqq \infty$ would be simply irrelevant: Any such function would be a Fourier multiplier and the more interesting Fourier multipliers à la Michlin-Hörmander are not covered. However for spaces of entire analytic functions the situation is different: On the basis of the above definition we obtain substantial assertions (in particular for $p < 1$). If one uses these spaces as buildings blocks in order to construct corresponding Besov spaces, then one obtains on this way multiplier assertions of Michlin-Hörmander type and beyond. Cf. [T, in particular 2.6] as far as the unweighted case is concerned. We give a second definition of a Fourier multiplier with less restrictive a priori assumptions for M.

Definition 2. *Let $\omega(x) \in \mathfrak{M}$ and $\varrho(x) \in R(\omega)$. Let Ω be a compact subset of R_n and let $0 < p \leqq \infty$. Let $M \in S'_\omega$ be essentially bounded on Ω. Then M is said to be a Fourier multiplier for $L_p^\Omega(\varrho, \mu_L)$ if there exists a constant c such that* (2) *holds for all $f \in S_\omega^\Omega$.*

Remark 2. Of course, S'_ω are the ultra-distributions from Definition 1.2.1/3 and S_ω^Ω has been defined in (1.4.2/1). Obviously, $F^{-1}M\,Ff$ with $f \in S_\omega^\Omega$ is well-defined, and only the values of $M(x)$ on Ω are of interest. Our hypothesis, that the restriction of $M \in S'_\omega$ on Ω is an essentially bounded function is not important but convenient for technical reasons. If one compares the two definitions then it is quite clear that the assumptions for M in the second definition are more natural than the ones in the first definition. The disadvantage of the second definition is the restriction $f \in S_\omega^\Omega$. But this is not a severe handicap. Let Λ be a bounded open set in R_n with the segment property, cf. 1.5.4., and let $\Omega = \bar{\Lambda}$. Let $0 < p < \infty$ and let M be a Fourier multiplier in the sense of the second definition. Then S_ω^Ω is dense in $L_p^\Omega(\varrho, \mu_L)$, and consequently, $F^{-1}MF$ can be extended by continuity from S_ω^Ω to $L_p^\Omega(\varrho, \mu_L)$. If $p = \infty$ then one has not this argument at hand, however one can use the approximation procedure which we described in 1.5.4. In some pathological cases the second definition is even meaningless, e.g. if $\Omega = \{0\}$ and $\varrho^{-1}(x) = \omega(x) = e^{|x|^\beta}$ with $0 < \beta < 1$, cf. 1.5.4. If one looks for sufficient conditions then the following point of view is reasonable: If Ω_1 and Ω_2 are compact subsets of R_n with $\Omega_1 \subset \Omega_2$ and if M is a Fourier multiplier for $L_p^{\Omega_2}(\varrho, \mu_L)$ in the sense of the second definition then it is also a Fourier multiplier for $L_p^{\Omega_1}(\varrho, \mu_L)$. If one chooses $\Omega_2 = Q_b$, cf. (1.5.5/1), then one has the above rather satisfactory situation (at least if $p < \infty$), because the interior of Q_b is a domain with the segment property. In order to include also $p = \infty$, one can postulate the following modification of Definition 2: *A linear and bounded operator T from $L_p^\Omega(\varrho, \mu_L)$ into itself is said to be a Fourier multiplier for $L_p^\Omega(\varrho, \mu_L)$ if the restriction of M on Ω is an essentially bounded function on Ω and if $T\varphi = F^{-1}MF\varphi$ holds for all $\varphi \in S_\omega^\Omega$.* For interesting concrete Fourier multipliers it is rather immaterial whether its definition is based on Definition 1, Definition 2 or the just-mentioned modification. Mostly for this reason we stop our meditations about the above definitions. We use the first definition in order to have a quick proof of a substantial Fourier multiplier assertion and we use the second definition in order to shed light on the lattice structure of Fourier multipliers.

Remark 3. We consider Fourier multipliers from $L_p^\Omega(\varrho, \mu_L)$ into itself. One can extend this task and ask for Fourier multipliers from $L_p^\Omega(\varrho, \mu_L)$ into $L_{\tilde{p}}^\Omega(\tilde{\varrho}, \mu_L)$. Our methods should be applicable also to this more general case.

1.7.2. Basic Multiplier Theorem

In this subsection a Fourier multiplier should be understood in the sense of Definition 1.7.1/1.

Theorem. *Let $\omega(x) \in \mathfrak{M}$ and $\varrho(x) \in R(\omega)$. Let Ω and Γ be compact subsets of R_n. Let $0 < p \leqq \infty$ and $\tilde{p} = \min(1, p)$. Then there exists a constant c such that*

$$\|\varrho F^{-1}MFf \mid L_p\| \leqq c\|e^\omega F^{-1}M \mid L_{\tilde{p}}\| \, \|\varrho f \mid L_p\| \tag{1}$$

holds for all $f \in L_p^\Omega(\varrho, \mu_L)$ and all $F^{-1}M \in L_{\tilde{p}}^\Gamma(e^\omega, \mu_L)$.

Proof. If $F^{-1}M \in L_{\tilde{p}}^\Gamma(e^\omega, \mu_L)$ then we have $F^{-1}M \in L_1^\Gamma(e^\omega, \mu_L)$, cf. (1.5.2/3). This shows that Definition 1.7.1./1 makes sense. If $p \geqq 1$ then (1) follows immediately from (1.7.1/3). Let $p < 1$. We have

$$(F^{-1}MFf)(x) = c \int_{R_n} (F^{-1}M)(y) f(x-y) \, dy, \quad x \in R_n. \tag{2}$$

For fixed $x \in R_n$, the integrand in (2) is considered as a function of y. In the same way as in the first step of the proof of Theorem 1.6.2 (cf. also [T, p. 25]) we have

$$\begin{aligned} \operatorname{supp} F[(F^{-1}M)(\cdot) f(x-\cdot)] &\subset \operatorname{supp} M + \operatorname{supp} F(f(x-\cdot)) \\ &= \operatorname{supp} M - \operatorname{supp} Ff \subset \Gamma - \Omega. \end{aligned} \tag{3}$$

We use again (1.4.1/2) and obtain that

$$\begin{aligned} \varrho(x) \|(F^{-1}M)(\cdot) f(x-\cdot) \mid L_p\| &\leqq c\|e^{\omega(\cdot)} (F^{-1}M)(\cdot) \varrho(x-\cdot) f(x-\cdot) \mid L_p\| \\ &\leqq c'\|e^\omega F^{-1}M \mid L_p\| \, \|\varrho f \mid L_\infty\| < \infty \end{aligned} \tag{4}$$

holds. We apply (1.5.2/3) (with $q = 1$) to $(F^{-1}M)(y) f(x-y) \in L_p^{\Gamma-\Omega}$ (unweighted space with respect to the Lebesgue measure). Then it follows from (2) and (4) that

$$\begin{aligned} \varrho^p(x) |(F^{-1}MFf)(x)|^p &\leqq c\varrho^p(x) \|(F^{-1}M)(\cdot) f(x-\cdot) \mid L_1\|^p \\ &\leqq c\varrho^p(x) \|(F^{-1}M)(\cdot) f(x-\cdot) \mid L_p\|^p \\ &\leqq c' \int_{R_n} e^{\omega(y)p} |(F^{-1}M)(y)|^p \varrho^p(x-y) |f(x-y)|^p dy \end{aligned} \tag{5}$$

holds. Integration with respect to $x \in R_n$ yields (1).

Remark 1. This is the counterpart of [T, Proposition 1.5.1]. Of course, only the case $p < 1$ is not obvious.

Remark 2. Because Ff has a compact support we always may assume without restriction of generality that $M = FF^{-1}M$ also has a compact support.

1.7.3. Convolution Algebras

If $f \in L_1$ and $g \in L_1$ are given functions then the convolution $f * g$ is defined by

$$(f * g)(x) = \int_{R_n} f(x-y) g(y) \, dy, \quad x \in R_n. \tag{1}$$

If c is an appropriate positive number then (1) can be rewritten as $f * g = cF^{-1}[FfFg]$. This reformulation fits very well in our framework and one can define on that way the convolution $f * g$ for other couples f and g (but this is not necessary for our purposes in this subsection). Let $\omega(x) \in \mathfrak{M}$ and $0 < p \leqq 1$. Let Ω be a compact subset of R_n and let $f \in L_p^\Omega(\mathrm{e}^\omega, \mu_L)$ and $g \in L_p^\Omega(\mathrm{e}^\omega, \mu_L)$. Then it follows from (1.5.2/3) that $f \in L_1^\Omega(\mathrm{e}^\omega, \mu_L)$ and $g \in L_1^\Omega(\mathrm{e}^\omega, \mu_L)$. In particular, (1) makes sense.

Theorem. *Let $\omega(x) \in \mathfrak{M}$ and $0 < p \leqq 1$. Let Ω be a compact subset of R_n. Then $L_p^\Omega(\mathrm{e}^\omega, \mu_L)$ is a convolution algebra (i.e. an algebra with the convolution as multiplication) and there exists a constant c such that*

$$\|\mathrm{e}^\omega (f * g) \mid L_p\| \leqq c\|\mathrm{e}^\omega f \mid L_p\| \, \|\mathrm{e}^\omega g \mid L_p\| \tag{2}$$

holds for all $f \in L_p^\Omega(\mathrm{e}^\omega, \mu_L)$ and all $g \in L_p^\Omega(\mathrm{e}^\omega, \mu_L)$.

Proof. As has been remarked above, $f * g$ is well-defined. Furthermore we have

$$\operatorname{supp} F(f * g) = \operatorname{supp} Ff \cdot Fg \subset \Omega \tag{3}$$

and (2) follows from (1.7.2/1) with $\varrho = \mathrm{e}^\omega$ and $f * g = cF^{-1}(FgFf)$. This proves that $L_p^\Omega(\mathrm{e}^\omega, \mu_L)$ is an algebra with the convolution as multiplication.

Remark 1. In a formal way the last theorem can be written as

$$L_p^\Omega(\mathrm{e}^\omega, \mu_L) * L_p^\Omega(\mathrm{e}^\omega, \mu_L) \subset L_p^\Omega(\mathrm{e}^\omega, \mu_L). \tag{4}$$

By (1.2.1/4), (1.2.1/5) and the equivalence assertions from Theorem 1.5.3(ii) (cf. also (1.4.1/13)) we have the following examples:

$$L_p^\Omega(|x|^d, \mu_L) \quad \text{with} \quad d \geqq 0 \quad \text{and} \quad 0 < p \leqq 1, \tag{5}$$

and

$$L_p^\Omega(\mathrm{e}^{|x|^\beta}, \mu_L) \quad \text{with} \quad 0 < \beta < 1 \quad \text{and} \quad 0 < p \leqq 1 \tag{6}$$

are convolution algebras. Formally, $d = 0$ in (5) is not covered by our approach. However it can be included immediately, cf. also [T, 1.5.3].

Remark 2. Let $0 < p \leqq 1$ and $d \geqq 0$. Then there exists a constant c such that

$$\| |x|^d f * g \mid L_p\| \leqq cb^{d+n\left(\frac{1}{p}-1\right)} \| |x|^d f \mid L_p\| \, \| |x|^d g \mid L_p\| \tag{7}$$

holds for all $b > 0$, all $f \in L_p^{Q_b}(|x|^d, \mu_L)$ and all $g \in L_p^{Q_b}(|x|^d, \mu_L)$, where Q_b stands for the cube from (1.5.5/1). This follows from $f\left(\frac{x}{b}\right) \in L_p^{Q_1}(|x|^d, \mu_L)$ if $f(x) \in L_p^{Q_b}(|x|^d, \mu_L)$ and a simple homogeneity argument.

Remark* 3. We followed essentially H. Triebel [2,5]. The unweighted case, i.e. $d = 0$ in (5), may be found in J. Peetre [4, p. 237] and [T, p. 28]. Inequalities of type (7) are useful in the theory of function spaces, cf. J. Peetre [4] and B. Jawerth [1].

1.7.4. Some Prerequisites: Besov-Sobolev Spaces

Effective criteria for Fourier multipliers $M(x)$ for the L_p-spaces with $1 < p < \infty$ are usually formulated via local smoothness properties of $M(x)$ and growth conditions. This can be done in the language of the spaces C^m (of m times continuously differentiable functions), W_2^m (Sobolev spaces) or $B_{2,q}^s$ (Besov spaces), at least locally, cf. e.g. L. Hörmander [1] and E. M. Stein [1, pp. 96, 109, 232]. This type of Michlin-Hörmander multiplier theorems can be extended to Hardy spaces, cf. C. Fefferman, E. M. Stein [2, p. 150]. A systematic treatment of corresponding multiplier theorems for the spaces $B_{p,q}^s(R_n)$ and $F_{p,q}^s(R_n)$ has been given in [T, in particular 2.6], cf. also

J. Peetre [4]. In the subsequent chapters of this book we describe respective multiplier theorems for different types of $B^s_{p,q} - F^s_{p,q}$-spaces: with dominating mixed smoothness properties (Chapter 2), periodic spaces (Chapter 3), weighted spaces (Section 5.1.), and abstract spaces (Chapter 6). In the next subsection we wish to formulate corresponding multiplier assertions for the spaces $L^\Omega_p(\varrho, \mu_L)$ (at least for some special weights $\varrho(x)$). For this purpose we need some facts from the theory of the spaces W^s_2 and $B^s_{2,q}$ with $s \geqq 0$ and $0 < q \leqq \infty$. A systematic treatment of the spaces $B^s_{p,q}(R_n)$ and $F^s_{p,q}(R_n)$ with $s \in R_1$, $0 < p < \infty$ and $0 < q \leqq \infty$ has been given in [T] (cf. also [I]). The special case $p = 2$ simplifies this theory essentially. We restrict ourselves to formulations. As for further details and proofs we refer to [T] (and [I]).

Let $s \geqq 0$. Then $H^s_2 = H^s_2(R_n)$ are the usual Bessel-potential spaces. The Sobolev spaces $W^m_2 = H^m_2$ with $m = 1, 2, \dots$ are special cases, $H^0_2 = L_2$. The spaces H^s_2 can be defined via

$$H^s_2 = \{f \mid f \in S', \|f \mid H^s_2\| = \|(1 + |x|^2)^{s/2} Ff \mid L_2\| < \infty\}. \tag{1}$$

If $s = m = 1, 2, \dots$ then simple properties of the Fourier transform yield

$$W^m_2 = \left\{f \mid f \in S', \|f \mid W^m_2\| = \Big(\sum_{|\alpha| \leqq m} \|D^\alpha f \mid L_2\|^2\Big)^{1/2} < \infty\right\} \tag{2}$$

(equivalent norms). This is the well-known definition of W^m_2. Let

$$K_0 = \{x \mid |x| \leqq 1\}, \qquad K_j = \{x \mid 2^{j-1} \leqq |x| \leqq 2^j\}, \tag{3}$$

where $j = 1, 2, 3, \dots$ Let $L_2(K_j)$ be the usual L_2-spaces on K_j. Then it follows immediately that

$$\|f \mid H^s_2\|^* = \left(\sum_{j=0}^\infty 2^{2js} \|Ff \mid L_2(K_j)\|^2\right)^{1/2} \tag{4}$$

is an equivalent norm on H^s_2. Let $0 < q < \infty$ and $s > 0$. Then we generalize (1) and (4) in the following way,

$$B^s_{2,q} = \left\{f \mid f \in S', \|f \mid B^s_{2,q}\|^* = \left(\sum_{j=0}^\infty 2^{sqj} \|Ff \mid L_2(K_j)\|^q\right)^{1/q} < \infty\right\}. \tag{5}$$

Similarly one defines $B^s_{2,\infty}$, however this is not of interest for us at this moment. We have

$$B^s_{2,2} = H^s_2, \quad s > 0. \tag{6}$$

$B^s_{2,q}$ are the usual Besov spaces. They are quasi-Banach spaces (Banach spaces if $q \geqq 1$) with $\|\cdot \mid B^s_{2,q}\|^*$ as quasi-norm (norm if $q \geqq 1$). In order to describe the counterpart of (2) we introduce the differences

$$(\Delta^1_h f)(x) = f(x + h) - f(x), \qquad \Delta^2_h = \Delta^1_h \Delta^1_h, \quad h \in R_n, x \in R_n. \tag{7}$$

Let $0 < s = [s] + \{s\}$, where $[s]$ is an integer and $0 < \{s\} < 1$. Then

$$\|f \mid B^s_{2,q}\| = \sum_{|\alpha| \leqq [s]} \|D^\alpha f \mid L_2\| + \sum_{|\alpha| = [s]} \left(\int_{R_n} |h|^{-\{s\}q} \|\Delta^1_h D^\alpha f \mid L_2\|^q \frac{dh}{|h|^n}\right)^{1/q} \tag{8}$$

is an equivalent quasi-norm on $B^s_{2,q}$. If s is a natural number then

$$\|f \mid B^s_{2,q}\| = \sum_{|\alpha| \leqq s-1} \|D^\alpha f \mid L_2\| + \sum_{|\alpha| = s-1} \left(\int_{R_n} |h|^{-q} \|\Delta^2_h D^\alpha f \mid L_2\|^q \frac{dh}{|h|^n}\right)^{1/q} \tag{8'}$$

is an equivalent quasi-norm, cf. [T, Theorem 2.5.12 and Remark 2.5.12/2]. If $1 \leqq q < \infty$ then we have a well-known classical assertion, cf. S. M. Nikol'skij [2], [I] or [T, 2.5.7 and 2.2.2]. If s is a natural number then we recall that $B^s_{2,2} = W^s_2$ holds and consequently, $\|f \mid W^s_2\|$ is also an equivalent norm.

Let Λ be a bounded C^∞-domain in R_n, or an open cube (or more general an open rectangle) in R_n. Then $B^s_{2,q}(\Lambda)$ with $s > 0$ and $0 < q < \infty$ is defined as the restriction of $B^s_{2,q} = B^s_{2,q}(R_n)$ on Λ (considered as a subspace of the distributions on Λ), quasi-normed via

$$\|f \mid B^s_{2,q}(\Lambda)\| = \inf \|g \mid B^s_{2,q}\| \tag{9}$$

where the infimum is taken over all $g \in B^s_{2,q} = B^s_{2,q}(R_n)$ with $g|_\Lambda = f$ (in the sense of distributions on Λ). These are the usual Besov-Sobolev spaces on domains, they are factor spaces of the corresponding spaces on R_n. Of course, if s is a natural number then $W^s_2(\Lambda) = B^s_{2,2}(\Lambda)$ are the usual Sobolev spaces on Λ. More generally, $B^s_{2,q}(\Lambda)$ are the familiar Besov spaces on domains, they can be normed (at least if $1 \leqq q < \infty$) in the usual way, cf. O. V. Besov, V. P. Il'in, S. M. Nikol'skij [1], [I] or [T, 3.4.2]. The spaces $B^s_{2,q}(\Lambda)$ have the extension property, cf. [T, 3.3.4]: Let $s > 0$ and $0 < q < \infty$. Then there exists a linear and bounded operator S (extension operator) from $B^s_{2,q}(\Lambda)$ into $B^s_{2,q}$ with

$$(Sf)(x) = f(x) \quad \text{if} \quad x \in \Lambda. \tag{10}$$

If Γ is another bounded C^∞-domain (or open cube or open rectangle) with $\bar{\Lambda} \subset \Gamma$, then we may assume that

$$\operatorname{supp} Sf \subset \Gamma \quad \text{for any} \quad f \in B^s_{2,q}(\Lambda). \tag{11}$$

This follows from the multiplication properties of the spaces $B^s_{2,q}(R_n)$, cf. [T, 2.8.2]. Finally we mention a few elementary topological embedding theorems which follow immediately from (5): If $s > 0$, $\varepsilon > 0$, $0 < q < \infty$ and $0 < q_0 \leqq q_1 < \infty$ then it follows that

$$B^{s+\varepsilon}_{2,q} \subset B^s_{2,q_0} \subset B^s_{2,q_1} \tag{12}$$

holds. In particular, we have

$$H^{s+\varepsilon}_2 \subset B^s_{2,q} \subset H^s_2 \quad \text{if} \quad 0 < q \leqq 2, s > 0 \quad \text{and} \quad \varepsilon > 0. \tag{13}$$

By the above procedure, we have the same embeddings for the respective spaces on Λ.

1.7.5. Special Multiplier Theorem

We wish to prove a Fourier multiplier theorem of Michlin-Hörmander-Fefferman-Stein type for the spaces $L^\Omega_p(\varrho, \mu_L)$. The basis is Theorem 1.7.2. For this purpose we are forced to specify the underlying weights $\omega(x) \in \mathfrak{M}$ and $\varrho(x) \in R(\omega)$. Let

$$\omega(x) = d \log(1 + |x|), \quad x \in R_n, \quad d > 0, \tag{1}$$

be the weights from (1.2.1/4). In other words: In this subsection we do not need ultra-distributions, the usual distributions are sufficient. Interesting examples of weights $\varrho(x) \in R(\omega)$ in the sense of Definition 1.4.1(i) are given by

$$\varrho(x) = (1 + |x|)^\lambda \quad \text{with} \quad |\lambda| \leqq d. \tag{2}$$

If $\lambda \neq 0$ is given, then the best results will be obtained if one chooses $d = |\lambda|$. The unweighted case $\lambda = 0$ is also covered by our approach, but we shall not stress this point, because this special case has been considered in [T, 1.5.2 and in particular 1.5.4].

Proposition. *Let $d \geqq 0$ and $0 < p \leqq 2$. Then there exists a positive number c such that*

$$\|(1 + |x|)^d \, Ff \mid L_p\| \leqq c \left\| f \mid B_{2,p}^{d+n\left(\frac{1}{p}-\frac{1}{2}\right)} \right\| \tag{3}$$

holds for all $f \in B_{2,p}^{d+n\left(\frac{1}{p}-\frac{1}{2}\right)}$.

Proof. We have

$$\begin{aligned} \|(1 + |x|)^d \, Ff \mid L_p\|^p &\leqq c \sum_{j=0}^{\infty} 2^{jdp} \|Ff \mid L_p(K_j)\|^p \\ &\leqq c' \sum_{j=0}^{\infty} 2^{jdp} \|Ff \mid L_2(K_j)\|^p \cdot 2^{jn\frac{2-p}{2}}, \end{aligned} \tag{4}$$

where K_j, $L_2(K_j)$ (and similarly $L_p(K_j)$) have the same meaning as in 1.7.4. Now, (3) follows from (1.7.4/5).

Remark* 1. Results of this type are known, cf. J. Peetre [1, pp. 9–11] and the references given there. (3) with $d = 0$ coincides with [T, (1.5.4/1)].

Let $0 < p \leqq \infty$ be given. Then we put $\tilde{p} = \min(1, p)$ and

$$\sigma_{d,p} = \begin{cases} d + \dfrac{n}{2} & \text{if} \quad 1 \leqq p \leqq \infty \quad \text{and} \quad d \geqq 0 \\ d + n\left(\dfrac{1}{p} - \dfrac{1}{2}\right) & \text{if} \quad 0 < p < 1 \quad \text{and} \quad d \geqq 0. \end{cases} \tag{5}$$

Theorem. *Let $d > 0$, $\omega(x) = d \log(1 + |x|)$ and $\varrho(x) \in R(\omega)$ (e.g. $\varrho(x) = (1 + |x|)^\lambda$ with $|\lambda| \leqq d$). Let Λ be a bounded C^∞-domain or an open cube (or open rectangle) in R_n. Let $\Omega = \bar{\Lambda}$ and let $0 < p \leqq \infty$. Then $M(x) \in B_{2,\tilde{p}}^{\sigma_{d,p}}(\Lambda)$ is a Fourier multiplier for $L_p^\Omega(\varrho, \mu_L)$ and there exists a constant c such that*

$$\|\varrho F^{-1} M F f \mid L_p\| \leqq c \|M \mid B_{2,\tilde{p}}^{\sigma_{d,p}}(\Lambda)\| \, \|\varrho f \mid L_p\| \tag{6}$$

holds for every $M \in B_{2,\tilde{p}}^{\sigma_{d,p}}(\Lambda)$ and every $f \in L_p^\Omega(\varrho, \mu_L)$.

Proof. Let S be the extension operator from 1.7.4. with (1.7.4/11). Then we have $SM \in B_{2,\tilde{p}}^{\sigma_{d,p}}$ and by (3) (with F^{-1} instead of F)

$$F^{-1} SM \in L_{\tilde{p}}^{\Gamma}(e^\omega, \mu_L). \tag{7}$$

We apply (1.7.2/1) with

$$\|(1 + |x|)^d \, F^{-1} SM \mid L_{\tilde{p}}\| \leqq c \|SM \mid B_{2,\tilde{p}}^{\sigma_{d,p}}\| \leqq c' \|M \mid B_{2,\tilde{p}}^{\sigma_{d,p}}(\Lambda)\|.$$

This proves (6).

Remark 2. The unweighted substitute of (6) reads as follows: *There exists a constant c such that*

$$\|F^{-1} M F f \mid L_p\| \leqq c \|M \mid B_{2,\tilde{p}}^{\sigma_{0,p}}(\Lambda)\| \, \|f \mid L_p\|$$

holds for every $M \in B_{2,\tilde{p}}^{\sigma_{0,p}}(\Lambda)$ and every $f \in L_p^\Omega$. Cf. [T, 1.5.4].

Remark* 3. The number $\frac{n}{2}$ is typical for Fourier multiplier theorems for L_p-spaces with $1 < p < \infty$ and for the Hardy space H_1, cf. E. M. Stein [1, p. 96], J. Peetre [1] and C. Fefferman, E. M. Stein [2, p. 150]. If $0 < p < 1$, then $n\left(\frac{1}{p} - \frac{1}{2}\right)$ is the natural counterpart, cf. [T, in particular 2.6].

Remark 4. If one dislikes the Besov spaces $B^{\sigma_{d,p}}_{2,\tilde{p}}(\Lambda)$ in (6) (in particular if $\tilde{p} = p < 1$) then one can use the embeddings (1.7.4/13) in order to formulate more handy (but also somewhat weaker) Fourier multiplier assertions: *Let $d > 0$, $\omega(x) = d \log(1 + |x|)$ and $\varrho(x) \in R(\omega)$. Let Λ be a bounded C^∞-domain or an open cube in R_n. Let $\Omega = \bar{\Lambda}$ and let $0 < p \leqq \infty$. Let k be an integer with $k > \sigma_{d,p}$,* cf. (5). *Then $M(x) \in W^k_2(\Lambda)$ is a Fourier multiplier for $L^\Omega_p(\varrho, \mu_L)$ and there exists a constant c such that*

$$\|\varrho F^{-1}MFf \mid L_p\| \leqq c \Big(\sum_{|\alpha| \leqq k} \|D^\alpha M \mid L_2(\Lambda)\|^2 \Big)^{1/2} \|\varrho f \mid L_p\| \tag{8}$$

holds for every $M \in W^k_2(\Lambda)$ and for every $f \in L^\Omega_p(\varrho, \mu_L)$. This follows immediately from the above theorem and (1.7.4/13).

1.7.6. Discrete Multiplier Theorem

Up to this moment Definition 1.7.1/1 was the basis of our theory of Fourier multipliers for $L^\Omega_p(\varrho, \mu_L)$. In this subsection a Fourier multiplier should be understood in the sense of Definition 1.7.1/2. This gives us a greater flexibility. We wish to show that not only the spaces $L^\Omega_p(\varrho, \mu_L)$ have a lattice-structure, but also their Fourier multipliers. Let Ω be a compact subset in R_n and let

$$Q_b = \{y \mid y = (y_1, \ldots, y_n) \in R_n, |y_j| \leqq b\}, \quad b > 0, \tag{1}$$

such that Ω is contained in the interior of Q_b, i.e.

$$\Omega \subset \{y \mid y = (y_1, \ldots, y_n) \in R_n, \ |y_j| < b\}. \tag{2}$$

Let $M(x)$ be a complex-valued essentially bounded function on Ω. We extend $M(x)$ outside of Ω by zero and denote also the extended function by $M(x)$. The lattice Z_n has the previous meaning, cf. (1.3.1/5). The Fourier coefficients of $M(x)$ with respect to Q_b are given by

$$\begin{aligned} M_k &= (2b)^{-n} \int_{Q_b} M(x)\, e^{i\frac{\pi}{b}kx}\, dx = (2b)^{-n} \int_{R_n} M(x)\, e^{i\frac{\pi}{b}kx}\, dx \\ &= (2\pi)^{n/2} (2b)^{-n} (F^{-1}M)\left(\frac{\pi}{b}k\right), \quad k \in Z_n. \end{aligned} \tag{3}$$

We have (L_2-convergence)

$$M(x) = \sum_{k \in Z_n} M_k\, e^{-i\frac{\pi}{b}kx}, \quad x \in Q_b. \tag{4}$$

We recall that $\|a_k \mid l_p\|$ has been defined in (1.3.1/6) and (1.3.1/7).

Theorem. *Let $\omega(x) \in \mathfrak{M}$ and $\varrho(x) \in R(\omega)$. Let Ω be a compact subset of R_n and let $0 < p \leqq \infty$. Let $M(x)$ be a complex-valued essentially bounded function on Ω (extended by zero outside of Ω). Let b in (2) be sufficiently large.*

(i) *$M(x)$ is a Fourier multiplier for $L^\Omega_p(\varrho, \mu_L)$ if and only if there exists a constant c such that*

$$\left\| \varrho\left(\frac{\pi}{b}k\right) \sum_{l \in Z_n} M_{k-l}\varphi\left(\frac{\pi}{b}l\right) \mid l_p \right\| \leqq c \left\| \varrho\left(\frac{\pi}{b}k\right) \varphi\left(\frac{\pi}{b}k\right) \mid l_p \right\| \tag{5}$$

holds for all $\varphi \in S^\Omega_\omega$.

(ii) *Let $\tilde{p} = \min(1, p)$ and*

$$\left\| M_k\, e^{\omega\left(\frac{\pi}{b}k\right)} \mid l_{\tilde{p}} \right\| < \infty. \tag{6}$$

Then $M(x)$ is a Fourier multiplier for $L_p^\Omega(\varrho, \mu_L)$ and

$$\|\varrho F^{-1} M F\varphi \mid L_p\| \leqq c \left\| M_k \, \mathrm{e}^{\omega\left(\frac{\pi}{b}k\right)} \mid l_{\tilde{p}} \right\| \|\varrho\varphi \mid L_p\|, \quad \varphi \in S_\omega^\Omega, \tag{7}$$

where c is independent both of φ and M.

Proof. Step 1. We prove (i). Let $\varphi(x) \in S_\omega^\Omega$. Then we have the following counterparts of (3) and (4),

$$(F\varphi)(x) = \sum_{k \in Z_n} a_k \, \mathrm{e}^{-\mathrm{i}\frac{\pi}{b}kx}, \quad x \in Q_b, \tag{8}$$

and

$$a_k = \left(\frac{\pi}{2}\right)^{n/2} b^{-n} \varphi\left(\frac{\pi}{b}k\right). \tag{9}$$

Then we have

$$\begin{aligned} M(x)(F\varphi)(x) &= \sum_{k \in Z_n} M(x) a_k \, \mathrm{e}^{-\mathrm{i}\frac{\pi}{b}kx} \\ &= \sum_{k \in Z_n} \mathrm{e}^{-\mathrm{i}\frac{\pi}{b}kx} \sum_{l \in Z_n} a_l M_{k-l}, \quad x \in Q_b. \end{aligned} \tag{10}$$

We recall that $\{a_k\}_{k \in Z_n}$ is a rapidly decreasing sequence. Then it is easy to see that (10) can be understood in the sense of an L_2-convergence. Then the counterpart of (3) yields

$$(F^{-1} M F\varphi)\left(\frac{\pi}{b}k\right) = b^n \left(\frac{2}{\pi}\right)^{n/2} \sum_{l \in Z_n} a_l M_{k-l} = \sum_{l \in Z_n} M_{k-l} \varphi\left(\frac{\pi}{b}l\right). \tag{11}$$

If b is sufficiently large then we can apply Theorem 1.5.3(ii), cf. also Remark 1.5.2/1. This proves (i).

Step 2. We prove (ii). Let (6) be satisfied and let $\varphi(x) \in S_\omega^\Omega$. We use (1.4.1/2). Then we have

$$\varrho\left(\frac{\pi}{b}k\right) \left| \sum_{l \in Z_n} M_{k-l} \varphi\left(\frac{\pi}{b}l\right) \right| \leqq c \sum_{l \in Z_n} \mathrm{e}^{\omega\left(\frac{\pi}{b}(k-l)\right)} M_{k-l} \varrho\left(\frac{\pi}{b}l\right) \left| \varphi\left(\frac{\pi}{b}l\right) \right|. \tag{12}$$

If $1 \leqq p \leqq \infty$ then we have

$$\left\| \varrho\left(\frac{\pi}{b}k\right) \sum_{l \in Z_n} M_{k-l} \varphi\left(\frac{\pi}{b}l\right) \mid l_p \right\| \leqq c \left\| \mathrm{e}^{\omega\left(\frac{\pi}{b}k\right)} M_k \mid l_1 \right\| \left\| \varrho\left(\frac{\pi}{b}k\right) \varphi\left(\frac{\pi}{b}k\right) \mid l_p \right\|. \tag{13}$$

If $0 < p < 1$ then we have

$$\begin{aligned} \left\| \varrho\left(\frac{\pi}{b}k\right) \sum_{k \in Z_n} M_{k-l} \varphi\left(\frac{\pi}{b}l\right) \mid l_p \right\|^p &\leqq c \sum_{k, l \in Z_n} \mathrm{e}^{p\omega\left(\frac{\pi}{b}(k-l)\right)} M_{k-l}^p \varrho^p\left(\frac{\pi}{b}l\right) \left| \varphi\left(\frac{\pi}{b}l\right) \right|^p \\ &= c \left\| \mathrm{e}^{\omega\left(\frac{\pi}{b}k\right)} M_k \mid l_p \right\|^p \left\| \varrho\left(\frac{\pi}{b}k\right) \varphi\left(\frac{\pi}{b}k\right) \mid l_p \right\|^p. \end{aligned} \tag{14}$$

The c's in (13) and (14) are independent of M and φ. Now (7) follows from (11) and (13), (14) on the basis of Theorem 1.5.3(ii) and Remark 1.5.2/1 (We recall that b is sufficiently large).

Remark. Part (i) is simply the discretization of Definition 1.7.1/2. Furthermore, part (ii) is essentially the discretization of Theorem 1.7.2. In [F, 1.5] we started with the above theorem and proved the other Fourier multiplier assertions on this basis. In particular, Theorem 1.7.2 is an easy consequence of (7).

The methods of this subsection are connected with Subsection 1.5.5., where we dealt with structure properties of special spaces $L_p^\Omega(\varrho, \mu_L)$. We discuss the above results from this point of view. Let M_k be given by (3) and let $\mathbf{M} = (M_{k-l})_{k,\, l \in Z_n}$ be interpreted as a matrix (at least for $n = 1$ matrices of this type are denoted as Toeplitz matrices). Let

$$l_p(d_k) = \{c \mid c = \{c_k\}_{k \in Z_n},\ \|d_k c_k \mid l_p\| < \infty\}, \tag{15}$$

where $\{d_k\}_{k \in Z_n}$ is a sequence of positive numbers. *If* $\mathbf{M}$ *generates a linear and bounded operator from* $l_p\left(\varrho\left(\frac{\pi}{b}k\right)\right)$ *into itself then* $M(x)$ *is a Fourier multiplier for* $L_p^\Omega(\varrho, \mu_L)$ (under the hypotheses of the above theorem). This follows immediately from the above theorem. We strengthen this assertion.

Proposition. *Let* $1 < p < \infty$, $\frac{1}{p} + \frac{1}{p'} = 1$ *and* $\varrho(x) = (1 + |x|)^d$ *with* $-\frac{1}{p} < d < \frac{1}{p'}$. *Let* $\Omega = Q_b$ *be given by* (1). *Let* $M(x)$ *be a complex-valued essentially bounded function on* Ω. *Let* $\mathbf{M} = (M_{k-l})_{k,l \in Z_n}$ *where the* M_k*'s are given by* (3). *Then* $M(x)$ *is a Fourier multiplier for* $L_p^\Omega(\varrho, \mu_L)$ *if and only if* $\mathbf{M}$ *generates a linear and bounded operator from* $l_p((1 + |k|)^d)$ *into itself.*

Proof. In this special case we can identify the b from the theorem with the above b. If $\mathbf{M}$ has the above property, then we have (5). By (1.5.5/11) and (11) follows that $M(x)$ is a Fourier multiplier. Conversely, if $M(x)$ is a Fourier multiplier in the sense of Definition 1.7.1/2 (and this is our point of view in this subsection) then we have (1.7.1/2) for all $f \in L_p^\Omega(\varrho, \mu_L)$. This follows from the density assertion from 1.5.4. By (1.5.5/11) we have (5) for all $\varphi \in L_p^\Omega(\varrho, \mu_L)$. Again by Theorem 1.5.5 this proves that $\mathbf{M}$ has the desired property.

1.7.7. Comments

We discuss briefly few questions connected with our definitions and results about Fourier multipliers for $L_p^\Omega(\varrho, \mu_L)$. More details are given in [F, 1.5.7].

(i) What about the a priori assumption in Definition 1.7.1/2 that $M(x)$ is essentially bounded on Ω? Let $M(x) \in L_1$ be a function with a support near the origin. We assume that $M(x)$ is a Fourier multiplier for $L_p^\Omega(\varrho, \mu_L)$, where Ω is, say, the unit ball and $1 < p < \infty$ (i.e. we have (1.7.1/2) at least for $f \in S_\omega^\Omega$). Then $M(x)$ is also a Fourier multiplier for $L_p(R_n, \varrho(x))$ (cf. (1.3.1/4) as far as the notations are concerned). By duality, $M(x)$ is a Fourier multiplier for $L_{p'}(R_n, \varrho^{-1}(x))$ with $\frac{1}{p} + \frac{1}{p'} = 1$. By interpolation, $M(x)$ is a Fourier multiplier for L_2, and, consequently, bounded.

(ii) What about $M(x) \in B_{2,p}^{\sigma_d,p}(\Lambda)$ in (1.7.5/6)? If $q > 1$ then $B_{2,q}^{n/2}$ contains essentially unbounded functions, cf. [T, 2.8.3] or better [F, p. 65/66] or [S, 2.6.2]. On this basis and with the help of (i) it is possible to prove that one cannot replace $\|M \mid B_{2,p}^{\sigma_d,p}(\Lambda)\|$ in (1.7.5/6) by $\|M \mid B_{2,q}^{n/2}(\Lambda)\|$ with $q > 1$ (even not in the unweighted case $d = 0$).

(iii) What about with (1.7.5/8) in comparison with Hörmander's multiplier theorem

$$\|F^{-1}MFf \mid L_p\| \leqq c \sup_{\substack{|\alpha| \leqq k \\ R > 0}} \left(R^{2|\alpha|-n} \int\limits_{R < |x| < 2R} |D^\alpha M(x)|^2 \,dx\right)^{1/2} \|f \mid L_p\| \tag{1}$$

for the usual L_p-spaces with $1 < p < \infty$ and $k > \frac{n}{2}$? In general, the multiplier factor on the right-hand side of (1.7.5/8) cannot be replaced by something like the multiplier factor in (1). E.g. if $n = 1$ $\Omega = [-b, b]$ with $b > 0, 0 < p < \infty$, $\varrho(x) = (1 + |x|)^d$ with $d \geqq 1 - \frac{1}{p}$ then such a replacement is not possible, cf. [F, pp. 66/67] for details.

1.8. Fourier Multipliers for Mixed $L_{\bar{p}}$-Spaces of Entire Analytic Functions

1.8.1. Definition and Basic Multiplier Theorem

Section 1.8. deals with Fourier multipliers for the spaces $L^\Omega_{\bar{p}}$ from Definition 1.6.1. On the one hand it is the generalization of [T, 1.5], where we considered Fourier multipliers for $L^\Omega_p = L^\Omega_{\bar{p}}$ with $\bar{p} = (p, p)$. On the other hand it is the mixed (and unweighted) counterpart of Section 1.7. However in contrast to 1.7. we restrict ourselves to those assertions which we need in Chapter 2. In this sense we are not interested in the mixed counterpart of Definition 1.7.1/2, but only in the mixed counterpart of Definition 1.7.1/1. All notations have the same meaning as in 1.6., cf. in particular the definition of the spaces $L^\Omega_{\bar{p}}$ in (1.6.1/1). We recall that $n = 2$: All considerations take place in R_2.

Definition. *Let Ω be a compact subset of R_2 and let $\bar{p} = (p_1, p_2)$ with $0 < p_1 \leqq \infty$ and $0 < p_2 \leqq \infty$. Let $M \in S'$ and $F^{-1}M \in L_1$. Then M is said to be a Fourier multiplier for $L^\Omega_{\bar{p}}$ if there exists a constant c such that*

$$\|F^{-1}MFf \mid L_{\bar{p}}\| \leqq \|f \mid L_{\bar{p}}\| \tag{1}$$

holds for all $f \in L^\Omega_{\bar{p}}$.

Remark 1. This is the counterpart of [T, Definition 1.5.1] (for the spaces $L^\Omega_p = L^\Omega_{\bar{p}}$ with $\bar{p} = (p, p)$) and of Definition 1.7.1/1. If $f \in L^\Omega_{\bar{p}}$, then we have $f \in L_\infty$, cf. Theorem 1.6.2. In the same way as in Remark 1.7.1 it follows that

$$(F^{-1}MFf)(x) = c \int_{R_2} (F^{-1}M)(y) f(x - y)\, dy, \quad x \in R_2, \tag{2}$$

holds. Furthermore, $M \in L_\infty$ is a continuous function.

Theorem. *Let Ω and Γ be compact subsets of R_2. Let $\bar{p} = (p_1, p_2)$ with $0 < p_1 \leqq \infty$ and $0 < p_2 \leqq \infty$. Let*

$$\tilde{p}_1 = \min(1, p_1) \quad \textit{and} \quad \tilde{p}_2 = \min(1, p_1, p_2). \tag{3}$$

Then there exists a constant c such that

$$\|F^{-1}MFf \mid L_{\bar{p}}\| \leqq c\|F^{-1}M \mid L_{(\tilde{p}_1, \tilde{p}_2)}\| \, \|f \mid L_{\bar{p}}\| \tag{4}$$

holds for all $f \in L^\Omega_{\bar{p}}$ and all $F^{-1}M \in L^\Gamma_{(\tilde{p}_1; \tilde{p}_2)}$.

Proof. By Theorem 1.6.2 we have $F^{-1}M \in L^\Gamma_1 = L^\Gamma_{(1,1)}$. In particular, the hypotheses of the above definition are satisfied and we have (2) and (1.7.2/3), where $x \in R_2$ in the latter formula is fixed. Furthermore, it follows that

$$\|(F^{-1}M)(\cdot) f(x - \cdot) \mid L_{(\tilde{p}_1, \tilde{p}_2)}\| \leqq c\|f \mid L_\infty\| \, \|F^{-1}M \mid L_{(\tilde{p}_1, \tilde{p}_2)}\| < \infty \tag{5}$$

and (again by Theorem 1.6.2 and (1.7.2/3)) that

$$\|(F^{-1}M)(\cdot) f(x - \cdot) \mid L_1\| \leqq c\|(F^{-1}M)(\cdot) f(x - \cdot) \mid L_{(\tilde{p}_1, \tilde{p}_2)}\| \tag{6}$$

holds. By (2) and (6) we have

$$|(F^{-1}MFf)(x)| \leqq c\left(\int_{R_1}\left(\int_{R_1}|(F^{-1}M)(y)|^{\tilde{p}_1}|f(x-y)|^{\tilde{p}_1}\,dy_1\right)^{\tilde{p}_2/\tilde{p}_1}dy_2\right)^{1/\tilde{p}_2}. \tag{7}$$

We apply $\|\cdot\mid L_{p_1|x_1}\|$ to (7) (we recall that $|\,x_1$ indicates the integration variable, cf. 1.3.2.). By the triangle inequality we have

$$\begin{aligned}&\|(F^{-1}MFf)(x)\mid L_{p_1|x_1}\|\\ &\leqq c\left\|\int_{R_1}\left(\int_{R_1}|(F^{-1}M)(y)|^{\tilde{p}_1}|f(x-y)|^{\tilde{p}_1}\,dy_1\right)^{\tilde{p}_2/\tilde{p}_1}dy_2 \mid L_{p_1/\tilde{p}_2|x_1}\right\|^{1/\tilde{p}_2}\\ &\leqq c\left(\int_{R_1}\|(F^{-1}M)(y)\mid L_{\tilde{p}_1|y_1}\|^{\tilde{p}_2}\,\|f(x_1,x_2-y_2)\mid L_{p_1|x_1}\|^{\tilde{p}_2}\,dy_2\right)^{1/\tilde{p}_2}.\end{aligned} \tag{8}$$

We apply $\|\cdot\mid L_{p_2|x_2}\|$ to (8) and obtain by the triangle inequality (4). This proves the theorem.

Remark 2. This is the mixed generalization of [T, Proposition 1.5.1] which dealt with $L_p^\Omega = L_{(p,p)}^\Omega$, and it is the mixed counterpart of Theorem 1.7.2. A somewhat weaker assertion (as far as the couples $(\tilde{p}_1, \tilde{p}_2)$ are concerned) may be found in B. Stöckert [2, Satz 1(ii)].

1.8.2. Convolution Algebras

Let $\bar{p} = (p_1, p_2)$ with $0 < p_1 \leqq 1$ and $0 < p_2 \leqq 1$. Let Ω be a compact subset of R_2 and let $f \in L_{\bar{p}}^\Omega$ and $g \in L_{\bar{p}}^\Omega$. By Theorem 1.6.2, both f and g are elements of L_1. In particular the convolution $f * g$ from (1.7.3/1) makes sense. The mixed counterpart of Theorem 1.7.3 reads as follows.

Theorem. *Let $0 < p_2 \leqq p_1 \leqq 1$ and let Ω be a compact subset of R_2. Then $L_{\bar{p}}^\Omega$ is a convolution algebra and there exists a constant c such that*

$$\|f * g \mid L_{\bar{p}}\| \leqq c\|f \mid L_{\bar{p}}\|\,\|g \mid L_{\bar{p}}\| \tag{1}$$

holds for all $f \in L_{\bar{p}}^\Omega$ and $g \in L_{\bar{p}}^\Omega$.

Proof. As has been remarked above, $f * g$ is well-defined. Furthermore, we have (1.7.3/3). Then (1) follows from Theorem 1.8.1 with $\Gamma = \Omega$, $g = F^{-1}M$, $\tilde{p}_1 = p_1$ and $\tilde{p}_2 = p_2$.

Remark. This is the mixed generalization of [T, 1.5.3] which dealt with $L_p^\Omega = L_{(p,p)}^\Omega$ and it is the mixed counterpart of Theorem 1.7.3.

1.8.3. Main Multiplier Theorem

Our next aim is to generalize the multiplier theorem of [T, 1.5.2] from the spaces $L_p^\Omega = L_{(p,p)}^\Omega$ to the mixed spaces $L_{\bar{p}}^\Omega$. In other words: We are looking for the mixed counterpart of Theorem 1.7.5. However in contrast to 1.7.5., where we used the Besov spaces $B_{2,\bar{p}}^\sigma$, we restrict ourselves in this subsection to the potential spaces $S_2^{\bar{r}}H$ (with dominating mixed smoothness properties). There is no doubt that our result can be strengthened if one replaces $S_2^{\bar{r}}H$ by appropriate spaces of Besov type. In this sense our approach is nearer to [T, 1.5.2] than to 1.7.5. First we introduce the spaces $S_2^{\bar{r}}H$, which are rather special cases of the spaces which we shall study in Chapter 2. However we dislike to refer at this moment to the next chapter.

Let $\bar{r} = (r_1, r_2)$ with $-\infty < r_1 < \infty$ and $-\infty < r_2 < \infty$. We recall that $S' = S'(R_2)$ and $L_2 = L_2(R_2)$ (all considerations take place in R_2). Then

$$S_2^{\bar{r}}H = \{f \mid f \in S', \|f \mid S_2^{\bar{r}}H\| = \|(1 + x_1^2)^{r_1/2} (1 + x_2^2)^{r_2/2} Ff \mid L_2\| < \infty\} \quad (1)$$

is a special potential space with dominating mixed smoothness. Spaces of this type have been introduced by P. I. Lizorkin, S. M. Nikol'skij [1]. If $\bar{r} = \bar{m} = (m_1, m_2)$, where m_1 and m_2 are non-negative integers then we have $S_2^{\bar{m}}H = S_2^{\bar{m}}W$ with

$$S_2^{\bar{m}}W = \{f \mid f \in L_2, \|f \mid S_2^{\bar{m}}W\| = \|f \mid L_2\| + \left\|\frac{\partial^{m_1}f}{\partial x_1^{m_1}} \mid L_2\right\| + \left\|\frac{\partial^{m_2}f}{\partial x_2^{m_2}} \mid L_2\right\| + \left\|\frac{\partial^{m_1+m_2}f}{\partial x_1^{m_1}\partial x_2^{m_2}} \mid L_2\right\| < \infty\Bigg\} \quad (2)$$

(equivalent norms). $S_2^{\bar{m}}W$ are Sobolev spaces with dominating mixed derivatives. Obviously we have

$$S_2^{\bar{r}}H \subset S_2^{(0,0)}H = L_2 \quad \text{if} \quad 0 \leqq r_1 < \infty \quad \text{and} \quad 0 \leqq r_2 < \infty. \quad (3)$$

Proposition. *Let $\bar{d} = (d_1, d_2)$ with $0 \leqq d_1 < \infty$, $0 \leqq d_2 < \infty$, and $\bar{p} = (p_1, p_2)$ with $0 < p_1 \leqq 2$, $0 < p_2 \leqq 2$. Let $\bar{r} = (r_1, r_2)$ with*

$$r_1 > \varkappa_1 = d_1 + \frac{1}{p_1} - \frac{1}{2} \quad \text{and} \quad r_2 > \varkappa_2 = d_2 + \frac{1}{p_2} - \frac{1}{2}. \quad (4)$$

Then there exists a positive number c such that

$$\|(1 + x_1^2)^{d_1/2} (1 + x_2^2)^{d_2/2} Ff \mid L_{\bar{p}}\| \leqq c\|f \mid S_2^{\bar{r}}H\| \quad (5)$$

holds for all $f \in S_2^{\bar{r}}H$.

Proof. Let

$$I_0 = [-1, 1] \quad \text{and}$$
$$I_j = [-2^j, -2^{j-1}] \cup [2^{j-1}, 2^j] \quad \text{if} \quad j = 1, 2, 3, \ldots,$$

and let χ_k with $k = 0, 1, 2, \ldots$ be the characteristic function of I_k. We apply Hölder's inequality with respect to $\frac{p_1}{2} + \frac{2-p_1}{2} = 1$ and $\frac{p_2}{2} + \frac{2-p_2}{2} = 1$. Then we have

$$\begin{aligned}
&\|(1 + x_1^2)^{d_1/2} (1 + x_2^2)^{d_2/2} Ff \mid L_{\bar{p}}\| \\
&= \left[\sum_{k=0}^{\infty} \int_{I_k} \left(\sum_{j=0}^{\infty} \int_{I_j} (1 + x_1^2)^{d_1p_1/2} (1 + x_2^2)^{d_2p_1/2} |(Ff)(x)|^{p_1} dx_1\right)^{p_2/p_1} dx_2\right]^{1/p_2} \\
&\leqq c_1 \left[\sum_{k=0}^{\infty} 2^{kd_2p_2} \int_{I_k} \left(\sum_{j=0}^{\infty} 2^{jd_1p_1} \int_{I_j} |(Ff)(x)|^{p_1} dx_1\right)^{p_2/p_1} dx_2\right]^{1/p_2} \\
&\leqq c_2 \left[\sum_{k=0}^{\infty} 2^{k\varkappa_2p_2} \left\|\chi_k(x_2) \left(\sum_{j=0}^{\infty} 2^{j\varkappa_1p_1} \|\chi_j(x_1)(Ff)(x) \mid L_{2\,|x_1}\|^{p_1}\right)^{1/p_1} \mid L_{2\,|x_2}\right\|^{p_2}\right]^{1/p_2} \\
&\leqq c_3 \left(\sum_{k=0}^{\infty} \sum_{j=0}^{\infty} 2^{2kr_2+2jr_1} \|\chi_k\chi_j Ff \mid L_2\|^2\right)^{1/2} \leqq c_4\|f \mid S_2^{\bar{r}}H\|,
\end{aligned} \quad (6)$$

where we used (4) in the last but one estimate. This proves the proposition.

Remark 1. This is a modification of Proposition 1.7.5.

Theorem. *Let $\bar{p} = (p_1, p_2)$ with $0 < p_1 \leqq \infty$ and $0 < p_2 \leqq \infty$. Let Ω be a compact subset of R_2 with*

$$\Omega \subset Q_{\bar{b}} = \{(x_1, x_2) \mid |x_1| \leqq b_1, |x_2| \leqq b_2\}. \tag{7}$$

Let

$$r_1 > \sigma_1 = \frac{1}{\min(1, p_1)} - \frac{1}{2} \quad \text{and} \quad r_2 > \sigma_2 = \frac{1}{\min(1, p_1, p_2)} - \frac{1}{2}. \tag{8}$$

Then $M(x) \in S_2^{\bar{r}}H$ is a Fourier multiplier for $L_{\bar{p}}^{\Omega}$. There exists a constant c such that

$$\|F^{-1}MFf \mid L_{\bar{p}}\| \leqq c\|M(b_1\cdot, b_2\cdot) \mid S_2^{\bar{r}}H\| \, \|f \mid L_{\bar{p}}\| \tag{9}$$

holds for all $M \in S_2^{\bar{r}}H$, all $\bar{b} = (b_1, b_2)$, all Ω with (7) and all $f \in L_{\bar{p}}^{\Omega}$.

Proof. Let $\psi \in S$ be a function with compact support and

$$\psi(x) = 1 \quad \text{if} \quad x \in \Omega. \tag{10}$$

Because $M \in L_2$ we have

$$[F(\psi M)](x) = c \int_{R_2} (F\psi)(x - y)(FM)(y)\,dy. \tag{11}$$

Because $r_1 \geqq 0$ and $r_2 \geqq 0$ it follows from (1) and (11) that

$$\|\psi M \mid S_2^{\bar{r}}H\| \leqq c\|M \mid S_2^{\bar{r}}H\| \tag{12}$$

holds. By (5) with $d_1 = d_2 = 0$, (8) and (12) we have

$$\|F^{-1}\psi M \mid L_{(\tilde{p}_1, \tilde{p}_2)}\| \leqq c\|M \mid S_2^{\bar{r}}H\| \tag{13}$$

with $\tilde{p}_1 = \min(1, p_1)$ and $\tilde{p}_2 = \min(1, p_1, p_2)$. We use (1.8.1/4) with ψM instead of M. By (10) and (13) it follows that

$$\|F^{-1}MFf \mid L_{\bar{p}}\| \leqq c\|M \mid S_2^{\bar{r}}H\| \, \|f \mid L_{\bar{p}}\| \tag{14}$$

holds for every $f \in L_{\bar{p}}^{\Omega}$. Finally, (9) follows from (14) and a homogeneity argument.

Remark 2. By a simple translation argument it follows that we can choose

$$b_1 = \max |x_1 - y_1|, \qquad b_2 = \max |x_2 - y_2| \tag{15}$$

in (9) where the maximum in (15) is taken over all $x \in \Omega$ and $y \in \Omega$.

Remark* 3. The above theorem with $M(x) = M_1(x_1)\,M_2(x_2)$ has been proved in H.-J. Schmeisser [4]. If $p_1 = p_2$ then we have

$$r = r_1 = r_2 > \frac{1}{\min(1, p)} - \frac{1}{2}.$$

By (14) we have

$$\|F^{-1}MFf \mid L_p\| \leqq c\|M \mid S_2^{(r,r)}H\| \, \|f \mid L_p\| \tag{16}$$

if $M \in S_2^{(r,r)}H$ and $f \in L_p^{\Omega}$. By (1) and (1.7.4/1) it follows that $H_2^{2r} \subset S_2^{(r,r)}H$. In other words, one can replace $\|M \mid S_2^{(r,r)}H\|$ in (16) by $\|M \mid H_2^{2r}\|$. Then one obtains the multiplier theorem from [T, 1.5.2]. Its mixed version (16) is somewhat sharper. The weighted counterpart may be found in 1.7.5. In B. Stöckert [2] one can find further multiplier assertions for weighted mixed spaces, inclusively discrete multiplier theorems in the sense of 1.7.6.

1.9. Vector-Valued Weighted L_p-Spaces

1.9.1. The Spaces $L_p^\Omega(R_n, \varrho(x), l_q)$

Section 1.9. has four subsections. This one deals with weighted spaces of $L_p(l_q)$-type of entire analytic functions on R_n. In the remaining three subsections we consider (somewhat in contrast to the title of this chapter) weighted $L_p(l_q)$-spaces and weighted L_p-spaces of usual functions on R_n. A brief description of some results of Section 1.9. has been given in [T, 7.1.1.–7.1.3.].

Let $\Omega = \{\Omega_k\}_{k=0}^\infty$ be a sequence of compact subsets of R_n. Let $d_k = \sup |x - y|$ be the diameter of Ω_k where the supremum is taken over all $x \in \Omega_k$ and $y \in \Omega_k$. We always assume that $d_k > 0$, i.e. Ω_k contains at least two different points. We recall that $\|\cdot \mid L_p(R_n, \varrho(x), l_q)\|$ has been defined in (1.3.1/8) and (1.3.1/9). Furthermore, $\omega(x) \in \mathfrak{M}$ and $\varrho(x) \in R(\omega)$ have been introduced in 1.2.1. and 1.4.1., respectively.

Definition. *Let $\omega(x) \in \mathfrak{M}$, $\varrho(x) \in R(\omega)$, $0 < p < \infty$, and $0 < q \leqq \infty$. Let $\Omega = \{\Omega_k\}_{k=0}^\infty$ be the above sequence of compact subsets of R_n. Then*

$$L_p^\Omega(R_n, \varrho(x), l_q) = \{f \mid f = \{f_k\}_{k=0}^\infty,\ f_k \in S'_\omega \text{ and } \operatorname{supp} Ff_k \subset \Omega_k \text{ if } k = 0, 1, 2, \ldots,\ \|f_k \mid L_p(R_n, \varrho(x), l_q)\| < \infty\}. \quad (1)$$

Remark 1. This is essentially the weighted counterpart of [T, Definition 1.6.1] (cf. also [T, Definition 7.1.1]) and the vector-valued counterpart of Definition 1.5.1 (restricted to the Lebesgue measure). Of course, the definition (and also the following proposition, but not the following theorem) can be extended to $p = \infty$.

Proposition. *$L_p^\Omega(R_n, \varrho(x), l_q)$ is a quasi-Banach space (under the hypotheses of the above definition).*

Proof. $L_p^{\Omega_k}(\varrho, \mu_L)$ (where μ_L stands for the Lebesgue measure) is a quasi-Banach space, cf. Theorem 1.5.3. Then the proposition follows by standard arguments.

Let D_ω be the space from Definition 1.2.1/2, where $\|\cdot\|_{\omega,\lambda}$ is given by (1.2.1/7). Furthermore, let $\Omega = \{\Omega_k\}_{k=0}^\infty$ be the above sequence of compact subsets of R_n with the diameters $d_k > 0$. If $\varphi = \{\varphi_k(x)\}_{k=0}^\infty \subset D_\omega$ then we put

$$A_{\lambda\omega}^{\varphi,\Omega} = \sup_k \|\varphi_k(d_k\cdot)\|_{\omega,\lambda} = \sup_k \int_{R_n} |(F\varphi_k(d_k\cdot))(x)| \, e^{\lambda\omega(x)} \, dx. \quad (2)$$

We wish to extend Theorem 1.7.2 (and the subsequent Theorem 1.7.5 via Proposition 1.7.5) from the scalar case to the vector-valued case. It is easy to see that $\|e^\omega F^{-1}M \mid L_{\tilde{p}}\|$ in (1.7.2/1) can be replaced by $\|M\|_{\omega,\lambda}$ or, more generally, by $\|e^{\lambda\omega}FM \mid L_r\|$, where r is a given number with $0 < r \leqq \infty$ and $\lambda = \lambda(r) \geqq 1$ is chosen in an appropriate way. Of course, these substitutes weaken Theorem 1.7.2, but nevertheless they are typical results. In this sense the following theorem is the vector-valued extension of Theorem 1.7.2 with, say, $\|M\|_{\omega,\lambda}$ instead of $\|e^\omega F^{-1}M \mid L_{\tilde{p}}\|$. We recall that $\|f_k \mid L_p(R_n, l_q)\|$ has been defined in 1.3.1.

Theorem. *Let $\omega(x) \in \mathfrak{M}$, $\varrho(x) \in R(\omega)$, $0 < p < \infty$, and $0 < q \leqq \infty$. Let $\Omega = \{\Omega_k\}_{k=0}^\infty$ be a sequence of compact subsets of R_n with the diameters $d_k \geqq 1$. Let $0 < r < \min(p,q)$ Then there exist two positive numbers c and λ such that*

$$\|F^{-1}\varphi_k Ff_k \mid L_p(R_n, \varrho(x), l_q)\| \leqq \left\| \sup_{z \in R_n} \varrho(z) \frac{|(F^{-1}\varphi_k Ff_k)(z)|}{1 + |d_k(x - z)|^{n/r}} \mid L_p(R_n, l_q) \right\| \leqq c A_{\lambda\omega}^{\varphi,\Omega} \|f_k \mid L_p(R_n, \varrho(x), l_q)\| \quad (3)$$

holds for all $\varphi = \{\varphi_k(x)\}_{k=0}^\infty \subset D_\omega$ and all $\{f_k\}_{k=0}^\infty \in L_p^\Omega(R_n, \varrho(x), l_q)$.

Proof. Step 1. The first inequality is obvious. Let the second inequality be valid and let $y^k \in R_n$. If one replaces f_k and Ω_k by $e^{ixy^k} f_k$ and

$$\Omega_k + y^k = \{z \mid z = y + y^k \text{ with } y \in \Omega_k\}$$

then this inequality remains unchanged, cf. (1.2.1/7). In other words, we may assume that $\Omega_k = \{y \mid |y| \leqq d_k\}$.

Step 2. In order to prove the second inequality in (3) we combine some ideas of the proof of Theorem 1.4.2 (scalar weighted case) and of the proof of [T, Theorem 1.6.2]. For sake of brevity we put

$$f_k^*(x) = \sup_{z \in R_n} \varrho(z) \frac{|f_k(z)|}{1 + |d_k(x - z)|^{n/r}}, \quad x \in R_n, \tag{4}$$

and

$$(F^{-1}\varphi_k F f_k)^* (x) = \sup_{z \in R_n} \varrho(z) \frac{|(F^{-1}\varphi_k F f_k)(z)|}{1 + |d_k(x - z)|^{n/r}}, \quad x \in R_n. \tag{5}$$

By (1.5.2/1) we have $f_k^*(x) < \infty$ a.e. We prove that

$$(F^{-1}\varphi_k F f_k)^* (x) \leqq c A_{\lambda\omega}^{\varphi, \Omega} f_k^*(x), \quad x \in R_n, k = 0, 1, 2, \ldots \tag{6}$$

holds, where c and λ are appropriate numbers. Let $x \in R_n$ and $y \in R_n$. By (1.4.1/2) we have

$$\begin{aligned}
&\varrho(x - y) \, |(F^{-1}\varphi_k F f_k)(x - y)| \\
&\leqq c \int_{R_n} |(F^{-1}\varphi_k)(x - y - z)| \, e^{\omega(x-y-z)} \, \varrho(z) \, |f_k(z)| \, dz \\
&\leqq c f_k^*(x) \int_{R_n} |(F^{-1}\varphi_k)(x - y - z)| \, e^{\omega(x-y-z)} \, (1 + |d_k(x - z)|)^{n/r} \, dz \\
&\leqq c'(1 + |d_k y|)^{n/r} f_k^*(x) \int_{R_n} |(F^{-1}\varphi_k)(x - y - z)| \, e^{\omega(x-y-z)} \\
&\qquad \times (1 + |d_k(x - y - z)|)^{n/r} \, dz.
\end{aligned} \tag{7}$$

We recall that

$$(F^{-1}\varphi_k(d_k \cdot))(\xi) = d_k^{-n}(F^{-1}\varphi_k)(d_k^{-1}\xi). \tag{8}$$

If we use (8), (1.2.1/2), $d_k \geqq 1$ and the monotonicity of ω then the last integral in (7) can be estimated from above by $A_{\lambda\omega}^{\varphi, \Omega}$, where λ is a appropriate number. Then (6) is an easy consequence of (7).

Step 3. Let Mf be the maximal function from (1.3.3/1). We prove that

$$f_k^*(x) \leqq c(M|\varrho f_k|^r)^{1/r}(x), \quad x \in R_n, k = 0, 1, 2, \ldots \tag{9}$$

holds, where c is independent of f_k. First we remark that

$$\operatorname{supp} F f_k(d_k^{-1} \cdot) \subset B = \{y \mid |y| \leqq 1\},$$

cf. (8) with d_k^{-1} instead of d_k. Because $d_k \geqq 1$ we have (1.4.1/2) with $\varrho(d_k^{-1} \cdot)$ instead of ϱ, but with the same c and $e^{\omega(x-y)}$. We apply Theorem 1.4.2 and Remark 1.4.2/2 and obtain

$$\begin{aligned}
\sup_{z \in R_n} \varrho(d_k^{-1}(x - z)) \frac{|f_k(d_k^{-1}(x - z))|}{1 + |z|^{n/r}} &\leqq c[M \mid (\varrho f_k)(d_k^{-1} \cdot)|^r (x)]^{1/r} \\
&\leqq c(M|\varrho f_k|^r)(d_k^{-1}x)^{1/r}, \quad x \in R_n,
\end{aligned} \tag{10}$$

where c is independent of $x \in R_n$ and $k = 0, 1, 2, \ldots$ In order to avoid technical difficulties we may assume that $f_k \in S_\omega^{\Omega_k}$. This justifies (10) and the following calculations. Finally one can apply the approximation procedure from 1.5.4. We shall not stress this point. But (10) proves (9). By (6) and (9) we have

$$\|(F^{-1}\varphi_k Ff_k)^* \mid L_p(R_n, l_q)\| \leqq cA_{\lambda\omega}^{\varphi,\Omega} \|(M|\varrho f_k|^r)^{1/r} \mid L_p(R_n, l_q)\| = cA_{\lambda\omega}^{\varphi,\Omega} \|M|\varrho f_k|^r \mid L_{p/r}(R_n, l_{q/r})\|^{1/r}. \tag{11}$$

We have $1 < \frac{p}{r} < \infty$ and $1 < \frac{q}{r} \leqq \infty$. Then it follows from Theorem 1.3.3/2 that

$$\|(F^{-1}\varphi_k Ff_k)^* \mid L_p(R_n, l_q)\| \leqq cA_{\lambda\omega}^{\varphi,\Omega} \||\varrho f_k|^r \mid L_{p/r}(R_n, l_{q/r})\|^{1/r} = cA_{\lambda\omega}^{\varphi,\Omega} \|\varrho f_k \mid L_p(R_n, l_q)\|. \tag{12}$$

This proves (3).

Remark 2. The above theorem is a combination of the scalar weighted theorems from 1.5.2. (maximal inequality) and 1.7.2. (Fourier multipliers) on the one hand and their vector-valued unweighted counterpart from [T, Theorem 1.6.2 and Theorem 1.6.3]. In this sense it is a sufficiently powerful basis in order to study weighted spaces of $B_{p,q}^s - F_{p,q}^s$ type, cf. 5.1. The above theorem has been proved in H. Triebel [6]. The number $A_{\lambda\omega}^{\varphi,\Omega}$ can be improved: At the end of Step 2 we used some rough estimates. But $A_{\lambda\omega}^{\varphi,\Omega}$ is a quite reasonable number, completely sufficient in order to study weighted spaces of $B_{p,q}^s - F_{p,q}^s$ type. We discussed some improvements of $A_{\lambda\omega}^{\varphi,\Omega}$ (and also some special cases) in H. Triebel [6]. What about the restriction $p < \infty$ in the theorem? The case $p = q = \infty$ is simple: There is a counterpart of (3). If $p = \infty$ and $0 < q < \infty$, then inequalities of type (3) are impossible, even in the unweighted case. We refer to [S, Theorem 2.1.4] where we constructed a (somewhat complicated) counter-example.

1.9.2. The Spaces $L_p(R_n, \varrho(x), l_q)$

Let $\omega(x) \in \mathfrak{M}$ and $\varrho(x) \in R(\omega)$, where $\mathfrak{M}$ and $R(\omega)$ have the same meaning as in 1.2.1. and 1.4.1. The Banach space $L_p(R_n, \varrho(x), l_q)$ with $1 < p < \infty$ and $1 < q < \infty$ has been defined in 1.3.1. We wish to prove a Fourier multiplier theorem which in some sense is the counterpart of Theorem 1.9.1, at least in the diagonal case. The matrix $\{\varphi_{k,j}(x)\}_{k,j=0}^{\infty}$ is called a diagonal matrix if

$$\left.\begin{aligned} &\varphi_{k,k}(x) = \varphi_k(x) \quad \text{and} \\ &\varphi_{k,j}(x) = 0 \quad \text{for } k = 0, 1, 2, \ldots \text{ and } j = 0, 1, 2, \ldots \text{ with } j \neq k, \end{aligned}\right\} \tag{1}$$

a vertical matrix if

$$\varphi_{k,0}(x) = \varphi_k(x) \quad \text{and} \quad \varphi_{k,j}(x) = 0 \quad \text{for} \quad k = 0, 1, 2, \ldots \text{ and } j = 1, 2, 3, \ldots, \tag{2}$$

and a horizontal matrix if

$$\varphi_{0,k}(x) = \varphi_k(x) \quad \text{and} \quad \varphi_{j,k}(x) = 0 \quad \text{for} \quad k = 0, 1, 2, \ldots \text{ and } j = 1, 2, 3, \ldots \tag{3}$$

In all three cases we put $\varphi = \{\varphi_k(x)\}_{k=0}^{\infty}$. Let D_ω be the space from Definition 1.2.1/2. If $\varphi = \{\varphi_k(x)\}_{k=0}^{\infty} \subset D_\omega$ and $\lambda > 0$ then we put

$$B_{\lambda\omega}^{\varphi} = \sup_k \int_{R_n} e^{\lambda\omega(x)} |(F^{-1}\varphi_k)(x)| \, dx + \sup_k \int_{R_n} \left| \frac{\partial^n \varphi_k}{\partial x_1 \ldots \partial x_n}(x) \right| dx. \tag{4}$$

This is the substitute of $A_{\lambda\omega}^{\varphi,\Omega}$ from (1.9.1/2). This subsection deals with the diagonal case. Under some restrictions for the admitted weights the vertical and horizontal cases are treated in the following subsection.

Theorem. *Let $\omega(x) \in \mathfrak{M}$, $\varrho(x) \in R(\omega)$, $1 < p < \infty$, and $1 < q < \infty$. There exist two positive numbers c and λ such that*

$$\|F^{-1}\varphi_k F f_k \mid L_p(R_n, \varrho(x), l_q)\| \leqq c B^{\varphi}_{\lambda\omega} \|f_k \mid L_p(R_n, \varrho(x), l_q)\| \tag{5}$$

holds for all systems $\varphi = \{\varphi_k(x)\}_{k=0}^{\infty} \subset D_\omega$ and all systems $\{f_k\}_{k=0}^{\infty} \in L_p(R_n, \varrho(x), l_q)$.

Proof. Step 1. First we assume $p = q$ and $f_k(x) \in S_\omega$. By (1.4.1/2) we have

$$\varrho(x) \, |(F^{-1}\varphi_k F f_k)(x)| \leqq c \int_{R_n} e^{\omega(y)} \, |(F^{-1}\varphi_k)(y)| \, \varrho(x-y) \, |f_k(x-y)| \, dy. \tag{6}$$

By the triangle inequality it follows that

$$\begin{aligned} \|\varrho F^{-1}\varphi_k F f_k \mid L_p\| &\leqq c \|\varrho f_k \mid L_p\| \int_{R_n} e^{\omega(y)} \, |(F^{-1}\varphi_k)(y)| \, dy \\ &\leqq c B^{\varphi}_{\omega} \|\varrho f_k \mid L_p\| . \end{aligned} \tag{7}$$

This proves (5) with $p = q$ for all $\{f_k(x)\}_{k=0}^{\infty} \subset S_\omega$ and by completion for all admissible $\{f_k(x)\}_{k=0}^{\infty}$.

Step 2. Next we recall an unweighted vector-valued Fourier multiplier theorem which is an easy consequence of P. I. Lizorkin [1, Theorem 5]. Let $1 < p < \infty$ and $1 < q < \infty$. Then exists a constant c such that

$$\begin{aligned} \|F^{-1}\varphi_k F f_k \mid L_p(R_n, l_q)\| &\leqq c \sup_{j=0,1,2,\dots} \int_{R_n} \left| \frac{\partial^n \varphi_j(x)}{\partial x_1 \dots \partial x_n} \right| dx \, \|f_k \mid L_p(R_n, l_q)\| \\ &\leqq c B^{\varphi}_{\omega} \|f_k \mid L_p(R_n, l_q)\| \end{aligned} \tag{8}$$

holds for all systems $\{\varphi_j\}$ and $\{f_k\}$. We interpolate (7) and (8). Let $[\cdot, \cdot]_\theta$ be Calderón's complex interpolation method. Let $\varrho_0(x)$ and $\varrho_1(x)$ be two weight functions. Let $1 < q < \infty$, $1 < p_0 < \infty$, and $1 < p_1 < \infty$. Then we have

$$[L_{p_0}(R_n, \varrho_0(x), l_q), L_{p_1}(R_n, \varrho_1(x), l_q)]_\theta = L_p(R_n, \varrho(x), l_q) \tag{9}$$

with $0 < \theta < 1$,

$$\frac{1}{p} = \frac{1-\theta}{p_0} + \frac{\theta}{p_1} \quad \text{and} \quad \varrho(x) = \varrho_0^{1-\theta}(x) \, \varrho_1^{\theta}(x), \quad x \in R_n, \tag{10}$$

cf. [I, Theorem 1.18.5]. In our case we have $p_0 = q$ and $\varrho_1(x) = 1$. Then $\varrho(x) = \varrho_0^{1-\theta}(x) \in R(\omega)$ if $\varrho_0(x) \in R(\omega)$, cf. (1.4.1/8). Then (5) with the above p, the above $\varrho(x)$ and $\lambda = 1$ follows from (7) and (8). Conversely, if p with $1 < p < \infty$ and $\varrho(x) \in R(\omega)$ are given (and q with $1 < q < \infty$) then we find appropriate numbers p_1 and θ with $1 < p_1 < \infty$ and $0 < \theta < 1$ such that (10) holds (again with $p_0 = q$ and $\varrho_1(x) = 1$), in particular $\varrho_0(x) = \varrho^{1/1-\theta}(x) \in R\left(\frac{1}{1-\theta}\omega\right)$ by (1.4.1/8). This proves (5) with $\lambda = \dfrac{1}{1-\theta}$.

Remark 1. Let $d_k \geqq 1$ if $k = 0, 1, 2, \dots$ By (1.9.1/8) and the monotonicity of $\omega(x)$ it follows that

$$\begin{aligned} B^{\varphi}_{\lambda\omega} &= \sup_k \int_{R_n} e^{\lambda\omega(d_k^{-1}x)} \, |(F^{-1}\varphi_k(d_k \cdot))(x)| \, dx + \sup_k \int_{R_n} \left| \frac{\partial^n \varphi_k(d_k x)}{\partial x_1 \dots \partial x_n} \right| dx \\ &\leqq \sup_k \int_{R_n} e^{\lambda\omega(x)} \, |(F^{-1}\varphi_k(d_k \cdot))(x)| \, dx + \sup_k \int_{R_n} \left| \frac{\partial^n \varphi_k(d_k x)}{\partial x_1 \dots \partial x_n} \right| dx . \end{aligned} \tag{11}$$

One can compare this expression with $A^{\varphi,\Omega}_{\lambda\omega}$ from (1.9.1/2). There is an additional but rather harmless term on the right-hand side of (11). In contrast to Theorem 1.9.1 we have now no restrictions for the supports of Ff_k, but the parameters are now restricted by $1 < p < \infty$ and $1 < q < \infty$. An extension of the above theorem to $p < 1$ is meaningless.

Remark* 2. The proof of the above theorem is rather simple. Essentially we reduced the general weighted case via the special weighted case with $p = q$ and interpolation to the unweighted case described in (8). The latter is Lizorkin's vector-valued Fourier multiplier theorem. One can replace this Fourier multiplier theorem by other vector-valued Fourier multiplier theorems of Michlin-Hörmander type. We mention two versions. Let $\psi(x) \in S_\omega$ with

$$0 \leqq \psi(x) \leqq 1, \qquad \operatorname{supp} \psi \subset \{y \mid \tfrac{1}{4} \leqq |y| \leqq 4\} \tag{12}$$

and

$$\psi(u) = 1 \quad \text{if} \quad \tfrac{1}{2} \leqq |u| \leqq 2. \tag{13}$$

Let $1 < p < \infty$ and $1 < q < \infty$. Then we have

$$\|F^{-1}\varphi_k Ff_k \mid L_p(R_n, l_q)\| \leqq c \sup \left(\sum_{l=0}^{\infty} \|\psi(\cdot)\, \varphi_l(2^j \cdot) \mid H_2^{\varkappa}(R_n)\|^2 \right)^{1/2} \|f_k \mid L_p(R_n, l_q)\| \tag{14}$$

where the supremum is taken over all integers j. Here $\varkappa$ is an arbitrary number with $\varkappa > \frac{n}{2}$ and $H_2^{\varkappa}(R_n)$ are the usual Bessel-potential spaces (Sobolev spaces if $\varkappa$ is a natural number), cf. 1.7.4. If $\varkappa > \frac{n}{2}$ is an integer then (14) can be rewritten in the more familiar form

$$\|F^{-1}\varphi_k Ff_k \mid L_p(R_n, l_q)\| \leqq c \sup \left(R^{2|\alpha|-n} \int_{R/2 \leqq |x| \leqq 2R} \sum_{l=0}^{\infty} |D^\alpha \varphi_l(x)|^2 \, dx \right)^{\frac{1}{2}} \|f_k \mid L_p(R_n, l_q)\|, \tag{15}$$

where the supremum is taken over all $R > 0$ and all multi-indices α with $0 \leqq |\alpha| \leqq \varkappa$. A proof of (15) (and also of (14) via appropriate modifications) may be found in [I, 2.2.4], cf. also [T, 2.4.8]. The scalar case of (15) is due to L. Hörmander [1]. As far as the scalar case of (14) is concerned we refer to J. Peetre [1] and the remarks in A. P. Calderón, A. Torchinsky [1, II, p. 168].

1.9.3. The Spaces $L_p(R_n, \varrho(x), l_2)$

This subsection is the continuation of 1.9.2. We prove the vertical and the horizontal counterpart of Theorem 1.9.2. We recall that we gave a description of vertical and horizontal matrices in 1.9.2, cf. (1.9.2/2) and (1.9.2/3), respectively. In comparison with Theorem 1.9.2 we have two essential restrictions. First we put $q = 2$. This is natural because one cannot expect vertical and horizontal Fourier multiplier theorems of Michlin-Hörmander type for the spaces $L_p(R_n, \varrho(x), l_q)$ with $q \neq 2$. Furthermore we deal only with the special weights $(1 + |x|)^d$ and $e^{d|x|^\beta}$ where $0<\beta<1$ and $d \in R_1$ (this corresponds to the usual distributions and the Gevrey ultra-distributions, respectively). Our method is qualitative and can also be applied to other weights. On the other hand it is too specific in order to cover all weights $\varrho(x) \in R(\omega)$ with $\omega(x) \in \mathfrak{M}$. However presumably the theorem below is valid for all these weights. But instead of trying to characterize those weights for which our method works we restrict ourselves to the above special weights. Next we introduce the counterparts of $A^{\varphi,\Omega}_{\lambda\omega}$ and $B^{\varphi}_{\lambda\omega}$ from (1.9.1/2) and (1.9.2/4), respectively. Let $\varphi = \{\varphi_k(x)\}_{k=0}^{\infty}$ be a sequence of infinitely differentiable functions. If $\varepsilon > 0$ then we put

$$C_\varepsilon^\varphi = \sup_{\substack{x \in R_n \\ 0 \leqq |\alpha| \leqq 1+[n/2]}} (1 + |x|)^{|\alpha|} \left(\sum_{k=0}^{\infty} |(D^\alpha \varphi_k)(x)|^2 \right)^{1/2} + \sup_{\substack{x \in R_n \\ |\alpha| > 1+[n/2]}} \frac{1}{(|\alpha|!)^{1+\varepsilon}} \left(\sum_{k=0}^{\infty} |(D^\alpha \varphi_k)(x)|^2 \right)^{1/2}. \tag{1}$$

We discuss (1). Let $\varphi_{k,j}(x)$ be given by (1.9.2/2) or (1.9.2/3) (vertical or horizontal case) and let $1 < p < \infty$. Then exists a positive constant c such that

$$\left\| \sum_{j=0}^{\infty} F^{-1}\varphi_{k,j}Ff_j \mid L_p(R_n, l_2)\right\| \leq c \sup_{\substack{R>0 \\ |\alpha| \leq 1+[n/2]}} \left(R^{2|\alpha|-n} \int_{R/2 \leq |x| \leq 2R} \sum_{l=0}^{\infty} |D^\alpha \varphi_l(x)|^2 \, dx \right)^{1/2} \|f_k \mid L_p(R_n, l_2)\| \quad (2)$$

holds for any system $\{\varphi_k(x)\}_{k=0}^{\infty}$ which belongs locally to $W_2^{1+[n/2]}(R_n)$ and any system $\{f_k(x)\}_{k=0}^{\infty} \in L_p(R_n, l_2)$. This is the l_2-version of Hörmander's multiplier theorem. A proof may be found in [I, pp. 161–165], cf. also [T, p. 88]. This is the same formulation as in (1.9.2/15). However the factor in front of $\|f_k \mid L_p(R_n, l_2)\|$ in (2) can be estimated from above by the first term on the right-hand side of (1). Hence if $C_\varepsilon^\varphi < \infty$ then we can apply the above unweighted l_2-Fourier multiplier theorem. The second term on the right-hand side of (1) is connected with the l_2-version of a well-known property of functions $\psi \in D_\omega$ from Definition 1.2.1/2 with $\omega(x) = |x|^\varkappa$, $0 < \varkappa < 1$: By (1.2.1/7) with $\omega(\xi) = |\xi|^\varkappa$ we have

$$\begin{aligned} |(D^\alpha\psi)(x)| &\leq \int_{R_n} |\xi|^{|\alpha|} \, |(F\psi)(\xi)| \, d\xi \\ &\leq (|\alpha|!)^{1/\varkappa} \int_{R_n} \left(\frac{|\xi|^{\varkappa|\alpha|}}{|\alpha|!} \right)^{1/\varkappa} |(F\psi)(\xi)| \, d\xi \\ &\leq (|\alpha|!)^{\frac{1}{\varkappa}} \int_{R_n} |(F\psi)(\xi)| \, e^{\frac{1}{\varkappa}|\xi|^\varkappa} \, d\xi = (|\alpha|!)^{\frac{1}{\varkappa}} \|\psi\|_{\omega, \frac{1}{\varkappa}} \end{aligned}$$

for any $x \in R_n$. The l_2-version is given by

$$\frac{\left(\sum_{k=0}^{\infty} |(D^\alpha\varphi_k)(x)|^2 \right)^{1/2}}{(|\alpha|!)^{1/\varkappa}} \leq \int_{R_n} \left(\sum_{k=0}^{\infty} |(F\varphi_k)(\xi)|^2 \right)^{\frac{1}{2}} e^{\frac{1}{\varkappa}|\xi|^\varkappa} \, d\xi \quad (3)$$

for any $x \in R_n$. In other words, one could replace the second term on the right-hand side of (1) by the right-hand side of (3) with $1 + \varepsilon = \frac{1}{\varkappa}$. We prefer the more general version given in (1) which will be of great service for us in the next subsection. As far as formulations of this type are concerned cf. also J.-L. Lions, E. Magenes [1, III, p. 2]. We formulate the theorem only for the weights $\varrho(x) = e^{d|x|^\beta}$ with $0 < \beta < 1$ and $d \in R_1$, $d \neq 0$. It will be clear that our method covers also the case $\varrho(x) = (1+|x|)^d$ with $d \in R_1$.

Theorem. *Let* $\varrho(x) = e^{d|x|^\beta}$ *with* $d \in R_1$, $d \neq 0$ *and* $0 < \beta < 1$. *Let* $1 < p < \infty$. *Let* $\{\varphi_{k,j}(x)\}_{k,j=0}^{\infty}$ *be either a vertical or a horizontal matrix in the sense of* (1.9.2/2) *and* (1.9.2/3), *respectively. Let* $1 < 1 + \varepsilon < \frac{1}{\beta}$. *Then exists a positive constant* c *such that*

$$\left\| \sum_{j=0}^{\infty} F^{-1}\varphi_{k,j}Ff_j \mid L_p(R_n, \varrho(x), l_2)\right\| \leq cC_\varepsilon^\varphi \|f_k \mid L_p(R_n, \varrho(x), l_2)\| \quad (4)$$

holds for all matrices $\{\varphi_{k,j}(x)\}_{k,j=0}^{\infty}$ *with* $C_\varepsilon^\varphi < \infty$ (*where* $\varphi = \{\varphi_k(x)\}_{k=0}^{\infty}$) *and all systems* $\{f_k\} \in L_p(R_n, \varrho(x), l_2)$.

Proof. Step 1. First we deal with the case $p = 2$. Then we have to prove that

$$\sum_{k=0}^{\infty} \|\varrho F^{-1}\varphi_k Ff \mid L_2\|^2 \leqq c(C_\varepsilon^\varphi)^2 \|\varrho f \mid L_2\|^2 \tag{5}$$

(vertical case) and

$$\left\|\varrho \sum_{k=0}^{\infty} F^{-1}\varphi_k Ff_k \mid L_2\right\|^2 \leqq c(C_\varepsilon^\varphi)^2 \sum_{k=0}^{\infty} \|\varrho f_k \mid L_2\|^2 \tag{6}$$

(horizontal case) hold. First we are interested in (6) with $\varrho(x) = e^{\frac{1}{2}|x|^\beta}$. We begin with some preliminaries. Let j be a natural number. Then we have

$$\begin{aligned}
&\left\|x_1^j \sum_{k=0}^{\infty} F^{-1}\varphi_k Ff_k \mid L_2\right\|^2 \\
&= \left\|\sum_{k=0}^{\infty} \frac{\partial^j}{\partial x_1^j}(\varphi_k Ff_k) \mid L_2\right\|^2 \leqq c\left\|\sum_{k=0}^{\infty}\sum_{l=0}^{j} (-i)^l \frac{j!}{l!(j-l)!} \frac{\partial^{j-l}\varphi_k}{\partial x_1^{j-l}} Fx_1^l f_k \mid L_2\right\|^2 \\
&\leqq c\left\|\sum_{l=0}^{j} \frac{j!}{l!(j-l)!}\left(\sum_{k=0}^{\infty}\left|\frac{\partial^{j-l}\varphi_k}{\partial x_1^{j-l}}(x)\right|^2\right)^{1/2}\left(\sum_{k=0}^{\infty}|(Fx_1^l f_k)(x)|^2\right)^{1/2} \mid L_2\right\|^2 .
\end{aligned} \tag{7}$$

Let $0 < \varepsilon < \varepsilon'$. Then (1) yields

$$\left\|x_1^j \sum_{k=0}^{\infty} F^{-1}\varphi_k Ff_k \mid L_2\right\|^2 \leqq c(C_\varepsilon^\varphi)^2 \sum_{l=0}^{j} \frac{j!^2(j-l)!^{2\varepsilon'}}{l!^2} \sum_{k=0}^{\infty} \|Fx_1^l f_k \mid L_2\|^2 . \tag{8}$$

We may choose $0 < \varepsilon' - \varepsilon$ as small as we want. Of course $\|Fx_1^l f_k \mid L_2\| = \|x_1^l f_k \mid L_2\|$. Furthermore we can replace x_1 in (8) by $x_2, \ldots, x_n$. Let $A \in SO(n)$ (special orthogonal group of orthogonal n-dimensional matrices of determinant 1). Then (8) and the subsequent remarks yield

$$\begin{aligned}
&\left\|\left(\sum_{r=1}^{n} |(Ax)_r|^{2j}\right)^{1/2} \sum_{k=0}^{\infty} F^{-1}\varphi_k Ff_k \mid L_2\right\|^2 \\
&\leqq c(C_\varepsilon^\varphi)^2 \sum_{k=0}^{\infty}\sum_{l=0}^{j} \frac{j!^2(j-l)!^{2\varepsilon'}}{l!^2} \left\|\left(\sum_{r=1}^{n} |(Ax)_r|^{2l}\right)^{1/2} f_k \mid L_2\right\|^2 .
\end{aligned} \tag{9}$$

Let μ be the normed Haar measure of the compact group $SO(n)$. Then we have

$$\int_{SO(n)} \sum_{r=1}^{n} (Ax)_r^{2l}\, d\mu = d_l |x|^{2l}, \quad l = 1, 2, 3, \ldots \tag{10}$$

We must estimate the positive numbers d_l. If $|x| = 1$ and $l = 1, 2, \ldots$ then we have

$$x_1^{2l} + \ldots + x_n^{2l} = (x_1^{2l} + \ldots + x_n^{2l})(x_1^2 + \ldots + x_n^2) \leqq K(x_1^{2l+2} + \ldots + x_n^{2l+2})$$

where K is independent of l (we used $ab \leqq a^\sigma + b^{\sigma'}$ with $a \geqq 0$, $b \geqq 0$, $1 < \sigma < \infty$, $\left.\frac{1}{\sigma} + \frac{1}{\sigma'} = 1\right)$. This proves $d_l \leqq K d_{l+1}$ and

$$d_l \leqq K^{j-l} d_j \quad \text{with} \quad j > l. \tag{11}$$

We remark that $K^{j-l} \leqq c(j-l)!^\eta$, where we may choose $\eta > 0$ arbitrarily small

(c is independent of $j - l$). Now (9)–(11) yield

$$\left\| |x|^j \sum_{k=0}^{\infty} F^{-1}\varphi_k F f_k \mid L_2 \right\|^2 \leqq c(C_\varepsilon^\varphi)^2 \sum_{k=0}^{\infty} \sum_{l=0}^{j} \frac{j!^2 (j-l)!^{2\varkappa}}{l!^2} \| |x|^l f_k \mid L_2 \|^2$$
$$\leqq c(C_\varepsilon^\varphi)^2 \sum_{k=0}^{\infty} \left\| \left(\sum_{l=0}^{j} \frac{j!^2 (j-l)!^{2\varkappa}}{l!^2} |x|^{2l} \right)^{1/2} f_k \mid L_2 \right\|^2 \tag{12}$$

with $0 < \varepsilon < \varkappa$. Interpolation of (12) and the unweighted case yields

$$\left\| |x|^{\frac{\beta}{2} j} \sum_{k=0}^{\infty} F^{-1}\varphi_k F f_k \mid L_2 \right\|^2$$
$$\leqq c(C_\varepsilon^\varphi)^2 \sum_{k=0}^{\infty} \sum_{l=0}^{j} \frac{j!^\beta (j-l)!^{\lambda\beta}}{l!^\beta} \left\| |x|^{\frac{\beta}{2} l} f_k \mid L_2 \right\|^2 \tag{13}$$

where $0 < \varepsilon < \lambda$. The constant c in (13) is independent of j. As far as corresponding interpolation methods are concerned we refer to [I, 1.18.5]. We use

$$e^{|x|^\beta} = \sum_{j=0}^{\infty} \frac{|x|^{\beta j}}{j!}.$$

Then (13) yields

$$\left\| e^{\frac{1}{2}|x|^\beta} \sum_{k=0}^{\infty} F^{-1}\varphi_k F f_k \mid L_2 \right\|^2$$
$$\leqq c(C_\varepsilon^\varphi)^2 \sum_{k=0}^{\infty} \sum_{l=0}^{\infty} \frac{1}{l!^\beta} \left\| |x|^{\frac{\beta}{2} l} f_k \mid L_2 \right\|^2 \sum_{j=l}^{\infty} \frac{j!^\beta (j-l)!^{\lambda\beta}}{j!}. \tag{14}$$

The sum $\sum_{j=l}^{\infty} \ldots$ in (14) begins with $l!^{\beta-1}$ and the quotient of two consecutive terms can be estimated from above by

$$\frac{(j+1-l)^{\lambda\beta}}{(j+1)^{1-\beta}} \leqq \frac{1}{(j+1)^{1-\beta(1+\lambda)}} \leqq q < 1$$

if $j \geqq 1$ and $\beta(1+\lambda) < 1$. However by assumption $\beta(1+\varepsilon) < 1$ and we may choose $\lambda - \varepsilon > 0$ as small as we want. Consequently,

$$\left\| e^{\frac{1}{2}|x|^\beta} \sum_{k=0}^{\infty} F^{-1}\varphi_k F f_k \mid L_2 \right\|^2 \leqq c(C_\varepsilon^\varphi)^2 \sum_{k=0}^{\infty} \| e^{\frac{1}{2}|x|^\beta} f_k \mid L_2 \|^2. \tag{15}$$

This proves (6) if $\varrho(x) = e^{\frac{1}{2}|x|^\beta}$.

Step 2. We prove (5) with $\varrho(x) = e^{\frac{1}{2}|x|^\beta}$. We replace $\left\| \sum_{k=0}^{\infty} \ldots \right\|^2$ in (7) by $\sum_{k=0}^{\infty} \| \ldots \|^2$ and f_k by f. Instead of (8) we obtain

$$\sum_{k=0}^{\infty} \| x_1^j F^{-1}\varphi_k F f \mid L_2 \|^2 \leqq c(C_\varepsilon^\varphi)^2 \sum_{l=0}^{j} \frac{j!^2 (j-l)!^{2\varepsilon'}}{l!^2} \| x_1^l f \mid L_2 \|^2. \tag{16}$$

Now we can follow the argument from the first step and arrive at

$$\sum_{k=0}^{\infty} \| e^{\frac{1}{2}|x|^\beta} F^{-1}\varphi_k F f \mid L_2 \|^2 \leqq c(C_\varepsilon^\varphi)^2 \, \| e^{\frac{1}{2}|x|^\beta} f \mid L_2 \|^2. \tag{17}$$

This is (5) with $\varrho(x) = e^{\frac{1}{2}|x|^\beta}$.

Step 3. Let $a > 0$ and $\varphi^a = \{\varphi_k(ax)\}_{k=0}^{\infty}$. Then we have $C_{\varepsilon'}^{\varphi^a} \leqq cC_{\varepsilon}^{\varphi}$ for any ε' with $\varepsilon < \varepsilon'$. This follows from (1). Now we use a homogeneity argument in order to prove the counterparts of (15) and (17) with $e^{d|x|^\beta}$, $d > 0$, instead of $e^{\frac{1}{2}|x|^\beta}$. Finally we obtain (5) and (6) with $\varrho(x) = e^{-d|x|^\beta}$ by duality from (6) and (5) with $\varrho(x) = e^{d|x|^\beta}$, respectively, $d > 0$. The proof is complete as far as the case $p = 2$ is concerned.

Step 4. By the considerations after formula (2) we obtain the unweighted case $\varrho(x) \equiv 1$ for all p with $1 < p < \infty$. Now we use (1.9.2/9) with $p_0 = q = 2$, $\varrho_0(x) = e^{d|x|^\beta}$, $d \in R_1$, $d \neq 0$ and $\varrho_1(x) \equiv 1$ in the same way as in the proof of Theorem 1.9.2. The proof is complete.

Remark 1. The above theorem extends (2) to the special weights $\varrho(x) = e^{d|x|^\beta}$ with $d \in R_1$ and $0<\beta<1$. By the same arguments one obtains (4) for the weight functions $\varrho(x) = (1 + |x|)^d$ with $d \in R_1$. The possibility to replace the second term on the right-hand side of (1) by corresponding terms from the right-hand side of (3) gives an idea how to generalize (4) to more general weights. We sketch a possibility. Let additionally $(F^{-1}\varphi_k)(x) \geqq 0$ and $f_k(x) \geqq 0$. Then we have

$$\begin{aligned}\varrho(x)\,(F^{-1}\varphi_k F f_k)(x) &= \varrho(x) \int_{R_n} (F^{-1}\varphi_k)(y)\, f_k(x-y)\,dy \\ &\leqq c \int_{R_n} e^{\omega(y)} (F^{-1}\varphi_k)(y)\,\varrho(x-y)\, f_k(x-y)\,dy\end{aligned}$$

and

$$\begin{aligned}\left\| \varrho(x) \sum_{k=0}^{\infty} F^{-1}\varphi_k F f_k \mid L_2 \right\| &\leqq c \left\| \int_{R_n} e^{\omega(y)} \left(\sum_{k=0}^{\infty} (F^{-1}\varphi_k)(y)^2 \right)^{\frac{1}{2}} \varrho(x-y) \left(\sum_{k=0}^{\infty} |f_k(x-y)|^2 \right)^{\frac{1}{2}} \mid L_2 \right\| \\ &\leqq c \int_{R_n} e^{\omega(y)} \left(\sum_{k=0}^{\infty} |(F\varphi_k)(y)|^2 \right)^{\frac{1}{2}} dy \left(\sum_{l=0}^{\infty} \|\varrho f_l \mid L_2\|^2 \right)^{\frac{1}{2}}.\end{aligned}$$

This is the counterpart of (6). Similarly one obtains the counterpart of (5). The assumptions $f_k(x) \geqq 0$ can be done without restriction of generality, i.e. we have the counterparts of (5) and (6) for all admissible systems $\{f_k(x)\}_{k=0}^{\infty}$. The rest is the same as in the above proof. This yields the counterpart of (4). The restrictions $(F^{-1}\varphi_k)(x) \geqq 0$ are more serious. In H. Triebel [6] we tried to overcome these disturbing assumptions, however our arguments have had some gaps, cf. also [T, Remark 7.1.2/2]. These difficulties are the main reason why we restricted the above proof to the weights $\varrho(x) = e^{d|x|^\beta}$ with $d \in R_1$ and $0 < \beta < 1$. On the other hand it is quite clear that the above theorem can be extended to some other weights. This follows from the essentially qualitative arguments of the proof, but also from subsequent manipulations with admissible weights such as summations, multiplications (via interpolation) and duality procedures, cf. H. Triebel [3] in this context.

Remark* 2. In Remark 1.3.3/4 we introduced Muckenhoupt's A_p-condition and described related maximal inequalities and some Fourier multiplier assertions in Remark 1.3.3/4 and in 1.5.4, respectively. It is not a surprise that one has inequalities of type (4) (and its scalar case) for these weights. This includes Littlewood-Paley theorems for the corresponding spaces $L_p(R_n, \varrho(x))$ (as we shall see in the next subsection). As we mentioned in Remark 1.3.3/4 we shall not deal with weights $\varrho(x)$ which obey the A_p-condition. However we give some references in this direction: S. I. Agalarov [1], V. M. Kokilašvili [1, 2], D. S. Kurtz [1], D. S. Kurtz, R. L. Wheeden [1, 2], B. Muckenhoupt, R. L. Wheeden, Wo-Sang Young [1, 2], B. Muckenhoupt, Wo-Sang-Young [1] and Wo-Sang Young [1]. Via interpolation etc. one can combine (4) with corresponding assertions for weights connected with the A_p-condition.

1.9.4. The Spaces $L_p(R_n, \varrho(x))$

The spaces $L_p(R_n, \varrho(x))$ and $L_p(R_n, \varrho(x), l_2)$ have been defined in 1.3.1. We prove a Littlewood-Paley theorem for the spaces $L_p(R_n, \varrho(x))$ under the assumption that the weights $\varrho(x)$ are the same as in the previous subsection. We restrict ourselves to $\varrho(x) = e^{d|x|^\beta}$ with $d \in R_1$ and $0 < \beta < 1$ although it is clear that corresponding assertions can be proved for $\varrho(x) = (1 + |x|)^d$ with $d \in R_1$, too. Let $\psi(x)$ be an infinitely differentiable function on R_n. Then

$$c_\varepsilon^\psi = \sup_{\substack{x \in R_n \\ 0 \leqq |\alpha| \leqq 1 + [n/2]}} (1 + |x|)^{|\alpha|} |D^\alpha \psi(x)| + \sup_{\substack{x \in R_n \\ |\alpha| > 1 + [n/2]}} \frac{|D^\alpha \psi(x)|}{(|\alpha|!)^{1+\varepsilon}} \tag{1}$$

is the scalar version of C_ε^φ from (1.9.3/1). Finally we recall that

$$[W(t) f](x) = (4\pi t)^{-n/2} \int_{R_n} e^{-|x-y|^2/4t} f(y)\, dy, \quad t > 0, \quad x \in R_n, \tag{2}$$

denotes the Gauss-Weiserstrass semi-group.

Theorem. *Let* $1 < p < \infty$ *and* $\varrho(x) = e^{d|x|^\beta}$ *with* $-\infty < d < \infty$ *and* $0 < \beta < 1$. *Let* $1 < 1 + \varepsilon < \frac{1}{\beta}$.

(i) *Let* $\{\varphi_k(x)\}_{k=0}^\infty$ *be a sequence of infinitely differentiable functions with* $\sup_k c_\varepsilon^{\varphi_k} < \infty$,

$$\sum_{k=0}^\infty \varphi_k(x) = 1 \quad \text{if} \quad x \in R_n, \tag{3}$$

$$\operatorname{supp} \varphi_0 \subset \{y \mid |y| \leqq 2\} \tag{4}$$

and

$$\operatorname{supp} \varphi_l \subset \{y \mid 2^{l-1} \leqq |y| \leqq 2^{l+1}\} \quad \text{if} \quad l = 1, 2, 3, \ldots \tag{5}$$

Then there exist two positive numbers c_1 *and* c_2 *such that*

$$c_1 \|f \mid L_p(R_n, \varrho(x))\| \leqq \|F^{-1}\varphi_k Ff \mid L_p(R_n, \varrho(x), l_2)\| \leqq c_2 \|f \mid L_p(R_n, \varrho(x))\| \tag{6}$$

holds for all $f \in L_p(R_n, \varrho(x))$.

(ii) *Let* $\varphi_0(x)$ *and* $\varphi_1(x)$ *be two infinitely differentiable real functions with* $c_\varepsilon^{\varphi_0} < \infty$ *and* $c_\varepsilon^{\varphi_1} < \infty$. *Let*

$$|\varphi_0(x)| > 0 \quad \text{if} \quad |x| \leqq 1 \quad \text{and} \quad |\varphi_1(y)| > 0 \quad \text{if} \quad \tfrac{1}{2} \leqq |y| \leqq 2. \tag{7}$$

Let $L > 1 + \left[\frac{n}{2}\right]$ *and*

$$\sup |(D^\alpha \varphi_1)(x)| (|x|^L + |x|^{-L}) < \infty \tag{8}$$

where the supremum is taken over all α *with* $0 \leqq |\alpha| \leqq 1 + \left[\frac{n}{2}\right]$ *and all* $x \in R_n - \{0\}$. *Let* $\varphi_k(x) = \varphi_1(2^{-k+1}x)$ *if* $k = 1, 2, 3, \ldots$ *Then there exist two positive numbers* c_1 *and* c_2 *such that* (6) *holds for all* $f \in L_p(R_n, \varrho(x))$.

(iii) *Let* $\varphi_0(x)$ *be an infinitely differentiable real function with* $c_\varepsilon^{\varphi_0} < \infty$ *and* $\varphi_0(0) \neq 0$. *Let L be a natural number with* $L > 1 + \left[\frac{n}{2}\right]$. *Then there exist two positive numbers* c_1 *and* c_2 *such that*

$$\begin{aligned} & c_1 \|f \mid L_p(R_n, \varrho(x))\| \\ & \leqq \|F^{-1}\varphi_0 Ff \mid L_p(R_n, \varrho(x))\| + \left\| \left(\int_0^1 \left| t^L \left(\frac{\partial^L}{\partial t^L} W(t) f \right)(\cdot) \right|^2 \frac{dt}{t} \right)^{1/2} \mid L_p(R_n, \varrho(x)) \right\| \\ & \leqq c_2 \|f \mid L_p(R_n, \varrho(x))\| \end{aligned} \tag{9}$$

holds for all $f \in L_p(R_n, \varrho(x))$.

Proof. Step 1. We prove (i). *Let* $\varphi = \{\varphi_k(x)\}_{k=0}^{\infty}$. By $\sup_k c_\varepsilon^{\varphi_k} < \infty$, (4) and (5) we have $C_\varepsilon^\varphi < \infty$, cf. (1.9.3/1). Then the second inequality in (6) is an immediate consequence of the vertical case of (1.9.3/4), cf. also (1.9.3/5). In order to prove the first inequality we choose a system $\psi = \{\psi_k(x)\}_{k=0}^{\infty}$ with $C_\varepsilon^\psi < \infty$ and $\psi_k(x) = 1$ if $x \in \operatorname{supp} \varphi_k$. There are no problems to construct systems of this type. By (3) and the horizontal case of (1.9.3/4) we have

$$\|f \mid L_p(R_n, \varrho(x))\| = \left\| \sum_{k=0}^{\infty} F^{-1}\psi_k F F^{-1}\varphi_k F f \mid L_p(R_n, \varrho(x)) \right\|$$
$$\leqq c C_\varepsilon^\psi \|F^{-1}\varphi_k F f \mid L_p(R_n, \varrho(x), l_2)\| . \tag{10}$$

The proof of (6) is complete.

Step 2. We prove (ii). We claim $C_\varepsilon^\varphi < \infty$ if $\varphi = \{\varphi_k(x)\}_{k=0}^{\infty}$ with $\varphi_l(x) = \varphi_1(2^{-l+1}x)$ where $l = 1, 2, 3, \ldots$ Let $k = 1, 2, 3, \ldots$ If $|\alpha| > 1 + \left[\frac{n}{2}\right]$ then the desired estimate follows from

$$|D^\alpha \varphi_k(x)| \leqq 2^{-k|\alpha|+|\alpha|} |(D^\alpha \varphi_1)(2^{-k+1}x)| \leqq c(|\alpha|!)^{1+\varepsilon} \cdot 2^{-(k-1)|\alpha|}. \tag{11}$$

Let $0 \leqq |\alpha| \leqq 1 + \left[\frac{n}{2}\right]$. If $|x| \leqq 1$ then (8) yields

$$|D^\alpha \varphi_k(x)| \leqq c \cdot 2^{-k|\alpha|} \cdot 2^{-kL}. \tag{12}$$

If $|x| \sim 2^j$, where j is a natural number, then (8) yields

$$2^{j|\alpha|} |D^\alpha \varphi_k(x)| \leqq c \cdot 2^{(j-k)|\alpha|} \cdot 2^{-|j-k|L}. \tag{13}$$

We recall $L > |\alpha|$. Then $C_\varepsilon^\varphi < \infty$ follows from (12) and (13). Now the second inequality in (6) is again a consequence of the vertical case of (1.9.3/4). In order to prove the first inequality in (6) we note that

$$\Phi(x) = \sum_{k=0}^{\infty} \varphi_k^2(x) \geqq c > 0 \quad \text{and} \quad c_{\varepsilon'}^{\Phi} < \infty \tag{14}$$

with $\varepsilon < \varepsilon'$ and $1 + \varepsilon' < \frac{1}{\beta}$. The latter assertion follows from the above estimates (11)–(13) with φ_k^2 instead of φ_k. Then $\Phi^{-1}(x)$ has the same properties. Now it follows from the scalar case of (1.9.3/4) that $\|F^{-1}\Phi F f \mid L_p(R_n, \varrho(x))\|$ is an equivalent norm in $L_p(R_n, \varrho(x))$. By the horizontal case of (1.9.3/4) we have now

$$\|f \mid L_p(R_n, \varrho(x))\| \leqq c\|F^{-1}\Phi F f \mid L_p(R_n, \varrho(x))\|$$
$$= \left\| \sum_{k=0}^{\infty} F^{-1}\varphi_k F F^{-1}\varphi_k F f \mid L_p(R_n, \varrho(x)) \right\| \leqq c\|F^{-1}\varphi_k F f \mid L_p(R_n, \varrho(x), l_2)\| .$$

The proof of (ii) is complete.

Step 3. In order to prove (iii) we choose

$$\varphi_1(x) = |x|^{2L} \mathrm{e}^{-|x|^2/2}. \tag{15}$$

We recall $(F \mathrm{e}^{-|x|^2/2})(\xi) = \mathrm{e}^{-|\xi|^2/2}$. Then it follows from (1.9.3/3) that $c_\varepsilon^{\varphi_1} < \infty$ is satisfied. Furthermore we have (8) and (7) (where $\varphi_0(0) \neq 0$ is sufficient in our case). Hence we can apply (ii). By (15) we have

$$\left(F^{-1}\varphi_1(\sqrt{2t}\,\cdot) F f\right)(x) = c t^L \Delta^L \int_{R_n} t^{-n/2} \mathrm{e}^{-|x-y|^2/4t} f(y)\, \mathrm{d}y$$
$$= c' t^L \Delta^L [W(t)f](x) = c' t^L \frac{\partial^L}{\partial t^L} [W(t)f](x), \tag{16}$$

the latter by well-known properties of the Gauss-Weierstrass semi-group. Then (ii) yields the discrete version of (9). The replacement of the discrete version by the continuous one will be proved in the next step.

Step 4. Let $\varphi_1(x)$ be given and let

$$\psi_1(x) = \int_{\frac{1}{2}}^{1} \varphi_1(tx) \frac{\mathrm{d}t}{t}. \tag{17}$$

Let $\varphi_k(x) = \varphi_1(2^{-k+1}x)$ and $\psi_k(x) = \psi_1(2^{-k+1}x)$, $k = 1, 2, \ldots$ Then we have

$$\begin{aligned} &\|F^{-1}\psi_k Ff \mid L_p(R_n, \varrho(x), l_2)\| \\ &\leqq \left\| \left(\int_{1/2}^{1} \sum_{k=1}^{\infty} |[F^{-1}\varphi_1(2^{-k+1}t\cdot)\, Ff]\,(\cdot)|^2 \frac{\mathrm{d}t}{t} \right)^{1/2} \mid L_p(R_n, \varrho(x)) \right\| \\ &= \left\| \left(\int_{0}^{1} |(F^{-1}\varphi_1(t\cdot)\, Ff)\,(x)|^2 \frac{\mathrm{d}t}{t} \right)^{1/2} \mid L_p(R_n, \varrho(x)) \right\|. \end{aligned} \tag{18}$$

If φ_1 is given by (15) then we can apply (ii) to ψ_1 from (17). Beside technical modifications this proves the first inequality in (9), where we used (16). In order to prove the second inequality in (9) we remark that $\varphi_1(tx)$ with $\frac{1}{2} \leqq t \leqq 1$ satisfies the conditions from (ii) uniformly. If $p = 2$ then the last equality in (18) and Fubini's theorem yield the desired estimate. In the unweighted case, i.e. $\varrho(x) \equiv 1$, estimates of this type are well-known, cf. e.g. [T, 2.12.2 and Remark 5.2.3/4] or E. M. Stein [1, IV.1] where in the latter book corresponding results for the Cauchy-Poisson semi-group are proved. The rest is now a matter of complex interpolation similar as in the second step of the proof of Theorem 1.9.2. This completes the proof.

Remark. This is a Littlewood-Paley theorem. Its proof is based on Theorem 1.9.3. This shows that this theorem can be extended to other weights, cf. Remark 1 and Remark 2 in 1.9.3. Littlewood-Paley theorems of type (9) with the Gauss-Weierstrass semi-group or the Cauchy-Poisson semi-group for unweighted L_p-spaces are known, cf. E. M. Stein [1, IV.1] and [T, 2.12.2 and Remark 5.2.3/4] A systematic study of assertions of this type for the unweighted spaces $B^s_{p,q}(R_n)$ and $F^s_{p,q}(R_n)$ of Besov-Hardy-Sobolev type has been given in the Subsections 2.5.15.–2.5.17. of the Russion edition of [T]. Littlewood-Paley theorems play an important role in Fourier analysis, cf. e.g. R. E. Edwards, G. I. Gaudry [1].

1.10. Vector-Valued Mixed $L_{\bar{p}}$-Spaces of Entire Analytic Functions

1.10.1. Definitions

The spaces $Sl_{\bar{q}}(L_{\bar{p}})$ and $L_{\bar{p}}(l_{\bar{q}})$ have the same meaning as in 1.3.2., cf. (1.3.2/6), (1.3.2/6′) and (1.3.2/7), (1.3.2/7′), respectively. We recall that all spaces are defined on R_2. We omit the suffix R_2 in the sequel.

Definition 1. *Let $\bar{p} = (p_1, p_2)$ and $\bar{q} = (q_1, q_2)$ with $0 < p_m \leqq \infty$ and $0 < q_m \leqq \infty$, $(m = 1, 2)$. Let $\Omega = \{\Omega_{j,k}\}_{j,k=0}^{\infty}$ be a sequence of compact subsets of R_2. Then*

$$\begin{aligned} Sl_{\bar{q}}(L^{\Omega}_{\bar{p}}) = \{f \mid f = \{f_{j,k}\}_{j,k=0}^{\infty},\ f_{j,k} \in S' \text{ and } \operatorname{supp} Ff_{j,k} \subset \Omega_{j,k} \\ \text{if } j, k = 0, 1, 2, \ldots,\ \|f_{j,k} \mid Sl_{\bar{q}}(L_{\bar{p}})\| < \infty\} \end{aligned} \tag{1}$$

and

$$L^{\Omega}_{\bar p}(l_{\bar q}) = \{f \mid f = \{f_{j,k}\}_{j,k=0}^{\infty},\ f_{j,k} \in S' \text{ and } \operatorname{supp} Ff_{j,k} \subset \Omega_{j,k} \text{ if } j, k = 0, 1, 2, \ldots,\ \|f_{j,k} \mid L_{\bar p}(l_{\bar q})\| < \infty\}. \tag{2}$$

Remark 1. This is the mixed counterpart of [T, Definition 1.6.1], cf. also with Definition 1.9.1 (weighted counterpart). By the Paley-Wiener-Schwartz theorem from 1.2.4. both $Sl_{\bar q}(L^{\Omega}_{\bar p})$ and $L^{\Omega}_{\bar p}(l_{\bar q})$ are spaces of sequences of entire analytic functions on R_2.

Proposition. *$Sl_{\bar q}(L^{\Omega}_{\bar p})$ and $L^{\Omega}_{\bar p}(l_{\bar q})$ are quasi-Banach spaces (under the hypotheses of the above definition).*

Proof. $L^{\Omega_{j,k}}_{\bar p}$ is a quasi-Banach space, cf. Theorem 1.6.3. Then the proposition follows by standard arguments.

Definition. 2. *Let $\bar p = (p_1, p_2)$ and $\bar q = (q_1, q_2)$ with $0 < p_m \leqq \infty$ and $0 < q_m \leqq \infty$, $(m = 1, 2)$. Let $\Omega = \{\Omega_{j,k}\}_{j,k=1}^{\infty}$ be a sequence of compact subsets of R_2. Let $M = \{M_{j,k}\}_{j,k=0}^{\infty} \subset S'$ and $\{F^{-1}M_{j,k}\}_{j,k=0}^{\infty} \subset L_1$.*

(i) *M is said to be a Fourier multiplier for $Sl_{\bar q}(L^{\Omega}_{\bar p})$ if there exists a constant c such that*

$$\|F^{-1}M_{j,k}Ff_{j,k} \mid Sl_{\bar q}(L_{\bar p})\| \leqq c\|f_{j,k} \mid Sl_{\bar q}(L_{\bar p})\| \tag{3}$$

holds for all $f = \{f_{j,k}\} \in Sl_{\bar q}(L^{\Omega}_{\bar p})$.

(ii) *M is said to be a Fourier multiplier for $L^{\Omega}_{\bar p}(l_{\bar q})$ if there exists a constant c such that*

$$\|F^{-1}M_{j,k}Ff_{j,k} \mid L_{\bar p}(l_{\bar q})\| \leqq c\|f_{j,k} \mid L_{\bar p}(l_{\bar q})\| \tag{4}$$

holds for all $f = \{f_{j,k}\} \in L^{\Omega}_{\bar p}(l_{\bar q})$.

Remark 2. This is the vector-valued counterpart of Definition 1.8.1. Cf. also with Remark 1.8.1/1 which explains this definition, in particular, (3) and (4) make sense.

1.10.2. Maximal Inequalities

Our aim is to extend the Fourier multiplier Theorem 1.8.3 from $L^{\Omega}_{\bar p}$ to $Sl_{\bar q}(L^{\Omega}_{\bar p})$ and $L^{\Omega}_{\bar p}(l_{\bar q})$. For that purpose it will be convenient to begin with some maximal inequalities. We specify the sequence $\Omega = \{\Omega_{j,k}\}_{j,k=0}^{\infty}$ of compact subsets R_2 from Definition 1.10.1/1 by

$$\Omega_{j,k} = \{x \mid x = (x_1, x_2), |x_1| \leqq a_j, |x_2| \leqq b_k\} \text{ with } a_j > 0, b_k > 0. \tag{1}$$

Theorem. *Let $\bar p = (p_1, p_2)$ and $\bar q = (q_1, q_2)$ with $0 < p_m < \infty$ and $0 < q_m \leqq \infty$; $(m = 1, 2)$. Let $\Omega = \{\Omega_{j,k}\}_{j,k=0}^{\infty}$ be the sequence (1) of compact subsets of R_2.*

(i) *Let*

$$0 < r_1 < p_1 \quad \text{and} \quad 0 < r_2 < \min(p_1, p_2, q_1). \tag{2}$$

Then there exists a positive constant c such that

$$\left\| \sup_{z \in R_2} \frac{|f_{j,k}(\cdot - z)|}{(1 + |a_j z_1|^{1/r_1})(1 + |b_k z_2|^{1/r_2})} \mid Sl_{\bar q}(L_{\bar p}) \right\| \leqq c\|f_{j,k} \mid Sl_{\bar q}(L_{\bar p})\| \tag{3}$$

bolds for all systems $\{f_{j,k}(x)\} \in Sl_{\bar q}(L^{\Omega}_{\bar p})$.

(ii) *Let*

$$0 < r_1 < \min(p_1, q_1, q_2) \quad \text{and} \quad 0 < r_2 < \min(p_1, p_2, q_1, q_2). \tag{4}$$

Then there exists a positive constant c such that

$$\left\| \sup_{z\in R_2} \frac{|f_{j,k}(\cdot - z)|}{(1+|a_j z_1|^{1/r_1})(1+|b_k z_2|^{1/r_2})} \mid L_{\bar p}(l_{\bar q}) \right\| \leq c\| f_{j,k} \mid L_{\bar p}(l_{\bar q})\| \tag{5}$$

holds for all systems $\{f_{j,k}(x)\} \in L^{\Omega}_{\bar p}(l_{\bar q})$.

Proof. Step 1. We prove (ii). We use (1.6.4/1) with $f_{j,k}(a_j^{-1}x_1, b_k^{-1}x_2)$ instead of $f(x_1, x_2)$ (then Ω from Theorem 1.6.4 is the unit cube). By simple transformations we obtain that

$$\sup_{z\in R_2} \frac{|f_{j,k}(x - z)|}{(1+|a_j z_1|^{1/r_1})(1+|b_k z_2|^{1/r_2})} \leq c(M_2(M_1\,|f_{j,k}|^{r_1})^{r_2/r_1})^{1/r_2}(x), \tag{6}$$

holds, where c is independent of $x \in R_2$, j, k, and $f_{j,k}$. Similarly as in (1.6.4/7) (and with the same notations) we have

$$\left\| \sup_{z\in R_2} \frac{|f_{j,k}(\cdot - z)|}{(1+|a_j z_1|^{1/r_1})(1+|b_k z_2|^{1/r_2})} \mid L_{\bar p}(l_{\bar q}) \right\|$$
$$\leq c\| M_2(M_1\,|f_{j,k}|^{r_1})^{r_2/r_1}\,|L_{p_1/r_2\,|\,x_1}(l_{(q_1/r_2,\,q_2/r_2)})|\; L_{p_2/r_2\,|\,x_2}\|^{1/r_2}. \tag{7}$$

By Theorem 1.3.3/1 we can estimate (7) from above by

$$c\|(M_1|f_{j,k}|^{r_1})^{r_2/r_1}\,|L_{p_1/r_2\,|\,x_1}(l_{(q_1/r_2,\,q_2/r_2)})|\; L_{p_2/r_2\,|\,x_2}\|^{1/r_2}$$
$$= c\| M_1\,|f_{j,k}|^{r_1}\,|L_{p_1/r_1\,|\,x_1}(l_{(q_1/r_1,\,q_2/r_1)})|\; L_{p_2/r_1\,|\,x_2}\|^{1/r_1}. \tag{8}$$

We apply again (1.3.3/5) to the inner x_1-integral. Then it follows that (7) can be estimated from above by

$$c\|\,|f_{j,k}|^{r_1}\,|L_{p_1/r_1\,|\,x_1}(l_{(q_1/r_1,\,q_2/r_1)})|\; L_{p_2/r_1\,|\,x_2}\|^{1/r_1} = c\| f_{j,k} \mid L_{\bar p}(l_{\bar q})\|.$$

This proves (5).

Step 2. The proof of (3) is similar. The left-hand side of (3) can be estimated from above by

$$c\| M_2(M_1|f_{j,k}|^{r_1})^{r_2/r_1}\,|l_{q_1/r_2\,|\,j}(L_{p_1/r_2\,|\,x_1})|\; l_{q_2/r_2\,|\,k}(L_{p_2/r_2\,|\,x_2})\|^{1/r_2}, \tag{9}$$

cf. the end of 1.3.2. as far as notations are concerned. We apply again Theorem 1.3.3/1 and estimate (9) from above by

$$c\| M_1\,|f_{j,k}|^{r_1}\,|l_{q_1/r_1\,|\,j}(L_{p_1/r_1\,|\,x_1})|\; l_{q_2/r_1\,|\,k}(L_{p_2/r_1\,|\,x_2})\|^{1/r_1}. \tag{10}$$

The latter can be estimated by the usual maximal inequality (1.3.3/3), which yields the right-hand side of (3). The proof is complete.

Remark* 1. The above theorem has been proved in H.-J. Schmeisser [5, Theorem 3.3]. It is the vector-valued version of Theorem 1.6.4(ii). The unmixed counterpart may be found in [T, Theorem 1.6.2]. The above assertion cannot be extended to $p_1 = \infty$ and/or $p_2 = \infty$. We refer to [T, Remark 1.6.2] and, in turn, to [S, 2.1.4.] for a discussion of this question (in the unmixed case). It is clear that the above theorem can be extended to other mixed spaces (with other combinations of l_q and L_p spaces).

Remark 2. The theorem can be extended to other sequences Ω of compact subsets of R_2. Of special interest is the following possibility. Let $\Omega_{j,k}$ be given by (1) and let $x^{j,k} \in R_2$. Then the theorem remains valid for the modified system $\Omega = \{\Omega_{j,k} + x^{j,k}\}_{j,k=0}^{\infty}$ (even with the same constants in (3) and (5) which, in particular, are independent of $x^{j,k}$). We prove this claim. Let $xx^{j,k}$ be the scalar product in R_2. Let $f_{j,k}(x)$ be given with $\operatorname{supp} Ff_{j,k} \subset \Omega_{j,k} + x^{j,k}$. If $h_{j,k}(x) = e^{-ixx^{j,k}} f_{j,k}(x)$ then we have $(Fh_{j,k})(x) = (Ff_{j,k})(x + x^{j,k})$ and $\operatorname{supp} Fh_{j,k} \subset \Omega_{j,k}$. This reduces the new case to the above theorem.

1.10.3. Fourier Multipliers

Fourier multipliers $M = \{M_{j,k}\}_{j,k=0}^{\infty}$ for the spaces $Sl_{\bar{q}}(L_{\bar{p}}^{\Omega})$ and $L_{\bar{p}}^{\Omega}(l_{\bar{q}})$ have been introduced in Definition 1.10.1/2. On the basis of the maximal inequalities of the preceding subsection we prove a Fourier multiplier theorem which we need later on. The spaces $S_2^{\bar{r}}H$ have been introduced in 1.8.3.

Theorem. *Let* $\bar{p} = (p_1, p_2)$ *and* $\bar{q} = (q_1, q_2)$ *with* $0 < p_m < \infty$ *and* $0 < q_m \leqq \infty$; $(m = 1, 2)$. *Let* $\Omega = \{\Omega_{j,k}\}_{j,k=0}^{\infty}$ *be the seqence* (1.10.2/1) *of compact subsets of* R_2.

(i) *Let* $\bar{r} = (r_1, r_2)$ *with*

$$r_1 > \frac{1}{p_1} + \frac{1}{2} \quad \text{and} \quad r_2 > \frac{1}{\min(p_1, p_2, q_1)} + \frac{1}{2}. \tag{1}$$

Then there exists a constant c such that

$$\begin{aligned} &\|F^{-1}M_{j,k}Ff_{j,k} \mid Sl_{\bar{q}}(L_{\bar{p}})\| \\ &\leqq c \sup_{m,n} \|M_{m,n}(a_m\cdot, b_n\cdot) \mid S_2^{\bar{r}}H\| \, \|f_{j,k} \mid Sl_{\bar{q}}(L_{\bar{p}})\| \end{aligned} \tag{2}$$

holds for all systems $\{f_{j,k}\} \in Sl_{\bar{q}}(L_{\bar{p}}^{\Omega})$ *and all systems* $M = \{M_{j,k}\} \subset S_2^{\bar{r}}H$.

(ii) *Let* $\bar{r} = (r_1, r_2)$ *with*

$$r_1 > \frac{1}{\min(p_1, q_1, q_2)} + \frac{1}{2} \quad \text{and} \quad r_2 > \frac{1}{\min(p_1, p_2, q_1, q_2)} + \frac{1}{2}. \tag{3}$$

Then there exists a constant c such that

$$\begin{aligned} &\|F^{-1}M_{j,k}Ff_{j,k} \mid L_{\bar{p}}(l_{\bar{q}})\| \\ &\leqq \sup_{m,n} \|M_{m,n}(a_m\cdot, b_n\cdot) \mid S_2^{\bar{r}}H\| \, \|f_{j,k} \mid L_{\bar{p}}(l_{\bar{q}})\| \end{aligned} \tag{4}$$

holds for all systems $\{f_{j,k}\} \in L_{\bar{p}}^{\Omega}(l_{\bar{q}})$ *and all systems* $M = \{M_{j,k}\} \subset S_2^{\bar{r}}H$.

Proof. By Proposition 1.8.3 we always have $F^{-1}M_{j,k} \in L_1$. This shows that Definition 1.10.1/2 makes sense. Furthermore it holds

$$(F^{-1}M_{j,k}Ff_{j,k})(x) = c \int_{R_2} (F^{-1}M_{j,k})(y) \, f_{j,k}(x-y) \, dy, \quad x \in R_2, \tag{5}$$

cf. Remark 1.8.1/1. For sake of brevity we put

$$f_{j,k}^*(x) = \sup_{z \in R_2} \frac{|f_{j,k}(x-z)|}{(1 + |a_j z_1|^{1/s_1})(1 + |b_k z_2|^{1/s_2})}, \tag{6}$$

where we choose s_1 and s_2 later on. Then we have

$$\begin{aligned} |(F^{-1}M_{j,k}Ff_{j,k})(x-z)| &\leqq c \int_{R_2} |(F^{-1}M_{j,k})(x-z-y)| \, |f_{j,k}(y)| \, dy \\ &\leqq c f_{j,k}^*(x) \int_{R_2} |(F^{-1}M_{j,k})(x-z-y)| \, (1 + |a_j(x_1-y_1)|^{1/s_1}) \\ &\quad \times (1 + |b_k(x_2-y_2)|^{1/s_2}) \, dy \\ &\leqq c f_{j,k}^*(x) (1 + |a_j z_1|^{1/s_1})(1 + |b_k z_2|^{1/s_2}) \\ &\quad \times \frac{1}{a_j b_k} \int_{R_2} \left| (F^{-1}M_{j,k}) \left(\frac{y_1}{a_j}, \frac{y_2}{b_k} \right) \right| (1 + |y_1|^{1/s_1})(1 + |y_2|^{1/s_2}) \, dy. \end{aligned} \tag{7}$$

We recall that

$$\frac{1}{a_j b_k}(F^{-1}M_{j,k})\left(\frac{y_1}{a_j}, \frac{y_2}{b_k}\right) = (F^{-1}M_{j,k}(a_j\cdot, b_k\cdot))(y)$$

holds. By the abbreviation (6) (with $F^{-1}M_{j,k}Ff_{j,k}$ instead of $f_{j,k}$), (7) and Proposition 1.8.3 $\left(\text{with } p_1 = p_2 = 1,\ d_1 = \frac{1}{s_1} \text{ and } d_2 = \frac{1}{s_2}\right)$ we have

$$(F^{-1}M_{j,k}Ff_{j,k})^*(x) \leqq cf^*_{j,k}(x)\, \|M_{j,k}(a_j\cdot, b_k\cdot) \mid S_2^{\bar{r}}H\| \tag{8}$$

provided that

$$r_1 > \frac{1}{s_1} + \frac{1}{2} \quad \text{and} \quad r_2 > \frac{1}{s_2} + \frac{1}{2}. \tag{9}$$

We note that

$$|(F^{-1}M_{j,k}Ff_{j,k})(x)| \leqq (F^{-1}M_{j,k}Ff_{j,k})^*(x), \quad x \in R_2. \tag{10}$$

If (1) is satisfied then we find numbers s_1 and s_2 with (9), $0 < s_1 < p_1$ and $0 < s_2 < \min(p_1, p_2, q_1)$. Now, (2) follows from Theorem 1.10.2(i) (with s_1 and s_2 instead of r_1 and r_2, respectively), (8) and (10). In the same way one proves (4) on the basis of Theorem 1.10.2(ii).

Remark 1. This theorem as well as some generalizations and modifications (for other types of mixed $L_{\bar{p}}$-spaces) have been proved in H.-J. Schmeisser [5]. The proof has been based on Theorem 1.10.2. In particular, by Remark 1.10.2/2 we can replace the above system Ω by the system $\{\Omega_{j,k} + x^{j,k}\}_{j,k=0}^{\infty}$ with $x^{j,k} \in R_2$ in the above theorem without any changes in the formulation. The most general case as far as sequences $\Omega = \{\Omega_{j,k}\}_{j,k=0}^{\infty}$ of compact subsets of R_2 are concerned reads as follows. Let

$$a_{j,k} = \max_{x,y \in \Omega_{j,k}} |x_1 - y_1| > 0 \quad \text{and} \quad b_{j,k} = \max_{x,y \in \Omega_{j,k}} |x_2 - y_2| > 0. \tag{11}$$

Then we have (1.10.2/6) with $a_{j,k}$ and $b_{j,k}$ instead of a_j and b_k, respectively. This modified inequality together with the translation argument $\Omega_{j,k} \to \Omega_{j,k} + x^{j,k}$ from Remark 1.10.2/2 was the basis of all our considerations. In other words, under the corresponding restrictions for the involved parameters we have the Fourier multiplier assertions (2) and (4) with $a_{m,n}$ and $b_{m,n}$ instead of a_m and b_n, respectively.

Remark 2. The above theorem generalizes Theorem 1.8.3 (mixed scalar case) and [T, Theorem 2.4.9/2] together with its forerunner [T, Theorem 1.6.3] (unmixed vector-valued case). If one compares the restrictions for the parameters in these theorems with (1) and (3) then it is quite clear that (1) and (3) can be improved. We have the following conjecture: *Part* (i) *of the theorem remains valid if* (1) *is replaced by*

$$r_1 > \frac{1}{\min(p_1, 1)} - \frac{1}{2} \quad \textit{and} \quad r_2 > \frac{1}{\min(p_1, p_2, q_1, 1)} - \frac{1}{2}. \tag{1'}$$

Part (ii) *of the theorem remains valid if* (3) *is replaced by*

$$r_1 > \frac{1}{\min(p_1, q_1, q_2, 1)} - \frac{1}{2} \quad \textit{and} \quad r_2 > \frac{1}{\min(p_1, p_2, q_1, q_2, 1)} - \frac{1}{2}. \tag{3'}$$

Maybe $q_1 < \infty$, $q_2 < \infty$ and some restrictions for the system Ω are necessary. In order to prove these conjectures one could try to follow the same way as in [T]: There we began with [T, Theorem 1.6.3] which is the direct counterpart of the above theorem (inclusively restrictions of type (1) and (3), and the proof via a maximal inequality). Afterwards we obtained [T, Theorem 2.4.9/2] on the basis of [T, Theorem 1.6.3], the sharp Michlin-Hörmander Fourier multiplier theorem (1.9.2/12) for $L_p(l_q)$ with $1 < p < \infty$ and $1 < q < \infty$, and the new complex interpolation method from [T,

2.4.9]. One can try to extend this method to the mixed spaces under consideration. Beside an extension of the complex interpolation method from the unmixed case to the mixed case one needs assertions of type (1.9.2/12) for the spaces $L_{\bar{p}}(l_{\bar{q}})$ where all involved parameters are strictly between 1 and ∞. At least partial results in this direction may be found in P. I. Lizorkin [1, Theorem 5] and P. I. Lizorkin [2]. However (at least at this moment) we have not tried to follow that way.

Remark 3. In H.-J. Schmeisser [1, II, Proposition 2] we proved the following result: Let $M_{j,k}(x) = M_j(x_1)\, N_k(x_2)$ with $j = 0, 1, 2, \ldots$ and $k = 0, 1, 2, \ldots$ Then (2) holds with $\bar{r} = (r_1, r_2)$, where r_1 and r_2 satisfy (1′) (instead of (1)), $0 < q_m \leqq \infty$ and $0 < p_m \leqq \infty$. Of course, this is a partial answer of the above conjecture. For the proof and further results of this type we refer to H.-J. Schmeisser [1, II].

2. Spaces with Dominating Mixed Smoothness Properties

2.1. Introduction

This chapter deals with the theory of spaces $S^{\bar{r}}_{\bar{p},\bar{q}}B(R_2)$, $SB^{\bar{r}}_{\bar{p},\bar{q}}(R_2)$, and $S^{\bar{r}}_{\bar{p},\bar{q}}F(R_2)$. Here $\bar{p} = (p_1, p_2)$, $\bar{q} = (q_1, q_2)$, $\bar{r} = (r_1, r_2)$ with $0 < p_i \leqq \infty$, $0 < q_i \leqq \infty$, $-\infty < r_i < \infty$; ($i = 1, 2$; $p_1 < \infty$ and $p_2 < \infty$ in the case of $S^{\bar{r}}_{\bar{p},\bar{q}}F(R_2)$). These spaces are treated in the framework of Fourier analysis on the basis of the results of Sections 1.6., 1.8., and 1.10. in analogy to the theory of the isotropic spaces $B^s_{p,q}$ and $F^s_{p,q}$ ($0 < p \leqq \infty$, $0 < q \leqq \infty$, $-\infty < s < \infty$) of Besov-Hardy-Sobolev type (cf. [T, Chapters 1, 2], [S], and [F, Chapter 2]).

Let us give a brief survey of the history of the theory of spaces with dominating mixed smoothness properties. In 1962/1963 S. M. Nikol'skij [3, 4] introduced the space of Sobolev type

$$S^{\bar{r}}_p W(R_2) = \Big\{ f \mid f \in L_p(R_2),\ \|f \mid S^{\bar{r}}_p W(R_2)\| = \|f \mid L_p\| + \left\| \frac{\partial^{r_1} f}{\partial x_1^{r_1}} \mid L_p \right\| + \left\| \frac{\partial^{r_2} f}{\partial x_2^{r_2}} \mid L_p \right\| + \left\| \frac{\partial^{r_1+r_2} f}{\partial x_1^{r_1} \partial x_2^{r_2}} \mid L_p \right\| < \infty \Big\}, \tag{1}$$

$1 < p < \infty$, $r_i = 0, 1, 2, \ldots$; ($i = 1, 2$). This space coincides with $S^{\bar{r}}_{\bar{p},\bar{2}}F(R_2)$, $\bar{p} = (p,p)$, $\bar{2} = (2, 2)$, in our context. In contrast to isotropic or anisotropic spaces (cf. 2.2.2.) which can be defined via "pure" derivatives in both x_1- and x_2-directions, the mixed derivative $\frac{\partial^{r_1+r_2} f}{\partial x_1^{r_1} \partial x_2^{r_2}}$ plays the dominant part here. Furthermore (1) represents the beginning of the theory. Following that, S. M. Nikol'skij [5] introduced and studied $S^{\bar{r}}_{\bar{p},\bar{\infty}}B(R_2)$, $\bar{p} = (p, p)$, $\bar{\infty} = (\infty, \infty)$, $0 \leqq r_1 < \infty$, $0 \leqq r_2 < \infty$. We refer also to N. S. Bachvalov [1]. Roughly speaking this space is constructed in the sense of (1) and Hölder-Zygmund spaces using mixed derivatives and mixed differences (cf. 2.3.4.). The detailed study of spaces of such type was advanced mainly by Soviet mathematicians. T. I. Amanov [1] and A. D. Džabrailov [1, 2] defined and considered $S^{\bar{r}}_{\bar{p},\bar{q}}B(R_2)$, $\bar{p} = (p, p)$, $\bar{q} = (q, q)$, $1 \leqq p \leqq \infty$, and $1 \leqq q \leqq \infty$. Instead of usual L_p-norms mixed $L_{\bar{p}}$-norms were used (Ja. S. Bugrov [2, 4], A. P. Uninskij [1, 2, 3], T. I. Amanov [2]). P. I. Lizorkin and S. M. Nikol'skij [1] extended $S^{\bar{r}}_p W$ with the help of fractional differentiation to $0 \leqq r_1 < \infty$, $0 \leqq r_2 < \infty$ and obtained spaces of Bessel-potential type. The main problems which have been investigated comprised the representation in terms of entire analytic functions of exponential type, connections with approximation theory, embedding theorems, and interpolation theorems. We refer also to Ja. S. Bugrov [1, 3, 5], A. D. Džabrailov [3, 4, 5], K. K. Golovkin [2], O. V. Besov [1], O. V. Besov, A. D. Džabrailov [1], M. K. Potapov [1–4], P. Grisvard [1], and G. Sparr [1]. A systematic treatment which covers the results of that period concerning spaces with dominating mixed smoothness properties can be found in the book of T. I. Amanov [3]. The progress in the theory of function spaces in the seventies, in particular the possibility to extend function spaces to values $p < 1$ using methods of Fourier analysis (C. Fefferman, E. M. Stein [2], J. Peetre [3, 4], and [F, S, T,]) gave new stimulations in order to deal with spaces with dominating mixed smoothness, too. The main goal was to find an approach which allows to treat both isotropic and anisotropic spaces and also spaces with dominating mixed smoothness from a common point of view. Properties of such general function spaces were studied in H. Triebel [4, III–IV], B. Stöckert, H. Triebel [1], and M. L. Gol'dman [4]. We refer also to [F, Chapter 2].

Then a new impetus came from the theory of integral operators. It is an immediate consequence of Hölder's inequality, that $\mathscr{K}$,

$$(\mathscr{K}f)(x_2) = \int_{R_1} K(x_1, x_2) f(x_1) \,\mathrm{d}x_1,$$

is a linear bounded operator from $L_{p_1'}(R_1)$ into $L_{p_2}(R_1)$ if the kernel $K = K(x_1, x_2)$ belongs to the space $L_{\bar{p}}(R_2)$, $\bar{p} = (p_1, p_2)$, $1 \leqq p_1 \leqq \infty$, $1 \leqq p_2 \leqq \infty$, $\dfrac{1}{p_1} + \dfrac{1}{p_1'} = 1$. New methods in the theory of eigenvalues of integral operators developed by A. Pietsch [1–3] raised the problem to find criterions for $K(x_1, x_2)$ which ensure that $\mathscr{K}$ maps between Besov spaces. This was solved by expressing smoothness properties of $K(x_1, x_2)$ via vector-valued Besov spaces $B^r_{p,q}(A)$, where $A = B^t_{u,v}$ is a Besov function space. From the standpoint of Fourier analysis the question was treated by H. Triebel [8]. It turned out that K must belong to $SB^{\bar{r}}_{\bar{p},\bar{q}}(R_2)$, which is a space with dominating mixed smoothness properties related to $S^{\bar{r}}_{\bar{p},\bar{q}}B(R_2)$ and $S^{\bar{r}}_{\bar{p},\bar{q}}F(R_2)$ (cf. Definition 2.2.1/2). For the exact definitions and more details we refer to Subsection 2.5.1. This is also the motivation to study spaces on the basis of mixed norms (quasi-norms). Properties of $SB^{\bar{r}}_{\bar{p},\bar{q}}(R_2)$ and related spaces have been investigated by H.-J. Schmeisser [1–6].

The aim of this chapter is to study the above mentioned families of spaces from a common point of view. We state the basic properties, describe equivalent representations, embeddings, and applications to integral operators for three types of spaces:

(i) the spaces $S^{\bar{r}}_{\bar{p},\bar{q}}B$ which cover the classical spaces of S. M. Nikol'skij [5] and T. I. Amanov [1–3],

(ii) the spaces $S^{\bar{r}}_{\bar{p},\bar{q}}F$ which generalize the spaces $S^{\bar{r}}_{\bar{p}}H$ of Bessel-potential type of P. I. Lizorkin, S. M. Nikol'skij [1],

(iii) the spaces $SB^{\bar{r}}_{\bar{p},\bar{q}}$ which are related to vector-valued Besov spaces (cf. P. Grisvard [1], G. Sparr [1], A. Pietsch [2, 3]).

2.2. Definitions and Basic Properties

2.2.1. Definitions

In Section 2.2. we give the definition of the spaces $S^{\bar{r}}_{\bar{p},\bar{q}}B(R_2)$, $SB^{\bar{r}}_{\bar{p},\bar{q}}(R_2)$, and $S^{\bar{r}}_{\bar{p},\bar{q}}F(R_2)$ and we prove some fundamental properties. We derive equivalent quasi-norms, elementary embeddings, Fourier multiplier assertions, and maximal inequalities which are crucial in our further investigations. Furthermore we intend to describe briefly other types of spaces. This is mainly in order to compare the definitions. In particular, we define the isotropic spaces $B^s_{p,q}$ and $F^s_{p,q}$ as well as the anisotropic spaces $B^{\bar{s}}_{p,q}$ and $F^{\bar{s}}_{p,q}$.

Recall that $|\alpha| = \sum_{j=1}^{n} \alpha_j$ if $\alpha = (\alpha_1, \ldots, \alpha_n)$ is a multi-index of non-negative integers.

Definition 1. *Let $\Phi(R_n)$ be the collection of all systems $\varphi = \{\varphi_j(x)\}_{j=0}^{\infty} \subset S(R_n)$ such that*

$$\left.\begin{aligned} &\operatorname{supp}\varphi_0 \subset \{x \mid |x| \leqq 2\} \\ &\operatorname{supp}\varphi_j \subset \{x \mid 2^{j-1} \leqq |x| \leqq 2^{j+1}\} \quad \textit{if} \quad j = 1, 2, \ldots; \end{aligned}\right\} \qquad (1)$$

for every multi-index α *there exists a positive number* c_α *such that*

$$2^{j|\alpha|}|D^\alpha \varphi_j(x)| \leqq c_\alpha \quad \text{for all} \quad j = 0, 1, 2, \ldots \text{ and all } x \in R_n, \tag{2}$$

and

$$\sum_{j=0}^{\infty} \varphi_j(x) = 1 \quad \text{for every} \quad x \in R_n. \tag{3}$$

Remark 1. Systems of test functions with the above properties (1)–(3) are standard in the modern theory of function spaces (cf. e.g. [T, Definition 2.3.1/1]). Of course $\Phi(R_n)$ is not empty. For sake of completeness we shall give a proof. Let $0 \leqq \psi(x) \in S(R_n)$ with

$$\operatorname{supp} \psi \subset \left\{x \,\middle|\, \frac{1}{2} \leqq |x| \leqq 2\right\} \quad \text{and} \quad \psi(x) > 0 \quad \text{if} \quad \frac{1}{\sqrt{2}} \leqq |x| \leqq \sqrt{2}. \tag{4}$$

Let $\varphi_j(x) = \psi(2^{-j}x)\left(\sum_{k=-\infty}^{\infty} \psi(2^{-k}x)\right)^{-1}$ if $j = 1, 2, \ldots$ and $\varphi_0(x) = 1 - \sum_{j=1}^{\infty} \varphi_j(x)$. Then it follows $\varphi = \{\varphi_j(x)\}_{j=0}^{\infty} \in \Phi(R_n)$.

We use the notations from 1.2.4. and 1.3.2. In particular, F and F^{-1} stand for the Fourier transform in $S'(R_2)$ and its inverse, respectively.

Definition 2. *Let* $-\infty < r_i < \infty$ *and* $0 < q_i \leqq \infty$; $(i = 1, 2)$. *Let* $\bar{r} = (r_1, r_2)$ *and* $\bar{q} = (q_1, q_2)$. *Let* $\varphi = \{\varphi_j(x)\}_{j=0}^{\infty} \in \Phi(R_1)$.

(i) *If* $\bar{p} = (p_1, p_2)$, $0 < p_1 \leqq \infty$, $0 < p_2 \leqq \infty$, *then*

$$\begin{aligned} S_{\bar{p},\bar{q}}^{\bar{r}}B(R_2) = \{f \mid f \in S'(R_2), \; &\|f \mid S_{\bar{p},\bar{q}}^{\bar{r}}B(R_2)\|^\varphi \\ &= \|2^{r_1 j + r_2 k} F^{-1}\varphi_j\varphi_k Ff \mid l_{\bar{q}}(L_{\bar{p}})\| < \infty\}. \end{aligned} \tag{5}$$

(ii) If $\bar{p} = (p_1, p_2)$, $0 < p_1 \leqq \infty$, $0 < p_2 \leqq \infty$, *then*

$$\begin{aligned} SB_{\bar{p},\bar{q}}^{\bar{r}}(R_2) = \{f \mid f \in S'(R_2), \; &\|f \mid SB_{\bar{p},\bar{q}}^{\bar{r}}(R_2)\|^\varphi \\ &= \|2^{r_1 j + r_2 k} F^{-1}\varphi_j\varphi_k Ff \mid Sl_{\bar{q}}(L_{\bar{p}})\| < \infty\}. \end{aligned} \tag{6}$$

(iii) *If* $\bar{p} = (p_1, p_2)$, $0 < p_1 < \infty$, $0 < p_2 < \infty$, *then*

$$\begin{aligned} S_{\bar{p},\bar{q}}^{\bar{r}}F(R_2) = \{f \mid f \in S'(R_2), \; &\|f \mid S_{\bar{p},\bar{q}}^{\bar{r}}F(R_2)\|^\varphi \\ &= \|2^{r_1 j + r_2 k} F^{-1}\varphi_j\varphi_k Ff \mid L_{\bar{p}}(l_{\bar{q}})\| < \infty\}. \end{aligned} \tag{7}$$

Remark 2. $F^{-1}\varphi_j\varphi_k Ff$ must be understood in the sense of Sections 1.8. and 1.10. $\varphi_j\varphi_k$ stands for $\varphi_j(\xi)\,\varphi_k(\eta)$, where $\xi \in R_1$, $\eta \in R_1$, $j = 0, 1, 2, \ldots$, and $k = 0, 1, 2, \ldots$ Then the Paley-Wiener-Schwartz theorem (cf. 1.2.4.) yields that $F^{-1}[\varphi_j(\xi)\,\varphi_k(\eta)\,Ff]$ is an entire analytic function of exponential type. Consequently, $(F^{-1}[\varphi_j(\xi)\,\varphi_k(\eta)\,Ff])\,(x_1, x_2)$ makes sense. It is an infinitely differentiable function of at most polynomial growth. For brevity we have written

$$F^{-1}\varphi_j\varphi_k Ff = (F^{-1}(\varphi_j(\xi)\,\varphi_k(\eta)\,Ff))\,(x_1, x_2), \; f \in S'(R_2). \tag{8}$$

If we take into consideration the definitions of $l_{\bar{q}}(L_{\bar{p}})$, $Sl_{\bar{q}}(L_{\bar{p}})$, $L_{\bar{p}}(l_{\bar{q}})$, and the corresponding quasi-norms then the expressions on the right-hand sides of (5), (6), and (7) are justified. Note that $L_{p_1} = L_{p_1 | x_1}$, $L_{p_2} = L_{p_2 | x_2}$, $l_{q_1} = l_{q_1 | j}$, and $l_{q_2} = l_{q_2 | k}$ (cf. the remarks at the end of 1.3.2.). In other words, in (5), (6), and (7) there are taken quasi-norms $\|\cdot \mid L_{p_1}\|$ with respect to x_1, $\|\cdot \mid L_{p_2}\|$ with respect to x_2, $\|\cdot \mid l_{q_1}\|$ with respect to j, and $\|\cdot \mid l_{q_2}\|$ with respect to k, but in different orders (cf. the definitions of $l_{\bar{q}}$, $L_{\bar{p}}$, $Sl_{\bar{q}}(L_{\bar{p}})$, $L_{\bar{p}}(l_{\bar{q}})$ and their quasi-norms in (1.3.2/4)–(1.3.2/7) and (1.3.2/4′)–(1.3.2/7′)). Of course, there are several other possibilities. We restrict ourselves to these three types of spaces, which seem to be the most important ones from the historical point of view and from the standpoint of applications to mapping properties of integral operators. By the way, the methods which shall be developed in the following sections, can also be applied to modifications of the spaces from Definition 2. This would be more or less routine work on the basis of this chapter.

Remark 3. We did not indicate "φ" in the notation of the spaces. In Section 2.2.3. it will be shown that $S^{\bar{r}}_{\bar{p},\bar{q}}B(R_2)$, $SB^{\bar{r}}_{\bar{p},\bar{q}}(R_2)$, and $S^{\bar{r}}_{\bar{p},\bar{q}}F(R_2)$ do not depend on the choice of the system $\varphi \in \Phi(R_1)$. This justifies our notations.

Remark 4. The definition of $S^{\bar{r}}_{\bar{p},\bar{q}}F(R_2)$ is restricted to $p_1 < \infty$ and $p_2 < \infty$. This is essential and also necessary. If $p_1 = \infty$ and/or $p_2 = \infty$ then Definition 2(iii) does not make any sense. In particular, corresponding spaces depend on the chosen system φ. For details (concerning isotropic spaces) we refer to [S, 2.1.4.]. In some cases one can overcome these difficulties (cf. [T, 2.3.4.]).

Remark 5. Let us remark, that the connection between $S^{\bar{r}}_{\bar{p},\bar{q}}F$ and the spaces $S^{\bar{r}}_{p}W$ described in (2.1/1) is indicated by the last inequality in (1.8.3/6) in the case $\bar{p} = (2,2)$ and $\bar{q} = (2,2)$. There characteristic functions have been used instead of systems $\{\varphi_j\}_{j=0}^{\infty} \in \Phi(R_1)$. But this is immaterial, as can be seen easily in that case.

2.2.2. Other Types of Spaces

We define the isotropic spaces $B^s_{p,q}(R_n)$, $F^s_{p,q}(R_n)$, and the anisotropic spaces $B^{\bar{s}}_{p,q}(R_n)$, $F^{\bar{s}}_{p,q}(R_n)$ in order to illustrate the differences between the three prominent types of function spaces (i.e. "isotropic", "anisotropic", and "dominating mixed"). Further we add some remarks on possible generalizations.

The Spaces $B^s_{p,q}(R_2)$ and $F^s_{p,q}(R_2)$:

We consider the case $n = 2$, only. The generalization to $n \neq 2$ is obvious. In the following L_p means $L_p(R_2) = L_{\bar{p}}(R_2)$ with $\bar{p} = (p, p)$, $0 < p \leqq \infty$. $L_p(l_q)$ is the space $L_p(R_2, \varrho(x), l_q)$ from (1.3.1/8, 9) with $\varrho(x) = 1$. We write $\{f_j(x)\}_{j=0}^{\infty} \in l_q(L_p)$ if

$$\|f_j \mid l_q(L_p)\| = \left(\sum_{j=0}^{\infty} \left(\int_{R_2} |f_j(x)|^p \,dx\right)^{q/p}\right)^{1/q} < \infty \tag{1}$$

(usual modification if $p = \infty$ or/and $q = \infty$). Let $-\infty < s < \infty$ and $0 < q \leqq \infty$. Let $\varphi = \{\varphi_j(x)\}_{j=0}^{\infty} \in \Phi(R_2)$.

(i) If $0 < p \leqq \infty$, then

$$B^s_{p,q}(R_2) = \{f \mid f \in S'(R_2),\ \|f \mid B^s_{p,q}(R_2)\| = \|2^{sj}F^{-1}\varphi_j Ff \mid l_q(L_p)\| < \infty\}. \tag{2}$$

(ii) If $0 < p < \infty$, then

$$F^s_{p,q}(R_2) = \{f \mid f \in S'(R_2),\ \|f \mid F^s_{p,q}(R_2)\| = \|2^{sj}F^{-1}\varphi_j Ff \mid L_p(l_q)\| < \infty\}. \tag{3}$$

The spaces $B^s_{p,q}(R_2)$ and $F^s_{p,q}(R_2)$ have been studied extensively, including applications of corresponding spaces on domains to elliptic differential operators. We refer to [T], [S], [I], and to J. Peetre [4] for the modern part of the theory. We have

$$f = \sum_{j=0}^{\infty} F^{-1}\varphi_j Ff = \sum_{j=0}^{\infty} f_j \quad \text{(convergence in } S'(R_2)) \tag{4}$$

with

$$\left.\begin{aligned} &\operatorname{supp} Ff_j \subset B_{2^{j+1}}(R_2) - B_{2^{j-1}}(R_2) = D_j, \quad j = 1, 2, \ldots, \\ &\operatorname{supp} Ff_0 \subset B_2(R_2) = D_0. \end{aligned}\right\} \tag{5}$$

Here, $B_b(R_2) = \{x \mid x = (x_1, x_2), |x| \leqq b\}$, $b > 0$, where $|x| = (x_1^2 + x_2^2)^{1/2}$ is the isotropic distance of $x \in R_2$ from the origin. We have $\bigcup_{j=0}^{\infty} D_j = R_2$, that means $\{D_j\}_{j=0}^{\infty}$ is a dyadic (isotropic) covering of R_2. Hence, the principle of construction is the following: We decompose $f \in S'(R_2)$ in a series of entire analytic functions of exponential type. The supports of their Fourier transforms are related to a dyadic (isotropic) covering of R_2 (cf. (4) and (5)). The rate of convergence of $f = \sum_{j=0}^{\infty} f_j$ is measured by (2) in the case of $B^s_{p,q}(R_2)$ and by (3) in the case of $F^s_{p,q}(R_2)$, i.e. by multiplying with the weights 2^{js} and afterwards by integration ($\|\cdot \mid L_p\|$) and summation ($\|\cdot \mid l_q\|$) in different orders. The outcoming families of spaces include many classical function spaces as special cases, in particular, the Hölder-Zygmund spaces, the Bessel-potential spaces and Sobolev spaces, the Besov spaces as well as the local counterparts of the Hardy spaces and of the space BMO. For example, if $1 < p < \infty$, and $r = 0, 1, 2, \ldots$, then

$$F^r_{p,2}(R_2) = W^r_p(R_2) = \Big\{ f \mid f \in L_p(R_2),\ \|f \mid W^r_p(R_2)\| = \|f \mid L_p\| + \left\| \frac{\partial^r f}{\partial x_1^r} \mid L_p \right\| + \left\| \frac{\partial^r f}{\partial x_2^r} \mid L_p \right\| < \infty \Big\}. \tag{6}$$

In contrast to (2.1/1) (i.e. to the space $S^{\bar{r}}_p W$, $\bar{r} = (r, r)$) the mixed derivative $\dfrac{\partial^{2r} f}{\partial x_1^r \, \partial x_2^r}$ need not belong to $L_p(R_2)$ and therefore it plays no role in the theory of the space $B^s_{p,q}(R_2)$ and $F^s_{p,q}(R_2)$. Further let us notice that the smoothness properties of a function belonging to $W^r_p(R_2)$ do not depend on the chosen coordinate-direction in R_2. For the details and for motivations we refer to [T, Chapter 2]. If we compare (2) and (3) with Definition 2.2.1/2 then we observe that both constructions follow the same principle. But we have specific decompositions and different measurements of the rate of approximation. This yields different smoothness properties, as we shall see in detail later on.

The Spaces $B^{\bar{s}}_{p,q}(R_2)$ and $F^{\bar{s}}_{p,q}(R_2)$:

We adopt some ideas from [T, 10.1.]. Let $1 < p < \infty$ and $\bar{r} = (r_1, r_2)$, where r_1 and r_2 are non-negative integers. Then

$$W^{\bar{r}}_p(R_2) = \Big\{ f \mid f \in L_p(R_2),\ \|f \mid W^{\bar{r}}_p(R_2)\| = \|f \mid L_p\| + \left\| \frac{\partial^{r_1} f}{\partial x_1^{r_1}} \mid L_p \right\| + \left\| \frac{\partial^{r_2} f}{\partial x_2^{r_2}} \mid L_p \right\| < \infty \Big\} \tag{7}$$

is a direct generalization of (6) (we have $W^r_p(R_2) = W^{\bar{r}}_p(R_2)$ if $\bar{r} = (r, r)$.) $W^{\bar{r}}_p(R_2)$ is usually called an anisotropic Sobolev space. Now the smoothness properties of $f \in W^{\bar{r}}_p(R_2)$ depend on the chosen direction in R_2. However, mixed derivatives do not play a dominating role. In general, $\dfrac{\partial^{r_1 + r_2}}{\partial x_1^{r_1} \, \partial x_2^{r_2}} f \notin L_p(R_2)$ if $f \in W^{\bar{r}}_p(R_2)$. Nevertheless, it is of interest which of the mixed derivatives belong to $W^{\bar{r}}_p(R_2)$. The answer is given by

$$W^{\bar{r}}_p(R_2) = \Big\{ f \mid f \in L_p(R_2),\ \|f \mid W^{\bar{r}}_p(R_2)\|^m = \sum_{0 \le \frac{\alpha_1}{r_1} + \frac{\alpha_2}{r_2} \le 1} \left\| \frac{\partial^{|\alpha|}}{\partial x_1^{\alpha_1} \, \partial x_2^{\alpha_2}} f \mid L_p \right\| < \infty \Big\}, \tag{8}$$

where $\|\cdot \mid W^{\bar{r}}_p(R_2)\|^m$ is an equivalent norm in $W^{\bar{r}}_p(R_2)$ (cf. S. M. Nikol'skij [2, 9.2.2]).

Similarly, using differences, anisotropic Besov spaces can be introduced. We refer to S. M. Nikol'skij [2]. The question arises whether we can find an approach in the spirit of (2) and (3) via the decomposition method. The first steps in this direction have been done in [F, 2.5.2.] and in B. Stöckert, H. Triebel [1, Section 4]. We shall give the definitions of the corresponding families of spaces $B^{\bar{s}}_{p,q}(R_2)$ and $F^{\bar{s}}_{p,q}(R_2)$. At first we have to generalize the set $\Phi(R_2)$ from Definition 2.2.1/1. Let

$$\bar{a} = (a_1, a_2) \in R_2 \quad \text{with} \quad a_1 > 0,\ a_2 > 0, \quad \text{and} \quad a_1 + a_2 = 2. \tag{9}$$

The anisotropic distance of $x = (x_1, x_2) \in R_2$ from the origin is defined by

$$|x|_{\bar{a}} = (|x_1|^{2/a_1} + |x_2|^{2/a_2})^{1/2}. \tag{10}$$

Obviously we have $|x|_{\bar{a}} = |x|$ if $a_1 = a_2 = 1$. The counterpart of $\Phi(R_2)$ reads as follows: $\Phi^{\bar{a}}(R_2)$ is the collection of all systems $\varphi = \{\varphi_j(x)\}_{j=0}^{\infty} \subset S(R_2)$ with

$$\operatorname{supp} \varphi_0 \subset \{x \mid |x|_{\bar{a}} \leqq 2\},\ \operatorname{supp} \varphi_j \subset \{x \mid 2^{j-1} \leqq |x|_{\bar{a}} \leqq 2^{j+1}\} \text{ if } j = 1, 2, \ldots; \tag{11}$$

for every multi-index $\alpha = (\alpha_1, \alpha_2)$ there exists a positive constant c_α such that

$$2^{ja_1\alpha_1 + ja_2\alpha_2}|D^{(\alpha_1, \alpha_2)}\varphi_j(x)| \leqq c_\alpha \tag{12}$$

for any $j = 0, 1, 2, \ldots$ and any $x \in R_2$; and

$$\sum_{j=0}^{\infty} \varphi_j(x) = 1 \quad \text{for any} \quad x \in R_2. \tag{13}$$

It is clear that $\Phi^{\bar{a}}(R_2) = \Phi(R_2)$ holds if $\bar{a} = (1, 1)$. Further we can show that $\Phi^{\bar{a}}(R_2)$ is not empty. Now we are prepared to define the anisotropic spaces $B^{\bar{s}}_{p,q}(R_2)$ and $F^{\bar{s}}_{p,q}(R_2)$.

Let $\bar{a} = (a_1, a_2)$ with $a_1 > 0$, $a_2 > 0$, and $a_1 + a_2 = 2$. Let $\varphi \in \Phi^{\bar{a}}(R_2)$. Let $-\infty < s < \infty$ and $\bar{s} = (s_1, s_2)$ with $s_1 = \dfrac{s}{a_1}$ and $s_2 = \dfrac{s}{a_2}$. Let $0 < q \leqq \infty$.

(i) If $0 < p \leqq \infty$, then

$$B^{\bar{s}}_{p,q}(R_2) = \{f \mid f \in S'(R_2),\ \|f \mid B^{\bar{s}}_{p,q}(R_2)\| = \|2^{sj}F^{-1}\varphi_j Ff \mid l_q(L_p)\| < \infty\}. \tag{14}$$

(ii) If $0 < p < \infty$, then

$$F^{\bar{s}}_{p,q}(R_2) = \{f \mid f \in S'(R_2),\ \|f \mid F^{\bar{s}}_{p,q}(R_2)\| = \|2^{sj}F^{-1}\varphi_j Ff \mid L_p(l_q)\| < \infty\}. \tag{15}$$

It follows immediately from the definition that $B^{\bar{s}}_{p,q}(R_2) = B^{s}_{p,q}(R_2)$ and $F^{\bar{s}}_{p,q}(R_2) = F^{s}_{p,q}(R_2)$ holds if $\bar{a} = (1, 1)$. Further we see that the definition follows the above principle of construction: instead of isotropic dyadic decompositions (coverings) anisotropic dyadic decompositions (coverings) are used. Then we have to proceed quite analogously. Let us mention that (14) and (15) are restricted to $\bar{s} = (s_1, s_2)$ such that $s_1 \cdot s_2 \geqq 0$. In other words both s_1 and s_2 must be either non-negative or non-positive. The case $s_1 \cdot s_2 < 0$ remains as an unsolved problem. If $\bar{s} = (s_1, s_2)$ with $s_1 > 0$, $s_2 > 0$ is given, then the above numbers s and $\bar{a} = (a_1, a_2)$ can be calculated as follows:

$$\frac{1}{s} = \frac{1}{2}\left(\frac{1}{s_1} + \frac{1}{s_2}\right) \quad \text{and} \quad a_1 = \frac{s}{s_1}, \quad a_2 = \frac{s}{s_2} \tag{16}$$

(similarly if $s_1 < 0$, $s_2 < 0$). s is called the mean smoothness and $\bar{a}$ the anisotropy. Let us remark that there does not exist a complete theory of the spaces $B^{\bar{s}}_{p,q}(R_2)$ and

$F^{\bar{s}}_{p,q}(R_2)$ for the full range of parameters $0 < p \leqq \infty$, $0 < q \leqq \infty$, and $-\infty < s < \infty$, as it has been given for the isotropic spaces e.g. in [T]. But many assertions can be carried over, e.g.

$$F^{\bar{r}}_{p,2}(R_2) = W^{\bar{r}}_p(R_2), \quad \text{if} \quad r_i = 0, 1, 2, \ldots (i = 1, 2),\ 1 < p < \infty\,. \tag{17}$$

We restrict ourselves to refer to some literature on this topic. Anisotropic spaces $B^{\bar{s}}_{p,q}(R_n)$ and $W^{\bar{s}}_p(R_n)$, with $s_1 > 0$, $s_2 > 0$, $1 \leqq p \leqq \infty$, $1 \leqq q \leqq \infty$ and their counterparts on domains have been extensively studied in S. M. Nikol'skij [2] and in O. V. Besov, V. P. Il'jin, S. M. Nikol'skij [1]. A treatment of these spaces from the standpoint of interpolation can be found in H.-J. Schmeisser, H. Triebel [1] and H.-J. Schmeisser [7]. For characterizations, embeddings, and traces of $B^{\bar{s}}_{p,q}(R_n)$ and $F^{\bar{s}}_{p,q}(R_n)$ with $1 \leqq p < \infty$, $1 \leqq q \leqq \infty$, we refer to M. L. Gol'dman [1–4] and G. A. Kaljabin [1–4]. Investigations which include also the case $p < 1$ and $q < 1$ have been made by B. Stöckert, H. Triebel [1] (cf. also [F, Chapter 2], and H. Triebel [4, V]). For traces on curves and applications to boundary value problems for semi-elliptic differential equations we refer to Chapter 4. Anisotropic Hardy spaces have been treated in A. P. Calderón, A. Torchinsky [1] and H. Triebel [7]. Surveys on anisotropic spaces and further literature can also be found in O. V. Besov, V. P. Il'jin, L. D. Kudrjavzev, P. I. Lizorkin, S. M. Nikol'skij [1], A. Kufner, O. John, S. Fučik [1, 8.5], and in [I, 2.13].

Further Types of Spaces:

One can try to generalize the described approach to the spaces above and to the spaces from Definition 2.2.1/2 in several directions. For example, if we look at the construction of the isotropic spaces (2) and (3) we observe as basic elements: the Euclidean n-space R_n, the covering $\{D_j\}_{j=0}^{\infty}$ of R_n, the space of tempered distributions $S'(R_n)$, the Fourier transform F, the spaces $l_q(L_p)$ and $L_p(l_q)$, and the sequence of weights $\{2^{js}\}_{j=0}^{\infty}$. Preserving the construction principle it is possible to replace each of these elements in an appropriate way. Besides anisotropic spaces (cf. (14), (15)) and spaces with dominating mixed smoothness properties (cf. Definition 2.2.1/2) we obtain a lot of useful spaces. Some work has been done in this direction. In particular, the investigations in the Chapters 3–6 can be considered from this point of view. We restrict ourselves here to some orientations.

Homogeneous Spaces. The homogeneous counterparts of $B^s_{p,q}(R_n)$ and $F^s_{p,q}(R_n)$, denoted by $\dot{B}^s_{p,q}(R_n)$ and $\dot{F}^s_{p,q}(R_n)$, respectively, are studied in [T, Chapter 5] and in [S, Chapter 3]. We refer also to J. Peetre [3, 4], P. I. Lizorkin [1, 4], and J. Bergh, J. Löfström [1, Chapter 6].

General Decomposition Spaces. On the basis of admissible coverings and corresponding systems of test functions (there must be exist a resolution of unity) general function spaces are constructed in the spirit of (2) and (3). One obtains a unified approach to isotropic and anisotropic spaces as well as to spaces with dominating mixed smoothness. Common basic properties can be proved. We refer to H. Triebel [4, II–V], [F, Chapter 2], B. Stöckert, H. Triebel [1], M. L. Gol'dman [4], and H. G. Feichtinger, P. Gröbner [1].

Modulation Spaces. Modulation spaces are a special case of the just mentioned general spaces based on uniform decompositions (decompositions in congruent cubes) of R_n. They are connected with spaces of Wiener type which are due to H. G. Feichtinger [1–4]. We refer to Section 5.2.

Weighted Spaces. We replace $l_q(L_p)$ by $l_q(L_p(R_n, \varrho(x)), L_p(l_q)$ by $L_p(R_n, \varrho(x), l_q)$, and $S'(R_n)$ by $S'_\omega(R_n)$ with $\omega \in \mathfrak{M}$, $\varrho \in R(\omega)$ (cf. 1.2.1., 1.3.1., and 1.4.1. as far as notations are concerned). This leads to weighted versions of $B^s_{p,q}(R_n)$ and $F^s_{p,q}(R_n)$. We refer to Section 5.1.

Periodic Spaces. The theory of the spaces $B^s_{p,q}(R_n)$, $F^s_{p,q}(R_n)$ has a full periodic counterpart. Entire analytic functions of exponential type in (4) are replaced by trigonometric polynomials the Fourier coefficients of which are related to a dyadic decomposition of Z_n (cf. (1.3.1/5)). We refer to Chapter 3.

Abstract Spaces. Several directions are possible. R_n and T_n (the n-torus) can be replaced by a locally compact abelian group. Instead of the Fourier transform F and $S'(R_n)$ we can use another appropriate integral transform and a corresponding space of distributions, respectively. Other orthogonal systems can play the role of the trigonometric system in the case of periodic spaces. Finally abstract spaces can be defined via spectral decompositions of suitable self-adjoint operators acting in a Hilbert space. As far as the latter possibility is concerned we refer to Chapter 6.

2.2.3. Equivalent Quasi-Norms and Elementary Embeddings

We return to the spaces $S^{\bar r}_{\bar p,\bar q}B(R_2)$, $SB^{\bar r}_{\bar p,\bar q}(R_2)$, and $S^{\bar r}_{\bar p,\bar q}F(R_2)$ which have been defined in Definition 2.2.1/2. For sake of brevity we omit the suffix R_2 and put $S^{\bar r}_{\bar p,\bar q}B = S^{\bar r}_{\bar p,\bar q}B(R_2)$, $SB^{\bar r}_{\bar p,\bar q} = SB^{\bar r}_{\bar p,\bar q}(R_2)$, and $S^{\bar r}_{\bar p,\bar q}F = S^{\bar r}_{\bar p,\bar q}F(R_2)$. It is clear that all these spaces are linear vector spaces. Further it is not difficult to see that $\|\cdot \mid S^{\bar r}_{\bar p,\bar q}B\|^\varphi$, $\|\cdot \mid SB^{\bar r}_{\bar p,\bar q}\|^\varphi$, and $\|\cdot \mid S^{\bar r}_{\bar p,\bar q}F\|^\varphi$ define quasi-norms (norms if $\min(p_1, p_2, q_1, q_2) \geqq 1$) in the corresponding spaces for each $\varphi \in \Phi(R_1)$. The notation of a quasi-norm has been described in 1.3.1/(i)–(iii). Recall that two quasi-norms $\|a\|_1$ and $\|a\|_2$ on a given linear vector space A are said to be equivalent if there exist two positive numbers c_1 and c_2 such that

$$c_1\|a\|_1 \leqq \|a\|_2 \leqq c_2\|a\|_1, \quad a \in A,$$

holds.

Proposition 1. *Let* $\varphi = \{\varphi_j(t)\}_{j=0}^\infty \in \Phi(R_1)$ *and* $\psi = \{\psi_j(t)\}_{j=0}^\infty \in \Phi(R_1)$. *Let* $\bar p = (p_1, p_2)$, $\bar q = (q_1, q_2)$, *and* $\bar r = (r_1, r_2)$.

(i) *If* $-\infty < r_m < \infty$, $0 < p_m \leqq \infty$, $0 < q_m \leqq \infty$; $(m = 1, 2)$, *then* $\|f \mid S^{\bar r}_{\bar p,\bar q}B\|^\varphi$ *and* $\|f \mid S^{\bar r}_{\bar p,\bar q}B\|^\psi$ *are equivalent quasi-norms in* $S^{\bar r}_{\bar p,\bar q}B$.

(ii) *If* $-\infty < r_m < \infty$, $0 < p_m \leqq \infty$, $0 < q_m \leqq 0$; $(m = 1, 2)$, *then* $\|f \mid SB^{\bar r}_{\bar p,\bar q}\|^\varphi$ *and* $\|f \mid SB^{\bar r}_{\bar p,\bar q}\|^\psi$ *are equivalent quasi-norms in* $SB^{\bar r}_{\bar p,\bar q}$.

(iii) *If* $-\infty < r_m < \infty$, $0 < p_m < \infty$, $0 < q_m \leqq \infty$; $(m = 1, 2)$, *then* $\|f \mid S^{\bar r}_{\bar p,\bar q}F\|^\varphi$ *and* $\|f \mid S^{\bar r}_{\bar p,\bar q}F\|^\psi$ *are equivalent quasi-norms in* $S^{\bar r}_{\bar p,\bar q}F$.

Proof. Step 1. To prove (i) we observe that

$$\varphi_j(t) = (\psi_{j-1}(t) + \psi_j(t) + \psi_{j+1}(t))\,\varphi_j(t), \quad t \in R_1, j = 0, 1, 2, \ldots,$$

with $\psi_{-1}(t) = 0$. This gives

$$F^{-1}\varphi_j\varphi_k Ff = \sum_{n=-1}^{1}\sum_{m=-1}^{1} F^{-1}\varphi_j\varphi_k F[F^{-1}\psi_{j+m}\psi_{k+n}Ff]. \tag{1}$$

By Theorem 1.8.3 (with $b_1 = 2^{j+2}$, $b_2 = 2^{k+2}$) we obtain

$$\begin{aligned}
&\|F^{-1}\varphi_j\varphi_k F[F^{-1}\psi_{j+m}\psi_{k+n}Ff] \mid L_{\bar p}\| \\
&\leqq c\|\varphi_j(2^{j+2}\cdot)\,\varphi_k(2^{k+2}\cdot) \mid S_2^{\bar l}W\|\ \|F^{-1}\psi_{j+m}\psi_{k+n}Ff \mid L_{\bar p}\| \\
&\leqq c'\|\varphi_j(2^{j+2}\cdot) \mid W_2^l(R_1)\|\ \|\varphi_k(2^{k+2}\cdot) \mid W_2^l(R_1)\|\ \|F^{-1}\psi_{j+m}\psi_{k+n}Ff \mid L_{\bar p}\|. \qquad (2)
\end{aligned}$$

Here $\bar l = (l, l)$ with a sufficient large natural number l. Recall that $S_2^{\bar l}W$ and $W_2^l(R_1)$ have been defined in (1.8.3/2) and (1.7.4/2), respectively. The properties of $\varphi \in \Phi(R_1)$, in particular (2.2.1/1) and (2.2.1/2), imply

$$\|\varphi_j(2^{j+2}\cdot) \mid W_2^l(R_1)\| \leqq c \quad \text{for} \quad j = 0, 1, 2, \ldots \qquad (3)$$

Consequently, (2) and (3) yield

$$\|F^{-1}\varphi_j\varphi_k F[F^{-1}\psi_{j+m}\psi_{k+n}Ff] \mid L_{\bar p}\| \leqq c\|F^{-1}\psi_{j+m}\psi_{k+n}Ff \mid L_{\bar p}\| \qquad (4)$$

for $j, k = 0, 1, 2, \ldots$; and $m, n = -1, 0, 1$. The constant c is independent of j, k, m, n, and $f \in S_{\bar p,\bar q}^{\bar r}B$. Combining (1), (4), and (2.2.1/5) (with φ and ψ, respectively) the quasi-norm $\|f \mid S_{\bar p,\bar q}^{\bar r}B\|^{\varphi}$ can be estimated from above bythe quasi-norm $\|f \mid S_{\bar p,\bar q}^{\bar r}B\|^{\psi}$. Changing the roles of φ and ψ we obtain part (i) of the theorem.

Step 2. The proof of (ii) proceeds analogously. Let us put

$$f_{j+m,k+n} = F^{-1}\psi_{j+m}\psi_{k+n}Ff; \quad j, k = 0, 1, 2, \ldots; \ m, n = -1, 0, 1.$$

It follows from (1)

$$\begin{aligned}
&\|2^{jr_1+kr_2}F^{-1}\varphi_j\varphi_k Ff \mid Sl_{\bar q}(L_{\bar p})\| \\
&\leqq c \sum_{n=-1}^{1} \sum_{m=-1}^{1} \|F^{-1}\varphi_j\varphi_k F[2^{(j+m)r_1+(k+n)r_2} f_{j+m,k+n}] \mid Sl_{\bar q}(L_{\bar p})\|. \qquad (5)
\end{aligned}$$

Now we apply Theorem 1.10.3(i) (with $a_j = 2^{j+2}$ and $b_k = 2^{k+2}$) and Remark 1.10.3/3 (if $p_1 = \infty$ and/or $p_2 = \infty$) to the right-hand side of (5). Together with (3) we obtain that $\|f \mid SB_{\bar p,\bar q}^{\bar r}\|^{\varphi}$ can be estimated from above by $c\|f \mid SB_{\bar p,\bar q}^{\bar r}\|^{\psi}$. Again we change the roles of φ and ψ to complete the proof of part (ii).

Step 3. The proof of part (iii) is the same as in Step 2 if one replaces $Sl_{\bar q}(L_{\bar p})$ in (5) by $L_{\bar p}(l_{\bar q})$ and uses Theorem 1.10.3(ii) instead of Theorem 1.10.3(i) and Remark 1.10.3/3.

Remark 1. It does not make any sense to admit $p_1 = \infty$ and/or $p_2 = \infty$ in the definition of $S_{\bar p,\bar q}^{\bar r}F$ (cf. (2.2.1/7)). One can show that in this case a corresponding assertion of part (iii) of the proposition would not be valid (cf. also Remark 1.10.2/1).

Remark 2. The assertions of the proposition can be generalized. In particular, the proof shows that

$$\|f \mid S_{\bar p,\bar q}^{\bar r}B\|^{(\varphi,\psi)} = \|2^{r_1 j+r_2 k}F^{-1}\varphi_j\psi_k Ff \mid l_{\bar q}(L_{\bar p})\|, \qquad (6)$$

where $\varphi \in \Phi(R_1)$ and $\psi \in \Phi(R_1)$, is also an equivalent quasi-norm in $S_{\bar p,\bar q}^{\bar r}B$. Analogously, we obtain equivalent quasi-norms $\|f \mid SB_{\bar p,\bar q}^{\bar r}\|^{(\varphi,\psi)}$ and $\|f \mid S_{\bar p,\bar q}^{\bar r}F\|^{(\varphi,\psi)}$ on $SB_{\bar p,\bar q}^{\bar r}$ and $S_{\bar p,\bar q}^{\bar r}F$, respectively, for the admitted parameters.

In the following propositions $A_1 \subset A_2$ always means that the quasi-normed linear vector space A_1 is continuously embedded in the quasi-normed linear vector space A_2, i.e. there exists a constant c such that $\|a \mid A_2\| \leqq c\|a \mid A_1\|$ for all $a \in A_1$.

Proposition 2. *Let* $\bar p = (p_1, p_2)$, $\bar q = (q_1, q_2)$, $\bar v = (v_1, v_2)$, $\bar r = (r_1, r_2)$, *and* $\bar\varepsilon = (\varepsilon_1, \varepsilon_2)$. *Let* $-\infty < r_m < \infty$, $0 < p_m \leqq \infty$ ($p_m < \infty$ *in the F-cases*); ($m = 1, 2$).

(i) *If* $0 < q_m \leqq v_m \leqq \infty$, *then*

$$S^{\bar r}_{\bar p,\bar q}B \subset S^{\bar r}_{\bar p,\bar v}B, \qquad SB^{\bar r}_{\bar p,\bar q} \subset SB^{\bar r}_{\bar p,\bar v}, \quad \text{and} \quad S^{\bar r}_{\bar p,\bar q}F \subset S^{\bar r}_{\bar p,\bar v}F. \tag{7}$$

(ii) *Let* $0 \leqq \varepsilon_m < \infty$, $0 < q_m \leqq \infty$ $(m = 1, 2)$. *Then*

$$S^{\bar r+\bar\varepsilon}_{\bar p,\bar q}B \subset S^{\bar r}_{\bar p,\bar v}B, \qquad S^{\bar r+\bar\varepsilon}_{\bar p,\bar q}B \subset SB^{\bar r}_{\bar p,\bar v}, \quad \text{and} \quad S^{\bar r+\bar\varepsilon}_{\bar p,\bar q}F \subset S^{\bar r}_{\bar p,\bar v}F \tag{8}$$

holds if $0 < v_1 \leqq \infty$, $0 < v_2 \leqq \infty$ *with* $v_m = q_m$ *in the case* $\varepsilon_m = 0$ $(m = 1, 2)$.

(iii) *Let* $0 < q_1 \leqq \infty$, $0 < q_2 \leqq \infty$. *Then we have*

$$S^{\bar r}_{\bar p,(\min(p_2,q_1),q_2)}B \subset SB^{\bar r}_{\bar p,\bar q} \subset S^{\bar r}_{\bar p,(\max(p_2,q_1),q_2)}B, \tag{9}$$

$$SB^{\bar r}_{\bar p,(\min(p_1,q_1),\min(p_1,p_2,q_2))} \subset S^{\bar r}_{\bar p,\bar q}F \subset SB^{\bar r}_{\bar p,(\max(p_1,q_1),\max(p_1,p_2,q_2))}, \tag{10}$$

$$S^{\bar r}_{\bar p,(\min(p_1,p_2,q_1),\min(p_1,p_2,q_2))}B \subset S^{\bar r}_{\bar p,\bar q}F \subset S^{\bar r}_{\bar p,(\max(p_1,p_2,q_1),\max(p_1,p_2,q_2))}B. \tag{11}$$

Proof. Step 1. Let $\varphi = \{\varphi_j(t)\}_{j=0}^{\infty} \in \Phi(R_1)$. We put

$$a_{j,k}(x_1, x_2) = 2^{r_1 j + r_2 k}(F^{-1}\varphi_j\varphi_k Ff)(x_1, x_2) = 2^{r_1 j + r_2 k} f_{j,k}(x_1, x_2). \tag{12}$$

It holds

$$\|c_j \mid l_v\| = (\sum |c_j|^v)^{\frac{q}{v}\frac{1}{q}} \leqq (\sum |c_j|^q)^{\frac{1}{q}} = \|c_j \mid l_q\| \tag{13}$$

if $0 < q \leqq v \leqq \infty$ (modification if $v = \infty$). Consequently,

$$\|f \mid S^{\bar r}_{\bar p,\bar v}B\|^{\varphi} = \|a_{j,k}(x) \,|L_{\bar p}|\, l_{v_1|j} \mid l_{v_2|k}\|$$
$$\leqq \|a_{j,k}(x) \,|L_{\bar p}|\, l_{q_1|j} \mid l_{q_2|k}\| = \|f \mid S^{\bar r}_{\bar p,\bar q}B\|^{\varphi}.$$

Similarly, we obtain the second and the third embedding of (i).

Step 2. Let $\bar\varepsilon = (\varepsilon_1, 0)$, $\varepsilon_1 > 0$. By (7) it is sufficient to prove (8) for the case $\bar q = (\infty, v_2)$, only. We have,

$$\|f \mid S^{\bar r}_{\bar p,\bar v}B\|^{\varphi} = \left(\sum_{k=0}^{\infty} 2^{r_2 k v_2}\left(\sum_{j=0}^{\infty} 2^{r_1 j v_1}\|f_{j,k} \mid L_{\bar p}\|^{v_1}\right)^{v_2/v_1}\right)^{1/v_2}$$
$$\leqq \left(\sum_{k=0}^{\infty} 2^{r_2 k v_2}\left(\sup_j 2^{(r_1+\varepsilon_1)j}\|f_{j,k} \mid L_{\bar p}\|\right)^{v_2}\right)^{1/v_2}\left(\sum_{j=0}^{\infty} 2^{-j\varepsilon_1 v_1}\right)^{1/v_1}$$
$$\leqq c\|f \mid S^{\bar r+\bar\varepsilon}_{\bar p,(\infty,v_2)}B\|^{\varphi}.$$

Analogously we proceed if $\bar\varepsilon = (0, \varepsilon_2)$, $\varepsilon_2 > 0$, and if $f \in S^{\bar r}_{\bar p,\bar q}F$ $(f \in SB^{\bar r}_{\bar p,\bar q})$.

Step 3. Part (iii) is a consequence of the generalized triangle (or Minkowski) inequality for Banach spaces. In particular it holds

$$\|a_j(t) \,|L_p|\, l_q\| = \left\|\int_{R_1} |a_j(t)|^p \,\mathrm{d}t \mid l_{q/p}\right\|^{1/p} \leqq \|a_j(t) \,|l_q|\, L_p\| \quad \text{if} \quad p \leqq q$$

and

$$\|a_j(t) \,|l_q|\, L_p\| = \left\|\sum_{j=0}^{\infty} |a_j(t)|^q \mid L_{p/q}\right\|^{1/q} \leqq \|a_j(t) \,|L_p|\, l_q\| \quad \text{if} \quad q \leqq p.$$

Therefore, using (12) and part (i) we can estimate

$$\|f \mid S^{\bar r}_{\bar p,(\max(p_2,q_1),q_2)}B\|^{\varphi} = \|a_{j,k}(x_1, x_2) \mid L_{p_1|x_1} \mid L_{p_2|x_2} \mid l_{\max(p_2,q_1)|j} \mid l_{q_2|k}\|$$
$$\leqq \|a_{j,k}(x_1, x_2) \mid L_{p_1|x_1} \mid l_{\max(p_2,q_1)|j} \mid L_{p_2|x_2} \mid l_{q_2|k}\|$$
$$\leqq \|f \mid SB^{\bar r}_{\bar p,\bar q}\|^{\varphi} \leqq \|a_{j,k}(x_1, x_2) \mid L_{p_1|x_1} \mid l_{\min(p_2,q_1)|j} \mid L_{p_2|x_2} \mid l_{q_2|k}\|$$
$$\leqq \|a_{j,k}(x_1, x_2) \mid L_{p_1|x_1} \mid L_{p_2|x_2} \mid l_{\min(p_2,q_1)|j} \mid l_{q_2|k}\|$$
$$= \|f \mid S^{\bar r}_{\bar p,(\min(p_2,q_1),q_2)}B\|^{\varphi}.$$

This yields (9). Similarly the embeddings (10) and (11) can be proved.

Remark 3. Proposition 2 states relations between the spaces $S^{\bar{r}}_{\bar{p},\bar{q}}B$, $SB^{\bar{r}}_{\bar{p},\bar{q}}$, and $S^{\bar{r}}_{\bar{p},\bar{q}}F$ in dependence on the parameters $\bar{p} = (p_1, p_2)$, $\bar{q} = (q_1, q_2)$ and $\bar{r} = (r_1, r_2)$. The question arises whether some spaces coincide (set-theoretical and topological). This problem was discussed in [T, 2.3.9.] for the isotropic spaces $B^s_{p,q}(R_n)$ and $F^s_{p,q}(R_n)$ (cf. also the references given there). We can apply the method of [T, 2.3.9.] to our spaces. The result reads as follows: Let $\bar{r} = (r_1, r_2)$, $\bar{p} = (p_1, p_2)$, and $\bar{q} = (q_1, q_2)$. Let $-\infty < r_m < \infty$, $0 < q_m \leqq \infty$; $(m = 1, 2)$. Then it holds (obviously)

$$S^{\bar{r}}_{\bar{p},\bar{q}}B = SB^{\bar{r}}_{\bar{p},\bar{q}} \quad \text{if} \quad p_2 = q_1, 0 < p_1 \leqq \infty, 0 < p_2 \leqq \infty, \tag{14}$$

$$SB^{\bar{r}}_{\bar{p},\bar{q}} = S^{\bar{r}}_{\bar{p},\bar{q}}F \quad \text{if} \quad p_1 = p_2 = q_1 = q_2, 0 < p_1 < \infty, 0 < p_2 < \infty, \tag{15}$$

$$S^{\bar{r}}_{\bar{p},\bar{q}}B = S^{\bar{r}}_{\bar{p},\bar{q}}F \quad \text{if} \quad p_1 = p_2 = q_1 = q_2, 0 < p_1 < \infty, 0 < p_2 < \infty. \tag{16}$$

These are the only cases in which two of the considered spaces $S^{\bar{r}}_{\bar{p},\bar{q}}B$, $S^{\bar{t}}_{\bar{u},\bar{v}}F$, and $SB^{\bar{\varrho}}_{\bar{\mu},\bar{v}}$ coincide.

Remark 4. Let $\bar{p} = (p_1, p_2)$, $\bar{q} = (q_1, q_2)$, and $\bar{r} = (r_1, r_2)$. Let $0 < p_m \leqq \infty$ ($p_m < \infty$ in the case $S^{\bar{r}}_{\bar{p},\bar{q}}F$), $0 < q_m \leqq \infty$, and $-\infty < r_m < \infty$; $(m = 1,2)$. As an immediate consequence of part (i) and part (iii) of the last proposition we obtain

$$S^{\bar{r}}_{\bar{p},\bar{q}}B,\ SB^{\bar{r}}_{\bar{p},\bar{q}},\ S^{\bar{r}}_{\bar{p},\bar{q}}F \subset S^{\bar{r}}_{\bar{p},(\infty,\infty)}B. \tag{17}$$

Proposition 3. *Let* $\bar{p} = (p_1, p_2)$, $\bar{q} = (q_1, q_2)$, *and* $\bar{r} = (r_1, r_2)$. *Let* $0 < q_1 \leqq \infty$, $0 < q_2 \leqq \infty$, *and*

$$r_1 > \frac{1}{\min(1, p_1)} - 1, \qquad r_2 > \frac{1}{\min(1, p_2)} - 1. \tag{18}$$

Then we have

$$S^{\bar{r}}_{\bar{p},\bar{q}}B \subset L_{\bar{p}} \cap L_{(\max(1,p_1), \max(1,p_2))}, \tag{19}$$

$$SB^{\bar{r}}_{\bar{p},\bar{q}} \subset L_{\bar{p}} \cap L_{(\max(1,p_1), \max(1,p_2))}, \tag{20}$$

if $0 < p_1 \leqq \infty$, $0 < p_2 \leqq \infty$, *and*

$$S^{\bar{r}}_{\bar{p},\bar{q}}F \subset L_{\bar{p}} \cap L_{(\max(1,p_1), \max(1,p_2))} \tag{21}$$

if $0 < p_1 < \infty$, $0 < p_2 < \infty$.

Proof. Step 1. By Proposition 2 (cf. Remark 4) it is sufficient to prove

$$S^{\bar{r}}_{\bar{p},(\infty,\infty)}B \subset L_{\bar{p}} \cap L_{(\max(1,p_1), \max(1,p_2))}. \tag{22}$$

Let $f \in S^{\bar{r}}_{\bar{p},(\infty,\infty)}B$ and let $\varphi = \{\varphi_j(t)\}_{j=0}^{\infty} \in \Phi(R_1)$. Then

$$f = \sum_{k=0}^{\infty} \sum_{j=0}^{\infty} F^{-1}\varphi_j\varphi_k Ff \quad \text{(convergence in } S'\text{)}.$$

Therefore, with the help of Nikol'skij's inequality (Theorem 1.6.2 with $\alpha = (0, 0)$) we can estimate

$$\begin{aligned} &\|f \mid L_{(\max(1,p_1), \max(1,p_2))}\| \\ &\leqq \sum_{k=0}^{\infty} \sum_{j=0}^{\infty} \|F^{-1}\varphi_j\varphi_k Ff \mid L_{(\max(1,p_1), \max(1,p_2))}\| \\ &\leqq c \sum_{k=0}^{\infty} \sum_{j=0}^{\infty} 2^{j\left(\frac{1}{\min(1,p_1)} - 1\right)} \cdot 2^{k\left(\frac{1}{\min(1,p_2)} - 1\right)} \|F^{-1}\varphi_j\varphi_k Ff \mid L_{\bar{p}}\| \\ &\leqq c\|f \mid S^{\bar{r}}_{\bar{p},(\infty,\infty)}B\|^{\varphi} \left(\sum_{j=0}^{\infty} 2^{-\varepsilon_1 j}\right) \left(\sum_{k=0}^{\infty} 2^{-\varepsilon_2 k}\right) \leqq c'\|f \mid S^{\bar{r}}_{\bar{p},(\infty,\infty)}B\|^{\varphi}. \end{aligned}$$

Hence, $f \in L_{(\max(1,p_1), \max(1,p_2))}$.

Step 2. We prove $f \in L_{\bar{p}}$. Step 1 shows that

$$\tilde{f}_{m,n} = \sum_{k=0}^{n} \sum_{j=0}^{m} F^{-1}\varphi_j\varphi_k Ff = \sum_{k=0}^{n} \sum_{j=0}^{m} f_{j,k} \to f \in L_{(\max(1,p_1),\max(1,p_2))} \tag{23}$$

(convergence in $L_{(\max(1,p_1),\max(1,p_2))}$) if $f \in S^{\bar{r}}_{\bar{p},(\infty,\infty)}B$. We claim that $\{\tilde{f}_{m,n}(x)\}^{\infty}_{m,n=0}$ is a fundamental sequence in $L_{\bar{p}}$. (Of course, only the case $\min(p_1, p_2) < 1$ is of interest.) Let $M > m$, $N > n$ be arbitrary natural numbers. Let $s = \min(1, p_1, p_2)$. Then, using (8), we have

$$\begin{aligned}
\|\tilde{f}_{M,N} - \tilde{f}_{m,n} \mid L_{\bar{p}}\| &\leqq c\left(\sum_{k=n+1}^{N}\sum_{j=0}^{M} \|f_{j,k} \mid L_{\bar{p}}\|^s\right)^{1/s} + c\left(\sum_{k=0}^{n}\sum_{j=m+1}^{M} \|f_{j,k} \mid L_{\bar{p}}\|^s\right)^{1/s} \\
&\leqq c\left(\sum_{k=n+1}^{N} 2^{-r_2sk}\right)^{1/s} \sup_k \left(\sum_{j=0}^{\infty} 2^{r_2sk}\, \|f_{j,k} \mid L_{\bar{p}}\|^s\right)^{1/s} \\
&\quad + c\left(\sum_{j=m+1}^{M} 2^{-r_1sj}\right)^{1/s} \left(\sum_{k=0}^{\infty}\left(\sup_j 2^{r_1j}\, \|f_{j,k} \mid L_{\bar{p}}\|\right)^s\right)^{1/s} \\
&\leqq c\left(\sum_{k=n+1}^{N} 2^{-r_2sk}\right)^{1/s} \|f \mid S^{(0,r_2)}_{\bar{p},(s,\infty)}B\|^{\varphi} \\
&\quad + c\left(\sum_{j=m+1}^{M} 2^{-r_1sj}\right)^{1/s} \|f \mid S^{(r_1 0)}_{\bar{p},(\infty,s)}B\|^{\varphi} \\
&\leqq c'\left(\sum_{k=n+1}^{N} 2^{-r_2sk} + \sum_{j=m+1}^{M} 2^{-r_1sj}\right)^{1/s} \|f \mid S^{\bar{r}}_{\bar{p},(\infty,\infty)}\|^{\varphi} \to 0
\end{aligned}$$

if $m, n \to \infty$.

Hence, $\{\tilde{f}_{m,n}\}$ is a fundamental sequence in $L_{\bar{p}}$ and $\tilde{f}_{m,n} \to g$ in $L_{\bar{p}}$. Consequently, because of (23) it holds $f = g$ a.e. Further we have

$$\begin{aligned}
\|f \mid L_{\bar{p}}\| &= \left\|\sum_{k=0}^{\infty}\sum_{j=0}^{\infty} F^{-1}\varphi_j\varphi_k Ff \mid L_{\bar{p}}\right\| \\
&\leqq \left\|\sum_{k=0}^{\infty}\sum_{j=0}^{\infty} |F^{-1}\varphi_j\varphi_k Ff| \mid L_{\bar{p}}\right\| \\
&= \|f \mid S^{\bar{0}}_{\bar{p},\bar{1}}F\|^{\varphi} \leqq c\|f \mid S^{\bar{r}}_{\bar{p},\bar{\infty}}B\|^{\varphi},
\end{aligned} \tag{24}$$

where $\bar{1} = (1, 1)$, $\bar{0} = (0, 0)$, and $\overline{\infty} = (\infty, \infty)$. This completes the proof.

Remark 5. In general, $S^{\bar{r}}_{\bar{p},\bar{q}}B(S^{\bar{r}}_{\bar{p},\bar{q}}F, SB^{\bar{r}}_{\bar{p},\bar{q}})$ is a space of tempered distributions. Proposition 3 shows that $f = f(x) \in S^{\bar{r}}_{\bar{p},\bar{q}}B(S^{\bar{r}}_{\bar{p},\bar{q}}F, SB^{\bar{r}}_{\bar{p},\bar{q}})$ is a regular distribution if

$$r_1 > \frac{1}{\min(1, p_1)} - 1 \quad \text{and} \quad r_2 > \frac{1}{\min(1, p_2)} - 1.$$

In particular, this means $r_1 > 0$, $r_2 > 0$ if $\min(p_1, p_2) \geqq 1$ and $r_1 > \frac{1}{p_1} - 1$, $r_2 > \frac{1}{p_2} - 1$ if $\max(p_1, p_2) \leqq 1$. On the other hand, it can easily be calculated that

$$\delta \in S^{\bar{r}}_{\bar{p},(\infty,\infty)}B \quad \text{if} \quad r_1 = \frac{1}{p_1} - 1,\ r_2 = \frac{1}{p_2} - 1. \tag{25}$$

Here δ stands for the δ-distribution. By Proposition 2 this assertion can be extended to the other spaces.

We shall finish this subsection with some further embeddings. Let us denote the Banach space of all complex-valued uniformly continuous bounded functions on R_2 by $C = C(R_2)$, endowed with the norm,

$$\|f \mid C\| = \sup_{x \in R_2} |f(x_1, x_2)|. \tag{26}$$

Proposition 4. *Let* $\bar{p} = (p_1, p_2)$, $1 \leqq p_1 < \infty$, $1 \leqq p_2 < \infty$. *Let* $\bar{1} = (1, 1)$, $\bar{0} = (0, 0)$, *and* $\overline{\infty} = (\infty, \infty)$. *Then we have*

$$S^{\bar{0}}_{\bar{p},\bar{1}}B \subset L_{\bar{p}} \subset S^{\bar{0}}_{\bar{p},\overline{\infty}}B, \tag{27}$$

$$S^{\bar{0}}_{\overline{\infty},\bar{1}}B \subset C \subset L_{\overline{\infty}} \subset S^{\bar{0}}_{\overline{\infty},\overline{\infty}}B. \tag{28}$$

Proof. Step 1. Let $\{\varphi_j\}_{j=0}^{\infty} \in \Phi(R_1)$. If $f \in S^{\bar{0}}_{\bar{p},\bar{1}}B$, $1 \leqq p_1 \leqq \infty$, $1 \leqq p_2 \leqq \infty$, then we can estimate as in Step 1 of the proof of Proposition 3

$$\|f \mid L_{\bar{p}}\| \leqq \sum_{k=0}^{\infty} \sum_{j=0}^{\infty} \|F^{-1}\varphi_j\varphi_k Ff \mid L_{\bar{p}}\| = \|f \mid S^{\bar{0}}_{\bar{p},\bar{1}}B\|^{\varphi}.$$

This proves the left-hand side of (27). Further we have

$$f = \sum_{k=0}^{\infty} \sum_{j=0}^{\infty} F^{-1}\varphi_j\varphi_k Ff(x) \quad \text{(convergence in } L_{\overline{\infty}})$$

if $\bar{p} = \overline{\infty}$. But, each function $F^{-1}\varphi_j\varphi_k Ff(x)$ is uniformly continuous. This follows from Theorem 1.6.2 (with $\alpha = (1, 0)$ and $\alpha = (0, 1)$). Hence, $f \in C(R_2)$ and the left-hand side of (28) is verified.

Step 2. Let $f \in L_{\bar{p}}$, $1 \leqq p_1 \leqq \infty$, $1 \leqq p_2 \leqq \infty$. In the same way as in the proof of Theorem 1.8.1 (note that we do not need that supp Ff is compact if $\min(p_1, p_2) \geqq 1$) we obtain

$$\begin{aligned} \|F^{-1}\varphi_j\varphi_k Ff \mid L_{\bar{p}}\| &\leqq \|(F^{-1}\varphi_j\varphi_k) \mid L_{\bar{1}}\| \, \|f \mid L_{\bar{p}}\| \\ &\leqq c\|f \mid L_{\bar{p}}\|. \end{aligned}$$

This proves the right-hand sides of (27) and (28).

2.2.4. Basic Properties

Let us recall that the topology of the locally convex space $S = S(R_2)$ is generated by

$$\|\varphi\|_N = \sup_{x \in R_2} (1 + |x|)^N \sum_{|\alpha| \leqq N} |D^\alpha \varphi(x)|, \quad N = 1, 2, \ldots, \tag{1}$$

where $\varphi \in S$. Here, $|x| = (x_1^2 + x_2^2)^{1/2}$ for $x \in R_2$, $|\alpha| = \alpha_1 + \alpha_2$, where $\alpha = (\alpha_1, \alpha_2)$ is a pair of non-negative integers. The topological dual S' of S is equipped with the strong topology. If A is a quasi-normed linear space then $S \subset A$ and $A \subset S'$ always mean continuous (topological) embeddings.

Theorem. (i) *Let* $\bar{p} = (p_1, p_2)$, $\bar{q} = (q_1, q_2)$, *and* $\bar{r} = (r_1, r_2)$ *with* $0 < p_m \leqq \infty$, $0 < q_m \leqq \infty$, *and* $-\infty < r_m < \infty$; $(m = 1, 2)$. *Then* $S^{\bar{r}}_{\bar{p},\bar{q}}B$ $(SB^{\bar{r}}_{\bar{p},\bar{q}})$ *is a quasi-Banach space (Banach space if* $\min(p_1, p_2, q_1, q_2) \geqq 1$) *and the quasi-norms* $\|f \mid S^{\bar{r}}_{\bar{p},\bar{q}}B\|^{\varphi}$ $(\|f \mid SB^{\bar{r}}_{\bar{p},\bar{q}}\|^{\varphi})$ *with* $\varphi \in \Phi(R_1)$ *and* $\|f \mid S^{\bar{r}}_{\bar{p},\bar{q}}B\|^{\psi}$ $(\|f \mid SB^{\bar{r}}_{\bar{p},\bar{q}}\|^{\psi})$ *with* $\psi \in \Phi(R_1)$ *are* *equivalent to each other. Furthermore it holds*

$$S \subset S^{\bar r}_{\bar p, \bar q}B \; (SB^{\bar r}_{\bar p, \bar q}) \subset S'. \tag{2}$$

If $\max(p_1, p_2, q_1, q_2) < \infty$, *then* S *is dense in* $S^{\bar r}_{\bar p, \bar q}B \; (SB^{\bar r}_{\bar p, \bar q})$.

(ii) *Let* $\bar p = (p_1, p_2)$, $\bar q = (q_1, q_2)$, *and* $\bar r = (r_1, r_2)$ *with* $0 < p_m < \infty$, $0 < q_m \leqq \infty$, $-\infty < r_m < \infty$; $(m = 1, 2)$. *Then,* $S^{\bar r}_{\bar p, \bar q}F$ *is a quasi-Banach space (Banach space if* $\min(p_1, p_2, q_1, q_2) \geqq 1$) *and the quasi-norms* $\|f \mid S^{\bar r}_{\bar p, \bar q}F\|^{\varphi}$ *with* $\varphi \in \Phi(R_1)$ *and* $\|f \mid S^{\bar r}_{\bar p, \bar q}F\|^{\psi}$ *with* $\psi \in \Phi(R_1)$ *are equivalent to each other. Furthermore it holds*

$$S \subset S^{\bar r}_{\bar p, \bar q}F \subset S'. \tag{3}$$

If $\max(p_1, p_2, q_1, q_2) < \infty$, *then* S *is dense in* $S^{\bar r}_{\bar p, \bar q}F$.

Proof. Step 1. The equivalence of the quasi-norms has been proved in Proposition 2.2.3/1. The completeness of the spaces follows by standard arguments if we use the embeddings in S' (cf. (2), (3)) and the Nikol'skij inequality from Theorem 1.6.2. We omit the details and refer to [T, Step 4 of the proof of Theorem 2.3.3].

Step 2. We prove (2) and (3). According to Proposition 2.2.3/2 (cf. also Remark 2.2.3/4) we may restrict ourselves to the proof of

$$S \subset S^{\bar r}_{\bar p, \bar\infty}B \subset S', \quad \bar\infty = (\infty, \infty) \tag{4}$$

for all $\bar p$, $\bar r$ with $0 < p_m \leqq \infty$, $-\infty < r_m < \infty$; $(m = 1, 2)$. Let $\{\varphi_j\} \in \Phi(R_1)$ and let $f \in S = S(R_2)$. If N_1, N_2, N_3, N are sufficiently large natural numbers then we have

$$\begin{aligned}
\|f \mid S^{\bar r}_{\bar p, \bar\infty}B\|^{\varphi} &= \sup_{j,k} 2^{r_1 j + r_2 k} \|F^{-1}\varphi_j\varphi_k Ff \mid L_{\bar p}\| \\
&\leqq c_1 \sup_{j,k} 2^{r_1 j + r_2 k} \|(1 + x_1^2)^{N_1} (1 + x_2^2)^{N_1} F^{-1}\varphi_j\varphi_k Ff \mid L_{\bar\infty}\| \\
&\leqq c_2 \sup_{j,k} 2^{r_1 j + r_2 k} \Big\| \sum_{|\alpha| \leqq 4N_1} \mathrm{i}^{|\alpha|} D^{\alpha}(\varphi_j\varphi_k Ff) \mid L_{\bar 1} \Big\| \\
&\leqq c_3 \Big\| (1 + x_1^2)^{N_2} (1 + x_2^2)^{N_2} \sum_{|\alpha| \leqq 4N_1} |D^{\alpha} Ff| \mid L_{\bar\infty} \Big\| \\
&\leqq c_4 \|Ff\|_{N_3} \leqq c_5 \|f\|_N .
\end{aligned} \tag{5}$$

The last inequality uses the fact that F is a continuous mapping from S onto itself. (5) proves the left-hand side of (4). It remains to prove the right-hand side of (4). If $f \in S^{\bar r}_{\bar p, \bar\infty}B$ and $\psi \in S$ then $f(\psi)$ denotes the value of the functional $f \in S'$ for the test function ψ. We put $\chi_j = \varphi_{j-1} + \varphi_j + \varphi_{j+1}$, $j = 0, 1, 2, \ldots$ (with $\varphi_{-1} = 0$), where again $\{\varphi_j\}_{j=0}^{\infty} \in \Phi(R_1)$. Because

$$f = \sum_{k=0}^{\infty} \sum_{j=0}^{\infty} F^{-1}\varphi_j\varphi_k Ff \quad \text{(convergence in } S')$$

we obtain

$$|f(\psi)| = \left| \sum_{k=0}^{\infty} \sum_{j=0}^{\infty} (F^{-1}\varphi_j\varphi_k Ff)(F\chi_j\chi_k F^{-1}\psi) \right|. \tag{6}$$

Now, it follows from $f \in S^{\bar r}_{\bar p, \bar\infty}B$ and from Theorem 1.6.2 that

$$\|F^{-1}\varphi_j\varphi_k Ff \mid L_{\bar\infty}\| \leqq c \cdot 2^{j\frac{1}{p_1} + k\frac{1}{p_2}} \|F^{-1}\varphi_j\varphi_k Ff \mid L_{\bar p}\| < \infty . \tag{7}$$

Furthermore, $F^{-1}\chi_j\chi_k F\psi$ belongs to S; $j, k = 0, 1, 2, \ldots$ Consequently, by (6) and (7) we have

$$
\begin{aligned}
|f(\psi)| &\leqq \sum_{k=0}^{\infty} \sum_{j=0}^{\infty} \|F^{-1}\varphi_j\varphi_k Ff \mid L_{\bar{\infty}}\| \, \|F\chi_j\chi_k F^{-1}\psi \mid L_{\bar{1}}\| \\
&\leqq c \sum_{k=0}^{\infty} \sum_{j=0}^{\infty} 2^{j\frac{1}{p_1}+k\frac{1}{p_2}} \|F^{-1}\varphi_j\varphi_k Ff \mid L_{\bar{p}}\| \, \|F\chi_j\chi_k F^{-1}\psi \mid L_{\bar{1}}\| \\
&\leqq c\|f \mid S^{\bar{r}}_{\bar{p},\bar{\infty}}B\|^{\varphi} \sum_{k=0}^{\infty} \sum_{j=0}^{\infty} 2^{j\left(\frac{1}{p_1}-r_1\right)+k\left(\frac{1}{p_2}-r_2\right)} \|F\chi_j\chi_k F^{-1}\psi \mid L_{\bar{1}}\| \\
&\leqq c'\|f \mid S^{\bar{r}}_{\bar{p},\bar{\infty}}B\|^{\varphi} \, \|\psi \mid S^{\bar{\varkappa}}_{\bar{1},\bar{1}}B\|^{\varphi},
\end{aligned} \tag{8}
$$

where $\bar{1} = (1, 1)$, $\bar{\varkappa} = (\varkappa_1, \varkappa_2)$, $\varkappa_1 = \dfrac{1}{p_1} - r_1$, $\varkappa_2 = \dfrac{1}{p_2} - r_2$. Using the left-hand side of (2) for $S^{\bar{\varkappa}}_{\bar{1},\bar{1}}B$ we obtain

$$|f(\psi)| \leqq c\|f \mid S^{\bar{r}}_{\bar{p},\bar{\infty}}B\|^{\varphi} \, \|\psi\|_N, \tag{9}$$

if N is a sufficiently large natural number. This completes the proof of (4).

Step 3. Next we prove the density of S in $S^{\bar{r}}_{\bar{p},\bar{q}}B$, $SB^{\bar{r}}_{\bar{p},\bar{q}}$, and $S^{\bar{r}}_{\bar{p},\bar{q}}F$ if $\max(p_1, p_2, q_1, q_2) < \infty$. Suppose that $f \in S^{\bar{r}}_{\bar{p},\bar{q}}B$. Let us put

$$f_N = \sum_{k=0}^{N} \sum_{j=0}^{N} F^{-1}\varphi_j\varphi_k Ff, \quad N = 0, 1, 2, \ldots,$$

where $\{\varphi_j\} \in \Phi(R_1)$. It is not difficult to see that $f_N \in S^{\bar{r}}_{\bar{p},\bar{q}}B \cap L^{\Omega}_{\bar{p}}$, with $\Omega = \{x \mid x = (x_1, x_2), |x_1| \leqq 2^{N+1}, |x_2| \leqq 2^{N+1}\}$. Further it is not hard to prove

$$\|f - f_N \mid S^{\bar{r}}_{\bar{p},\bar{q}}B\| \to 0, \quad N \to \infty.$$

Hence, by the definition of $S^{\bar{r}}_{\bar{p},\bar{q}}B$, the proof of the density is established if we are able to approximate $f_N(x)$ in, say, $L^{\Omega_1}_{\bar{p}}$, $\Omega_1 = \{x \mid x = (x_1, x_2), |x_1| \leqq 2^{N+2}, |x_2| \leqq 2^{N+2}\}$. For this purpose we apply the approximation procedure from Proposition 1.2.2/2(iii) (cf. also Step 1 of the proof of Theorem 1.5.2). Let $\psi \in S$, with $\psi(0) = 1$ and $\operatorname{supp} F\psi \subset \{x \mid x = (x_1, x_2), |x_1| \leqq 1, |x_2| \leqq 1\}$. Then, $\psi(\varepsilon x) f_N(x) \in L^{\Omega_1}_{\bar{p}} \cap S(R_2)$ and

$$\psi(\varepsilon x) f_N(x) \to f_N(x), \quad \varepsilon \to 0,$$

pointwise (cf. (1.2.2./4) with $\omega(x) = \log(1 + |x|)$). Lebesgue's bounded convergence theorem yields

$$\|\psi(\varepsilon x) f_N(x) - f_N(x) \mid L_{\bar{p}}\| \to 0, \quad \varepsilon \to 0.$$

Hence, the density of S in $S^{\bar{r}}_{\bar{p},\bar{q}}B$ is proved. The other cases can be treated similarly.

Remark 1. It can be proved that S is not dense in $S^{\bar{r}}_{\bar{p},\bar{q}}B$ ($SB^{\bar{r}}_{\bar{p},\bar{q}}$, $S^{\bar{r}}_{\bar{p},\bar{q}}F$) if $\max(p_1, p_2, q_1, q_2) = \infty$.

Remark 2. Because of the equivalence of the quasi-norms $\|f \mid \ldots\|^{\varphi}$ in the corresponding spaces we shall not indicate "φ" in the sequel. In this sense we write simply $\|f \mid \ldots\|$ instead of $\|f \mid \ldots\|^{\varphi}$.

Remark 3. The proof of the Theorem and the proofs in Subsection 2.2.3. show that the Nikol'skij inequality of Theorem 1.6.2, and the Fourier multiplier theorems for $L^{\Omega}_{\bar{p}}$, $Sl_{\bar{q}}(L^{\Omega}_{\bar{p}})$, and $L^{\Omega}_{\bar{p}}(l_{\bar{q}})$ (cf. Theorem 1.8.3, Theorem 1.10.3) hold a key position in our approach. As we shall see in the sequel many other properties can be derived by these methods.

2.2.5. Fourier Multipliers

Our method allows a comparatively easy approach to state criterions for Fourier multipliers.

Definition. *Let $m(x_1, x_2)$ be a complex-valued infinitely differentiable function of at most polynomial growth on R_2. A stands for $S^{\bar{r}}_{\bar{p},\bar{q}}B$, $SB^{\bar{r}}_{\bar{p},\bar{q}}$, and $S^{\bar{r}}_{\bar{p},\bar{q}}F$, respectively, with $\bar{p} = (p_1, p_2)$, $\bar{q} = (q_1, q_2)$, $\bar{r} = (r_1, r_2)$, $0 < p_m \leqq \infty$, $0 < q_m \leqq \infty$, $-\infty < r_m < \infty$; $(m = 1, 2)$. We suppose $\max(p_1, p_2) < \infty$ if $A = S^{\bar{r}}_{\bar{p},\bar{q}}F$. Then $m(x)$ is said to be a Fourier multiplier for A if there exists a positive constant c such that*

$$\|F^{-1}mFf \mid A\| \leqq c\|f \mid A\| \tag{1}$$

holds for all $f \in A$.

Recall that the spaces $S^{\bar{\sigma}}_2H$ and $S^{\bar{\sigma}}_2W$ have been defined in 1.8.3/(1), (2).

Theorem. *Let $\{\psi_j(t)\}_{j=0}^{\infty} \subset S(R_1)$ be a sequence of functions satisfying*

$$0 \leqq \psi_0(t) \leqq 1, \operatorname{supp} \psi_0 \subset [-4, 4], \psi_0(t) = 1 \quad \text{if} \quad t \in [-2, 2], \tag{2}$$

$$\psi_j(t) = \psi(2^{-j}t), \quad j = 1, 2, \ldots, \text{ where} \tag{3}$$

$$0 \leqq \psi(t) \leqq 1, \operatorname{supp} \psi \subset \{t \mid \tfrac{1}{4} \leqq |t| \leqq 4\}, \psi(t) = 1 \quad \text{if} \quad \tfrac{1}{2} \leqq |t| \leqq 2. \tag{4}$$

Let $m(x)$ be a complex-valued infinitely differentiable function of at most polynomial growth on R_2. Let $\bar{p} = (p_1, p_2)$, $\bar{q} = (q_1, q_2)$, $\bar{r} = (r_1, r_2)$, and $\bar{\sigma} = (\sigma_1, \sigma_2)$. For given $\bar{\sigma}$ and $m(x)$ we put

$$M_{\bar{\sigma}} = \sup_{\substack{j=0,1,2,\ldots \\ k=0,1,2,\ldots}} \|\psi_j(2^jx_1)\,\psi_k(2^kx_2)\,m(2^jx_1, 2^kx_2) \mid S^{\bar{\sigma}}_2H\|. \tag{5}$$

(i) *Let $0 < p_m \leqq \infty$, $0 < q_m \leqq \infty$, and $-\infty < r_m < \infty$; $(m = 1, 2)$. If*

$$\sigma_1 > \frac{1}{\min(1, p_1)} - \frac{1}{2}, \qquad \sigma_2 > \frac{1}{\min(1, p_1, p_2)} - \frac{1}{2}$$

then there exists a positive constant c such that

$$\|F^{-1}mFf \mid S^{\bar{r}}_{\bar{p},\bar{q}}B\| \leqq cM_{\bar{\sigma}}\|f \mid S^{\bar{r}}_{\bar{p},\bar{q}}B\| \tag{6}$$

holds for all $f \in S^{\bar{r}}_{\bar{p},\bar{q}}B$.

(ii) *Let $0 < p_m < \infty$, $0 < q_m \leqq \infty$, $-\infty < r_m < \infty$; $(m = 1, 2)$. If*

$$\sigma_1 > \frac{1}{p_1} + \frac{1}{2}, \qquad \sigma_2 > \frac{1}{\min(p_1, p_2, q_1)} + \frac{1}{2} \tag{7}$$

then there exists a positive constant c such that

$$\|F^{-1}mFf \mid SB^{\bar{r}}_{\bar{p},\bar{q}}\| \leqq cM_{\bar{\sigma}}\|f \mid SB^{\bar{r}}_{\bar{p},\bar{q}}\| \tag{8}$$

holds for all $f \in SB^{\bar{r}}_{\bar{p},\bar{q}}$.

(iii) *Let $0 < p_m < \infty$, $0 < q_m \leqq \infty$, $-\infty < r_m < \infty$; $(m = 1, 2)$. If*

$$\sigma_1 > \frac{1}{\min(p_1, q_1, q_2)} + \frac{1}{2}, \qquad \sigma_2 > \frac{1}{\min(p_1, p_2, q_1, q_2)} + \frac{1}{2} \tag{9}$$

then there exists a positive constant c such that

$$\|F^{-1}mFf \mid S^{\bar{r}}_{\bar{p},\bar{q}}F\| \leqq cM_{\bar{\sigma}}\|f \mid S^{\bar{r}}_{\bar{p},\bar{q}}F\| \tag{10}$$

holds for all $f \in S^{\bar{r}}_{\bar{p},\bar{q}}F$.

Proof. The assertions of the theorem are consequences of the multiplier theorems for $L^{\Omega}_{\bar{p}}$, $Sl_{\bar{q}}(L^{\Omega}_{\bar{p}})$, and $L^{\Omega}_{\bar{p}}(l_{\bar{q}})$ from Chapter 1. Indeed, if $f \in S'$ and $\{\varphi_j\}_{j=0}^{\infty} \in \Phi(R_1)$ then

$$F^{-1}\varphi_j(\xi_1)\,\varphi_k(\xi_2)\,F(F^{-1}mFf) = F^{-1}m_{j,k}Ff_{j,k},$$

where

$$f_{j,k}(x) = [F^{-1}(\varphi_j(\xi_1)\,\varphi_k(\xi_2)\,Ff)]\,(x)$$

and

$$m_{j,k}(x) = \psi_j(x_1)\,\psi_k(x_2)\,m(x).$$

Suppose $f \in S^{\bar{r}}_{\bar{p},\bar{q}}F$. This implies that $\{2^{jr_1+kr_2}f_{j,k}\} \in L^{\Omega}_{\bar{p}}(l_{\bar{q}})$ with $\Omega = \{\Omega_{j,k}\}_{j,k=0}^{\infty}$, $\Omega_{j,k} = \{x \mid x = (x_1, x_2),\ |x_1| \leqq 2^{j+1}, |x_2| \leqq 2^{k+1}\}$. Applying Theorem 1.10.3(ii) we obtain according to (9) and (5)

$$\begin{aligned}
&\|F^{-1}mFf \mid S^{\bar{r}}_{\bar{p},\bar{q}}F\| \\
&= \|F^{-1}m_{j,k}F(2^{jr_1+kr_2}f_{j,k}) \mid L_{\bar{p}}(l_{\bar{q}})\| \\
&\leqq cM_{\bar{\sigma}}\|2^{jr_1+kr_2}f_{j,k}(x) \mid L_{\bar{p}}(l_{\bar{q}})\| = cM_{\bar{\sigma}}\|f \mid S^{\bar{r}}_{\bar{p},\bar{q}}F\|.
\end{aligned}$$

In complete analogy (8) and (6) follow from Theorem 1.10.3(i) and Theorem 1.8.3, respectively.

Remark 1. If $m(x) = m_1(x_1)\,m_2(x_2)$ then part (ii) of the theorem can be improved concerning the assumptions on $\bar{\sigma}$. This can be done on the basis of Remark 1.10.3/3. For the proof we refer to H.-J. Schmeisser [1, II, Theorem 3(i)]. The result reads as follows: Let $\bar{p} = (p_1, p_2)$, $\bar{q} = (q_1, q_2)$, $\bar{r} = (r_1, r_2)$, $\bar{\sigma} = (\sigma_1, \sigma_2)$ with $0 < p_m \leqq \infty$, $0 < q_m \leqq \infty$, $-\infty < r_m < \infty$; $(m = 1, 2)$. Let

$$\sigma_1 > \frac{1}{\min(1, p_1)} - \frac{1}{2}, \quad \sigma_2 > \frac{1}{\min(1, p_1, p_2, q_1)} - \frac{1}{2}. \tag{7'}$$

Let $m(x) = m_1(x_1)\,m_2(x_2)$ be a complex-valued infinitely differentiable function of at most polynomial growth on R_2. Then there exists a positive constant c such that

$$\|F^{-1}mFf \mid SB^{\bar{r}}_{\bar{p},\bar{q}}\| \leqq cM_{\bar{\sigma}}\|f \mid SB^{\bar{r}}_{\bar{p},\bar{q}}\| \tag{8'}$$

holds for all $f \in SB^{\bar{r}}_{\bar{p},\bar{q}}$. Of course, $M_{\bar{\sigma}}$ has the meaning of (5). Note that in (8′) the restriction $\max(p_1, p_2) < \infty$ is removed. A corresponding improvement of part (iii) of the theorem should be possible and (8′) should be true for arbitrary $m(x)$. This depends on sharp multiplier theorems for $Sl_{\bar{q}}(L^{\Omega}_{\bar{p}})$ and $L^{\Omega}_{\bar{p}}(l_{\bar{q}})$, respectively (cf. Remark 1.10.3/2).

The theorem can be used to derive the following multiplier criterion of Michlin-Hörmander type. Let $\sigma, \sigma > 0$, be a real number. Recall $\sigma = [\sigma] + \{\sigma\}$, where $[\sigma]$ is a non-negative integer and $0 \leqq \{\sigma\} < 1$.

Corollary. *Let $m(x)$ be a complex-valued infinitely differentiable function of at most polynomial growth on R_2. If $\bar{N} = (N_1, N_2)$ with $N_1, N_2 = 0, 1, 2, \ldots$, then we put*

$$m_{\bar{N}} = \sup_{\substack{0 \leqq \alpha_1 \leqq N_1 \\ 0 \leqq \alpha_2 \leqq N_2}} \sup_{x \in R_2} (1 + x_1^2)^{\alpha_1/2}\,(1 + x_2^2)^{\alpha_2/2}\,|D^{\alpha}m(x_1, x_2)|. \tag{11}$$

Let $\bar{p} = (p_1, p_2)$, $\bar{q} = (q_1, q_2)$, and $\bar{r} = (r_1, r_2)$.

(i) *If* $0 < p_m \leqq \infty$, $0 < q_m \leqq \infty$, $-\infty < r_m < \infty$; $(m = 1, 2)$ *and if*

$$N_1 \geqq \left[\frac{1}{\min(1, p_1)} - \frac{1}{2}\right] + 1, \qquad N_2 \geqq \left[\frac{1}{\min(1, p_1, p_2)} - \frac{1}{2}\right] + 1 \tag{12}$$

then there exists a positive constant c such that

$$\|F^{-1}mFf \mid S^{\bar{r}}_{\bar{p},\bar{q}}B\| \leqq cm_{\bar{N}}\|f \mid S^{\bar{r}}_{\bar{p},\bar{q}}B\| \tag{13}$$

holds for all $f \in S^{\bar{r}}_{\bar{p},\bar{q}}B$.

(ii) *If* $0 < p_m < \infty$, $0 < q_m \leqq \infty$, *and* $-\infty < r_m < \infty$; $(m = 1, 2)$ *and if*

$$N_1 \geqq \left[\frac{1}{p_1} + \frac{1}{2}\right] + 1, \qquad N_2 \geqq \left[\frac{1}{\min(p_1, p_2, q_1)} + \frac{1}{2}\right] + 1 \tag{14}$$

then there exists a positive constant c such that

$$\|F^{-1}mFf \mid SB^{\bar{r}}_{\bar{p},\bar{q}}\| \leqq cm_{\bar{N}}\|f \mid SB^{\bar{r}}_{\bar{p},\bar{q}}\| \tag{15}$$

holds for all $f \in SB^{\bar{r}}_{\bar{p},\bar{q}}$.

(iii) *If* $0 < p_m < \infty$, $0 < q_m \leqq \infty$, *and* $-\infty < r_m < \infty$; $(m = 1, 2)$ *and if*

$$N_1 \geqq \left[\frac{1}{\min(p_1, q_1, q_2)} + \frac{1}{2}\right] + 1, \qquad N_2 \geqq \left[\frac{1}{\min(p_1, p_2, q_1, q_2)} + \frac{1}{2}\right] + 1 \tag{16}$$

then there exists a positive constant c such that

$$\|F^{-1}mFf \mid S^{\bar{r}}_{\bar{p},\bar{q}}F\| \leqq cm_{\bar{N}}\|f \mid S^{\bar{r}}_{\bar{p},\bar{q}}F\| \tag{17}$$

holds for all $f \in S^{\bar{r}}_{\bar{p},\bar{q}}F$.

Proof. $M_{\bar{N}}$, $m_{\bar{N}}$ have the meaning of (5) and (11), respectively. It is not complicated to verify that $M_{\bar{N}} \leqq cm_{\bar{N}}$, where c is a positive constant which is independent of $m(x)$. Hence, the assertions of the corollary are immediate consequences of the preceding theorem.

Remark 2. As in Remark 1 we can improve part (ii) of the corollary if $m(x) = m_1(x_1)\, m_2(x_2)$. In this case (15) remains true if

$$N_1 \geqq \left[\frac{1}{\min(1, p_1)} - \frac{1}{2}\right] + 1, \qquad N_2 \geqq \left[\frac{1}{\min(1, p_1, p_2, q_1)} - \frac{1}{2}\right] + 1 \tag{14'}$$

(cf. (7′) and (8′)).

The remainder of this subsection will be devoted to the translation invariance of our spaces. If $h = (h_1, h_2) \in R_2$ then the translation operator T_h is defined by $(T_h f)(x) = f(x + h)$ for any regular distribution. In general, we put

$$(T_h f)(\varphi) = f(\varphi(\cdot - h)), \quad f \in S', \varphi \in S. \tag{18}$$

Then T_h is a continuous one-to-one mapping from S' onto itself.

Proposition. *Let* $\bar{p} = (p_1, p_2)$, $\bar{q} = (q_1, q_2)$, $\bar{r} = (r_1, r_2)$ *with* $0 < p_m \leqq \infty$, $0 < q_m \leqq \infty$, $-\infty < r_m < \infty$; $(m = 1, 2)$. *Then* T_h *yields an isomorphic mapping from* $S^{\bar{r}}_{\bar{p},\bar{q}}B$ $(SB^{\bar{r}}_{\bar{p},\bar{q}})$ *onto itself. If additionally* $\max(p_1, p_2) < \infty$ *then* T_h *is an isomorphic mapping from* $S^{\bar{r}}_{\bar{p},\bar{q}}F$ *onto itself.*

Proof. Let $\{\varphi_j\}_{j=0}^{\infty} \in \Phi(R_1)$. If $f \in S'$ then it holds

$$(F^{-1}\varphi_j\varphi_k F T_h f)(x) = (F^{-1}\varphi_j\varphi_k Ff)(x+h). \tag{19}$$

Now the proposition is a direct consequence of the translation invariance of the spaces $L_{\bar{p}}$.

Remark 3. In (19) we used

$$FT_h f = e^{ihx} Ff \quad \text{and} \quad F^{-1}(e^{ih\cdot} f) = T_h(F^{-1}f).$$

Thus, the proposition implies that $m(x) = e^{ihx}$ is a Fourier multiplier for $S^{\bar{r}}_{\bar{p},\bar{q}}B$, $(SB^{\bar{r}}_{\bar{p},\bar{q}}, S^{\bar{r}}_{\bar{p},\bar{q}}F)$ in the sense of (1).

2.2.6. Lifting Property and Related Equivalent Quasi-Norms

Let $\bar{\varrho} = (\varrho_1, \varrho_2)$ with $-\infty < \varrho_1 < \infty$, $-\infty < \varrho_2 < \infty$. We define the so-called lifting operator $I_{\bar{\varrho}}$ by

$$I_{\bar{\varrho}} f = F^{-1}(1 + x_1^2)^{\varrho_1/2}(1 + x_2^2)^{\varrho_2/2} Ff, \quad f \in S'. \tag{1}$$

Clearly $I_{\bar{\varrho}}$ is a continuous one-to-one mapping from $S'(R_2)$ onto itself. We have $I_{\bar{\varrho}}^{-1} = I_{-\bar{\varrho}}$, $-\bar{\varrho} = (-\varrho_1, -\varrho_2)$. If A, $A \subset S'$, is a linear space, then the restriction of $I_{\bar{\varrho}}$ on A is denoted by the same symbol $I_{\bar{\varrho}}$. Obviously $I_{\bar{\varrho}}$ is also an one-to-one mapping from S onto itself. Further let us write $\bar{r} - \bar{\varrho} = (r_1 - \varrho_1, r_2 - \varrho_2)$ if $\bar{r} = (r_1, r_2)$ and $\bar{\varrho} = (\varrho_1, \varrho_2)$.

Theorem 1. *Let $\bar{q} = (q_1, q_2)$, $\bar{r} = (r_1, r_2)$, $\bar{\varrho} = (\varrho_1, \varrho_2)$ with $0 < q_m \leqq \infty$, $-\infty < r_m < \infty$, and $-\infty < \varrho_m < \infty$; $(m = 1, 2)$.*

(i) *Let $\bar{p} = (p_1, p_2)$ with $0 < p_1 \leqq \infty$ and $0 < p_2 \leqq \infty$. Then $I_{\bar{\varrho}}$ maps $S^{\bar{r}}_{\bar{p},\bar{q}}B$ $(SB^{\bar{r}}_{\bar{p},\bar{q}})$ isomorphically onto $S^{\bar{r}-\bar{\varrho}}_{\bar{p},\bar{q}}B$ $(SB^{\bar{r}-\bar{\varrho}}_{\bar{p},\bar{q}})$ and $\|I_{\bar{\varrho}}f \mid S^{\bar{r}-\bar{\varrho}}_{\bar{p},\bar{q}}B\|$ $(\|I_{\bar{\varrho}}f \mid SB^{\bar{r}-\bar{\varrho}}_{\bar{p},\bar{q}}\|)$ is an equivalent quasi-norm in $S^{\bar{r}}_{\bar{p},\bar{q}}B$ $(SB^{\bar{r}}_{\bar{p},\bar{q}})$.*

(ii) *Let $\bar{p} = (p_1, p_2)$ with $0 < p_1 < \infty$ and $0 < p_2 < \infty$. Then $I_{\bar{\varrho}}$ maps $S^{\bar{r}}_{\bar{p},\bar{q}}F$ isomorphically onto $S^{\bar{r}-\bar{\varrho}}_{\bar{p},\bar{q}}F$ and $\|I_{\bar{\varrho}}f \mid S^{\bar{r}-\bar{\varrho}}_{\bar{p},\bar{q}}F\|$ is an equivalent quasi-norm in $S^{\bar{r}}_{\bar{p},\bar{q}}F$.*

Proof. We prove (ii). Let $\{\varphi_j\}_{j=0}^{\infty} \in \Phi(R_1)$ and let $f \in S^{\bar{r}}_{\bar{p},\bar{q}}F$. We put $\chi_j = \varphi_{j-1} + \varphi_j + \varphi_{j+1}$ with $\varphi_{-1} = 0$. Then we have

$$\|I_{\bar{\varrho}}f \mid S^{\bar{r}-\bar{\varrho}}_{\bar{p},\bar{q}}F\| = \|F^{-1}M_{j,k}F(2^{r_1 j + r_2 k} f_{j,k}) \mid L_{\bar{p}}(l_{\bar{q}})\|$$

where

$$M_{j,k} = [2^{-j\varrho_1}(1 + x_1^2)^{\varrho_1/2}\chi_j(x_1)]\,[2^{-k\varrho_2}(1 + x_2^2)^{\varrho_2/2}\chi_k(x_2)]$$

and $f_{j,k} = F^{-1}\varphi_j\varphi_k Ff$. Now we argue as in the proof of Theorem 2.2.5 and (ii) follows from Theorem 1.10.3(ii) and the fact that $I_{\bar{\varrho}}^{-1} = I_{-\bar{\varrho}}$. In the same way we obtain part (i) as a consequence of Theorem 1.8.3 and Theorem 1.10.3(i), respectively. If $\max(p_1, p_2) = \infty$ then we have to use the multiplier result from Remark 1.10.3/3 (in the case $SB^{\bar{r}}_{\bar{p},\bar{q}}$).

We discuss the same problem in the special case that $\bar{\varrho} = \bar{l} = (l_1, l_2)$ is a pair of non-negative integers. One has the following result as a consequence of Theorem 2.2.5 and the above theorem.

Theorem 2. *Let $\bar{q} = (q_1, q_2)$, $\bar{r} = (r_1, r_2)$ with $0 < q_m \leqq \infty$ and $-\infty < r_m < \infty$; $(m = 1, 2)$. Let $\bar{l} = (l_1, l_2)$ be a pair of non-negative integers.*

(i) *Let* $\bar{p} = (p_1, p_2)$ *with* $0 < p_1 < \infty$ *and* $0 < p_2 < \infty$. *Then*

$$\sum_{\alpha_1=0}^{l_1} \sum_{\alpha_2=0}^{l_2} \|D^\alpha f \mid S^{\bar{r}-\bar{l}}_{\bar{p},\bar{q}}F\| \tag{2}$$

and

$$\|f \mid S^{\bar{r}-\bar{l}}_{\bar{p},\bar{q}}F\| + \|D^{(l_1,0)}f \mid S^{\bar{r}-\bar{l}}_{\bar{p},\bar{q}}F\| + \|D^{(0,l_2)}f \mid S^{\bar{r}-\bar{l}}_{\bar{p},\bar{q}}F\| + \|D^{\bar{l}}f \mid S^{\bar{r}-\bar{l}}_{\bar{p},\bar{q}}F\| \tag{3}$$

are equivalent quasi-norms on $S^{\bar{r}}_{\bar{p},\bar{q}}F$.

(ii) *Let* $\bar{p} = (p_1, p_2)$ *with* $0 < p_1 \leqq \infty$ *and* $0 < p_2 \leqq \infty$. *Then* (2) *and* (3) *with* $S^{\bar{r}-\bar{l}}_{\bar{p},\bar{q}}B$ $(SB^{\bar{r}-\bar{l}}_{\bar{p},\bar{q}})$ *instead of* $S^{\bar{r}-\bar{l}}_{\bar{p},\bar{q}}F$ *yield equivalent quasi-norms in* $S^{\bar{r}}_{\bar{p},\bar{q}}B$ $(SB^{\bar{r}}_{\bar{p},\bar{q}})$.

Proof. Step 1. Let $f \in S^{\bar{r}}_{\bar{p},\bar{q}}F$. By Theorem 2.2.5 (or Corollary 2.2.5)

$$m(x) = x_1^{\alpha_1} x_2^{\alpha_2} (1 + x_1^2)^{-l_1/2} (1 + x_2^2)^{-l_2/2} \tag{4}$$

is a Fourier multiplier for $S^{\bar{r}-\bar{l}}_{\bar{p},\bar{q}}F$ if $0 \leqq \alpha_1 \leqq l_1$ and $0 \leqq \alpha_2 \leqq l_2$. Using this fact and the lifting property of Theorem 1 then we obtain

$$\begin{aligned} \|D^\alpha f \mid S^{\bar{r}-\bar{l}}_{\bar{p},\bar{q}}F\| &= \|F^{-1} x_1^{\alpha_1} x_2^{\alpha_2} Ff \mid S^{\bar{r}-\bar{l}}_{\bar{p},\bar{q}}F\| \\ &= \|F^{-1} m(x)\, F(I_{\bar{l}} f) \mid S^{\bar{r}-\bar{l}}_{\bar{p},\bar{q}}F\| \leqq c\|f \mid S^{\bar{r}}_{\bar{p},\bar{q}}F\|. \end{aligned} \tag{5}$$

Hence, (2) and (3) can be estimated from above by $c\|f \mid S^{\bar{r}}_{\bar{p},\bar{q}}F\|$.

Step 2. Let $f \in S'$ such that (3) is finite. We choose an odd infinitely differentiable function $\varrho(t)$, $t \in R_1$, with

$$\varrho(t) = 0 \quad \text{if} \quad 0 \leqq t \leqq \tfrac{1}{2} \quad \text{and}$$
$$\varrho(t) = 1 \quad \text{if} \quad 1 \leqq t < \infty.$$

By Theorem 2.2.5 (Corollary 2.2.5) the functions

$$\left.\begin{aligned} m_1(x) &= \frac{(1 + x_1^2)^{l_1/2} (1 + x_2^2)^{l_2/2}}{(1 + \varrho^{l_1}(x_1)\, x_1^{l_1})(1 + \varrho^{l_2}(x_2)\, x_2^{l_2})}, \\ m_2(x) &= \varrho^{l_1}(x_1), \quad m_3(x) = \varrho^{l_2}(x_2), \quad m_4(x) = \varrho^{l_1}(x_1)\, \varrho^{l_2}(x_2) \end{aligned}\right\} \tag{6}$$

are Fourier multipliers for $S^{\bar{r}-\bar{l}}_{\bar{p},\bar{q}}F$. For sake of convenience let us temporarily put $A = S^{\bar{r}-\bar{l}}_{\bar{p},\bar{q}}F$. Using the multipliers from (6) and the lifting property (cf. Theorem 1) we obtain

$$\begin{aligned} &\|f \mid S^{\bar{r}}_{\bar{p},\bar{q}}F\| \\ &\leqq c\|F^{-1}(1 + x_1^2)^{l_1/2} (1 + x_2^2)^{l_2/2}\, Ff \mid A\| \\ &\leqq c'(\|f \mid A\| + \|F^{-1} m_2 F D^{(l_1,0)} f \mid A\| + \|F^{-1} m_3 F D^{(0,l_2)} f \mid A\| \\ &\qquad + \|F^{-1} m_4 F D^{\bar{l}} f \mid A\|) \\ &\leqq c''(\|f \mid A\| + \|D^{(l_1,0)} f \mid A\| + \|D^{(0,l_2)} f \mid A\| + \|D^{\bar{l}} f \mid A\|). \end{aligned} \tag{7}$$

Hence part (i) is proved if we take into account that obviously (3) can be estimated from above by (2).

Step 3. The proof of part (ii) follows the same line. Note that the functions from (4) and (6) are Fourier multipliers on $S^{\bar{r}-\bar{l}}_{\bar{p},\bar{q}}B$ and $SB^{\bar{r}-\bar{l}}_{\bar{p},\bar{q}}$ by Theorem 2.2.5 (or Corollary 2.2.5) and Remark 2.2.5/1. Further these spaces have the lifting property. Thus, the theorem is proved.

Remark 1. Of peculiar interest is the case $\bar{r} = \bar{l}$ in the theorem. Then (2) and (3) read as follows: The quasi-norm $\|f \mid S^{\bar{l}}_{\bar{p},\bar{q}}F\|$ is equivalent to

$$\sum_{\alpha_1=0}^{l_1} \sum_{\alpha_2=0}^{l_2} \|D^\alpha f \mid S^{\bar{0}}_{\bar{p},\bar{q}}F\| \tag{8}$$

and to

$$\|f \mid S^{\bar{0}}_{\bar{p},\bar{q}}F\| + \|D^{(l_1,0)}f \mid S^{\bar{0}}_{\bar{p},\bar{q}}F\| + \|D^{(0,l_2)}f \mid S^{\bar{0}}_{\bar{p},\bar{q}}F\| + \|D^{\bar{l}}f \mid S^{\bar{0}}_{\bar{p},\bar{q}}F\|. \tag{9}$$

Here $\bar{0} = (0, 0)$ and $\bar{l}, \bar{p}, \bar{q}$ satisfy the assumptions of the preceding theorem.

Remark 2. In connection with the embeddings from Proposition 2.2.3/3, Theorem 2 sheds some light upon the smoothness of $f \in S^{\bar{r}}_{\bar{p},\bar{q}}B$ $(S^{\bar{r}}_{\bar{p},\bar{q}}F, SB^{\bar{r}}_{\bar{p},\bar{q}})$. Let $\varkappa, \varkappa > 0$, be a real number. Let us decompose $\varkappa = [\varkappa]^- + \{\varkappa\}^+$, where $0 < \{\varkappa\}^+ \leqq 1$, and $[\varkappa]^-$ is an integer. Let $\bar{p}, \bar{q}$ satisfy the assumptions of Theorem 2. Then the following assertion holds true: If $f \in S^{\bar{r}}_{\bar{p},\bar{q}}B$ $(S^{\bar{r}}_{\bar{p},\bar{q}}F, SB^{\bar{r}}_{\bar{p},\bar{q}})$ then

$$D^\alpha f \in L_{(\max(1,p_1),\max(1,p_2))} \cap L_{\bar{p}} \tag{10}$$

for all $\alpha = (\alpha_1, \alpha_2)$ with $0 \leqq \alpha_1 \leqq \left[r_1 - \left(\frac{1}{\min(1, p_1)} - 1\right)\right]^-$ and $0 \leqq \alpha_2 \leqq \left[r_2 - \left(\frac{1}{\min(1, p_2)} - 1\right)\right]^-$. In particular, f and the described derivatives are regular distributions.

2.2.7. Maximal Inequalities and Equivalent Quasi-Norms

As we have seen in Chapter 1 (cf. 1.3.3., 1.6.4., 1.10.) maximal inequalities are useful tools in the theory of the "basic" spaces $L^\Omega_{\bar{p}}$, $Sl_{\bar{q}}(L^\Omega_{\bar{p}})$, and $L^\Omega_{\bar{p}}(l_{\bar{q}})$. It turns out that these ideas can be carried over to the investigation of the spaces with dominating mixed smoothness properties $S^{\bar{r}}_{\bar{p},\bar{q}}B$, $SB^{\bar{r}}_{\bar{p},\bar{q}}$, and $S^{\bar{r}}_{\bar{p},\bar{q}}F$. However, it is not our aim to develop this topic in its full generality. We restrict ourselves to a reformulation of the results of 1.6.4. and 1.10.2. in the language of the spaces $S^{\bar{r}}_{\bar{p},\bar{q}}B$, $SB^{\bar{r}}_{\bar{p},\bar{q}}$, and $S^{\bar{r}}_{\bar{p},\bar{q}}F$. This is sufficient for the later purposes.

Definition. *Let $\{\Omega_j\}_{j=0}^\infty$ be a sequence of compact subsets of R_1 such that there exists a natural number N with*

$$\Omega_0 \subset [-2^N, 2^N], \quad \Omega_j \subset \{t \mid t \in R_1, 2^{j-N} \leqq |t| \leqq 2^{j+N}\}; \quad j = 1, 2, \ldots \tag{1}$$

Let $\{\psi_{j,k}\}_{j,k=0}^\infty \subset S(R_2)$ with

$$\operatorname{supp} \psi_{j,k} \subset \Omega_j \times \Omega_k; \quad j, k = 0, 1, 2, \ldots \tag{2}$$

If $f \in S'(R_2)$, $a_1 > 0$, and $a_2 > 0$ then we put

$$(\psi^*_{j,k}f)(x) = \sup_{y \in R_2} \frac{|(F^{-1}\psi_{j,k}Ff)(x-y)|}{(1 + |2^j y_1|^{a_1})(1 + |2^k y_2|^{a_2})}, \quad x \in R_2; \quad j, k = 0, 1, 2, \ldots \tag{3}$$

Remark 1. By the Paley-Wiener-Schwartz theorem (cf. 1.2.4.), $\{f_{j,k} = F^{-1}\psi_{j,k}Ff\}_{j,k=0}^\infty$ is a sequence of entire analytic functions. Hence $\psi^*_{j,k}f$ is a special case of the maximal functions from 1.6.4. or 1.10.2. In the following considerations $(\psi^*_{j,k}f)(x) = \infty$ will be excluded. Therefore we do not stress this possibility.

Recall that we write $s = [s] + \{s\}$, $[s]$ integer, $0 \leqq \{s\} < 1$, if s is a positive real number.

Theorem. *Let $\bar{q} = (q_1, q_2)$, $\bar{r} = (r_1, r_2)$ with $0 < q_m \leqq \infty$ and $-\infty < r_m < \infty$; $(m = 1, 2)$. Let $\{\Omega_{j,k} = \Omega_j \times \Omega_k\}_{j,k=0}^\infty$ and $\{\psi_{j,k}(x)\}_{j,k=0}^\infty$ be defined as in (1) and (2).*

If $\bar{L} = (L_1, L_2)$ *then*

$$C_{\varphi,\bar{L}} = \sup_{j,k} \sup_{x \in R_2} \sum_{\alpha_1=0}^{L_1} \sum_{\alpha_2=0}^{L_2} (1 + x_1^2)^{\alpha_1/2} (1 + x_2^2)^{\alpha_2/2} |D^\alpha \psi_{j,k}(x)|. \tag{4}$$

(i) *Let* $\bar{p} = (p_1, p_2)$ *with* $0 < p_1 \leqq \infty$ *and* $0 < p_2 < \infty$. *Let*

$$\left.\begin{aligned} &a_1 > \frac{1}{p_1}, \qquad a_2 > \frac{1}{\min(p_1, p_2)}, \\ &L_1 > \left[\frac{1}{\min(1, p_1)} - \frac{1}{2}\right], \qquad L_2 > \left[\frac{1}{\min(1, p_1, p_2)} - \frac{1}{2}\right]. \end{aligned}\right\} \tag{5}$$

If $C_{\varphi,\bar{L}} < \infty$ *then there exists a positive constant c such that*

$$\|2^{r_1 j + r_2 k} \psi_{j,k}^* f \mid l_{\bar{q}}(L_{\bar{p}})\| \leqq c C_{\varphi,\bar{L}} \|f \mid S_{\bar{p},\bar{q}}^{\bar{r}} B\| \tag{6}$$

holds for all $f \in S_{\bar{p},\bar{q}}^{\bar{r}} B$.

(ii) *Let* $\bar{p} = (p_1, p_2)$ *with* $0 < p_1 < \infty$ *and* $0 < p_2 < \infty$. *Let*

$$\left.\begin{aligned} &a_1 > \frac{1}{p_1}, \qquad a_2 > \frac{1}{\min(p_1, p_2, q_1)}, \\ &L_1 > \left[\frac{1}{p_1} + \frac{1}{2}\right], \qquad L_2 > \left[\frac{1}{\min(p_1, p_2, q_1)} + \frac{1}{2}\right]. \end{aligned}\right\} \tag{7}$$

If $C_{\varphi,\bar{L}} < \infty$ *then there exists a positive constant c such that*

$$\|2^{r_1 j + r_2 k} \psi_{j,k}^* f \mid S l_{\bar{q}}(L_{\bar{p}})\| \leqq c C_{\varphi,\bar{L}} \|f \mid S B_{\bar{p},\bar{q}}^{\bar{r}}\| \tag{8}$$

holds for all $f \in S B_{\bar{p},\bar{q}}^{\bar{r}}$.

(iii) *Let* $\bar{p} = (p_1, p_2)$ *with* $0 < p_1 < \infty$ *and* $0 < p_2 < \infty$. *Let*

$$\left.\begin{aligned} &a_1 > \frac{1}{\min(p_1, q_1, q_2)}, \qquad a_2 > \frac{1}{\min(p_1, p_2, q_1, q_2)}, \\ &L_1 > \left[\frac{1}{\min(p_1, q_1, q_2)} + \frac{1}{2}\right], \qquad L_2 > \left[\frac{1}{\min(p_1, p_1, q_1, q_2)} + \frac{1}{2}\right]. \end{aligned}\right\} \tag{9}$$

If $C_{\varphi,\bar{L}} < \infty$ *then there exists a positive constant c such that*

$$\|2^{r_1 j + r_2 k} \psi_{j,k}^* f \mid L_{\bar{p}}(l_{\bar{q}})\| \leqq c C_{\varphi,\bar{L}} \|f \mid S_{\bar{p},\bar{q}}^{\bar{r}} F\| \tag{10}$$

holds for all $f \in S_{\bar{p},\bar{q}}^{\bar{r}} F$.

Proof. Let $f \in S_{\bar{p},\bar{q}}^{\bar{r}} F$. Let $\{\varphi_j\}_{j=0}^\infty \in \Phi(R_1)$. By Theorem 1.10.3(ii) we have

$$\begin{aligned} &\|2^{r_1 j + r_2 k} F^{-1} \psi_{j,k} F f \mid L_{\bar{p}}(l_{\bar{q}})\| \\ &\leqq c \sum_{m,n} \|F^{-1} \psi_{j,k} F(2^{r_1 j + r_2 k} F^{-1} \varphi_{j+m} \varphi_{k+n} F f) \mid L_{\bar{p}}(l_{\bar{q}})\| \\ &\leqq c' C_{\varphi,\bar{L}} \|f \mid S_{\bar{p},\bar{q}}^{\bar{r}} F\|. \end{aligned} \tag{11}$$

(11) and the maximal inequality (1.10.2/5) yield (10). Analogously we prove (6) and (8) using Theorem 1.8.3, Theorem 1.6.4(ii), and 1.10.3(i), Theorem 1.10.2(i), respectively.

Corollary. *Let* $\bar{q} = (q_1, q_2)$, $\bar{r} = (r_1, r_2)$ *with* $0 < q_m \leqq \infty$ *and* $-\infty < r_m < \infty$; $(m = 1, 2)$. *Let* $\varphi = \{\varphi_j\}_{j=0}^\infty \in \Phi(R_1)$. *Let*

$$\varphi_{j,k}(x_1, x_2) = \varphi_j(x_1)\, \varphi_k(x_2), \quad x \in R_2, j = 0, 1, 2, \ldots, k = 0, 1, 2, \ldots \tag{12}$$

(i) *Let* $\bar{p} = (p_1, p_2)$ *with* $0 < p_1 \leqq \infty$ *and* $0 < p_2 < \infty$. *Let* $a_1 > \frac{1}{p_1}$ *and* $a_2 > \frac{1}{\min(p_1, p_2)}$. *Then*

$$\|2^{r_1 j + r_2 k} \varphi^*_{j,k} f \mid l_{\bar{q}}(L_{\bar{p}})\| \tag{13}$$

is an equivalent quasi-norm in $S^{\bar{r}}_{\bar{p},\bar{q}}B$.

(ii) *Let* $\bar{p} = (p_1, p_2)$ *with* $0 < p_1 < \infty$ *and* $0 < p_2 < \infty$. *Let* $a_1 > \frac{1}{p_1}$ *and* $a_2 > \frac{1}{\min(p_1, p_2, q_1)}$. *Then*

$$\|2^{r_1 j + r_2 k} \varphi^*_{j,k} f \mid Sl_{\bar{q}}(L_{\bar{p}})\| \tag{14}$$

is an equivalent quasi-norm in $SB^{\bar{r}}_{\bar{p},\bar{q}}$.

(iii) *Let* $\bar{p} = (p_1, p_2)$ *with* $0 < p_1 < \infty$ *and* $0 < p_2 < \infty$. *Let* $a_1 > \frac{1}{\min(p_1, q_1, q_2)}$ *and* $a_2 > \frac{1}{\min(p_1, p_2, q_1, q_2)}$. *Then*

$$\|2^{r_1 j + r_2 k} \varphi^*_{j,k} f \mid L_{\bar{p}}(l_{\bar{q}})\| \tag{15}$$

is an equivalent quasi-norm in $S^{\bar{r}}_{\bar{p},\bar{q}}F$.

Proof. If $\{\varphi_j\}_{j=0}^{\infty} \in \Phi(R_1)$ then $C_{\varphi,\bar{L}}$ (cf. (4)) with $\{\varphi_{j,k}\} = \{\varphi_j(x_1)\, \varphi_k(x_2)\}$ is finite for all pairs $\bar{L} = (L_1, L_2)$. Hence the equivalence of the quasi-norms follows from the preceding theorem and the trivial inequalities

$$|F^{-1} \varphi_{j,k} Ff(x)| \leqq (\varphi^*_{j,k} f)(x).$$

Remark* 2. Maximal inequalities in the sense of (6), (8), and (10) have been considered for various types of function spaces (cf. B. Stöckert, H. Triebel [1], [F, 2.3.2.], [T, 2.3.6.] and the references given there). The use of maximal functions has its origin in the fundamental paper by C. Fefferman, E. M. Stein [1]. Then the idea was modified by J. Peetre [3] in the framework of homogeneous $F^s_{p,q}$-spaces.

Remark 3. The assumptions concerning the sequence $\{\psi_{j,k}\}_{j,k=0}^{\infty}$ in (3) and in the theorem can be essentially weakened. In particular, the strong restrictions with respect to the supports of $\psi_{j,k}$ are not necessary. We can replace them by appropriated growth conditions on $\psi_{j,k}$. We refer to [T, Definition 2.3.6/1 and Remark 2.3.6/4], H. Triebel [9], and to H.-J. Schmeisser [1, I]. In the latter paper more general maximal inequalities for $SB^{\bar{r}}_{\bar{p},\bar{q}}$ have been proved.

2.3. Representations and Special Cases

2.3.1. Bessel-Potential and Sobolev Spaces with a Dominating Mixed Derivative

Until now we have introduced and studied the spaces $S^{\bar{r}}_{\bar{p},\bar{q}}B$, $SB^{\bar{r}}_{\bar{p},\bar{q}}$, and $S^{\bar{r}}_{\bar{p},\bar{q}}F$ in the framework of Fourier analysis. In particular, we dealt with equivalent quasi-norms which are related to our approach via the decomposition method and we established fundamental properties. The Fourier multiplier theorems and the maximal inequalities from Chapter 1 (cf. Subsections 1.3.3., 1.6.4., 1.8.3., 1.10.2., and 1.10.3.) have been the basic tools. It is the advantage of this method that we are able to treat the above spaces without unnatural restrictions concerning the range of the parameters $\bar{p}$, $\bar{q}$, and $\bar{r}$. In particular, the cases $0 < p_m < 1$, $0 < q_m < 1$, and $-\infty < r_m < 0$; $(m = 1, 2)$ are included in our considerations. We called $S^{\bar{r}}_{\bar{p},\bar{q}}B$, $SB^{\bar{r}}_{\bar{p},\bar{q}}$, and

$S^{\bar{r}}_{\bar{p},\bar{q}}F$ spaces with dominating mixed smoothness properties, and we explained this notation in Subsection 2.2.6. (cf. Theorem 2.2.6/2 and Remark 2.2.6/1). In other words, we classified functions (distributions) according to a prescribed smoothness. However, this can also be done (in a more obvious way) via differences and derivatives or via approximation procedures. There exists a lot of corresponding spaces with dominating mixed smoothness properties (cf. the literature mentioned in Section 2.1.). It is our aim to give a description of the relations between the just mentioned possibilities on the one hand and the spaces $S^{\bar{r}}_{\bar{p},\bar{q}}B$, $SB^{\bar{r}}_{\bar{p},\bar{q}}$, and $S^{\bar{r}}_{\bar{p},\bar{q}}F$ on the other hand.

This subsection deals with Bessel-potential and Sobolev spaces with a dominating mixed derivative, that means with spaces of type $S^{\bar{r}}_2H$ and $S^{\bar{r}}_2W$ which we already used in connection with Fourier multiplier criterions (cf. 1.8.3., 1.10.3., and also (2.1/1)).

Recall, that we defined the spaces $L_{\bar{p}} = L_{\bar{p}}(R_2)$ in (1.3.2/5). We restrict ourselves to the case $\bar{p} = (p_1, p_2)$ with $1 < p_1 < \infty$ and $1 < p_2 < \infty$.

Definition. *Let* $\bar{p} = (p_1, p_2)$ *with* $1 < p_1 < \infty$ *and* $1 < p_2 < \infty$.

(i) *Let* $\bar{r} = (r_1, r_2)$ *with* $-\infty < r_1 < \infty$ *and* $-\infty < r_2 < \infty$. *We put*

$$S^{\bar{r}}_{\bar{p}}H = \{f \mid f \in S',\ \|f \mid S^{\bar{r}}_{\bar{p}}H\| = \|F^{-1}(1+x_1^2)^{r_1/2}(1+x_2^2)^{r_2/2}Ff \mid L_{\bar{p}}\| < \infty\}. \tag{1}$$

(ii) *Let* $\bar{l} = (l_1, l_2)$ *with the non-negative integers* l_1, l_2. *We put*

$$\begin{aligned} S^{\bar{l}}_{\bar{p}}W = \{f \mid f \in L_{\bar{p}},\ &\|f \mid S^{\bar{l}}_{\bar{p}}W\| \\ &= \|f \mid L_{\bar{p}}\| + \|D^{(l_1,0)}f \mid L_{\bar{p}}\| + \|D^{(0,l_2)}f \mid L_{\bar{p}}\| + \|D^{(l_1,l_2)}f \mid L_{\bar{p}}\| < \infty\}. \end{aligned} \tag{2}$$

Remark* 1. Of course, the derivatives in (2) have to be understood in the sense of distributions. As already mentioned in 2.1. the spaces $S^{\bar{l}}_{\bar{p}}W$, $\bar{p} = (p, p)$, $1 < p < \infty$, were the starting point of the theory of spaces with dominating mixed smoothness properties. We refer to S. M. Nikol'skij [3.4]. The generalization $S^{\bar{r}}_{\bar{p}}H$ is due to P. I. Lizorkin, S. M. Nikol'skij [1]. In this paper properties of $S^{\bar{r}}_{\bar{p}}H$, $\bar{p} = (p, p)$, $1 < p < \infty$, and related spaces have been studied extensively in the framework of fractional derivatives. Maybe there is some confusion with the notation. In the Russian literature $S^{\bar{r}}_{\bar{p}}H$ stands for $S^{\bar{r}}_{\bar{p},\bar{\infty}}B$, $\bar{\infty} = (\infty, \infty)$, and $S^{\bar{r}}_{\bar{p}}$ or $S^{\bar{r}}_{\bar{p}}L$ stand for $S^{\bar{r}}_{\bar{p}}H$ (in our context). In H. Triebel [4, III, IV] we have $S^{\bar{r}}_{\bar{p}}H = H^{g(x)}_p$, $g(x) = (1+x_1^2)^{r_1/2}(1+x_2^2)^{r_2/2}$, $\bar{p} = (p, p)$, $1 < p < \infty$. We used the notation in the sense of [T] for isotropic spaces. Concerning properties of $S^{\bar{r}}_{\bar{p}}H$ we refer also to H. Triebel [4, III–V], G. Sparr [1], J. A. Bryčkov [1], and D. L. Fernandez [3, 5].

The key to the solution of our problem is a theorem of Littlewood-Paley type for the spaces $L_{\bar{p}}$ which we shall state here without a detailed proof.

Proposition. *Let* $\bar{p} = (p_1, p_2)$ *with* $1 < p_1 < \infty$, $1 < p_2 < \infty$ *and let* $\varphi = \{\varphi_j\}_{j=0}^{\infty} \in \Phi(R_1)$. *Then there exist two positive constants* c_1 *and* c_2 *such that*

$$c_1\|f \mid L_{\bar{p}}\| \leq \left\| \left(\sum_{k=0}^{\infty} \sum_{j=0}^{\infty} |(F^{-1}\varphi_j(\xi_1)\,\varphi_k(\xi_2)\,Ff)(x)|^2 \right)^{1/2} \Big| L_{\bar{p}} \right\| \leq c_2\|f \mid L_{\bar{p}}\| \tag{3}$$

holds for all $f \in L_{\bar{p}}$.

Remark 2. We quote this Littlewood-Paley theorem without proof although there is no reference where we can find (3) in this form. However, we add few remarks about the proof.

(i) Using the Fourier multiplier theorem for the spaces $L_{\bar{p}}(l_{\bar{2}})$, $\bar{2} = (2, 2)$ of P. I. Lizorkin [2, Lemma 2], (3) can be reduced to $\varphi_j(t) = \chi_j(t)$, $t \in R_1$, $j = 0, 1, 2, \ldots,$ where $\{\chi_j(t)\}_{j=0}^{\infty}$ is the sequence of characteristic functions from the proof of Proposition 1.8.3. O. V. Besov, V. P. Il'in, S. M. Nikol'skij [1] proved (3) with χ_j instead of φ_j, $j = 0, 1, 2, \ldots$, in the periodic case (F, F^{-1} must be replaced by the discrete Fourier transform and its inverse, respectively). This result can be extended to the non-periodic case by the method of S. M. Nikol'skij [2, 1.5.6.]. Then one obtains (3).

(ii) P. I. Lizorkin [2, Theorem 2 and Lemma 2] proved (3) for homogeneous decompositions of $R_2 - \{x \mid x = (x_1, x_2), x_1 x_2 = 0\}$. The crucial point is to use a theorem on vector-valued convolution operators by P. Krée [1] (cf. also A. Benedek, A. P. Calderón, R. Panzone [1]) in contrast to the case $\bar{p} = (p, p)$ (cf. P. I. Lizorkin [3]) where the corresponding theorem due to J. T. Schwartz [1] is sufficient. If we deal with non-homogeneous decompositions (as in our situation) the proof of (3) can be reduced to horizontal and vertical Fourier multipliers for $L_{\bar{p}}(l_{\bar{2}})$, $\bar{2} = (2, 2)$ (cf. (1.9.2/3) and (1.9.2/2)) by the method of Theorem 1.9.4. It is possible to obtain the desired multiplier criterion for $L_{\bar{p}}(l_{\bar{2}})$ if one follows the proof of [I, Theorem 2.2.4(a)] taking into consideration the already quoted result by P. Krée [1].

Remark 3. It is clear that $S^{\bar{r}}_{\bar{p}}H$ and $S^{\bar{l}}_{\bar{p}}W$ equipped with the respective norms are Banach spaces of tempered distributions. Note that in our context the proposition reads as $S^{\bar{0}}_{\bar{p},\bar{2}}F = L_{\bar{p}}$, $\bar{2} = (2, 2)$, $\bar{0} = (0, 0)$.

Theorem. *Let* $\bar{p} = (p_1, p_2)$ *with* $1 < p_1 < \infty$, $1 < p_2 < \infty$, *and let* $\bar{2} = (2, 2)$.

(i) *Let* $\bar{r} = (r_1, r_2)$ *with* $-\infty < r_1 < \infty$ *and* $-\infty < r_2 < \infty$. *Then we have*

$$S^{\bar{r}}_{\bar{p},\bar{2}}F = S^{\bar{r}}_{\bar{p}}H, \tag{4}$$

where the the corresponding norms are equivalent to each other.

(ii) *Let* $\bar{l} = (l_1, l_2)$ *with non-negative integers* l_1 *and* l_2. *Then we have*

$$S^{\bar{l}}_{\bar{p}}H = S^{\bar{l}}_{\bar{p}}W, \tag{5}$$

where $\|f \mid S^{\bar{l}}_{\bar{p}}W\|$ *and*

$$\|f \mid S^{\bar{l}}_{\bar{p}}W\|^* = \sum_{\alpha_1=0}^{l_1} \sum_{\alpha_2=0}^{l_2} \|D^\alpha f \mid L_{\bar{p}}\| \tag{6}$$

are equivalent norms in $S^{\bar{l}}_{\bar{p}}H$.

Proof. (4) is an immediate consequence of the preceding proposition and the lifting property of Theorem 2.2.6/1. Part (ii) of the theorem follows obviously from part (i), Theorem 2.2.6/2, and the proposition.

Remark 4. Part (ii) of the theorem shows that there exists a positive constant c such that

$$\|D^\alpha f \mid L_{\bar{p}}\| \leqq c(\|f \mid L_{\bar{p}}\| + \|D^{(l_1,0)}f \mid L_{\bar{p}}\| + \|D^{(0,l_2)}f \mid L_{\bar{p}}\| + \|D^{(l_1,l_2)}f \mid L_{\bar{p}}\|) \tag{7}$$

holds if $0 \leqq \alpha_1 \leqq l_1$ and $0 \leqq \alpha_2 \leqq l_2$. Inequalities of this type (and more general ones) are often used in the theory of function spaces. We refer to O. V. Besov, V. P. Il'in, S. M. Nikol'skij [1, § 13] and the literature cited there. Note that by Theorem 2.2.6/2 we can replace the norm $L_{\bar{p}}$ in (7) by the quasi-norms $S^{\bar{r}}_{\bar{p},\bar{q}}B$, $SB^{\bar{r}}_{\bar{p},\bar{q}}$, and $S^{\bar{r}}_{\bar{p},\bar{q}}F$, respectively.

Remark 5. Combining Propositions 2.2.3/3, 4, Theorem 2.2.6/1, and the above theorem we may obtain embeddings in $S^{\bar{r}}_{\bar{p}}H$ or $S^{\bar{l}}_{\bar{p}}W$. For example.

$$S^{\bar{r}}_{\bar{p},\bar{q}}B\,(SB^{\bar{r}}_{\bar{p},\bar{q}}) \subset S^{\bar{l}}_{\bar{p}}W \quad \text{if} \quad r_1 > l_1, r_2 > l_2 \tag{8}$$

$$S^{\bar{l}}_{\bar{p},\bar{1}}B \subset S^{\bar{l}}_{\bar{p}}W \subset S^{\bar{l}}_{\bar{p},\overline{\infty}}B, \quad \bar{1} = (1, 1), \overline{\infty} = (\infty, \infty), \tag{9}$$

if $\bar{l} = (l_1, l_2)$, $l_m = 0, 1, 2, \ldots$; $(m = 1, 2)$.

2.3.2. Representations of Nikol'skij Type

Let $\bar{p} = (p_1, p_2)$ with $0 < p_1 \leqq \infty$ and $0 < p_2 \leqq \infty$. We put

$$\mathfrak{A}_{\bar{p}} = \{a \mid a = \{a_{j,k}(x)\}_{j,k=0}^{\infty},\ a_{j,k}(x) \in S' \cap L_{\bar{p}},$$
$$\operatorname{supp} Fa_{j,k} \subset \{y \mid y = (y_1, y_2), |y_1| \leqq 2^{j+1}, |y_2| \leqq 2^{k+1}\},$$
$$\text{for } j, k = 0, 1, 2, \ldots\} \tag{1}$$

Obviously, $L_{\bar{p}} \subset S'$ if $\min(p_1, p_2) \geqq 1$. Recall that $a_{j,k} \subset S' \cap L_{\bar{p}}$ makes sense by the Paley-Wiener-Schwartz theorem (cf. 1.2.4.). In particular, $a \in \mathfrak{A}_{\bar{p}}$ is a system of entire analytic functions of exponential type.

Theorem. *Let $\bar{q} = (q_1, q_2)$ with $0 < q_1 \leqq \infty$ and $0 < q_2 \leqq \infty$.*

(i) *Let $\bar{p} = (p_1, p_2)$ with $0 < p_1 \leqq \infty$ and $0 < p_2 \leqq \infty$. Let $\bar{r} = (r_1, r_2)$ with*

$$r_1 > \frac{1}{\min(1, p_1)} - 1, \qquad r_2 > \frac{1}{\min(1, p_1, p_2)} - 1. \tag{2}$$

Then,

$$S_{\bar{p},\bar{q}}^{\bar{r}}B = \{f \mid f \in S', \exists \{a_{j,k}\}_{j,k=0}^{\infty} \in \mathfrak{A}_{\bar{p}} \text{ such that } f = \sum_{j,k=0}^{\infty} a_{j,k} \text{ in } S' \text{ and}$$
$$\|f \mid S_{\bar{p},\bar{q}}^{\bar{r}}B\|^{a,N} = \|2^{r_1 j + r_2 k} a_{j,k}(x) \mid l_{\bar{q}}(L_{\bar{p}})\| < \infty\} \tag{3}$$

holds. Furthermore,

$$\|f \mid S_{\bar{p},\bar{q}}^{\bar{r}}B\|^{N} = \inf_{a} \|f \mid S_{\bar{p},\bar{q}}^{\bar{r}}B\|^{a,N} \tag{4}$$

is an equivalent quasi-norm in $S_{\bar{p},\bar{q}}^{\bar{r}}B$, where the infimum is taken over all admissible representations in (3).

(ii) *Let $\bar{p} = (p_1, p_2)$ with $0 < p_1 \leqq \infty$ and $0 < p_2 \leqq \infty$. Let $\bar{r} = (r_1, r_2)$ with*

$$r_1 > \frac{1}{\min(1, p_1)} - 1, \qquad r_2 > \frac{1}{\min(1, p_1, p_2, q_1)} - 1. \tag{5}$$

Then

$$SB_{\bar{p},\bar{q}}^{\bar{r}} = \{f \mid f \in S', \exists \{a_{j,k}\}_{j,k=0}^{\infty} \in \mathfrak{A}_{\bar{p}} \text{ such that } f = \sum_{j,k=0}^{\infty} a_{j,k} \text{ in } S' \text{ and}$$
$$\|f \mid SB_{\bar{p},\bar{q}}^{\bar{r}}\|^{a,N} = \|2^{r_1 j + r_2 k} a_{j,k}(x) \mid Sl_{\bar{q}}(L_{\bar{p}})\| < \infty\} \tag{6}$$

holds. Furthermore,

$$\|f \mid SB_{\bar{p},\bar{q}}^{\bar{r}}\|^{N} = \inf_{a} \|f \mid SB_{\bar{p},\bar{q}}^{\bar{r}}\|^{a,N} \tag{7}$$

is an equivalent quasi-norm in $SB_{\bar{p},\bar{q}}^{\bar{r}}$, where the infimum is taken over all admissible representations in (6).

(iii) *Let $\bar{p} = (p_1, p_2)$ with $0 < p_1 < \infty$ and $0 < p_2 < \infty$. Let $\bar{r} = (r_1, r_2)$ with*

$$r_1 > \frac{1}{\min(p_1, q_1, q_2)}, \qquad r_2 > \frac{1}{\min(p_1, p_2, q_1, q_2)}. \tag{8}$$

Then

$$S^{\bar r}_{\bar p,\bar q}F = \{f \mid f \in S', \quad \exists \{a_{j,k}\}^{\infty}_{j,k=0} \in \mathfrak{A}_{\bar p} \text{ such that } f = \sum_{j,k=0}^{\infty} a_{j,k} \text{ in } S' \text{ and}$$

$$\|f \mid S^{\bar r}_{\bar p,\bar q}F\|^{a,N} = \|2^{r_1 j + r_2 k} a_{j,k}(x) \mid L_{\bar p}(l_{\bar q})\| < \infty\} \tag{9}$$

holds. Furthermore,

$$\|f \mid S^{\bar r}_{\bar p,\bar q}F\|^{N} = \inf_a \|f \mid S^{\bar r}_{\bar p,\bar q}F\|^{a,N} \tag{10}$$

is an equivalent quasi-norm in $S^{\bar r}_{\bar p,\bar q}F$, where the infimum is taken over all admissible representations in (9).

Proof. Step 1. Let $\varphi = \{\varphi_j\}^{\infty}_{j=0} \in \Phi(R_1)$. Let $f \in S^{\bar r}_{\bar p,\bar q}F$. We put $a_{j,k}(x) = (F^{-1}\varphi_j(\xi_1) \times \varphi_k(\xi_2)\, Ff)(x)$, $j, k = 0, 1, 2, \ldots$ Then, $\{a_{j,k}(x)\}^{\infty}_{j,k=0} \in \mathfrak{A}_{\bar p}$, $f = \sum_{k=0}^{\infty}\sum_{j=0}^{\infty} a_{j,k}(x)$ in S' and $\|f \mid S^{\bar r}_{\bar p,\bar q}F\|^{a,N} = \|f \mid S^{\bar r}_{\bar p,\bar q}F\|$. This proves one direction of part (iii). Analogously we argue in part (i) and part (ii).

Step 2. We show the converse direction. Suppose that $f \in S'$ belongs to the right-hand side of (9), where (8) is satisfied, i.e. we have an admissible system $a = \{a_{j,k}(x)\}^{\infty}_{j,k=0} \in \mathfrak{A}_{\bar p}$, such that $f = \sum_{k=0}^{\infty}\sum_{j=0}^{\infty} a_{j,k}(x)$ in S' and $\|f \mid S^{\bar r}_{\bar p,\bar q}F\|^{a,N} < \infty$. Let a sequence $\{\varphi_j\}^{\infty}_{j=0} \in \Phi(R_1)$ be given. We put $\varphi_{j,k}(x) = \varphi_j(x_1)\,\varphi_k(x_2)$, if $j, k = 0, 1, \ldots$ Then

$$(F^{-1}\varphi_{j,k}Ff)(x) = \sum_{n=k-1}^{\infty} \sum_{m=j-1}^{\infty} (F^{-1}\varphi_{j,k}Fa_{m,n})(x) \quad \text{in} \quad S' \tag{11}$$

(with $a_{-1,n} = a_{m,-1} = 0$). We claim that the right-hand side of (11) converges also pointwise to $(F^{-1}\varphi_{j,k}Ff)(x)$, $x \in R_2$. Let s, $0 < s < \infty$, be an arbitrary positive real number. Applying Nikol'skij's inequality (Theorem 1.6.2) and Theorem 1.8.1 we obtain

$$\begin{aligned}
&\sum_{n=k-1}^{\infty} \sum_{m=j-1}^{\infty} |F^{-1}\varphi_{j,k}Fa_{m,n}(x)|^s \\
&\leqq c \cdot 2^{j\frac{s}{p_1} + k\frac{s}{p_2}} \sum_{n=k-1}^{\infty} \sum_{m=j-1}^{\infty} \|F^{-1}\varphi_{j,k}Fa_{m,n} \mid L_{\bar p}\|^s \\
&\leqq c' \cdot 2^{j\frac{s}{p_1} + k\frac{s}{p_2}} \sum_{n=k-1}^{\infty} \sum_{m=j-1}^{\infty} 2^{\sigma_1 sm + \sigma_2 sn} \|a_{m,n} \mid L_{\bar p}\|^s \\
&\leqq c' \cdot 2^{j\frac{s}{p_1} + k\frac{s}{p_2}} (\|f \mid S^{\bar\sigma}_{\bar p,\bar s}B\|^{a,N})^s
\end{aligned} \tag{12}$$

where $\bar\sigma = (\sigma_1, \sigma_2)$ with $\sigma_1 = \dfrac{1}{\min(1, p_1)} - 1$, $\sigma_2 = \dfrac{1}{\min(1, p_1, p_2)} - 1$, and $\bar s = (s, s)$ with $0 < s < \infty$. By the same arguments as in the proof of Proposition 2.2.3/2 we can verify

$$\|f \mid S^{\bar\sigma}_{\bar p,\bar s}B\|^{a,N} \leqq c\|f \mid S^{\bar r}_{\bar p,\bar q}F\|^{a,N} < \infty\,. \tag{13}$$

Hence, the convergence in (11) can be understood in $L_{\overline{\infty}}$, $\overline{\infty} = (\infty, \infty)$, and therefore pointwise. Moreover, if $s = \min(p_1, p_2, q_1, q_2, 1)$ then

$$|(F^{-1}\varphi_{j,k}Ff)(x)| \leqq \left(\sum_{n=0}^{\infty}\sum_{m=0}^{\infty}|(F^{-1}\varphi_{j,k}Fa_{m+j-1,n+k-1})(x)|^s\right)^{1/s} \tag{14}$$

pointwise. Now the following calculations are justified. Starting with (14) and taking into consideration that

$$\sup_{j,k}\|\varphi_{j,k}(2^{m+j}\cdot, 2^{n+k}\cdot) \mid S_2^{\bar{\varrho}}H\| \leqq c\cdot 2^{m(\varrho_1-\frac{1}{2})+n(\varrho_2-\frac{1}{2})}, \tag{15}$$

$\bar{\varrho} = (\varrho_1, \varrho_2)$, where c is independent of j, k, we obtain with the help of Theorem 1.10.3(ii)

$$\begin{aligned}
\|f \mid S_{\bar{p},\bar{q}}^{\bar{r}}F\| &= \|2^{r_1 j + r_2 k}(F^{-1}\varphi_{j,k}Ff) \mid L_{\bar{p}}(l_{\bar{q}})\| \\
&\leqq c\left\|\left(\sum_{n=0}^{\infty}\sum_{m=0}^{\infty}2^{-mr_1 s-nr_2 s}|F^{-1}\varphi_{j,k}F(2^{(m+j-1)r_1+(n+k-1)r_2}a_{m+j-1,n+k-1})|^s\right)^{1/s} \mid L_{\bar{p}}(l_{\bar{q}})\right\| \\
&\leqq c\left(\sum_{n=0}^{\infty}\sum_{m=0}^{\infty}2^{-mr_1 s-nr_2 s}\|F^{-1}\varphi_{j,k}F(2^{(m+j-1)r_1+(n+k-1)r_2}a_{m+j-1,n+k-1}) \mid L_{\bar{p}}(l_{\bar{q}})\|^s\right)^{1/s} \\
&\leqq c'\left(\sum_{m=0}^{\infty}2^{-m(r_1-\varrho_1+\frac{1}{2})s}\right)^{1/s}\left(\sum_{n=0}^{\infty}2^{-n(r_2-\varrho_2+\frac{1}{2})s}\right)^{1/s}\|f \mid S_{\bar{p},\bar{q}}^{\bar{r}}F\|^{a,N}.
\end{aligned} \tag{16}$$

Here $\bar{\varrho} = (\varrho_1, \varrho_2)$ comes from the right-hand sides of (1.10.3/3). Because of (8) it is possible to choose ϱ_1, ϱ_2 such that

$$\begin{aligned}
r_1 + \frac{1}{2} > \varrho_1 > \frac{1}{\min(p_1, q_1, q_2)} + \frac{1}{2} \quad \text{and} \\
r_2 + \frac{1}{2} > \varrho_2 > \frac{1}{\min(p_1, p_2, q_1, q_2)} + \frac{1}{2}.
\end{aligned} \tag{17}$$

Thus, the sums on the right-hand side of (16) are finite and

$$\|f \mid S_{\bar{p},\bar{q}}^{\bar{r}}F\| \leqq c\|f \mid S_{\bar{p},\bar{q}}^{\bar{r}}F\|^{a,N}. \tag{18}$$

The above procedure, in particular the constant c in (18) does not depend on the special choice of the admissible system a. Therefore,

$$\|f \mid S_{\bar{p},\bar{q}}^{\bar{r}}F\| \leqq c\|f \mid S_{\bar{p},\bar{q}}^{\bar{r}}F\|^{N}$$

by taking $\inf_a$ on both sides of (18). This completes the proof of part (iii) of the theorem. The proofs of part (i) and part (ii) can be finished on exactly the same line. (13) remains valid in the cases $S_{\bar{p},\bar{q}}^{\bar{r}}B$ and $SB_{\bar{p},\bar{q}}^{\bar{r}}$. Instead of Theorem 1.10.3(ii) we have to apply Theorem 1.8.3, and Theorem 1.10.3(i) (Remark 1.10.3/3) in (16). Hence the theorem is proved.

Remark* 1. Representations of this type are due to S. M. Nikol'skij. They have been extensively used in isotropic, anisotropic, and spaces with dominating mixed smoothness properties in order to derive embedding theorems for the „classical" spaces originally defined with the help of differences and derivatives (cf. 2.3.1., 2.3.4.). We refer to S. M. Nikol'skij [2, 5.5., 5.6.] for isotropic and anisotropic spaces. Representations as in part (i) go back to S. M. Nikol'skij [5] for $S_{\bar{p},\bar{\infty}}^{\bar{r}}B$, $\bar{p} = (p, p)$, $1 \leqq p \leqq \infty$, $\bar{\infty} = (\infty, \infty)$ and T. I. Amanov [1] for spaces $S_{\bar{p},\bar{q}}^{\bar{r}}B$, $\bar{p} = (p, p)$, $\bar{q} = (q, q)$, $1 \leqq p \leqq \infty$, $1 \leqq q \leqq \infty$. The extension to mixed norms was done by J. S. Bugrov [2, 4] and T. I. Amanov [2]. We proved part (i) of the theorem in H.-J. Schmeisser [4] and part (ii) in H.-J. Schmeisser [1, II].

Remark 2. The above representations require restrictions concerning $\bar{r} = (r_1, r_2)$. Part (iii) could be improved if we would have at hand a sharp multiplier theorem for $L_{\bar{p}}^{\Omega}(l_{\bar{q}})$ (cf. Remark 1.10.3/2).

Remark 3. Let us note that we can replace $\mathfrak{A}_{\bar{p}}$ by

$$\mathfrak{B}_{\bar{p}} = \{b \mid b = \{b_{j,k}(x)\}_{j,k=0}^{\infty},\ b_{j,k}(x) \in S' \cap L_{\bar{p}},$$
$$\operatorname{supp} Fb_{j,k} \subset K_j \times K_k,\ K_0 = [-2, 2],\ K_j = [-2^{j+1}, -2^{j-1}] \cup [2^{j-1}, 2^{j+1}],$$
$$j = 1, 2, \ldots\} \tag{19}$$

in the above theorem. Then the restrictions with respect to $\bar{r} = (r_1, r_2)$ are not necessary. The modified assertions are valid for all $\bar{r}$, $-\infty < r_1 < \infty$ and $-\infty < r_2 < \infty$. This leads to equivalent definitions of $S^{\bar{r}}_{\bar{p},\bar{q}}B$, $SB^{\bar{r}}_{\bar{p},\bar{q}}$, and $S^{\bar{r}}_{\bar{p},\bar{q}}F$.

Remark 4. We shall finish this subsection with some thoughts on so-called Lizorkin representations implicitly used in the proof of Proposition 1.8.3. Let $\{\chi_j(t)\}_{j=0}^{\infty}$ be the same sequence as in the proof of Proposition 1.8.3, i.e. let $\chi_j(t)$ be the characteristic function of I_j ($I_0 = [-1, 1]$, $I_j = [-2^j, -2^{j-1}] \cup [2^{j-1}, 2^j]$; $j = 1, 2, \ldots$). By P. I. Lizorkin [2, Lemma 2] the functions $\chi_{j,k}(x) = \chi_j(x_1)\chi_k(x_2)$, $x \in R_2$, are Fourier multipliers for $L_{\bar{p}}$, $\bar{p} = (p_1, p_2)$ with $1 < p_1 < \infty$ and $1 < p_2 < \infty$, where the multiplier constant does not depend on j and k ($j, k = 0, 1, 2, \ldots$). By these results and on the basis of Theorem 1.8.3 we can prove that

$$\|f \mid S^{\bar{r}}_{\bar{p},\bar{q}}B\|^L = \|2^{r_1 j + r_2 k} F^{-1}\chi_{j,k}Ff \mid l_{\bar{q}}(L_{\bar{p}})\| \tag{20}$$

is an equivalent quasi-norm in $S^{\bar{r}}_{\bar{p},\bar{q}}B$ if $\bar{p} = (p_1, p_2)$, $\bar{q} = (q_1, q_2)$, $\bar{r} = (r_1, r_2)$ with $1 < p_m < \infty$, $0 < q_m \leqq \infty$ and $-\infty < r_m < \infty$; ($m = 1, 2$). We call representations associated with a quasi-norm of type (20) a Lizorkin representation. We refer to H.-J. Schmeisser [4]. Further, by P. I. Lizorkin [2, Lemma 2] the sequence $\{\chi_{j,k}\}_{j,k=0}^{\infty}$ generates a Fourier multiplier in $L_{\bar{p}}(l_{\bar{2}})$, $\bar{2} = (2, 2)$, $\bar{p} = (p_1, p_2)$, $1 < p_1 < \infty$, $1 < p_2 < \infty$. This yields together with Theorem 2.3.1 and Theorem 1.10.3 that

$$\|f \mid S^{\bar{r}}_{\bar{p}}H\|^L = \|2^{r_1 j + r_2 k} F^{-1}\chi_{j,k}Ff \mid L_{\bar{p}}(l_{\bar{2}})\| \tag{21}$$

is an equivalent norm in $S^{\bar{r}}_{\bar{p}}H$ $(= S^{\bar{r}}_{\bar{p},2}F$, $\bar{2} = (2, 2))$ if $\bar{p} = (p_1, p_2)$, $\bar{r} = (r_1, r_2)$ with $1 < p_m < \infty$ and $-\infty < r_m < \infty$; ($m = 1, 2$). The question arises whether $\{\chi_{j,k}\}$ is a Fourier multiplier in $L_{\bar{p}}(l_{\bar{q}})$ and $Sl_{\bar{q}}(L_{\bar{p}})$ (cf. 1.3.2/(6), (7)) if $\min(p_1, p_2, q_1, q_2) > 1$. An affirmative answer would lead to Lizorkin representations for $S^{\bar{r}}_{\bar{p},\bar{q}}F$ and $SB^{\bar{r}}_{\bar{p},\bar{q}}$. Note that $S^{\bar{r}}_{\bar{p},\bar{q}}F$ admits such a representation if $\bar{p} = (p, p)$, $\bar{q} = (q, q)$ with $1 < p < \infty$ and $1 < q < \infty$. This is implied by the multiplier theorem in P. I. Lizorkin [1] for the spaces $L_{\bar{p}}(l_{\bar{q}})$ with $\bar{p} = (p, p)$, $\bar{q} = (q, q)$, $1 < p < \infty$, $1 < q < \infty$.

2.3.3. Characterizations of $S^{\bar{r}}_{\bar{p},\bar{q}}F$ by Differences

In this subsection and in the next one we return in some sense to the starting point of the theory of spaces with dominating mixed smoothness properties. We shall establish equivalent quasi-norms of those types as they have been used by S. M. Nikol'skij [5], T. I. Amanov [1], and A. D. Džhabrailov [1, 2] in order to introduce the spaces $S^{\bar{r}}_{\bar{p},\bar{q}}B$, $\bar{p} = (p, p)$, $\bar{q} = (q, q)$, $1 \leqq p \leqq \infty$, $1 \leqq q \leqq \infty$. We show that all spaces $S^{\bar{r}}_{\bar{p},\bar{q}}B$, $SB^{\bar{r}}_{\bar{p},\bar{q}}$, and $S^{\bar{r}}_{\bar{p},\bar{q}}F$, where $\bar{p}, \bar{q}, \bar{r}$ satisfy the conditions from the preceding subsection, allow representations of that type. In particular, the "classical cases" can be extended to values $p_1, p_2, q_1, q_2 < 1$. At first we concentrate our attention on $S^{\bar{r}}_{\bar{p},\bar{q}}F$.

Let us start with some notations. If $f(x) = f(x_1, x_2)$ is a function defined on R_2 and if $m = 1, 2, \ldots$; $h = (h_1, h_2) \in R_2$, we put

$$\left.\begin{aligned} \Delta^1_{h_1,1}f(x) &= f(x_1 + h_1, x_2) - f(x_1, x_2), & \Delta^0_{h_1,1}f(x) &= f(x),\\ \Delta^1_{h_2,2}f(x) &= f(x_1, x_2 + h_2) - f(x_1, x_2), & \Delta^0_{h_2,2}f(x) &= f(x), \end{aligned}\right\} \tag{1}$$

$$\Delta^m_{h_i,i}f(x) = \Delta^1_{h_i,i}(\Delta^{m-1}_{h_i,i}f)(x); \quad i = 1, 2. \tag{2}$$

If $\bar{m} = (m_1, m_2)$; $m_i = 0, 1, 2, \ldots$; $(i = 1, 2)$, we define the mixed differences by

$$\Delta_h^{\bar{m}} f(x) = \Delta_{h_2,2}^{m_2}(\Delta_{h_1,1}^{m_1} f)(x) = \Delta_{h_1,1}^{m_1}(\Delta_{h_2,2}^{m_2} f)(x). \tag{3}$$

Obviously, $\Delta_h^{(m_1,0)} f = \Delta_{h_1,1}^{m_1} f$ and $\Delta_h^{(0,m_2)} f = \Delta_{h_2,2}^{m_2} f$. Further we have

$$\left.\begin{aligned} \Delta_{h_1,1}^{m_1} f(x) &= \sum_{\mu=0}^{m_1} c_\mu^{m_1} f(x_1 + \mu h_1, x_2),\ c_\mu^{m_1} = (-1)^{m_1-\mu}\binom{m_1}{\mu}, \quad \mu = 0, \ldots, m_1; \\ \Delta_{h_2,2}^{m_2} f(x) &= \sum_{\nu=0}^{m_2} c_\nu^{m_2} f(x_1, x_2 + \nu h_2),\ c_\nu^{m_2} = (-1)^{m_2-\nu}\binom{m_2}{\nu}, \quad \nu = 0, \ldots, m_2; \end{aligned}\right\} \tag{4}$$

and

$$\Delta_h^{\bar{m}} f(x) = \sum_{\mu=0}^{m_1} \sum_{\nu=0}^{m_2} c_{\mu,\nu}^{\bar{m}} f(x_1 + \mu h_1, x_2 + \nu h_2),\ c_{\mu,\nu}^{\bar{m}} = c_\mu^{m_1} c_\nu^{m_2}, \tag{5}$$

$$\mu = 0, \ldots, m_1; \qquad \nu = 0, \ldots, m_2.$$

We need maximal functions of the type used in 1.6.4., 1.10.2., and 2.2.7., Let $f \in S'$ with

$$\operatorname{supp} Ff \subset Q_{\bar{b}} = \{y \mid y = (y_1, y_2),\ |y_1| \leqq b_1, |y_2| \leqq b_2\},$$

where $\bar{b} = (b_1, b_2)$ with $0 < b_1 < \infty$ and $0 < b_2 < \infty$. Let a_1 and a_2 be positive real numbers. Then we put

$$f_{\bar{b}}^*(x) = \sup_{y \in R_2} \frac{|f(x - y)|}{(1 + |b_1 y_1|^{a_1})(1 + |b_2 y_2|^{a_2})}, \quad x \in R_2. \tag{6}$$

Recall that $L_{\bar{p}}^{Q_{\bar{b}}}$ has been introduced in Definition 1.6.1.

Lemma. *Let $\bar{p} = (p_1, p_2)$, $\bar{m} = (m_1, m_2)$, $\bar{b} = (b_1, b_2)$, $\bar{a} = (a_1, a_2)$ with $m_i = 1, 2, \ldots$, $0 < b_i < \infty$, $0 < a_i \leqq m_i$, $0 < p_i \leqq \infty$; $(i = 1, 2)$ and let $h = (h_1, h_2) \in R_2$. If $f \in L_{\bar{p}}^{Q_{\bar{b}}}$ then there exists a positive constant c independent of $f(x)$, $\bar{b}$, $\bar{a}$, and h such that*

$$|\Delta_h^{\bar{m}} f(x)| \leqq c \min(|b_1 h_1|^{m_1}, |b_1 h_1|^{a_1}) \min(|b_2 h_2|^{m_2}, |b_2 h_2|^{a_2}) f_{\bar{b}}^*(x). \tag{7}$$

Proof. We observe that

$$\left.\begin{aligned} &\Delta_h^{\bar{m}} f(x) = \Delta_{h_1,1}^{m_1} g(x),\ g(x) = \Delta_{h_2,2}^{m_2} f(x) \in L_{\bar{p}}, \quad \text{and} \\ &\operatorname{supp} Fg = \operatorname{supp}(e^{iy_2h_2} - 1)^{m_2} Ff \subset Q_{\bar{b}}. \end{aligned}\right\} \tag{8}$$

By (4) it follows easily

$$|\Delta_{h_1,1}^{m_1} g(x)| \leqq c \max(1, |b_1 h_1|^{a_1}) \sup_{y_1 \in R_1} \frac{|g(x_1 - y_1, x_2)|}{1 + |b_1 y_1|^{a_1}}. \tag{9}$$

On the other hand repeated application of the mean value theorem shows that

$$\begin{aligned} |\Delta_{h_1,1}^{m_1} g(x)| &\leqq c|h_1|^{m_1} \sup_{|y_1| \leqq c_1|h_1|} \left| \frac{\partial^{m_1}}{\partial x_1^{m_1}} g(x_1 - y_1, x_2) \right| \\ &\leqq c|h_1|^{m_1} \sup_{|y_1| \leqq c_1|h_1|} \left[\frac{\left| \frac{\partial^{m_1}}{\partial x_1^{m_1}} g(x_1 - y_1, x_2) \right|}{1 + |b_1 y_1|^{a_1}} (1 + |b_1 y_1|^{a_1}) \right] \\ &\leqq c'|h_1|^{m_1} \max(1, |b_1 h_1|^{a_1}) \sup_{y_1 \in R_1} \frac{\left| \frac{\partial^{m_1}}{\partial x_1^{m_1}} g(x_1 - y_1, x_2) \right|}{1 + |b_1 y_1|^{a_1}}. \end{aligned}$$

On the right-hand side we use the one-dimensional version of the first inequality in (1.6.4/1). With the help of a homogeneity argument this leads to

$$|\Delta_{h_1,1}^{m_1} g(x)| \leqq c|b_1 h_1|^{m_1} \max(1, |b_1 h_1|^{a_1}) \sup_{y_1 \in R_1} \frac{|g(x_1 - y_1, x_2)|}{1 + |b_1 h_1|^{a_1}}. \tag{10}$$

Combining (9) and (10) we obtain

$$|\Delta_{h_1,1}^{m_1} g(x)| \leqq c \min(|b_1 h_1|^{m_1}, |b_1 h_1|^{a_1}) \sup_{y_1 \in R_1} \frac{|g(x_1 - y_1, x_2)|}{1 + |b_1 y_1|^{a_1}}. \tag{11}$$

Now, we estimate

$$g(x_1 - y_1, x_2) = \Delta_{h_2,2}^{m_2} f(x_1 - y_1, x_2)$$

by the same procedure. Then (8) and (11) yield the desired inequality (7).

We need some further preparations. We shall use constructions from S. M. Nikol'-skij [2, 5.2.1.] and [5, § 2]. Let $g(t) \in S(R_1)$ be a function with the properties

$$g(t) \geqq 0, \qquad \int_{R_1} g(t)\,dt = 1, \tag{12}$$

$$\operatorname{supp} F_1 g \subset [-1, 1], \tag{13}$$

where F_1 denotes the one-dimensional Fourier transform. Let $\bar{b} = (b_1, b_2)$, with $0 < b_1 < \infty$ and $0 < b_2 < \infty$. If $f \in L_{\bar{p}} = L_{\bar{p}}(R_2)$, $\bar{p} = (p_1, p_2)$ with $1 \leqq p_1 \leqq \infty$ and $1 \leqq p_2 \leqq \infty$, then the function

$$\begin{aligned} a(x_1, x_2) &= \int_{R_2} g(h_1)\,g(h_2)\,f\left(x_1 - \frac{h_1}{b_1}, x_2 - \frac{h_2}{b_2}\right) dh_1\,dh_2 \\ &= cF^{-1}((F[g(\cdot)\,g(\cdot)])\,(b_1^{-1}\cdot, b_2^{-1}\cdot)\,Ff)\,(x_1, x_2) \end{aligned} \tag{14}$$

is an entire analytic function of exponential type with

$$\operatorname{supp} Fa \subset \{y \mid y = (y_1, y_2), \quad |y_1| \leqq b_1, |y_2| \leqq b_2\}. \tag{15}$$

Formula (4) yields

$$f(x_1, x_2) = (-1)^{m_1+1} \sum_{\mu=1}^{m_1} c_\mu^{m_1} f(x_1 + \mu h_1, x_2) + (-1)^{m_1} \Delta_{h_1,1}^{m_1} f(x). \tag{16}$$

Substituting each summand on the right-hand side by an analogous identity with respect to x_2, h_2, m_2 (cf. (4)) and integrating the resulting equality with respect to $g(h_1)\,g(h_2)\,dh$ then we obtain $f(x) = f_1(x) + f_2(x) + f_3(x) + f_4(x)$, where

$$\left.\begin{aligned} f_1(x) &= (-1)^{m_1+m_2} \sum_{\mu=1}^{m_1} \sum_{\nu=1}^{m_2} c_{\mu,\nu}^{\bar{m}} \int_{R_2} f(x_1 + \mu h_1, x_2 + \nu h_2)\,g(h_1)\,g(h_2)\,dh, \\ f_2(x) &= -(-1)^{m_1+m_2} \sum_{\mu=1}^{m_1} c_\mu^{m_1} \int_{R_2} \Delta_{h_2,2}^{m_2} f(x_1 + \mu h_1, x_2)\,g(h_1)\,g(h_2)\,dh, \\ f_3(x) &= -(-1)^{m_1+m_2} \sum_{\nu=1}^{m_2} c_\nu^{m_2} \int_{R_2} \Delta_{h_1,1}^{m_1} f(x_1, x_2 + \nu h_2)\,g(h_1)\,g(h_2)\,dh, \\ f_4(x) &= (-1)^{m_1+m_2} \int_{R_2} \Delta_h^{\bar{m}} f(x)\,g(h_1)\,g(h_2)\,dh. \end{aligned}\right\} \tag{17}$$

In (17) we use the indentities

$$\Delta_{h_i,i}^{m_i} f = \Delta_{h_i 2^{-r-1},i}^{m_i} f + \sum_{j=0}^{r} [\Delta_{h_i 2^{-j},i}^{m_i} - \Delta_{h_i 2^{-j-1},i}^{m_i}] f; \quad i = 1, 2;$$

$$\Delta_h^{\bar{m}} f = -\Delta^{\bar{m}}_{(h_1 2^{-r-1}, h_2 2^{-s-1})} f + \Delta^{\bar{m}}_{(h_1, h_2 2^{-s-1})} f + \Delta^{\bar{m}}_{(h_1 2^{-r-1}, h_2)} f$$
$$+ \sum_{j=0}^{r} \sum_{k=0}^{s} [\Delta^{\bar{m}}_{(h_1 2^{-j-1}, h_2 2^{-k-1})} - \Delta^{\bar{m}}_{(h_1 2^{-j-1}, h_2 2^{-k})}$$
$$- \Delta^{\bar{m}}_{(h_1 2^{-j}, h_2 2^{-k-1})} + \Delta^{\bar{m}}_{(h_1 2^{-j}, h_2 2^{-k})}] f,$$

where $r = 0, 1, 2, \ldots$ and $s = 0, 1, 2, \ldots$ Then the limit process $r \to \infty$, $s \to \infty$ shows $f(x) = (-1)^{m_1+m_2} \sum_{j=0}^{\infty} \sum_{k=0}^{\infty} a_{j,k}(x)$, where

$$\left.\begin{aligned}
a_{0,0}(x) &= (-1)^{m_1+m_2} f_1(x), \\
a_{0,k}(x) &= \sum_{\mu=1}^{m_1} c_\mu^{m_1} \int_{R_2} g(h)\, [\Delta^{m_2}_{h_2 2^{-k-1},2} - \Delta^{m_2}_{h_2 2^{-k},2}]\, f(x_1 + \mu h_1, x_2)\, \mathrm{d}h, \\
a_{j,0}(x) &= \sum_{\nu=1}^{m_2} c_\nu^{m_2} \int_{R_2} g(h)\, [\Delta^{m_1}_{h_1 2^{-j-1},1} - \Delta^{m_1}_{h_1 2^{-j},1}]\, f(x_1, x_2 + \nu h_2)\, \mathrm{d}h, \\
a_{j,k}(x) &= \int_{R_2} g(h)\, [\Delta^{\bar{m}}_{(h_1 2^{-j-1}, h_2 2^{-k-1})} - \Delta^{\bar{m}}_{(h_1 2^{-j-1}, h_2 2^{-k})} \\
&\qquad - \Delta^{\bar{m}}_{(h_1 2^{-j}, h_2 2^{-k-1})} + \Delta^{\bar{m}}_{(h_1 2^{-j}, h_2 2^{-k})}]\, f(x)\, \mathrm{d}h,
\end{aligned}\right\} \tag{18}$$

with $j = 1, 2, \ldots$; $k = 1, 2, \ldots$, and $g(h) = g(h_1)\, g(h_2)$ (convergence in $L_{\bar{p}}$ and hence in S'). By means of (14) and (15) it is not difficult to see that

$$\operatorname{supp} Fa_{j,k} \subset \{y \mid y = (y_1, y_2),\ |y_1| \leqq c \cdot 2^j,\ |y_2| \leqq c \cdot 2^k\} \tag{19}$$

holds for all $j = 0, 1, 2, \ldots$ and $k = 0, 1, 2, \ldots$, where c is independent of j and k (in the construction of $a_{j,k}(x)$ appear only summands of type (14)).

Before we state our main theorem let us introduce some notations. If $0 < q < \infty$ we put

$$\left.\begin{aligned}
\|\varphi(t) \mid L_q^*\| &= \left(\int_{R_1} |\varphi(t)|^q \frac{\mathrm{d}t}{|t|}\right)^{1/q}, \quad \text{and} \\
\|\varphi(t) \mid L_\infty^*\| &= \operatorname*{ess\text{-}sup}_{t \in R_1} |\varphi(t)|.
\end{aligned}\right\} \tag{20}$$

If $\bar{q} = (q_1, q_2)$ with $0 < q_1 \leqq \infty$ and $0 < q_2 \leqq \infty$ then we write

$$\begin{aligned}
\|\varphi(t_1, t_2) \mid L_{\bar{q}}^*\| &= \|\varphi(t_1, t_2) \mid L^*_{q_1 \mid t_1} \mid L^*_{q_2 \mid t_2}\| \\
&= \left(\int_{R_1} \left(\int_{R_1} |\varphi(t_1, t_2)|^{q_1} \frac{\mathrm{d}t_1}{|t_1|}\right)^{q_2/q_1} \frac{\mathrm{d}t_2}{|t_2|}\right)^{1/q_2}
\end{aligned} \tag{21}$$

(modification if $\max(q_1, q_2) = \infty$). Further pure and mixed moduli of continuity appear. Let $\bar{m} = (m_1, m_2)$ with $m_i = 1, 2, \ldots$ $(i = 1, 2)$. Then we put

$$\left.\begin{aligned}
\omega_t^{\bar{m}} f(x) &= \sup_{|h_1| \leqq t_1} \sup_{|h_2| \leqq t_2} |\Delta_h^{\bar{m}} f(x)|, \quad x \in R_2,\ t \in R_2, \\
\omega_{t_i,i}^{m_i} f(x) &= \sup_{|h_i| \leqq t_i} |\Delta_{h_i,i}^{m_i} f(x)|; \quad i = 1, 2.
\end{aligned}\right\} \tag{22}$$

In the following $\|\cdot \mid L_{q_i}^*\|$ means $\|\cdot \mid L^*_{q_i \mid h_i}\|$; $i = 1, 2$, and $\|\cdot \mid L_{\bar{p}}\|$ means $\|\cdot \mid L_{\bar{p} \mid x}\|$.

Theorem. *Let* $\bar{p} = (p_1, p_2)$, $\bar{q} = (q_1, q_2)$, $\bar{r} = (r_1, r_2)$ *with* $0 < p_i < \infty$, $0 < q_i \leqq \infty$; $(i = 1, 2)$ *and*

$$r_1 > \frac{1}{\min(p_1, q_1, q_2)}, \qquad r_2 > \frac{1}{\min(p_1, p_2, q_1, q_2)}.$$

If $\bar{m} = (m_1, m_2)$ is a pair of natural numbers such that $m_1 > r_1$ and $m_2 > r_2$, then

$$\|f \mid S^{\bar{r}}_{\bar{p},\bar{q}}F\|^{\Delta} = \|f \mid L_{\bar{p}}\| + \| |h_1|^{-r_1} \Delta^{m_1}_{h_1,1} f(x) \,|L^*_{q_1}|\, L_{\bar{p}}\| + \| |h_2|^{-r_2} \Delta^{m_2}_{h_2,2} f(x) \,|L^*_{q_2}|\, L_{\bar{p}}\| + \| |h_1|^{-r_1} |h_2|^{-r_2} \Delta^{\bar{m}}_h f(x) \,|L^*_{\bar{q}}|\, L_{\bar{p}}\| \tag{23}$$

and

$$\|f \mid S^{\bar{r}}_{\bar{p},\bar{q}}F\|^{\omega} = \|f \mid L_{\bar{p}}\| + \| |h_1|^{-r_1} \omega^{m_1}_{h_1,1} f(x) \,|L^*_{q_1}|\, L_{\bar{p}}\| + \| |h_2|^{-r_2} \omega^{m_2}_{h_2,2} f(x) \,|L^*_{q_2}|\, L_{\bar{p}}\| + \| |h_1|^{-r_1} |h_2|^{-r_2} \omega^{\bar{m}}_h f(x) \,|L^*_{\bar{q}}|\, L_{\bar{p}}\| \tag{24}$$

are equivalent quasi-norms in $S^{\bar{r}}_{\bar{p},\bar{q}}F$.

Proof. Step 1. Let $f \in S^{\bar{r}}_{\bar{p},\bar{q}}F$, where $\bar{r}, \bar{p}, \bar{q}$ satisfy the assumptions of the theorem. Because of Proposition 2.2.3/3, f belongs to $L_{\bar{p}}$ and

$$\|f \mid L_{\bar{p}}\| \leqq c\|f \mid S^{\bar{r}}_{\bar{p},\bar{q}}F\|. \tag{25}$$

We prove that there exists a positive constant c such that

$$\| |h_1|^{-r_1} |h_2|^{-r_2} \omega^{\bar{m}}_h f(x) \,|L^*_{\bar{q}}|\, L_{\bar{p}}\| \leqq c\|f \mid S^{\bar{r}}_{\bar{p},\bar{q}}F\|. \tag{26}$$

Let $I_\mu = [-2^\mu, -2^{\mu-1}] \cup [2^{\mu-1}, 2^\mu]$; $\mu = 0, \pm 1, \pm 2, \ldots$ Then

$$\| |h_1|^{-r_1} |h_2|^{-r_2} \omega^{\bar{m}}_h f(x) \mid L^*_{\bar{q}}\| = \left(\sum_{\nu=-\infty}^{\infty} \int_{I_\nu} \left(\sum_{\mu=-\infty}^{\infty} \int_{I_\mu} [|h_1|^{-r_1} |h_2|^{-r_2} \omega^{\bar{m}}_h f(x)]^{q_1} \frac{dh_1}{|h_1|} \right)^{q_2/q_1} \frac{dh_2}{|h_2|} \right)^{1/q_2}$$
$$\leqq c \left\| 2^{r_1\mu + r_2\nu} \sup_{|h_1| \leqq 2^{-\mu}} \sup_{|h_2| \leqq 2^{-\nu}} |\Delta^{\bar{m}}_h f(x)| \mid l^*_{\bar{q}} \right\|, \tag{27}$$

where $l^*_{\bar{q}}$ means $l^*_{\bar{q}|(\mu,\nu)}$ and stands for the mixed $l_{\bar{q}}$ quasi-norm with respect to (μ, ν) by summing from $-\infty$ to $+\infty$. Let $\varphi = \{\varphi_j\}_{j=0}^{\infty} \in \Phi(R_1)$. We put $\varphi_{j,k}(x) = \varphi_j(x_1)\varphi_k(x_2)$; $j, k = 0, 1, 2, \ldots$ Furthermore, let $(\varphi^*_{j,k}f)(x)$; $j, k = 0, 1, 2, \ldots$, be the maximal functions from (2.2.7/3). The preceding lemma and inequality (7) yield

$$\sup_{|h_1| \leqq 2^{-\mu}} \sup_{|h_2| \leqq 2^{-\nu}} |\Delta^{\bar{m}}_h F^{-1} \varphi_{j+\mu,k+\nu} Ff|\,(x)$$
$$\leqq c \min(2^{jm_1}, 2^{ja_1}) \min(2^{km_2}, 2^{ka_2})\, \varphi^*_{j+\mu,k+\nu} f(x) \tag{28}$$

if $j, k, \mu, \nu = 0, \pm 1, \pm 2, \ldots$ and $(\varphi_{j,k}f)(x) = 0$ for $j \cdot k < 0$. Here a_1 and a_2 are the numbers from the definition of $\varphi^*_{j,k}f$ (cf. (2.2.7/3)). We use

$$f = \sum_{j=-\infty}^{\infty} \sum_{k=-\infty}^{\infty} F^{-1} \varphi_{j+\mu,k+\nu} Ff,$$

where μ and ν are arbitrary integers. Taking into account the elementary inequality $(\sum |b_j|)^s \leqq \sum |b_j|^s$, $0 < s \leqq 1$, then (28) yields

$$\left\| 2^{r_1\mu} \sup_{|h_1| \leqq 2^{-\mu}} \sup_{|h_2| \leqq 2^{-\nu}} |\Delta^{\bar{m}}_h f(x)| \mid l^*_{q_1|\mu} \right\|$$
$$\leqq c \left\| \sum_{k=-\infty}^{\infty} \min(2^{km_2}, 2^{ka_2})^s \sum_{j=-\infty}^{\infty} [2^{r_1\mu} \min(2^{jm_1}, 2^{ja_1})\, \varphi^*_{j+\mu,k+\nu} f(x)]^s \mid l^*_{\frac{q_1}{s}|\mu} \right\|^{\frac{1}{s}}$$
$$\leqq c \left(\sum_{k=-\infty}^{\infty} \min(2^{km_2}, 2^{ka_2})^s \sum_{j=-\infty}^{\infty} \min(2^{jm_1}, 2^{ja_1})^s \cdot 2^{-jr_1 s} \times \|2^{(j+\mu)r_1} \varphi^*_{j+\mu,k+\nu} f(x) \mid l^*_{q_1|\mu}\|^s \right)^{\frac{1}{s}}. \tag{29}$$

Here $s = \min(1, q_1, q_2)$. If $m_1 > r_1 > a_1$, then

$$\sum_{j=-\infty}^{\infty} \min(2^{jm_1}, 2^{ja_1})^s \cdot 2^{-jr_1 s} < \infty .$$

Furthermore,

$$\|2^{(j+\mu)r_1}\varphi^*_{j+\mu,k+\nu} f \mid l^*_{q_1|\mu}\| = \|2^{\mu r_1}\varphi^*_{\mu,k+\nu} f \mid l_{q_1|\mu}\| .$$

Hence (29) yields

$$\left\| 2^{r_1\mu} \sup_{|h_1| \leqq 2^{-\mu}} \sup_{|h_2| \leqq 2^{-\nu}} |\Delta_h^{\bar{m}} f(x)| \mid l^*_{q_1|\mu} \right\|$$
$$\leqq c \left(\sum_{k=-\infty}^{\infty} \min(2^{km_2}, 2^{ka_2})^s \, \|2^{\mu r_1}\varphi^*_{\mu,k+\nu} f(x) \mid l_{q_1|\mu}\|^s \right)^{1/s} . \tag{30}$$

Multiplying (30) with $2^{r_2\nu}$ and taking afterwards the quasi-norm $\| \cdot \mid l^*_{q_2|\nu}\|$ we obtain by the same arguments as in (29)

$$\left\| 2^{r_1\mu + r_2\nu} \sup_{|h_1| \leqq 2^{-\mu}} \sup_{|h_2| \leqq 2^{-\nu}} |\Delta_h^{\bar{m}} f(x)| \mid l^*_{\bar{q}|(\mu,\nu)} \right\|$$
$$\leqq c \|2^{r_1\mu + r_2\nu}\varphi^*_{\mu,\nu} f(x) \mid l_{\bar{q}|(\mu,\nu)}\| \tag{31}$$

if $m_2 > r_2 > a_2$. (31) and (27) show

$$\| |h_1|^{-r_1} |h_2|^{-r_2} \omega_h^{\bar{m}} f(x) |L^*_{\bar{q}}| L_{\bar{p}}\| \leqq c \, \|2^{r_1 j + r_2 k}\varphi^*_{j,k} f(x) \mid L_{\bar{p}}(l_{\bar{q}})\| . \tag{32}$$

By our assumptions we can choose a_1 and a_2 such that

$$r_1 > a_1 > \frac{1}{\min(p_1, q_1, q_2)} \quad \text{and} \quad r_2 > a_2 > \frac{1}{\min(p_1, p_2, q_1, q_2)} .$$

Consequently, (32) and Theorem 2.2.7(iii) (cf. also Corollary 2.2.7) yield the desired inequality (26).

Step 2. If we choose $s = \min(1, q_1)$, then by the same arguments as in Step 1 it can be proved

$$\| |h_1|^{-r_1} \omega^{m_1}_{h_1,1} f(x) |L^*_{q_1}| L_{\bar{p}}\| \leqq c\|f \mid S^{(r_1,0)}_{\bar{p},(q_1,s)}F\| \tag{33}$$

and

$$\| |h_2|^{-r_2} \omega^{m_2}_{h_2,2} f(x) |L^*_{q_2}| L_{\bar{p}}\| \leqq c\|f \mid S^{(0,r_2)}_{\bar{p},(1,q_2)}F\| . \tag{34}$$

The embeddings of Proposition 2.2.3/2(ii) show that the right-hand sides of (33) and (34) can be estimated from above by $c\|f \mid S^{\bar{r}}_{\bar{p},\bar{q}}F\|$. Now, together with (25) and (32) we obtain that there exists a positive constant c such that

$$\|f \mid S^{\bar{r}}_{\bar{p},\bar{q}}F\|^{\omega} \leqq c\|f \mid S^{\bar{r}}_{\bar{p},\bar{q}}F\|$$

for all $f \in S^{\bar{r}}_{\bar{p},\bar{q}}F$.

Step 3. Obviously the proof of the theorem is complete if there exists a positive number c such that

$$\|f \mid S^{\bar{r}}_{\bar{p},\bar{q}}F\| \leqq c\|f \mid S^{\bar{r}}_{\bar{p},\bar{q}}F\|^{\Delta}$$

holds for all $f \in S^{\bar{r}}_{\bar{p},\bar{q}}F$. By Theorem 2.3.2(iii) it is sufficient to find a system

$a = \{a_{j,k}\}_{j,k=0}^{\infty} \in \mathfrak{A}_{\bar{p}}$ such that $f = \sum_{j,k} a_{j,k}(x)$ in S' and

$$\|f \mid S_{\bar{p},\bar{q}}^{\bar{r}}F\|^{a,N} = \|2^{r_1 j + r_2 k} a_{j,k}(x) \mid L_{\bar{p}}(l_{\bar{q}})\| \leqq c\|f \mid S_{\bar{p},\bar{q}}^{\bar{r}}F\|^{\Delta}. \tag{35}$$

Let $f \in S_{\bar{p},\bar{q}}^{\bar{r}}F$. Let $\{a_{j,k}\}_{j,k=0}^{\infty}$ be the system of entire analytic functions of exponential type considered in (18). Then $f = \sum_{j,k} a_{j,k}(x)$ in S'. The first identity (4) can be reformulated by

$$\sum_{\mu=1}^{m_1} c_{\mu}^{m_1} f(x_1 + \mu h_1, x_2) = (-1)^{m_1+1} f(x) + \Delta_{h_1,1}^{m_1} f(x).$$

From the second identity in (4) and from (5) we can deduce analogous formulas for $\Delta_{h_2,2}^{m_2} f$ and $\Delta_h^{\bar{m}} f$. Using these equations and (12) we find

$$\begin{aligned} |a_{0,0}(x)| &= \left| \int_{R_2} g(h) \left(\sum_{\mu=1}^{m_1} \sum_{\nu=1}^{m_2} c_{\mu}^{m_1} c_{\nu}^{m_2} f(x_1 + \mu h_1, x_2 + \nu h_2) \right) dh \right| \\ &\leqq |f(x)| + \int_{R_1} g(h_1) |\Delta_{h_1,1}^{m_1} f(x)| \, dh_1 + \int_{R_1} g(h_2) |\Delta_{h_2,2}^{m_2} f(x)| \, dh_2 \\ &\quad + \int_{R_2} g(h_1) g(h_2) |\Delta_h^{\bar{m}} f(x)| \, dh \end{aligned} \tag{36}$$

and

$$\begin{aligned} |a_{0,k}(x)| &= \left| \int_{R_2} g(h) [\Delta_{h_2 2^{-k-1},2}^{m_2} - \Delta_{h_2 2^{-k},2}^{m_2}] \sum_{\mu=1}^{m_1} c_{\mu}^{m_1} f(x_1 + \mu h_1, x_2) \, dh \right| \\ &\leqq \int_{R_1} g(h_2) |\Delta_{h_2 2^{-k-1},2}^{m_2} f(x)| \, dh_2 + \int_{R_1} g(h_2) |\Delta_{h_2 2^{-k},2}^{m_2} f(x)| \, dh_2 \\ &\quad + \int_{R_2} g(h) |\Delta_{(h_1, h_2 2^{-k-1})}^{\bar{m}} f(x)| \, dh + \int_{R_2} g(h) |\Delta_{(h_1, h_2 2^{-k})}^{\bar{m}} f(x)| \, dh. \end{aligned} \tag{37}$$

An analogous estimate is valid for $|a_{j,0}(x)|$. Obviously, $|a_{j,k}(x)|$ allows an inequality of this type, too. This yields

$$\begin{aligned} &\|2^{r_1 j + r_2 k} a_{j,k}(x) \mid l_{\bar{q}}\| \\ &= \left(\sum_{k=0}^{\infty} \left(\sum_{j=0}^{\infty} |2^{r_1 j + r_2 k} a_{j,k}(x)|^{q_1} \right)^{q_2/q_1} \right)^{1/q_2} \\ &\leqq c \left[|f(x)| + \left(\sum_{j=0}^{\infty} 2^{r_1 j q_1} \left(\int_{R_1} g(h_1) |\Delta_{h_1 2^{-j},1}^{m_1} f(x)| \, dh_1 \right)^{q_1} \right)^{1/q_1} \right. \\ &\quad + \left(\sum_{k=0}^{\infty} 2^{r_2 k q_2} \left(\int_{R_1} g(h_2) |\Delta_{h_2 2^{-k},2}^{m_2} f(x)| \, dh_2 \right)^{q_2} \right)^{1/q_2} \\ &\quad + \left(\sum_{k=0}^{\infty} \left(\sum_{j=0}^{\infty} 2^{r_1 j q_1 + r_2 k q_1} \right. \right. \\ &\quad \left. \left. \left. \times \left(\int_{R_2} g(h_1) g(h_2) |\Delta_{(h_1 2^{-j}, h_2 2^{-k})}^{\bar{m}} f(x)| \, dh_1 \, dh_2 \right)^{q_1} \right)^{q_2/q_1} \right)^{1/q_2} \right]. \end{aligned} \tag{38}$$

To prove (35) it is our task to estimate the $L_{\bar{p}|x}$-quasi-norm of the left hand-side of (38) from above by $c\|f \mid S_{\bar{p},\bar{q}}^{\bar{r}}F\|^{\Delta}$. We have (with $I_{\mu} = [-2^{\mu+1}, -2^{\mu}] \cup [2^{\mu}, 2^{\mu+1}]$,

$\mu = 0, \pm 1, \pm 2, \ldots)$

$$\int_{R_2} g(h_1)\, g(h_2)\, |\Delta^{\bar{m}}_{(h_1 2^{-j},\, h_2 2^{-k})} f(x)|\, dh_1\, dh_2$$

$$= \sum_{\nu=-\infty}^{\infty} \sum_{\mu=-\infty}^{\infty} \int_{I_\nu} \int_{I_\mu} g(h_1)\, g(h_2)\, |\Delta^{\bar{m}}_{(h_1 2^{-j},\, h_2 2^{-k})} f(x)|\, dh_1\, dh_2$$

$$= \sum_{\nu=-\infty}^{\infty} \sum_{\mu=-\infty}^{\infty} 2^{\mu+\nu} \int_{I_0} \int_{I_0} g(2^\mu h_1)\, g(2^\nu h_2)\, |\Delta^{\bar{m}}_{(h_1 2^{\mu-j},\, h_2 2^{\nu-k})} f(x)|\, dh_1\, dh_2$$

$$\leqq c \sum_{\nu=-\infty}^{\infty} \sum_{\mu=-\infty}^{\infty} \min(1, 2^{-\mu d_1}) \min(1, 2^{-\nu d_2}) \int_{I_0} \int_{I_0} |\Delta^{\bar{m}}_{(h_1 2^{\mu-j},\, h_2 2^{\nu-k})} f(x)|\, dh_1\, dh_2 \tag{39}$$

where $d_1 > 0$ and $d_2 > 0$ are at our disposal. Let us abbreviate

$$B_{j,k}(x) = \int_{R_2} g(h_1)\, g(h_2)\, |\Delta^{\bar{m}}_{(h_1 2^{-j},\, h_2 2^{-k})} f(x)|\, dh_1\, dh_2 \tag{40}$$

and

$$b_{j,k}(x) = \int_{I_0} \int_{I_0} |\Delta^{\bar{m}}_{(h_1 2^{j},\, h_2 2^{k})} f(x)|\, dh_1\, dh_2 \tag{41}$$

for $j, k = 0, \pm 1, \pm 2, \ldots$ Let $s = \min(1, q_1, q_2)$. Then it follows from (39)

$$B_{j,k}(x) \leqq c \left(\sum_{\nu=-\infty}^{\infty} \sum_{\mu=-\infty}^{\infty} \min(1, 2^{-\mu d_1 s}) \min(1, 2^{-\nu d_2 s})\, b^s_{\mu-j,\nu-k}(x) \right)^{1/s}.$$

Therefore, if $d_1 > r_1$ and $d_2 > r_2$, then we obtain

$$\|2^{r_1 j + r_2 k} B_{j,k} \mid l_{\bar{q}|(j,k)}\|$$

$$\leqq c \left(\sum_{\nu=-\infty}^{\infty} \sum_{\mu=-\infty}^{\infty} 2^{\mu r_1 s} \min(1, 2^{-\mu d_1 s}) \cdot 2^{\nu r_2 s} \min(1, 2^{-\nu d_2 s}) \right.$$

$$\left. \times \|2^{(j-\mu) r_1 + (k-\nu) r_2} b_{\mu-j,\nu-k} \mid l_{\bar{q}|(j,k)}\|^s \right)^{1/s} \leqq c' \|2^{r_1 j + r_2 k} b_{-j,-k} \mid l^*_{\bar{q}}\|, \tag{42}$$

where $\|\cdot \mid l^*_{\bar{p}}\|$ has the same meaning as in Step 1 of the proof (see formula (27)). Temporarily we put

$$f_{j,k} = f_{j,k}(h, x) = \Delta^{\bar{m}}_{(h_1 2^{-j},\, h_2 2^{-k})} f(x); \quad j, k = 0, \pm 1, \pm 2, \ldots$$

Then, with the help of Hölder's inequality for $\dfrac{s}{q_1} + \dfrac{q_1 - s}{q_1} = 1$ and $s = \min(1, q_1, q_2)$, we have

$$b_{-j,-k}(x) = \int_{I_0} \int_{I_0} |f_{j,k}(h, x)|\, dh_1\, dh_2$$

$$\leqq \left[\int_{I_0} \int_{I_0} |f_{j,k}(h, x)|^s\, dh_1\, dh_2 \right] \sup_{\substack{h_1 \in I_0, \\ h_2 \in I_0}} |f_{j,k}(h, x)|^{1-s}$$

$$\leqq c \int_{I_0} \left(\int_{I_0} |f_{j,k}(h, x)|^{q_1}\, dh_1 \right)^{s/q_1} dh_2 \sup_{\substack{h_1 \in I_0, \\ h_2 \in I_0}} |f_{j,k}(h, x)|^{1-s}. \tag{43}$$

Applying Hölder's inequality for series with $s + (1 - s) = 1$ this inequality yields

$$\begin{aligned}
&\|2^{r_1 j} b_{-j,-k}(x) \mid l^*_{q_1}\| \\
&= \left(\sum_{j=-\infty}^{\infty} (2^{r_1 j} b_{-j,-k}(x))^{q_1}\right)^{1/q_1} \\
&\leqq c\left(\sum_{j=-\infty}^{\infty} 2^{r_1 j s q_1}\left[\int_{I_0}\left(\int_{I_0} |f_{j,k}(h,x)|^{q_1}\,\mathrm{d}h_1\right)^{s/q_1}\mathrm{d}h_2\right]^{q_1}\right. \\
&\qquad \left.\times\left[2^{r_1 j q_1}\sup_{h_1,h_2\in I_0}|f_{j,k}(h,x)|^{q_1}\right]^{1-s}\right)^{1/q_1} \\
&\leqq c\left\|\int_{I_0}\left(\int_{I_0} 2^{r_1 j q_1}|f_{j,k}(h,x)|^{q_1}\,\mathrm{d}h_1\right)^{s/q_1}\mathrm{d}h_2 \mid l^*_{(q_1/s)\,|j}\right\| \\
&\qquad \times \|2^{r_1 j}\sup|f_{j,k}(h,x)| \mid l^*_{q_1\,|j}\|^{1-s} \\
&\leqq c\int_{I_0}\left(\sum_{j=-\infty}^{\infty}\int_{I_0} 2^{r_1 j q_1}|f_{j,k}(h,x)|^{q_1}\,\mathrm{d}h_1\right)^{s/q_1}\mathrm{d}h_2\|2^{r_1 j}\sup|f_{j,k}(h,x)| \mid l^*_{q_1\,|j}\|^{1-s},
\end{aligned} \tag{44}$$

where sup stands for $\sup_{h_1\in I_0}\sup_{h_2\in I_0}$. Next we estimate the first factor on the right-hand side by Hölder's inequality with $\frac{s}{q_2}+\frac{q_2-s}{q_2}=1$. Afterwards we take on both sides of (44) the quasi-norm $\|\cdot \mid l^*_{q_2\,|k}\|$ and repeat the procedure of (44). In this way we obtain

$$\begin{aligned}
&\|2^{r_1 j + r_2 k} b_{-j,-k}(x) \mid l^*_{\bar q}\| \\
&\leqq \left(\sum_{k=-\infty}^{\infty}\int_{I_0}\left(\sum_{j=-\infty}^{\infty}\int_{I_0}|2^{r_1 j + r_2 k} f_{j,k}(h,x)|^{q_1}\,\mathrm{d}h_1\right)^{q_2/q_1}\mathrm{d}h_2\right)^{s/q_2} \\
&\quad \times\left(\sum_{k=-\infty}^{\infty}\left(\sum_{j=-\infty}^{\infty}\left(2^{r_1 j + r_2 k}\sup_{\substack{h_1\in I_0,\\ h_2\in I_0}}|f_{j,k}(h,x)|\right)^{q_1}\right)^{q_2/q_1}\right)^{(1-s)/q_2}.
\end{aligned} \tag{45}$$

Taking into consideration the meaning of $f_{j,k}(h, x)$ we see as in (27) that the first factor on the right-hand side can be estimated from above by

$$c'\||h_1|^{-r_1}\,|h_2|^{-r_2}\,\Delta_h^{\bar m} f(x) \mid L^*_{\bar q\,|h}\|^s.$$

On the other hand,

$$\begin{aligned}
&\sum_{j=-\infty}^{\infty} 2^{r_1 j q_1}\sup_{h_1\in I_0}\left(\sup_{h_2\in I_0}|\Delta^{\bar m}_{(h_1 2^{-j},\,h_2 2^{-k})} f(x)|^{q_1}\right) \\
&\leqq c\sum_{j=-\infty}^{\infty}\int_{I_0} 2^{r_1 j q_1}\sup_{0\leqq |h_1|\leqq 2^{-j+1}|t_1|}\left(\sup_{h_2\in I_0}|\Delta^{\bar m}_{(h_1,\,h_2 2^{-k})} f(x)|^{q_1}\right)\mathrm{d}t_1 \\
&\leqq c'\int_{R_1}|t_1|^{-r_1 q_1}\sup_{0\leqq|h_1|\leqq|t_1|}\left(\sup_{h_2\in I_0}|\Delta^{\bar m}_{(h_1,\,h_2 2^{-k})} f(x)|^{q_1}\right)\frac{\mathrm{d}t_1}{|t_1|}.
\end{aligned}$$

Analogously we deal with $\sup_{h_2\in I_0}$ and $\sum_{k=-\infty}^{\infty}$ in the second factor on the right-hand side of (45). Hence, it can be estimated from above by

$$c''\||t_1|^{-r_1}\,|t_2|^{-r_2}\,\omega_t^{\bar m} f(x) \mid L^*_{\bar q\,|t}\|^{1-s}.$$

Thus, putting together (42) and (45) one obtains

$$\|2^{r_1 j+r_2 k} B_{j,k}(x) \mid l_{\bar{q}}\| \\ \leqq c \| |h_1|^{-r_1} |h_2|^{-r_2} \Delta_h^{\bar{m}} f(x) \mid L^*_{\bar{q}|h}\|^s \, \| |h_1|^{-r_1} |h_2|^{-r_2} \omega_h^{\bar{m}} f(x) \mid L^*_{\bar{q}|h}\|^{1-s}. \tag{46}$$

Recall $s = \min(1, q_1, q_2) \leqq 1$. Then (46) and Hölder's inequality yield

$$\|2^{r_1 j+r_2 k} B_{j,k}(x) \mid L_{\bar{p}}(l_{\bar{q}})\| \leqq c(\|f \mid S^{\bar{r}}_{\bar{p},\bar{q}}F\|^{\Delta})^s \, (\|f \mid S^{\bar{r}}_{\bar{p},\bar{q}}F\|^{\omega})^{1-s}. \tag{47}$$

Now we return to (38). If we have in mind the definition of $B_{j,k}(x)$ (cf. (40)) inequality (47) says that the last summand on the right-side of (38) has been estimated. But, estimates of type (47) are true for the remaining summands of (38), too. Clearly,

$$\|f(x) \mid L_{\bar{p}}\| = \|f(x) \mid L_{\bar{p}}\|^s \, \|f(x) \mid L_{\bar{p}}\|^{1-s} \\ \leqq (\|f \mid S^{\bar{r}}_{\bar{p},\bar{q}}F\|^{\Delta})^s \, (\|f \mid S^{\bar{r}}_{\bar{p},\bar{q}}F\|^{\omega})^{1-s}. \tag{48}$$

Similar as in the proof of (46) we have

$$\left\| 2^{r_i j} \int_{R_1} g(h_i) \, |\Delta^{m_i}_{h_i 2^{-j}, i} f(x)| \, dh_i \mid l_{q_i|j} \right\| \\ \leqq c \| |h_i|^{-r_i} \Delta^{m_i}_{h_i, i} f(x) \mid L^*_{q_i|h_i}\|^s \, \| |h_i|^{-r_i} \omega^{m_i}_{h_i, i} f(x) \mid L^*_{q_i|h_i}\|^{1-s}; \tag{49}$$

$i = 1, 2$, and afterwards as in (47)

$$\left\| 2^{r_i j} \int_{R_1} g(h_i) \, |\Delta^{m_i}_{h_i 2^{-j}, i} f(x)| \, dh_i \, |l_{q_i|j}| \, L_{\bar{p}|x} \right\| \\ \leqq c(\|f \mid S^{\bar{r}}_{\bar{p},\bar{q}}F\|^{\Delta})^s \, (\|f \mid S^{\bar{r}}_{\bar{p},\bar{q}}F\|^{\omega})^{1-s}. \tag{50}$$

Finally, (38), (47), (48), (50) as well as the results from Step 1 and Step 2 yield

$$\|f \mid S^{\bar{r}}_{\bar{p},\bar{q}}F\|^{a,N} \leqq c(\|f \mid S^{\bar{r}}_{\bar{p},\bar{q}}F\|^{\Delta})^s \, (\|f \mid S^{\bar{r}}_{\bar{p},\bar{q}}F\|^{\omega})^{1-s} \\ \leqq c'(\|f \mid S^{\bar{r}}_{\bar{p},\bar{q}}F\|^{\Delta})^s \, \|f \mid S^{\bar{r}}_{\bar{p},\bar{q}}F\|^{1-s}, \tag{51}$$

$s = \min(1, q_1, q_2)$, for all $f \in S^{\bar{r}}_{\bar{p},\bar{q}}F$. We obtain the desired estimate (35) by Theorem 2.3.2(iii). Thus, the proof of the theorem is complete.

Remark 1. As an immediate consequence of Theorem 2.3.1 and the above theorem we can describe the spaces $S^{\bar{r}}_{\bar{p}}H$ and $S^{\bar{l}}_{\bar{p}}W$ via differences. For example, if $\bar{p} = (p_1, p_2)$, $\bar{l} = (l_1, l_2)$ with $1 < p_1 < \infty$, $1 < p_2 < \infty$, $l_1 = 1, 2, \ldots$, and $l_2 = 1, 2, \ldots$ then

$$\|f \mid S^{\bar{l}}_{\bar{p}}W\|^{\Delta} = \|f \mid L_{\bar{p}}\| \\ + \left\| \left(\int_{R_1} |h_1|^{-2l_1} |\Delta^{m_1}_{h_1,1} f(x)|^2 \frac{dh_1}{|h_1|} \right)^{1/2} \mid L_{\bar{p}} \right\| \\ + \left\| \left(\int_{R_1} |h_2|^{-2l_2} |\Delta^{m_2}_{h_2,2} f(x)|^2 \frac{dh_2}{|h_2|} \right)^{1/2} \mid L_{\bar{p}} \right\| \\ + \left\| \left(\int_{R_2} |h_1|^{-2l_1} |h_2|^{-2l_2} |\Delta^{\bar{m}}_h f(x)|^2 \frac{dh_1}{|h_1|} \frac{dh_2}{|h_2|} \right)^{1/2} \mid L_{\bar{p}} \right\| \tag{52}$$

is an equivalent norm in $S^{\bar{l}}_{\bar{p}}W$. Here $\bar{m} = (m_1, m_2)$, where m_1 and m_2 are natural numbers with $m_1 > l_1$ and $m_2 > l_2$.

Remark 2. In (23) and (24) we can replace some differences by derivatives. One possibility is to use the equivalent quasi-norms of Theorem 2.2.6/2. However, if we compare the quasi-norms obtained in such a way with those of the classical spaces $S^{\bar{r}}_{\bar{p},\bar{q}}B$ (Banach space case) in S. M. Nikol'skij [5] or T. I. Amanov [1–3] it turns out that this method does not yield satisfactory results. We sketch a second possibility. Let $\bar{p} = (p_1, p_2)$, $\bar{q} = (q_1, q_2)$ with $0 < p_i < \infty$ and $0 < q_i \leqq \infty$ $(i = 1, 2)$. Let $\bar{m} = (m_1, m_2)$ and $\bar{l} = (l_1, l_2)$ be pairs of non-negative integers satisfying

$$m_1 > r_1 - l_1 > \frac{1}{\min(p_1, q_1, q_2)}, \qquad m_2 > r_2 - l_2 > \frac{1}{\min(p_1, p_2, q_1, q_2)}. \tag{53}$$

Then, we claim that

$$\begin{aligned} \|f \mid S^{\bar{r}}_{\bar{p},\bar{q}}F\|^{\omega}_{\bar{m},\bar{l}} = {} & \|f \mid L_{\bar{p}}\| \\ & + \left\| |h_1|^{-r_1+l_1}\, \omega^{m_1}_{h_1,1} \frac{\partial^{l_1}}{\partial x_1^{l_1}} f(x) \,|L^*_{q_1}|\, L_{\bar{p}} \right\| \\ & + \left\| |h_2|^{-r_2+l_2}\, \omega^{m_2}_{h_2,2} \frac{\partial^{l_2}}{\partial x_2^{l_2}} f(x) \,|L^*_{q_2}|\, L_{\bar{p}} \right\| \\ & + \left\| |h_1|^{-r_1+l_1}\, |h_2|^{-r_2+l_2}\, \omega^{\bar{m}}_{h} \frac{\partial^{l_1+l_2}}{\partial x_1^{l_1}\,\partial x_2^{l_2}} f(x) \,|L^*_{\bar{q}}|\, L_{\bar{p}} \right\| \end{aligned} \tag{54}$$

is an equivalent quasi-norm in $S^{\bar{r}}_{\bar{p},\bar{q}}F$. Further $\|f \mid L_{\bar{p}}\|$ can be replaced by $\|f \mid S^{\bar{l}}_{\bar{p}}W\|$. The proof follows from

$$\|f \mid S^{\bar{r}}_{\bar{p},\bar{q}}F\|^{\omega}_{\bar{m},\bar{l}} \leqq c_1 \|f \mid S^{\bar{r}}_{\bar{p},\bar{q}}F\| \leqq c_2 \|f \mid S^{\bar{r}}_{\bar{p},\bar{q}}F\|^{\Delta} \leqq c_3 \|f \mid S^{\bar{r}}_{\bar{p},\bar{q}}F\|^{\omega}_{\bar{m},\bar{l}}. \tag{55}$$

The first estimate can be obtained by the same arguments as in Step 1 and Step 2 of the proo fof the above theorem. Here we have to use Lemma 2.3.4 for $D^{\bar{l}}f (D^{(l_1,0)}f, D^{(0,l_2)}f)$ and Theorem 1.6.4 in order to estimate $\varphi^*_{j,k}(D^{\bar{l}}f)$. The second inequality has been proved in the theorem and the third one follows by iterated use of

$$|\Delta^{m_i}_{h_i,i} f(x)| \leqq c|h_i|\, \omega^{m_i-1}_{ch_i,i} \frac{\partial f}{\partial x_1}(x). \tag{56}$$

Remark* 3. Characterizations of spaces of Sobolev and Bessel-potential type (or F-type) by differences are not new. For the isotropic spaces $F^s_{p,q}(R_n)$ we refer to [T, 2.5.10.] and the literature cited there. Descriptions which include also the anisotropic case can be found in G. A. Kaljabin [2, 4].

2.3.4. Characterizations of $SB^{\bar{r}}_{\bar{p},\bar{q}}$ and $S^{\bar{r}}_{\bar{p},\bar{q}}B$ by Differences

This subsection contains the counterpart of Theorem 2.3.3 for the spaces $S^{\bar{r}}_{\bar{p},\bar{q}}B$ and $SB^{\bar{r}}_{\bar{p},\bar{q}}$. Moreover, we discuss the connections with classical spaces of this type.

Recall that $\Delta^{\bar{m}}_h f$, $\Delta^{m_i}_{h_i,i} f$, $\omega^{\bar{m}}_h f$, and $\omega^{m_i}_{h_i,i} f$ have the meaning of 2.3.3/(2), (3), and (22). Further the quasi-norms $\|\cdot \mid L^*_q\|$ and $\|\cdot \mid L^*_{\bar{q}}\|$, $0 < q \leqq \infty$, $\bar{q} = (q_1, q_2)$ with $0 < q_i \leqq \infty$ $(i = 1, 2)$ have been defined in 2.3.3/(20), (21).

Theorem 1. *Let* $\bar{q} = (q_1, q_2)$, $\bar{r} = (r_1, r_2)$ *with* $0 < q_i \leqq \infty$, $0 < r_i < \infty$; $(i = 1, 2)$. *Let* $\bar{m} = (m_1, m_2)$, $m_i = 1, 2, \ldots$; $(i = 1, 2)$.

(i) *If* $\bar{p} = (p_1, p_2)$ *with* $0 < p_1 \leqq \infty$, $0 < p_2 < \infty$ *and if*

$$m_1 > r_1 > \frac{1}{p_1}, \qquad m_2 > r_2 > \frac{1}{\min(p_1, p_2)} \tag{1}$$

then

$$\begin{aligned}\|f \mid S^{\bar r}_{\bar p,\bar q}B\|^{\Delta} = {} & \|f \mid L_{\bar p}\| \\ & + \||h_1|^{-r_1}\,\Delta^{m_1}_{h_1,1}f(x)\,|L_{\bar p}|\,L^*_{q_1}\| + \||h_2|^{-r_2}\,\Delta^{m_2}_{h_2,2}f(x)\,|L_{\bar p}|\,L^*_{q_2}\| \\ & + \||h_1|^{-r_1}\,|h_2|^{-r_2}\,\Delta^{\bar m}_{h}f(x)\,|L_{\bar p}|\,L^*_{\bar q}\| \end{aligned} \tag{2}$$

and

$$\begin{aligned}\|f \mid S^{\bar r}_{\bar p,\bar q}B\|^{\omega} = {} & \|f \mid L_{\bar p}\| \\ & + \||h_1|^{-r_1}\,\omega^{m_1}_{h_1,1}f(x)\,|L_{\bar p}|\,L^*_{q_1}\| + \||h_2|^{-r_2}\,\omega^{m_2}_{h_2,2}f(x)\,|L_{\bar p}|\,L^*_{q_2}\| \\ & + \||h_1|^{-r_1}\,|h_2|^{-r_2}\,\omega^{\bar m}_{h}f(x)\,|L_{\bar p}|\,L^*_{\bar q}\| \end{aligned} \tag{3}$$

are equivalent quasi-norms in $S^{\bar r}_{\bar p,\bar q}B$.

(ii) *If* $\bar p = (p_1, p_2)$ *with* $0 < p_1 < \infty$ *and* $0 < p_2 < \infty$ *and if*

$$m_1 > r_1 > \frac{1}{p_1}, \qquad m_2 > r_2 > \frac{1}{\min(p_1, p_2, q_1)} \tag{4}$$

then

$$\begin{aligned}\|f \mid SB^{\bar r}_{\bar p,\bar q}\|^{\Delta} = {} & \|f \mid L_{\bar p}\| \\ & + \||h_1|^{-r_1}\,\Delta^{m_1}_{h_1,1}f(x)\,|L_{p_1}|\,L^*_{q_1} \mid L_{p_2}\| \\ & + \||h_2|^{-r_2}\,\Delta^{m_2}_{h_2,2}f(x)\,|L_{\bar p}|\,L^*_{q_2}\| \\ & + \||h_1|^{-r_1}\,|h_2|^{-r_2}\,\Delta^{\bar m}_{h}f(x)\,|L_{p_1}|\,L^*_{q_1}\,|L_{p_2}|\,L^*_{q_2}\| \end{aligned} \tag{5}$$

and

$$\begin{aligned}\|f \mid SB^{\bar r}_{\bar p,\bar q}\|^{\omega} = {} & \|f \mid L_{\bar p}\| \\ & + \||h_1|^{-r_1}\,\omega^{m_1}_{h_1,1}f(x)\,|L_{p_1}|\,L^*_{q_1} \mid L_{p_2}\| \\ & + \||h_2|^{-r_2}\,\omega^{m_2}_{h_2,2}f(x)\,|L_{\bar p}|\,L^*_{q_2}\| \\ & + \||h_1|^{-r_1}\,|h_2|^{-r_2}\,\omega^{\bar m}_{h}f(x)\,|L_{p_1}|\,L^*_{q_1}|L_{p_2}|\,L^*_{q_2}\| \end{aligned} \tag{6}$$

are equivalent quasi-norms in $SB^{\bar r}_{\bar p,\bar q}$. *(In* (2), (3), (5), *and* (6) *the quasi-norms* $\|\cdot \mid L_{p_i}\|$ *and* $\|\cdot \mid L^*_{q_i}\|$ $(i = 1, 2)$ *are taken with respect to* x_i *and* h_i, *respectively.)*

Proof. We follow the method of the proof of Theorem 2.3.3. With the help of Lemma 2.3.3 and Theorem 2.2.7(i, ii) one obtains as in Step 1 and Step 2 of the proof of Theorem 2.3.3 that

$$\|f \mid S^{\bar r}_{\bar p,\bar q}B\|^{\omega} \leqq c\|f \mid S^{\bar r}_{\bar p,\bar q}B\| \tag{7}$$

and

$$\|f \mid SB^{\bar r}_{\bar p,\bar q}\|^{\omega} \leqq c\|f \mid SB^{\bar r}_{\bar p,\bar q}\| \tag{8}$$

if the assumptions of part (i) and part (ii) are satisfied, respectively. We omit the details because the modifications are more or less obvious. Then, the method of Step 3 of the proof of Theorem 2.3.3 is applicable in order to prove

$$\|f \mid S^{\bar r}_{\bar p,\bar q}B\|^{N} \leqq c\|f \mid S^{\bar r}_{\bar p,\bar q}B\|^{\Delta} \quad \text{for} \quad f \in S^{\bar r}_{\bar p,\bar q}B \tag{9}$$

and

$$\|f \mid SB^{\bar r}_{\bar p,\bar q}\|^{N} \leqq c\|f \mid SB^{\bar r}_{\bar p,\bar q}\|^{\Delta} \quad \text{for} \quad f \in SB^{\bar r}_{\bar p,\bar q}. \tag{10}$$

Again, we do not go into the details. Then Theorem 2.3.2(iii) completes the proof.

Remark 1. Let $f \in L_{\bar{p}}$, $\bar{p} = (p_1, p_2)$ with $0 < p_1 \leqq \infty$ and $0 < p_2 \leqq \infty$. As usually, $\bar{m} = (m_1, m_2)$, $m_i = 1, 2, \ldots$; $(i = 1, 2)$. Let us denote

$$\left.\begin{aligned} \omega_h^{\bar{m}} f(x)_{\bar{p}} &= \sup_{|t_1| \leqq |h_1|} \sup_{|t_2| \leqq |h_2|} \|\Delta_t^{\bar{m}} f(x) \mid L_{\bar{p}}\| \\ \omega_{h_i, i}^{m_i} f(x)_{\bar{p}} &= \sup_{|t_i| \leqq |h_i|} \|\Delta_{t_i, i}^{m_i} f(x) \mid L_{\bar{p} \mid x}\|; \quad i = 1,2 \end{aligned}\right\} \tag{11}$$

As an immediate consequence of part (i) of the above theorem we obtain (under the same assumptions for $\bar{p}$, $\bar{q}$, $\bar{r}$, and $\bar{m}$ as in the theorem) that

$$\begin{aligned} \|f \mid S_{\bar{p},\bar{q}}^{\bar{r}} B\|^M &= \|f \mid L_{\bar{p}}\| \\ &+ \| |h_1|^{-r_1} \omega_{h_1, 1}^{m_1} f(x)_{\bar{p}} \mid L_{q_1}^*\| + \| |h_2|^{-r_2} \omega_{h_2, 2}^{m_2} f(x)_{\bar{p}} \mid L_{q_2}^*\| \\ &+ \| |h_1|^{-r_1} |h_2|^{-r_2} \omega_h^{\bar{m}} f(x)_{\bar{p}} \mid L_{\bar{q}}^*\| \end{aligned} \tag{12}$$

is an equivalent quasi-norm in $S_{\bar{p},\bar{q}}^{\bar{r}} B$. A similar assertion holds in the case $SB_{\bar{p},\bar{q}}^{\bar{r}}$. In the Banach space case (12) coincides with the classical norm of the spaces $S_{\bar{p},\bar{q}}^{\bar{r}} B$ as it has been introduced by S. M. Nikol'skij [5], T. I. Amanov [1–3], J. S. Bugrov [2, 4], and A. D. Džabrailov [1, 2]. However, the restrictions $r_1 > \frac{1}{p_1}$ and $r_2 > \frac{1}{\min(p_1, p_2)}$ are unnatural. Next we shall show how these restrictions can be relaxed in the case $\min(p_1, p_2, q_1, q_2) \geqq 1$.

Recall that $L_{\bar{p}}^{\Omega}$ and $Sl_{\bar{q}}(L_{\bar{p}}^{\Omega})$ have been defined in (1.6.1/1) and (1.10.1/1), respectively. Further $\Delta_h^{\bar{m}} f$ and $\Delta_{h_i, i}^{m_i} f$ $(i = 1, 2)$ have the meaning of 2.3.3/(2), (3).

Lemma. *Let* $\bar{m} = (m_1, m_2)$, $m_i = 0, 1, 2, \ldots$; $(i = 1, 2)$. *Let* $h = (h_1, h_2) \in R_2$ *with* $h_1 \cdot h_2 \neq 0$.

(i) *Let* $\bar{p} = (p_1, p_2)$ *with* $0 < p_1 \leqq \infty$ *and* $0 < p_2 \leqq \infty$. *Let* $\bar{b} = (b_1, b_2)$ *with* $0 < b_1 < \infty$ *and* $0 < b_2 < \infty$. *Let* $\Omega = \{x \mid x = (x_1, x_2),\ |x_1| \leqq b_1, |x_2| \leqq b_2\}$. *Then there exists a positive constant c such that*

$$\|\Delta_h^{\bar{m}} f \mid L_{\bar{p}}\| \leqq c \min(1, |b_1 h_1|^{m_1}) \min(1, |b_2 h_2|^{m_2}) \|f \mid L_{\bar{p}}\| \tag{13}$$

holds for all $f \in L_{\bar{p}}^{\Omega}$.

(ii) *Let* $\bar{p} = (p_1, p_2)$, $\bar{q} = (q_1, q_2)$ *with* $0 < p_i \leqq \infty$, $0 < q_i \leqq \infty$. *Let* $\{a_j\}_{j=0}^{\infty}$ *and* $\{b_j\}_{j=0}^{\infty}$ *be sequences of positive numbers. Let* $\Omega = \{\Omega_{j,k}\}_{j,k=0}^{\infty}$ *wit h*$\Omega_{j,k} = \{x \mid x = (x_1, x_2),$ $|x_1| \leqq a_j$, $|x_2| \leqq b_k\}$, $j, k = 0, 1, 2, \ldots$ *Then there exists a positive constant c such that*

$$\|\Delta_h^{\bar{m}} f_{j,k} \mid Sl_{\bar{q}}(L_{\bar{p}})\| \leqq c \|\min(1, |a_j h_1|^{m_1}) \min(1, |b_k h_2|^{m_2}) f_{j,k} \mid Sl_{\bar{q}}(L_{\bar{p}})\| \tag{14}$$

holds for all $\{f_{j,k}\} \in Sl_{\bar{q}}(L_{\bar{p}}^{\Omega})$.

Proof. Step 1. Let $g \in L_{\bar{p}}^{\Omega}$, where Ω has the meaning of (i). We choose a function $\psi \in S(R_1)$ such that

$$\psi(t) = 1 \quad \text{if} \quad |t| \leqq 1, \ \operatorname{supp} \psi \subset [-2, 2].$$

Obviously,

$$\Delta_{h_1, 1}^{m_1} g(x) = (F^{-1}(e^{i\xi_1 h_1} - 1)^{m_1} Fg)(x).$$

Hence,

$$\begin{aligned} &\|\Delta_{h_1, 1}^{m_1} g(x) \mid L_{\bar{p}}\| \\ &= \left\| F^{-1} \left[\frac{(e^{ih_1 \cdot} - 1)^{m_1}}{(h_1 \cdot)^{m_1}} \psi\left(\frac{\cdot}{b_1}\right) \psi\left(\frac{\cdot}{b_2}\right) F(F^{-1} \xi_1^{m_1} F h_1^{m_1} g) \right] \mid L_{\bar{p}} \right\|. \end{aligned} \tag{15}$$

Let $|b_1 h_1| \leqq 1$. We claim

$$\left\| \frac{(e^{ih_1 b_1 x_1} - 1)^{m_1}}{(h_1 b_1 x_1)^{m_1}} \psi(x_1)\, \psi(x_2) \mid S_2^{\bar{\varkappa}} H \right\| \leqq c, \quad \bar{\varkappa} = (\varkappa_1, \varkappa_2) \tag{16}$$

with $0 < \varkappa_1 < \infty$ and $0 < \varkappa_2 < \infty$, where c is independent of h_1, b_1, and b_2: Recall that $H_2^{\varkappa_1}$ has been defined in (1.7.4/1). Let $\|f \mid \dot{H}_2^{\varkappa_1}\| = \| |x_1|^{\varkappa_1} F_1 f \mid L_2\|$. Then we have

$$\begin{aligned}
&\left\| \frac{(e^{ih_1 b_1 x_1} - 1)^{m_1}}{(h_1 b_1 x_1)^{m_1}} \psi(x_1)\, \psi(x_2) \mid S_2^{\bar{\varkappa}} H \right\| \\
&\leqq c + c|h_1 b_1|^{\varkappa_1 - 1/2} \left\| \frac{(e^{ix_1} - 1)^{m_1}}{x_1^{m_1}} \psi\left(\frac{x_1}{h_1 b_1}\right) \mid \dot{H}_2^{\varkappa_1} \right\| \\
&\leqq c + c_1 |h_1 b_1|^{\varkappa_1 - 1/2} \left\| \psi\left(\frac{x_1}{h_1 b_1}\right) \mid H_2^{\varkappa_1} \right\| \\
&\quad \times \int_{R_1} (1 + y_1^2)^{\varkappa_1/2} \left| F_1 \left(\frac{(e^{i\xi_1} - 1)^{m_1}}{\xi_1^{m_1}}\right)(y_1) \right| dy_1 .
\end{aligned} \tag{17}$$

It holds

$$F_1 \left(\frac{(e^{i\xi_1} - 1}{\xi_1}\right)(x_1) = c\chi_{[0,1]}(x_1), \tag{18}$$

where $\chi_{[0,1]}(x_1)$ is the characteristic function of the interval [0, 1]. Therefore,

$$\operatorname{supp} F \left(\frac{(e^{i\xi_1} - 1)^{m_1}}{\xi_1^{m_1}}\right) = \operatorname{supp}\, [\chi_{[0,1]} * \dots * \chi_{[0,1]}]$$

is compact and the right-hand side of (17) can be estimated from above by

$$c + c_2 |h_1 b_1|^{\varkappa_1 - 1/2} \left\| \psi\left(\frac{x_1}{h_1 b_1}\right) \mid H_2^{\varkappa_1} \right\| \leqq c + c_3 \|\psi \mid H_2^{\varkappa_1}\| \leqq c_4 .$$

We apply Theorem 1.8.3 to (15). Because of (16) we obtain

$$\begin{aligned}
\|\Delta_{h_1,1}^{m_1} g(x) \mid L_{\bar{p}}\| &\leqq c \|F^{-1} \xi_1^{m_1} F(h_1^{m_1} g)(x) \mid L_{\bar{p}}\| \\
&= c \|F^{-1} (b_1^{-1} \xi_1)^{m_1} \psi(b_1^{-1} \xi_1)\, \psi(b_2^{-1} \xi_2)\, F((b_1 h_1)^{m_1} g)(x) \mid L_{\bar{p}}\| .
\end{aligned}$$

Using again Theorem 1.8.3 we get

$$\begin{aligned}
\|\Delta_{h_1,1}^{m_1} g(x) \mid L_{\bar{p}}\| &\leqq c \|x_1^{m_1} \psi(x_1)\, \psi(x_2) \mid S_2^{\bar{\varkappa}}\|\, \|(b_1 h_1)^{m_1} g(x) \mid L_{\bar{p}}\| \\
&\leqq c' |b_1 h_1|^{m_1} \|g(x) \mid L_{\bar{p}}\|
\end{aligned} \tag{19}$$

under the assumption $|b_1 h_1| \leqq 1$. On the other hand the translation invariance of $L_{\bar{p}}$ yields

$$\|\Delta_{h_1,1}^{m_1} g(x) \mid L_{\bar{p}}\| \leqq c \|g(x) \mid L_{\bar{p}}\| . \tag{20}$$

Thus we have

$$\|\Delta_{h_1,1}^{m_1} g(x) \mid L_{\bar{p}}\| \leqq c \min(1, |b_1 h_1|^{m_1}) \|g(x) \mid L_{\bar{p}}\| \tag{21}$$

and (13) is proved for $\bar{m} = (m_1, 0)$. The case $\bar{m} = (0, m_2)$ can be treated analogously. In the general case we put

$$\Delta_h^{\bar{m}} f = \Delta_{h_1,1}^{m_1} g, \qquad g = \Delta_{h_2,2}^{m_2} f \in L_{\bar{p}}^{\Omega},$$

if $f \in L_{\bar{p}}^{\Omega}$. Now, (13) for $\bar{m} = (m_1, m_2)$, $m_1 > 0$, $m_2 > 0$, follows from (21) by iteration.

Step 2. We prove (ii). At first let us show that there exists a positive constant c such that

$$\|\Delta^{m_1}_{h_1,1} g_{j,k}(x) \mid L_{p_1 \mid x_1}\| \leqq c \min(1, |a_j h_1|^{m_1}) \|g_{j,k}(x) \mid L_{p_1 \mid x_1}\| \tag{22}$$

holds for all $\{g_{j,k}(x)\} \in Sl_{\bar{q}}(L^{\Omega}_{\bar{p}})$. Here Ω has the meaning of (ii) and c is independent of x_2, j, and k. By the approximation procedure of Proposition 1.2.2/2(iii) we may additionally assume that $g_{j,k}(x) \in S(R_2)$ (cf. Step 1 of the proof of Theorem. 1.5.2). If $x_2 \in R_1$ is fixed then $g_{j,k}(\cdot, x_2)$ belongs to $S(R_1)$. Moreover,

$$\operatorname{supp} F_1 g_{j,k}(\cdot, x_2) \subset \{y_1 \mid |y_1| \leqq a_j\},\ x_2 \in R_1. \tag{23}$$

Here F_1, F_1^{-1} denote the one-dimensional Fourier transform and its inverse with respect to x_1. F_1 and F_1^{-1} are one-to-one mappings. Hence,

$$\Delta^{m_1}_{h_1,1} g_{j,k} = F^{-1}\left(\frac{(e^{i\xi_1 h_1} - 1)^{m_1}}{(\xi_1 h_1)^{m_1}} \psi(a_j^{-1}\xi_1)\, F_1(F_1^{-1}\xi_1^{m_1} F_1 h_1^{m_1} g_{j,k})\right). \tag{24}$$

We obtain (22) similarly as in Step 1 using the one-dimensional version of Theorem 1.8.3 (cf. also [T, Theorem 1.5.2]). Next we claim that

$$\begin{aligned} &\|\Delta^{m_2}_{h_2,2} g_{j,k} \mid L_{p_2 \mid x_2}(l_{q_1 \mid j}(L_{p_1 \mid x_1}))\| \\ &\leqq c \min(1, |b_k h_2|^{m_2}) \|g_{j,k} \mid L_{p_2 \mid x_2}(l_{q_1 \mid j}(L_{p_1 \mid x_1}))\| \end{aligned} \tag{25}$$

holds for all $\{g_{j,k}(x)\} \in Sl_{\bar{q}}(L^{\Omega}_{\bar{p}})$, where c is independent of $k = 0, 1, 2, \dots$ This can be obtained by repeating the arguments from Step 1 applying the multipliers of Remark 1.10.3/3 (cf. also H.-J. Schmeisser [1, II], where the proof is given without the use of maximal techniques). Then (22) and (25) yield (14). Thus, the Lemma is proved.

Theorem 2. *Let* $\bar{p} = (p_1, p_2), \bar{q} = (q_1, q_2), \bar{r} = (r_1, r_2)$ *with* $1 \leqq p_i \leqq \infty, 1 \leqq q_i \leqq \infty$, *and* $0 < r_i < \infty$; $(i = 1, 2)$. *Let* $\bar{m} = (m_1, m_2), m_i = 1, 2, \dots; (i = 1, 2)$.

(i) *If* $m_1 > r_1$ *and* $m_2 > r_2$ *then*

$$\begin{aligned} &\|f \mid S^{\bar{r}}_{\bar{p},\bar{q}}B\|^{\Delta} \\ &= \|f \mid L_{\bar{p}}\| + \||h_1|^{-r_1} \Delta^{m_1}_{h_1,1} f(x) \,|L_{\bar{p}}|\, L^*_{q_1}\| + \||h_2|^{-r_2} \Delta^{m_2}_{h_2,2} f(x) \,|L_{\bar{p}}|\, L^*_{q_2}\| \\ &\quad + \||h_1|^{-r_1} |h_2|^{-r_2} \Delta^{\bar{m}}_h f(x) \,|L_{\bar{p}}|\, L^*_{\bar{q}}\| \end{aligned} \tag{26}$$

and

$$\begin{aligned} &\|f \mid S^{\bar{r}}_{\bar{p},\bar{q}}B\|^{M} \\ &= \|f \mid L_{\bar{p}}\| + \||h_1|^{-r_1} \omega^{m_1}_{h_1,1} f(x)_{\bar{p}} \mid L^*_{q_1}\| + \||h_2|^{-r_2} \omega^{m_2}_{h_2,2} f(x)_{\bar{p}} \mid L^*_{q_2}\| \\ &\quad + \||h_1|^{-r_1} |h_2|^{-r_2} \omega^{\bar{m}}_h f(x)_{\bar{p}} \mid L^*_{\bar{q}}\| \end{aligned} \tag{27}$$

are equivalent norms in $S^{\bar{r}}_{\bar{p},\bar{q}}B$. *Further,*

$$\begin{aligned} S^{\bar{r}}_{\bar{p},\bar{q}}B &= \{f \mid f \in L_{\bar{p}}, \|f \mid S^{\bar{r}}_{\bar{p},\bar{q}}B\|^{\Delta} < \infty\} \\ &= \{f \mid f \in L_{\bar{p}}, \|f \mid S^{\bar{r}}_{\bar{p},\bar{q}}B\|^{M} < \infty\}. \end{aligned} \tag{28}$$

(ii) *If* $m_1 > r_1$ *and* $m_2 > r_2$ *then*

$$\begin{aligned} &\|f \mid SB^{\bar{r}}_{\bar{p},\bar{q}}\|^{\Delta} \\ &= \|f \mid L_{\bar{p}}\| + \||h_1|^{-r_1} \Delta^{m_1}_{h_1,1} f(x) |L_{p_1}|\, L^*_{q_1} \mid L_{p_2}\| + \||h_2|^{-r_2} \Delta^{m_2}_{h_2,2} f(x) \,|L_{\bar{p}}|\, L^*_{q_2}\| \\ &\quad + \||h_1|^{-r_1} |h_2|^{-r_2} \Delta^{\bar{m}}_h f(x) \,|L_{p_1}|\, L^*_{q_1} |L_{p_2}|\, L^*_{q_2}\| \end{aligned} \tag{29}$$

and

$$\|f \mid SB^{\bar{r}}_{\bar{p},\bar{q}}\|^M$$
$$= \|f \mid L_{\bar{p}}\| + \left\| |h_1|^{-r_1} \sup_{|t_1| \leq |h_1|} \|\Delta^{m_1}_{t_1,1} f(x) \mid L_{p_1}\| \, |L^*_{q_1}| \, L_{p_2} \right\|$$
$$+ \left\| |h_2|^{-r_2} \sup_{|t_2| \leq |h_2|} \|\Delta^{m_2}_{t_2,2} f(x) \mid L_{\bar{p}}\| \mid L^*_{q_2} \right\|$$
$$+ \left\| |h_2|^{-r_2} \sup_{|t_2| \leq |h_2|} \left\| \left\| |h_1|^{-r_1} \sup_{|t_1| \leq |h_1|} \|\Delta^{\bar{m}}_t f(x) \mid L_{p_1}\| \mid L^*_{q_1} \right\| \mid L_{p_2} \right\| \mid L^*_{q_2} \right\| \tag{30}$$

are equivalent norms in $SB^{\bar{r}}_{\bar{p},\bar{q}}$. *Further,*

$$SB^{\bar{r}}_{\bar{p},\bar{q}} = \{f \mid f \in L_{\bar{p}}, \|f \mid SB^{\bar{r}}_{\bar{p},\bar{q}}\|^\Delta < \infty\}$$
$$= \{f \mid f \in L_{\bar{p}}, \|f \mid SB^{\bar{r}}_{\bar{p},\bar{q}}\|^M < \infty\}. \tag{31}$$

Proof. Step 1. Let $f \in SB^{\bar{r}}_{\bar{p},\bar{q}}(S^{\bar{r}}_{\bar{p},\bar{q}}B)$, where $\bar{p}$, $\bar{q}$, and $\bar{r}$ satisfy the assumptions of the theorem. In order to show that

$$\|f \mid SB^{\bar{r}}_{\bar{p},\bar{q}}\|^M \leq c\|f \mid SB^{\bar{r}}_{\bar{p},\bar{q}}\|, \quad f \in SB^{\bar{r}}_{\bar{p},\bar{q}}, \tag{32}$$

and

$$\|f \mid S^{\bar{r}}_{\bar{p},\bar{q}}B\|^M \leq c\|f \mid S^{\bar{r}}_{\bar{p},\bar{q}}B\|, \quad f \in S^{\bar{r}}_{\bar{p},\bar{q}}B, \tag{33}$$

we follow again Step 1 and Step 2 of the proof of Theorem 2.3.3. We restrict ourselves to (32). The counterpart of (2.3.3/27) reads as follows,

$$J = \left(\int_{R_1} |h_2|^{-r_2 q_2} \left(\sup_{|t_2| \leq |h_2|} \left\| \left(\int_{R_1} |h_1|^{-r_1 q_1} \right. \right. \right. \right.$$
$$\left. \left. \left. \left. \times \left(\sup_{|t_1| \leq |h_1|} \|\Delta^{\bar{m}}_t f(x) \mid L_{p_1}\| \right)^{q_1} \frac{dh_1}{|h_1|} \right)^{1/q_1} \mid L_{p_2} \right\| \right)^{q_2} \frac{dh_2}{|h_2|} \right)^{1/q_2}$$
$$\leq c \left(\sum_{\nu=-\infty}^{\infty} 2^{r_2 \nu q_2} \right.$$
$$\left. \times \sup_{|h_2| \leq 2^{-\nu}} \left\| \left(\sum_{\mu=-\infty}^{\infty} 2^{r_1 \mu q_1} \sup_{|h_1| \leq 2^{-\mu}} \|\Delta^{\bar{m}}_h f(x) \mid L_{p_1}\|^{q_1} \right)^{1/q_1} \mid L_{p_2} \right\|^{q_2} \right)^{1/q_2}. \tag{34}$$

Let $\varphi = \{\varphi_j\}_{j=0}^{\infty} \in \Phi(R_1)$. We put $\varphi_j = 0$ if $j = -1, -2, \ldots$, and $\varphi_{j,k}(x) = \varphi_j(x_1) \times \varphi_k(x_2)$ for all integers j, k. Then,

$$f(x) = \sum_{k=-\infty}^{\infty} \sum_{j=-\infty}^{\infty} (F^{-1}\varphi_{j,k} Ff)(x) = \sum_{k=-\infty}^{\infty} \sum_{j=-\infty}^{\infty} f_{j,k}(x).$$

According to (22) we can estimate

$$\sup_{|h_1| \leq 2^{-\mu}} \|\Delta^{\bar{m}}_h f(x) \mid L_{p_1}\|$$
$$\leq \sum_{k=-\infty}^{\infty} \sum_{j=-\infty}^{\infty} \|\Delta^{\bar{m}}_h f_{j,k}(x) \mid L_{p_1}\|$$
$$\leq c \sum_{k=-\infty}^{\infty} \sum_{j=-\infty}^{\infty} \min(1, 2^{(j-\mu)m_1}) \|\Delta^{m_2}_{h_2,2} f_{j,k}(x) \mid L_{p_1}\|$$
$$= c \sum_{k=-\infty}^{\infty} \sum_{j=-\infty}^{\infty} \min(1, 2^{jm_1}) \|\Delta^{m_2}_{h_2,2} f_{j+\mu,k}(x) \mid L_{p_1}\|. \tag{35}$$

Hence,

$$\left(\sum_{\mu=-\infty}^{\infty} 2^{r_1\mu q_1} \sup_{|h_1|\leqq 2^{-\mu}} \|\Delta_h^{\bar{m}} f(x) \mid L_{p_1}\|^{q_1}\right)^{1/q_1}$$

$$\leqq c\left\|\sum_{k=-\infty}^{\infty}\sum_{j=-\infty}^{\infty} 2^{-r_1 j}\min(1, 2^{jm_1})\,[2^{r_1(j+\mu)}\,\|\Delta_{h_2,2}^{m_2} f_{j+\mu,k} \mid L_{p_1}\|] \mid l_{q_1|\mu}^{*}\right\|$$

$$\leqq c\sum_{k=-\infty}^{\infty}\left(\sum_{j=-\infty}^{\infty} 2^{-r_1 j}\min(1, 2^{jm_1})\right)\|2^{r_1\mu}\,\Delta_{h_2,2}^{m_2} f_{\mu,k}(x)\,|L_{p_1}|\, l_{q_1|\mu}\|$$

$$\leqq c'\sum_{k=-\infty}^{\infty}\|2^{r_1\mu}\,\Delta_{h_2,2}^{m_2} f_{\mu,k}(x)\,|L_{p_1}|\, l_{q_1|\mu}\| \tag{36}$$

because $m_1 > r_1 > 0$. Here $\|\cdot \mid l_{q_1}^{*}\|$ means $\left(\sum_{-\infty}^{\infty} |\ldots|^{q_1}\right)^{1/q_1}$. Consequently,

$$J \leqq c\left(\sum_{\nu=-\infty}^{\infty} 2^{r_2\nu q_2} \sup_{|h_2|\leqq 2^{-\nu}}\left\|\sum_{k=-\infty}^{\infty}\|2^{r_1\mu}\,\Delta_{h_2,2}^{m_2} f_{\mu,k}(x)\,|L_{p_1}|\, l_{q_1|\mu}\| \mid L_{p_2}\right\|^{q_2}\right)^{1/q_2}. \tag{37}$$

Now, we repeat the steps in (35) and (36) using (25) instead of (22). Because $m_2 > r_2$ (37) yields

$$J \leqq c\,\|2^{r_1\mu+r_2\nu} f_{\mu,\nu}(x) \mid Sl_{\bar{q}}(L_{\bar{p}})\| = c\|f \mid SB_{\bar{p},\bar{q}}^{\bar{r}}\| \tag{38}$$

for all $f\in SB_{\bar{p},\bar{q}}^{\bar{r}}$. Analogously we proceed in the case $S_{\bar{p},\bar{q}}^{\bar{r}}B$ applying part (i) of the above Lemma instead of (22) and (25). The remaining terms on the right-hand sides of (27)and (30) can be estimated similarly as in Step 2 of the proof of Theorem 2.3.3, using (22), (25), and part (i) of the Lemma as well as Proposition 2.2.3/2.

Step 2. Obviously, by Theorem 2.3.2/(i, ii) the proof of the theorem is complete if we are able to show that

$$\|f \mid SB_{\bar{p},\bar{q}}^{\bar{r}}\|^{N} \leqq c\|f \mid SB_{\bar{p},\bar{q}}^{\bar{r}}\|^{\Delta} \tag{39}$$

holds if $f\in L_{\bar{p}}$ and $\|f \mid SB_{\bar{p},\bar{q}}^{\bar{r}}\|^{\Delta} < \infty$, and

$$\|f \mid S_{\bar{p},\bar{q}}^{\bar{r}}B\|^{N} \leqq c\|f \mid S_{\bar{p},\bar{q}}^{\bar{r}}B\|^{\Delta} \tag{40}$$

holds if $f\in L_{\bar{p}}$ and $\|f \mid S_{\bar{p},\bar{q}}^{\bar{r}}B\|^{\Delta} < \infty$. Again we follow the method of the proof of Theorem 2.3.3. Let $f\in L_{\bar{p}}$ such that $\|f \mid SB_{\bar{p},\bar{q}}^{\bar{r}}\|^{\Delta} < \infty$. Let $a = \{a_{j,k}\}_{j,k=0}^{\infty}$ be the system of entire analytic functions of exponential type constructed in (2.3.3/18). Then $a\in\mathfrak{A}_{\bar{p}}$ (beside minor technical changes) and $f = \sum_j\sum_k a_{j,k}$ in S' (even in $L_{\bar{p}}$). We prove

$$\|f \mid SB_{\bar{p},\bar{q}}^{\bar{r}}\|^{a,N} \leqq c\|f \mid SB_{\bar{p},\bar{q}}^{\bar{r}}\|^{\Delta}. \tag{41}$$

The counterpart of (2.3.3/38) reads as follows,

$$\|2^{r_1 j+r_2 k} a_{j,k}(x) \mid Sl_{\bar{q}}(L_{\bar{p}})\|$$

$$= \left(\sum_{k=0}^{\infty}\left\|\left(\sum_{j=0}^{\infty} 2^{r_1 jq_1+r_2 kq_1}\|a_{j,k}(x) \mid L_{p_1|x_1}\|^{q_1}\right)^{1/q_1} \mid L_{p_2|x_2}\right\|^{q_2}\right)^{1/q_2}$$

$$\leq c\|f \mid L_{\bar{p}}\|$$

$$+ c\left\|\left(\sum_{j=0}^{\infty} 2^{r_1 j q_1}\left\|\int_{R_1} g(h_1)\,|\Delta^{m_1}_{h_1 2^{-j},1} f(x)|\,\mathrm{d}h_1 \mid L_{p_1|x_1}\right\|^{q_1}\right)^{1/q_1} \mid L_{p_2|x_2}\right\|$$

$$+ c\left(\sum_{k=0}^{\infty} 2^{r_2 k q_2}\left\|\int_{R_1} g(h_2)\,|\Delta^{m_2}_{h_2 2^{-k},2} f(x)|\,\mathrm{d}h_2 \mid L_{\bar{p}|x}\right\|^{q_2}\right)^{1/q_2}$$

$$+ c\left(\sum_{k=0}^{\infty} 2^{r_2 k q_2}\left\|\left(\sum_{j=0}^{\infty} 2^{r_1 j q_1}\right.\right.\right.$$

$$\left.\left.\left.\times\left\|\int_{R_2} g(h_1)\,g(h_2)\,|\Delta^{\bar{m}}_{(h_1 2^{-j}, h_2 2^{-k})} f(x)|\,\mathrm{d}h \mid L_{p_1|x_1}\right\|^{q_1}\right)^{1/q_1} \mid L_{p_2|x_2}\right\|^{q_2}\right)^{1/q_2}$$

$$= c(\|f \mid L_{\bar{p}}\| + A_1 + A_2 + A_3). \tag{42}$$

If $B_{j,k}(x)$ and $b_{j,k}(x)$ ($j, k = 0, \pm 1, \pm 2, \ldots$) have the meaning of 2.3.3/(40), (41) then we can reformulate

$$A_3 = \left(\sum_{k=0}^{\infty} 2^{r_2 k q_2}\left\|\left(\sum_{j=0}^{\infty} 2^{r_1 j q_1}\|B_{j,k}(x) \mid L_{p_1|x_1}\|^{q_1}\right)^{1/q_1} \mid L_{p_2|x_2}\right\|^{q_2}\right)^{1/q_2} \tag{43}$$

and

$$B_{j,k}(x) \leq c\left(\sum_{\nu=-\infty}^{\infty}\sum_{\mu=-\infty}^{\infty} \min(1, 2^{-\mu d}) \min(1, 2^{-\nu d})\, b_{\mu-j,\nu-k}(x)\right), \tag{44}$$

where $d > 0$ is at our disposal. With the help of (44) the right-hand side of (43) can be estimated similarly as in (2.3.3/42). We obtain

$$A_3 \leq c\left(\sum_{k=-\infty}^{\infty} 2^{r_2 k q_2}\left\|\left(\sum_{j=-\infty}^{\infty} 2^{r_1 j q_1}\|b_{-j,-k}(x) \mid L_{p_1|x_1}\|^{q_1}\right)^{1/q_1} \mid L_{p_2|x_2}\right\|^{q_2}\right)^{1/q_2}. \tag{45}$$

Using the generalized triangle-inequality and Hölder's inequality, (45) yields

$$A_3 \leq c\left(\sum_{k=-\infty}^{\infty} 2^{r_2 k q_2}\right.$$

$$\left.\times \int_{I_0}\left\|\left(\sum_{j=-\infty}^{\infty} 2^{r_1 j q_1}\int_{I_0}\|\Delta^{\bar{m}}_{(h_1 2^{-j}, h_2 2^{-k})} f(x) \mid L_{p_1|x_1}\|^{q_1}\,\mathrm{d}h_1\right)^{1/q_1} \mid L_{p_2|x_2}\right\|^{q_2}\mathrm{d}h_2\right)^{1/q_2}$$

$$\leq c'\left(\int_{R_1} |h_2|^{-r_2 q_2}\left\|\left(\int_{R_1} |h_1|^{-r_1 q_1}\|\Delta^{\bar{m}}_h f(x) \mid L_{p_1|x_1}\|^{q_1}\frac{\mathrm{d}h_1}{|h_1|}\right)^{1/q_1} \mid L_{p_2|x_2}\right\|^{q_2}\frac{\mathrm{d}h_2}{|h_2|}\right)^{1/q_2}$$

$$= c'\||h_1|^{-r_1}\,|h_2|^{-r_2}\,\Delta^{\bar{m}}_h f(x)\,|L_{p_1|x_1}|\,L^*_{q_1|h_1}|L_{p_2|x_2}|\,L^*_{q_2|h_2}\|. \tag{46}$$

Here $I_0 = [-2, -1] \cup [1, 2]$. By similar arguments we have

$$A_1 \leq c\||h_1|^{-r_1}\,\Delta^{m_1}_{h_1,1} f(x)\,|L_{p_1|x_1}|\,L^*_{q_1|h_1} \mid L_{p_2|x_2}\| \tag{47}$$

and

$$A_2 \leq c\||h_2|^{-r_2}\,\Delta^{m_2}_{h_2,2} f(x)\,|L_{\bar{p}|x}|\,L^*_{q_2|h_2}\|. \tag{48}$$

Thus, (42), (46), (47), and (48) yield the desired estimates (41) and (39). Similarly we prove (40). The proof of the theorem is complete.

Remark 2. It would be desirable to extend the assertions of Theorem 2 to $\bar{p} = (p_1, p_2)$, $\bar{q} = (q_1, q_2)$ with $0 < p_i \leqq \infty$ and $0 < q_i \leqq \infty$; $(i = 1, 2)$. A natural condition with respect to $\bar{r} = (r_1, r_2)$ is

$$r_1 > \frac{1}{\min(1, p_1)} - 1 \quad \text{and} \quad r_2 > \frac{1}{\min(1, p_1, p_2)} - 1$$

in the case of $S^{\bar{r}}_{\bar{p},\bar{q}}B$ and

$$r_1 > \frac{1}{\min(1, p_1)} - 1 \quad \text{and} \quad r_2 > \frac{1}{\min(1, p_1, p_2, q_1)} - 1$$

in the case of $SB^{\bar{r}}_{\bar{p},\bar{q}}$ (cf. Theorem 2.3.2/(i, ii)). One direction can be proved by the methods of Step 1 of the proofs of Theorem 2.3.3 and the preceding Theorem 2 applying the last Lemma which holds true in these cases. The converse direction remains open.

Remark 3. Of peculiar interest is the space $S^{\bar{r}}_{\overline{\infty},\overline{\infty}}B = SB^{\bar{r}}_{\overline{\infty},\overline{\infty}}$, where $\overline{\infty} = (\infty, \infty)$ and $\bar{r} = (r_1, r_2)$ with $0 < r_1 < \infty$ and $0 < r_2 < \infty$. Let $C(R_2)$ be the space of all complex-valued uniformly continuous bounded functions on R_2. Equipped with the norm

$$\|f \mid C\| = \sup_{x \in R_2} |f(x)| \tag{49}$$

$C(R_2)$ is a Banach space. Let $\bar{m} = (m_1, m_2)$, $m_i = 1, 2, \ldots$ $(i = 1, 2)$ with $m_1 > r_1 > 0$ and $m_2 > r_2 > 0$. We define

$$S^{\bar{r}}\mathscr{C} = \Big\{ f \mid f \in C(R_2),\ \|f \mid S^{\bar{r}}\mathscr{C}\| = \|f \mid C\| + \sup_{h_1 \in R_1} \sup_{x \in R_2} |h_1|^{-r_1} |\Delta^{m_1}_{h_1,1} f(x)| + \sup_{h_2 \in R_1} \sup_{x \in R_2} |h_2|^{-r_2} |\Delta^{m_2}_{h_2,2} f(x)| + \sup_{h \in R_2} \sup_{x \in R_2} |h_1|^{-r_1} |h_2|^{-r_2} |\Delta^{\bar{m}}_h f(x)| < \infty \Big\}. \tag{50}$$

Then part (i) of Theorem 2 shows

$$S^{\bar{r}}_{\overline{\infty},\overline{\infty}}B = S^{\bar{r}}\mathscr{C}. \tag{51}$$

In particular, $S^{\bar{r}}\mathscr{C}$ does not depend on the chosen pair $\bar{m} = (m_1, m_2)$. $S^{\bar{r}}\mathscr{C}$ is a space of Hölder-Zygmund type with dominating mixed smoothness properties.

Remark 4. We discuss another interesting special case. Let A be a Banach space and let $L_p(R_1, A)$, $1 \leqq p \leqq \infty$ be the L_p-space of A-valued functions with the usual norm

$$\|f \mid L_p(R_1, A)\| = \Big(\int_{R_1} \|f(t) \mid A\|^p \,dt\Big)^{1/p}, \qquad 1 \leqq p < \infty$$

$$\|f \mid L_\infty(R_1, A)\| = \operatorname*{ess\text{-}sup}_{t \in R_1} \|f(t) \mid A\|.$$

Let $1 \leqq p \leqq \infty$ and $1 \leqq q \leqq \infty$. Let $m = 1, 2, \ldots$ and $0 < r < \infty$ with $m > r$. We define the Besov (Lipschitz) space of A-valued functions

$$B^r_{p,q}(R_1, A) = \Big\{ f \mid f \in L_p(R_1, A),\ \|f \mid B^r_{p,q}(R_1, A)\| = \|f \mid L_p(R_1, A)\| + \Big(\int_{R_1} |h|^{-rq} \|\Delta^m_h f(t) \mid L_p(R_1, A)\|^q \frac{dh}{|h|} \Big)^{1/q} < \infty \Big\} \tag{52}$$

(with the usual modification if $q = \infty$). Of course,

$$\Delta^m_h f(t) = \Delta^1_h(\Delta^{m-1}_h f)(t), \qquad \Delta^1_h f(t) = f(t + h) - f(t).$$

By the same procedure as in the scalar case one shows that the definition of $B^r_{p,q}(R_1, A)$ is independent of the choice of $m > r$ (cf. K. K. Golovkin [1, 3]). Further $B^r_{p,q}(R_1, A)$ is a Banach space. If $A = C$ is the space of all complex numbers then we obtain the usual Besov or Lipschitz spaces $B^r_{p,q}(R_1)$. Now, let $\bar{p} = (p_1, p_2)$, $\bar{q} = (q_1, q_2)$, $\bar{r} = (r_1, r_2)$ with $1 \leqq p_i < \infty$, $1 \leqq q_i < \infty$, and $0 < r_i < \infty$; $(i = 1, 2)$. Then we have by part (ii) of Theorem 2 and (52)

$$B^{r_2}_{p_2,q_2}(R_1, B^{r_1}_{p_1,q_1}(R_1)) = SB^{\bar{r}}_{\bar{p},\bar{q}}. \tag{53}$$

We used

$$L_{p_2}(R_1, B^{r_1}_{p_1,q_1}(R_1)) \subset L_{p_2}(R_1, L_{p_1}(R_1)) = L_{\bar{p}}.$$

Hence, by iteration of (52) we obtain

$$\|f \mid SB^{\bar{r}}_{\bar{p},\bar{q}}\|^{\Delta} \sim \|f \mid B^{r_2}_{p_2,q_2}(R_1, B^{r_1}_{p_1,q_1}(R_1))\|.$$

(53) shows that in the Banach space case $SB^{\bar{r}}_{\bar{p},\bar{q}}$ can be introduced by means of iteration of vector-valued Besov spaces. This procedure is often used in the literature. We refer to P. Grisvard [1], G. Sparr [1], and A. Pietsch [2–4].

Remark 5. As in the case $S^{\bar{r}}_{\bar{p},\bar{q}}F$ (cf. Remark 2.3.3/2) in (26), (27), (29), and (30) some differences can be replaced by derivatives. We omit the details.

2.4. Embedding Theorems

2.4.1. Embedding Theorems for Different Metrics

The proof of direct and inverse embedding theorems for spaces of functions is one of the standard problems of the theory of function spaces. Two types are of interest: embeddings for different metrics and traces on hyperplanes. For classical spaces with dominating mixed smoothness we refer to S. M. Nikol'skij [5], T. I. Amanov [1–3], A. P. Uninskij [1, 3], J. S. Bugrov [3] and M. K. Potapov [3–5]. The main tools are representations as in Definition 2.2.1/2 or in Subsection 2.3.2. and inequalities of Nikol'skij type for different metrics (cf. Theorem 1.6.2) and different dimensions. In the following we shall be concerned with embeddings for different metrics. The next subsection is devoted to the trace problem.

The spaces $S^{\bar{r}}_{\bar{p},\bar{q}}B$ can be treated with the help of Theorem 1.6.2. In order to investigate $SB^{\bar{r}}_{\bar{p},\bar{q}}$ and $S^{\bar{r}}_{\bar{p},\bar{q}}F$ we need the counterparts for $Sl_{\bar{q}}(L^{\Omega}_{\bar{p}})$ and $L^{\Omega}_{\bar{p}}(l_{\bar{q}})$. Recall that $Sl_{\bar{q}}(L^{\Omega}_{\bar{p}})$ and $L^{\Omega}_{\bar{p}}(l_{\bar{q}})$ have been defined in (1.10.1/1) and (1.10.1/2), respectively.

Proposition. *Let $\Omega = \{\Omega_{j,k}\}^{\infty}_{j,k=0}$ be the sequence of compact subsets of R_2,*

$$\Omega_{j,k} = \{y \mid y = (y_1, y_2), |y_1| \leqq a_j, |y_2| \leqq b_k\}; \quad j, k = 0, 1, 2, \ldots, \tag{1}$$

where $\{a_j\}^{\infty}_{j=0}$ and $\{b_j\}^{\infty}_{j=0}$ are sequences of positive numbers.

(i) *Let $\bar{p} = (p_1, p_2)$, $\bar{q} = (q_1, q_2)$, $\bar{u} = (u_1, u_2)$ with $0 < p_i \leqq u_i \leqq \infty$ and $0 < q_i \leqq \infty$; $(i = 1, 2)$. Then there exists a positive constant c such that*

$$\left\| a_j^{\frac{1}{u_1}-\frac{1}{p_1}} b_k^{\frac{1}{u_2}-\frac{1}{p_2}} f_{j,k}(x) \mid Sl_{\bar{q}}(L_{\bar{u}}) \right\| \leqq c \|f_{j,k}(x) \mid Sl_{\bar{q}}(L_{\bar{p}})\| \tag{2}$$

holds for all $\{f_{j,k}(x)\} \in Sl_{\bar{q}}(L^{\Omega}_{\bar{p}})$.

(ii) *Let $\bar{p} = (p_1, p_2)$, $\bar{u} = (u_1, u_2)$, $\bar{q} = (q_1, q_2)$, and $\bar{v} = (v_1, v_2)$ with $0 < p_i < u_i < \infty$, $0 < q_i \leqq \infty$, $0 < v_i \leqq \infty$; $(i = 1, 2)$. Let additionally $1 < a \leqq \dfrac{a_{j+1}}{a_j} \leqq A < \infty$; $j = 0, 1, 2, \ldots$, and $1 < b \leqq \dfrac{b_{k+1}}{b_k} \leqq B < \infty$; $k = 0, 1, 2, \ldots$ Then there exists a* *positive constant c such that*

$$\left\| a_j^{\frac{1}{u_1}-\frac{1}{p_1}} b_k^{\frac{1}{u_2}-\frac{1}{p_2}} f_{j,k}(x) \mid L_{\bar{u}}(l_{\bar{v}}) \right\| \leqq c \| f_{j,k}(x) \mid L_{\bar{p}}(l_{\bar{q}}) \| \tag{3}$$

holds for all $\{f_{j,k}(x)\} \in L_{\bar{p}}^{\Omega}(l_{\bar{q}})$.

Proof. Step 1. We prove part (i). Obviously it holds

$$\left\| a_j^{\frac{1}{u_1}-\frac{1}{p_1}} b_k^{\frac{1}{u_2}-\frac{1}{p_2}} f_{j,k}(x) \mid Sl_{\bar{q}}(L_{\bar{u}}) \right\| = \left\| a_j^{-\frac{1}{p_1}} b_k^{-\frac{1}{p_2}} f_{j,k}\left(\frac{x_1}{a_j}, \frac{x_2}{b_k}\right) \mid Sl_{\bar{q}}(L_{\bar{p}}) \right\|. \tag{4}$$

Let $\{f_{j,k}(x)\} \in Sl_{\bar{q}}(L_{\bar{p}}^{\Omega})$. By the properties of the Fourier transform we have

$$\operatorname{supp}\left[Ff_{j,k}\left(\frac{\cdot}{a_j}, \frac{\cdot}{b_k}\right)\right](y) \subset [-1, 1] \times [-1, 1].$$

We choose a function $\psi(x_1, x_2) \in S(R_2)$ such that supp $F\psi$ is compact and $(F\psi)(y_1, y_2) = 1$ if $y \in [-1, 1] \times [-1, 1]$. As in Step 1 of the proof of Theorem 1.6.2 we have

$$f_{j,k}\left(\frac{x_1}{a_j}, \frac{x_2}{b_k}\right) = c \int_{R_2} f_{j,k}\left(\frac{y_1}{a_j}, \frac{y_2}{b_k}\right) \psi(x-y)\, \mathrm{d}y. \tag{5}$$

If $x \in R_2$ is fixed then

$$\operatorname{supp} F\left[f_{j,k}\left(\frac{\cdot}{a_j}, \frac{\cdot}{b_k}\right) \psi(x-\cdot)\right] \subset [-1, 1] \times [-1, 1] - \operatorname{supp} F\psi$$

is compact and the support does not depend on $x \in R_2$, $j = 0, 1, 2, \ldots$, and $k = 0, 1, 2, \ldots$ Consequently, we obtain from (5) and Theorem 1.6.2 (or [T, p. 22])

$$\left| f_{j,k}\left(\frac{x_1}{a_j}, \frac{x_2}{b_k}\right) \right| \leqq c \left(\int_{R_2} \left| f_{j,k}\left(\frac{y_1}{a_j}, \frac{y_2}{b_k}\right) \psi(x-y) \right|^r \mathrm{d}y \right)^{1/r}, \tag{6}$$

where $0 < r \leqq 1$. Now we proceed as in Step 2 and Step 3 of the proof of Theorem 1.6.2. We choose $\psi(x) = \varphi(x_1)\varphi(x_2)$ and $r = \min(1, p_1, p_2, q_1)$. Then (6) and Hölder's inequality yield in the same way as in (1.6.2/6)

$$\left\| a_j^{-\frac{1}{p_1}} f_{j,k}\left(\frac{x_1}{a_j}, \frac{x_2}{b_k}\right) |L_{p_1|x_1}| \, l_{q_1|j} \right\|$$

$$\leqq c \left\| a_j^{-\frac{1}{p_1}} f_{j,k}\left(\frac{x_1}{a_j}, \frac{x_2}{b_k}\right) |L_{p_1|x_1}| \, l_{q_1|j} \mid L_{p_2|x_2} \right\|. \tag{7}$$

Hence,

$$\left\| a_j^{-\frac{1}{p_1}} b_k^{-\frac{1}{p_2}} f_{j,k}\left(\frac{x_1}{a_j}, \frac{x_2}{b_k}\right) |Sl_{\bar{q}}(L_{\bar{u}}) \right\| \leqq c \| f_{j,k}(x) \mid Sl_{\bar{q}}(L_{\bar{p}}) \| \tag{8}$$

if $\bar{u} = (p_1, \infty)$. Similar as in (1.6.2/8) we obtain (8) for $\bar{u} = (p_1, u_2)$ with $p_2 \leqq u_2 \leqq \infty$. Now let $\bar{u} = (u_1, u_2)$ as in part (i). Following the steps in 1.6.2/(10)–(13) we obtain

$$\left\| a_j^{-\frac{1}{p_1}} b_k^{-\frac{1}{p_2}} f_{j,k}\left(\frac{x_1}{a_j}, \frac{x_2}{b_k}\right) \mid Sl_{\bar{q}}(L_{\bar{u}}) \right\|$$

$$\leqq c \left\| a_j^{-\frac{1}{p_1}} b_k^{-\frac{1}{p_2}} f_{j,k}\left(\frac{x_1}{a_j}, \frac{x_2}{b_k}\right) |L_{p_1}| \, l_{q_1} |L_{u_2}| \, l_{q_2} \right\|. \tag{9}$$

Then (4), (8), and (9) yield the inequality (2).

Step 2. We prove part (ii). Let $\bar{p}$, $\bar{u}$, $\bar{q}$, and $\bar{v}$ be as supposed in part (ii). At first we claim that it is sufficient to prove

$$\left\| a_j^{\frac{1}{u_1}-\frac{1}{p_1}} b_k^{\frac{1}{u_2}-\frac{1}{p_2}} f_{j,k}(x) \,|l_{v_1\,|\,j}|\, L_{u_1\,|\,x_1}|\, l_{v_2\,|\,k}|\, L_{u_2\,|\,x_2} \right\|$$
$$\leqq c\|f_{j,k}(x)\,|l_{\infty\,|\,j}|\, L_{p_1\,|\,x_1}|l_{\infty\,|\,k}|\, L_{p_2\,|\,x_2}\| \tag{10}$$

for all $\{f_{j,k}(x)\} \in L_{\bar{p}}^{\Omega}(l_{\bar{q}})$, This is a consequence of the elementary estimates

$$\|\ldots \mid L_{\bar{u}}(l_{\bar{v}})\| \leqq \|\ldots |l_{v_1}|\, L_{u_1}|l_{\min(v_2,u_1)}|\, L_{u_2}\|$$

and

$$\|\ldots |l_\infty|\, L_{p_1}|l_\infty|\, L_{p_2}\| \leqq \|\ldots \mid L_{\bar{p}}(l_{\bar{q}})\|\,.$$

In order to prove (10) we proceed as in Step 1 by iteration. We prove

$$\left\| b_k^{\frac{1}{u_2}-\frac{1}{p_2}} f_{j,k}(x)\,|L_{p_1}(l_\infty)|\, L_{u_2}(l_{v_2})\right\| \leqq c\|f_{j,k}(x)\,|L_{p_1}(l_\infty)|\, L_{p_2}(l_\infty)\|\,. \tag{11}$$

Without loss of generality we may assume

$$\|f_{j,k}(x)\,|L_{p_1}(l_\infty)|\, L_{p_2}(l_\infty)\| = 1\,.$$

We choose $r = \min(1, p_1)$ and use again (4) and (6). Similar as in (7) and (8) it follows

$$b_k^{-\frac{1}{p_2}} \|f_{j,k}(x)\,|L_{p_1\,|\,x_1}(l_{\infty\,|\,j})|\, L_{\infty\,|\,x_2}\|$$
$$\leqq c\|f_{j,k}(x)\,|L_{p_1\,|\,x_1}(l_{\infty\,|\,j})|\, L_{p_2\,|\,x_2}\|$$
$$\leqq c\|f_{j,\nu}(x)\,|L_{p_1\,|\,x_1}(l_{\infty\,|\,j})|\, L_{p_2\,|\,x_2}(l_{\infty\,|\,\nu})\| = c\,. \tag{12}$$

We split

$$\left\| b_k^{\frac{1}{u_2}-\frac{1}{p_2}} f_{j,k}(x)\,|L_{p_1}(l_\infty)|\, l_{v_2\,|\,k}\right\|^{v_2}$$
$$= \sum_{k=0}^{L} b_k^{\left(\frac{1}{u_2}-\frac{1}{p_2}\right)v_2} \|f_{j,k}(x) \mid L_{p_1}(l_\infty)\|^{v_2} + \sum_{k=L+1}^{\infty} b_k^{\left(\frac{1}{u_2}-\frac{1}{p_2}\right)v_2} \|f_{j,k}(x) \mid L_{p_1}(l_\infty)\|^{v_2}$$

with $L = 0, 1, 2, \ldots$ (12) yields

$$\sum_{k=0}^{L} b_k^{\left(\frac{1}{u_2}-\frac{1}{p_2}\right)v_2} \|f_{j,k}(x) \mid L_{p_1}(l_\infty)\|^{v_2} \leqq c\sum_{k=0}^{L} b_k^{\frac{v_2}{u_2}} \leqq c_1 b_L^{\frac{v_2}{u_2}}\,. \tag{13}$$

We used $\dfrac{b_k}{b_{k+1}} \leqq \dfrac{1}{b} < 1$; $k = 0, 1, 2, \ldots$ On the other hand because of $\dfrac{1}{u_2} - \dfrac{1}{p_2} < 0$ we have

$$\sum_{k=L+1}^{\infty} b_k^{\left(\frac{1}{u_2}-\frac{1}{p_2}\right)v_2} \|f_{j,k}(x) \mid L_{p_1}(l_\infty)\|^{v_2}$$
$$= b_L^{\left(\frac{1}{u_2}-\frac{1}{p_2}\right)v_2} \sum_{k=L+1}^{\infty} \left(\frac{b_k}{b_L}\right)^{\left(\frac{1}{u_2}-\frac{1}{p_2}\right)v_2} \|f_{j,k}(x) \mid L_{p_1}(l_\infty)\|^{v_2}$$
$$\leqq c_2 b_L^{\left(\frac{1}{u_2}-\frac{1}{p_2}\right)v_2} \|f_{j,k}(x)\,|L_{p_1}(l_\infty)|\, l_{\infty\,|\,k}\|^{v_2}. \tag{14}$$

Note that the constants c_1 and c_2 are independent of L and x_2. Next we recall the well-known formula

$$\|g(x_2) \mid L_{u_2}\|^{u_2} = u_2 \int_0^\infty t^{u_2-1} \, |\{x_2 \mid |g(x_2)| > t\}| \, dt. \tag{15}$$

Here $|\{\ldots\}|$ stands for the Lebesgue measure of $\{\ldots\}$. Using this representation one obtains

$$\left\| b_k^{\frac{1}{u_2}-\frac{1}{p_2}} f_{j,k}(x) \mid L_{p_1}(l_\infty) \mid L_{u_2}(l_{v_2}) \right\|^{u_2}$$

$$= u_2 \int_0^\infty t^{u_2-1} \left| \left\{ x_2 \mid \left\| b_k^{\frac{1}{u_2}-\frac{1}{p_2}} f_{j,k}(x) \, |L_{p_1}(l_\infty)| \, l_{v_2} \right\| > t \right\} \right| dt$$

$$= u_2 \int_0^{(2c_1)^{1/v_2} b_0} (\ldots) \, dt + u_2 \int_{(2c_1)^{1/v_2} b_0}^\infty (\ldots) \, dt$$

$$= u_2 J_0 + u_2 J_\infty, \tag{16}$$

where c_1 is the constant on the right-hand side of (13). In order to estimate J_0 we use

$$\left\| b_k^{\frac{1}{u_2}-\frac{1}{p_2}} f_{j,k}(x) \, |L_{p_1}(l_\infty)| \, l_{v_2 \mid k} \right\| \leqq c_3 \| f_{j,k}(x) \, |L_{p_1}(l_\infty)| \, l_{\infty \mid k} \| .$$

This follows easily by $\dfrac{1}{u_2} - \dfrac{1}{p_2} < 0$ and the properties of $\{b_k\}_{k=0}^\infty$. Furthermore, there exists a constant c_4, such that $t^{u_2} < c_4 t^{p_2}$ if $t \in (0, (2c_1)^{1/v_2} b_0)$. Consequently, by (15) we have

$$J_0 \leqq c_4 \int_0^\infty t^{p_2-1} \left| \left\{ x_2 \mid \| f_{j,k}(x) \mid L_{p_1}(l_\infty) \mid l_\infty \| > \frac{t}{c_3} \right\} \right| dt = c_5 . \tag{17}$$

Next we split

$$J_\infty = \sum_{L=0}^\infty \int_{I_L} t^{u_2-1} \left| \left\{ x_2 \mid \sum_{k=0}^\infty b_k^{\left(\frac{1}{u_2}-\frac{1}{p_2}\right) v_2} \| f_{j,k}(x) \mid L_{p_1}(l_\infty) \|^{v_2} > t^{v_2} \right\} \right| dt, \tag{18}$$

where

$$I_L = \left((2c_1)^{\frac{1}{v_2}} b_L^{\frac{1}{u_2}}, (2c_1)^{\frac{1}{v_2}} b_{L+1}^{\frac{1}{u_2}} \right); \quad L = 0, 1, 2, \ldots$$

(13) and (14) imply for $t \in I_L$; $L = 0, 1, 2, \ldots$,

$$\left| \left\{ x_2 \mid \sum_{k=0}^\infty b_k^{\left(\frac{1}{u_2}-\frac{1}{p_2}\right) v_2} \| f_{j,k}(x) \mid L_{p_1}(l_\infty) \|^{v_2} > t^{v_2} \right\} \right|$$

$$\leqq \left| \left\{ x_2 \mid \sum_{k=L+1}^\infty b_k^{\left(\frac{1}{u_2}-\frac{1}{p_2}\right) v_2} \| f_{j,k}(x) \mid L_{p_1}(l_\infty) \|^{v_2} > c_1 b_L^{\frac{v_2}{u_2}} \right\} \right|$$

$$\leqq \left| \left\{ x_2 \mid \| f_{j,k}(x) \, |L_{p_1}(l_\infty)| \, l_\infty \|^{v_2} > \frac{c_1}{c_2} b_L^{\frac{v_2}{p_2}} \right\} \right|$$

$$\leqq \left| \left\{ x_2 \mid \| f_{j,k}(x) \, |L_{p_1}(l_\infty)| \, l_\infty \| > c_6 t^{\frac{u_2}{p_2}} \right\} \right|. \tag{19}$$

In the last estimate in (19) we used $\dfrac{b_{k+1}}{b_k} \leqq B$; $k = 0, 1, 2, \ldots$ If $t \in I_L$ this yields

$$\frac{c_1}{c_2} b_L^{\frac{v_2}{p_2}} \geqq \frac{c_1}{c_2} B^{-\frac{v_2}{p_2}} b_{L+1}^{\frac{v_2}{p_2}} \geqq \left(c_6 t^{\frac{u_2}{p_2}}\right)^{v_2}.$$

We substitute (19) in the summand of the right-hand side of (18). With the help of the transform $c_6 t^{u_2/p_2} = \tau$ and (15) it follows

$$J_\infty \leqq c_7 \int_0^\infty \tau^{p_2-1} |\{x_2 \mid \|f_{j,k}(x) \,|L_{p_1}(l_\infty)|\, l_\infty\| > \tau\}| \,d\tau = c_8. \tag{20}$$

We put together (16), (17), and (20). The proof of (11) is complete. It remains to show

$$\left\| a_j^{\frac{1}{u_1}-\frac{1}{p_1}} b_k^{\frac{1}{u_2}-\frac{1}{p_2}} f_{j,k}(x) \,|L_{u_1}(l_{v_1})|\, L_{u_2}(l_{v_2})\right\|$$
$$\leqq c \left\| b_k^{\frac{1}{u_2}-\frac{1}{p_2}} f_{j,k}(x) \,|L_{p_1}(l_\infty)|\, L_{u_2}(l_{v_2})\right\|. \tag{21}$$

Let for brevity

$$g_{j,k}(x) = b_k^{\frac{1}{u_2}-\frac{1}{p_2}} f_{j,k}(x); \quad j, k = 0, 1, 2, \ldots$$

We have

$$\left\| a_j^{-\frac{1}{p_1}} g_{j,k}(x) \mid L_{\infty \mid x_1} \right\| \leqq c \|g_{j,k}(x) \mid L_{p_1 \mid x_1}\| \leqq \|g_{\mu,k}(x) \,|L_{p_1 \mid x_1}|\, l_{\infty \mid \mu}\| \tag{22}$$

for all $j, k = 0, 1, 2, \ldots$, and $x_2 \in R_2$. This is a Nikol'skij type inequality with respect to x_1, where x_2 is fixed. It can be proved as in Proposition 1.4.3 or Theorem 1.6.2 provided that $g_{j,k}(x) \in S(R_2)$. Then the approximation procedure of Step 2 of the proof of Theorem 1.5.2 yields (22) for arbitrary $\{g_{j,k}(x)\}$. (22) is the counterpart of (12). Repeating the arguments from (12) to (20) we can verify that (22) yields

$$\left\| a_j^{\frac{1}{u_1}-\frac{1}{p_1}} g_{j,k}(x) \mid L_{u_1 \mid x_1}(l_{v_1 \mid j}) \right\| \leqq c \|g_{j,k}(x) \mid L_{p_1 \mid x_1}(l_{\infty \mid j})\| \tag{23}$$

for all $k = 0, 1, 2, \ldots$ and $x_2 \in R_1$. Taking the quasi-norm in $L_{u_2}(l_{v_2})$ on both sides with respect to k and x_2 we obtain (21). (11) and (21) yield (10). Thus the proof is complete.

Remark* 1. The principle of the proof in Step 2 is due to B. Jawerth [1] (cf. also [T, Theorem 2.7.1]). Part (i) extends the famous Nikol'skij inequality of 1.6.2. to the spaces $Sl_{\bar{q}}(L_{\bar{p}}^{\Omega})$. We refer also to H.-J. Schmeisser [3].

Theorem. (i) *Let* $\bar{p} = (p_1, p_2)$, $\bar{u} = (u_1, u_2)$, $\bar{q} = (q_1, q_2)$, $\bar{r} = (r_1, r_2)$, $\bar{t} = (t_1, t_2)$ *with* $0 < p_i \leqq u_i \leqq \infty$, $0 < q_i \leqq \infty$, *and* $-\infty < t_i \leqq r_i < \infty$; $(i = 1, 2)$. *Then*

$$S_{\bar{p},\bar{q}}^{\bar{r}} B \subset S_{\bar{u},\bar{q}}^{\bar{t}} B \tag{24}$$

and

$$SB_{\bar{p},\bar{q}}^{\bar{r}} \subset SB_{\bar{u},\bar{q}}^{\bar{t}} \tag{25}$$

if $r_1 - \dfrac{1}{p_1} = t_1 - \dfrac{1}{u_1}$ *and* $r_2 - \dfrac{1}{p_2} = t_2 - \dfrac{1}{u_2}$.

(ii) *Let* $\bar{p} = (p_1, p_2)$, $\bar{u} = (u_1, u_2)$, $\bar{q} = (q_1, q_2)$, $\bar{v} = (v_1, v_2)$, $\bar{r} = (r_1, r_2)$, $\bar{t} = (t_1, t_2)$ *with* $0 < p_i < u_i < \infty$, $0 < q_i \leqq \infty$, $0 < v_i \leqq \infty$, *and* $-\infty < t_i < r_i < \infty$;

$(i = 1, 2)$. *Then*

$$S^{\bar{r}}_{\bar{p},\bar{q}}F \subset S^{\bar{t}}_{\bar{u},\bar{v}}F \tag{26}$$

if $r_1 - \frac{1}{p_1} = t_1 - \frac{1}{u_1}$ *and* $r_2 - \frac{1}{p_2} = t_2 - \frac{1}{u_2}$.

Proof. Let $\varphi = \{\varphi_j\}_{j=0}^{\infty} \in \Phi(R_1)$. We put

$$f_{j,k}(x) = 2^{r_1 j + r_2 k}(F^{-1}\varphi_j(\xi_1)\,\varphi_k(\xi_2)\,Ff)(x); \quad j, k = 0, 1, 2, \ldots,$$

and $a_j = 2^j$, $b_k = 2^k$; $j, k = 0, 1, 2, \ldots$ Then obviously

$$2^{t_1 j + t_2 k}(F^{-1}\varphi_j\varphi_k Ff)(x) = a_j^{\frac{1}{u_1} - \frac{1}{p_1}}\, b_k^{\frac{1}{u_2} - \frac{1}{p_2}} f_{j,k}(x);$$

$j, k = 0, 1, 2, \ldots$ Hence (25) is an immediate consequence of part (i) of the preceding proposition and (26) follows from part (ii). Finally, (24) is a consequence of Theorem 1.6.2.

Remark 2. Recall the spaces $S^{\bar{r}}\mathscr{C} = S^{\bar{r}}_{\bar{\infty},\bar{\infty}}B$, $\bar{\infty} = (\infty, \infty)$, from Remark 2.3.4/3. As a consequence of the above theorem and Proposition 2.2.3/2 (cf. also (2.2.3/17)) we have

$$S^{\bar{r}}_{\bar{p},\bar{q}}B \subset S^{\bar{t}}\mathscr{C}, \tag{27}$$

$$SB^{\bar{r}}_{\bar{p},\bar{q}} \subset S^{\bar{t}}\mathscr{C}, \tag{28}$$

and

$$S^{\bar{r}}_{\bar{p},\bar{q}}F \subset S^{\bar{t}}\mathscr{C} \tag{29}$$

if $0 < p_i \leqq \infty$ ($p_i < \infty$ in (29)), $0 < q_i \leqq \infty$, and $0 < t_i \leqq r_i < \infty$ with $r_i = t_i + \frac{1}{p_i}$; $(i = 1, 2)$. Furthermore we proved in Proposition 2.2.3/4 (cf. (2.2.3/28))

$$S^{\bar{0}}_{\bar{\infty},\bar{1}}B \subset C(R_2)$$

$\bar{\infty} = (\infty, \infty)$, $\bar{1} = (1, 1)$, $\bar{0} = (0, 0)$. Here $C(R_2)$ denotes the space of all uniformly continuous bounded functions on R_2. Then Theorem 2.2.6/2(ii) yields

$$S^{\bar{l}}_{\bar{\infty},\bar{1}}B \subset S^{\bar{l}}C, \tag{30}$$

where $\bar{l} = (l_1, l_2)$; $l_i = 0, 1, 2, \ldots$ $(i = 1, 2)$ and

$$S^{\bar{l}}C = \left\{ f \mid f \in C(R_2),\ \|f \mid S^{\bar{l}}C\| = \sum_{\alpha_1=0}^{l_1} \sum_{\alpha_2=0}^{l_2} \|D^{\alpha}f \mid C\| < \infty \right\}. \tag{31}$$

Now it follows from (30) and (24)

$$S^{\bar{r}}_{\bar{p},\bar{q}}B \subset S^{\bar{l}}C \tag{32}$$

if $0 < p_i \leqq \infty$, $0 < q_i \leqq 1$, and $r_i = l_i + \frac{1}{p_i}$ with $l_i = 0, 1, 2, \ldots$ $(i = 1, 2)$. Again, Proposition 2.2.3/2 gives similar embeddings for $SB^{\bar{r}}_{\bar{p},\bar{q}}$ and $S^{\bar{r}}_{\bar{p},\bar{q}}F$. (32) states that $f \in S^{\bar{r}}_{\bar{p},\bar{q}}B$ has uniformly continuous bounded derivatives $D^{(\alpha_1,\alpha_2)}f$ if $0 \leqq \alpha_i \leqq l_i$ with $r_i = l_i + \frac{1}{p_i}$ and $0 < q_i \leqq 1$ $(i = 1, 2)$.

Remark 3. In connection with 2.3.1. the theorem describes embeddings in $L_{\bar{p}}$ or in spaces of Sobolev-Lebesgue type. On the other hand we obtain embeddings of $S^{\bar{r}}_{\bar{p}}W$ and $S^{\bar{r}}_{\bar{p}}H$ in $S^{\bar{t}}\mathscr{C}$ etc. In particular it holds:

$$S^{\bar{r}}_{\bar{p},\bar{q}}F \subset L_{\bar{u}} \tag{33}$$

if $1 < u_i < \infty$, $0 < p_i < u_i < \infty$, $0 < q_i \leqq \infty$ and $r_i = \frac{1}{p_i} - \frac{1}{u_i}$; $(i = 1, 2)$,

$$S^{\bar{r}}_{\bar{p}}H \subset S^{\bar{t}}\mathscr{C} \tag{34}$$

if $1 < p_i < \infty$, $r_i - \dfrac{1}{p_i} = t_i > 0$; $(i = 1, 2)$,

$$S^{\bar{r}}_{\bar{p}}H \subset S^{\bar{l}}C, \quad \bar{l} = (l_1, l_2), \tag{35}$$

if $1 < p_i < \infty$, $l_i = 0, 1, 2, \ldots, r_i - \dfrac{1}{p_i} > l_i \geqq 0$; $(i = 1, 2)$.

2.4.2. Traces

We deal with so-called direct embedding theorems (or traces) and inverse embedding theorems (or extensions) for the spaces $S^{\bar{r}}_{\bar{p},\bar{q}}B$, $SB^{\bar{r}}_{\bar{p},\bar{q}}$, and $S^{\bar{r}}_{\bar{p},\bar{q}}F$. Let us briefly sketch the problem. If $A(R_2)$ stands for one of these spaces then we know that $f \in A(R_2)$ can be approximated by functions of the form $F^{-1}\varphi Ff$ where $\varphi \in S(R_2)$ has a compact support. By the Paley-Wiener-Schwartz theorem (cf. 1.2.4.) $F^{-1}\varphi Ff$ is an infinitely differentiable function. For such functions we can define the natural traces on $\{x \mid x = (x_1, 0), x_1 \in R_1\}$ and $\{x \mid x = (0, x_2), x_2 \in R_1\}$ by

$$\left.\begin{aligned} &(\mathfrak{R}_1 f)(x_1) = f(x_1, 0) \\ \text{and}\quad &(\mathfrak{R}_2 f)(x_2) = f(0, x_2), \end{aligned}\right\} \tag{1}$$

respectively. The aim of this subsection is to determine a space $A_i(R_1)$ such that $\mathfrak{R}_i$ $(i = 1, 2)$ can be extended to a retraction from $A(R_2)$ onto $A_i(R_1)$ and to find the corresponding coretraction $\mathfrak{S}_i$ $(i = 1, 2)$. Let us recall the concept of retractions and coretractions. A linear continuous mapping $\mathfrak{R}$ from $A(R_2)$ onto $A_0(R_1)$ is said to be a retraction if there exists a linear continuous operator $\mathfrak{S}$ from $A_0(R_1)$ into $A(R_2)$ such that

$$\mathfrak{R}\mathfrak{S} = E \quad (\text{identity in } A_0(R_1)).$$

$\mathfrak{S}$ is called the corresponding coretraction. We refer also to [I, 2.9.] and [T, 2.7.2.] where direct and inverse embedding theorems are formulated in this language.

We need the one-dimensional spaces $B^s_{p,q}(R_1)$ and $F^s_{p,q}(R_1)$ of Besov-Sobolev-Hardy type (cf. 2.2.2/(2), (3), where we have to replace $\Phi(R_2)$ by $\Phi(R_1)$, $S'(R_2)$ by $S'(R_1)$ and L_p by $L_p(R_1)$).

Theorem. *Let $\mathfrak{R}_1$ and $\mathfrak{R}_2$ be given by (1). Let $\bar{p} = (p_1, p_2)$, $\bar{q} = (q_1, q_2)$, and $\bar{r} = (r_1, r_2)$.*

(i) *Let $0 < p_i \leqq \infty$, $0 < q_i \leqq \infty$, and $-\infty < r_i < \infty$; $(i = 1, 2)$. If $r_i > \dfrac{1}{p_i}$ then $\mathfrak{R}_i$ is a retraction from*

$$S^{\bar{r}}_{\bar{p},\bar{q}}B(R_2) \quad \textit{onto} \quad B^{r_i}_{p_i,q_i}(R_1) \tag{2}$$

and from

$$SB^{\bar{r}}_{\bar{p},\bar{q}}(R_2) \quad \textit{onto} \quad B^{r_i}_{p_i,q_i}(R_1); \quad (i = 1, 2). \tag{3}$$

(ii) *Let $0 < p_i < \infty$, $0 < q_i \leqq \infty$, and $-\infty < r_i < \infty$; $(i = 1, 2)$. If $r_i > \dfrac{1}{p_i}$ then $\mathfrak{R}_i$ is a retraction from*

$$S^{\bar{r}}_{\bar{p},\bar{q}}F(R_2) \quad \textit{onto} \quad F^{r_i}_{p_i,q_i}(R_1); \quad (i = 1, 2). \tag{4}$$

Proof. Step 1. Let $f \in S'(R_2)$ such that supp Ff is compact. We denote the Fourier transform and its inverse in $S'(R_1)$ by F_1 and F_1^{-1}, respectively. We choose a sequence $\varphi = \{\varphi_j(t)\}_{j=0}^{\infty} \in \Phi(R_1)$. We put $\varphi_{j,k}(x) = \varphi_j(x_1)\varphi_k(x_2)$; $j, k = 0, 1, 2, \ldots$ and

$\varphi^n_{j,k}(x) = \varphi_{j+n}(x_1)\,\varphi_{j,k}(x)$ where $n = -1, 0, 1$ with $\varphi_{-1} = 0$. Then we have

$$[F_1^{-1}\varphi_j F_1 f(\cdot, 0)]\,(x_1) = (F_1^{-1}\varphi_j F_1 f)\,(x_1, 0) = (F^{-1}\varphi_j Ff)\,(x_1, 0)$$
$$= \sum_{n=-1,0,1} \sum_{k=0}^{\infty} (F^{-1}\varphi^n_{j,k} Ff)\,(x_1, 0) \tag{5}$$

and similarly

$$[F_1^{-1}\varphi_k F_1 f(0, \cdot)]\,(x_2) = \sum_{n=-1,0,1} \sum_{j=0}^{\infty} (F^{-1}\varphi^n_{j,k} Ff)\,(0, x_2). \tag{6}$$

The sums on the right-hand sides of (5) and (6) are finite. Now, let $f \in S^{\bar r}_{\bar p,\bar q}B$. Using (5), Theorem 1.6.2 with $\alpha = (0, 0)$, $u_1 = p_1$, $u_2 = \infty$, and Theorem 1.8.3 then we obtain

$$\|f(x_1, 0) \mid B^{r_1}_{p_1,q_1}(R_1)\|$$
$$\leqq \left\| \left(\sum_{k=0}^{\infty} \sum_{n=-1,0,1} \|2^{r_1 j}(F^{-1}\varphi^n_{j,k} Ff)\,(x_1, 0) \mid L_{p_1 \mid x_1}\|^s \right)^{\frac{1}{s}} \mid l_{q_1 \mid j} \right\|$$
$$\leqq c \left\| 2^{r_1 j + \frac{1}{p_2} k} (F^{-1}\varphi_{j,k} Ff)\,(x) \,|L_{\bar p}|\, l_{s \mid k} \mid l_{q_1 \mid j} \right\|$$
$$\leqq c \left\| 2^{r_1 j + \frac{1}{p_2} k} (F^{-1}\varphi_{j,k} Ff)\,(x) \,|L_{\bar p}|\, l_{q_1 \mid j} \mid l_{s \mid k} \right\|, \tag{7}$$

where $s = \min(1, p_1, q_1)$. Hence, because $r_2 > \dfrac{1}{p_2}$ we have

$$\|f(x_1, 0) \mid B^{r_1}_{p_1,q_1}(R_1)\| \leqq c \left\| f \mid S^{\left(r_1, \frac{1}{p_2}\right)}_{\bar p,(q_1, s)} B \right\| \leqq c' \|f \mid S^{\bar r}_{\bar p,\bar q}B\|. \tag{8}$$

In the same way with $s = \min(1, p_1)$ and (6) we obtain

$$\|f(0, x_2) \mid B^{r_2}_{p_2,q_2}(R_1)\|$$
$$\leqq c \left\| 2^{\frac{1}{p_1} j + r_2 k} (F^{-1}\varphi_{j,k} Ff)\,(x) \,|L_{\bar p}|\, l_{s \mid j} \mid l_{q_2 \mid k} \right\|$$
$$= c \left\| f \mid S^{\left(\frac{1}{p_1}, r_2\right)}_{\bar p,(s, q_2)} B \right\| \leqq c' \|f \mid S^{\bar r}_{\bar p,\bar q}B\|. \tag{9}$$

If $f \in SB^{\bar r}_{\bar p,\bar q}$ we use Proposition 2.4.1(i) with $u_1 = p_1$, $u_2 = \infty$, and Theorem 1.10.3(i). As in (7) we obtain

$$\|f(x_1, 0) \mid B^{r_1}_{p_1,q_1}(R_1)\| \leqq c \left\| f \mid SB^{\left(r_1, \frac{1}{p_2}\right)}_{\bar p,(q_1, s)} \right\| \leqq c' \|f \mid SB^{\bar r}_{\bar p,\bar q}\|, \tag{10}$$

where $s = \min(1, p_1, q_1)$. Similarly,

$$\|f(0, x_2) \mid B^{r_2}_{p_2,q_2}(R_1)\| \leqq c \left\| f \mid SB^{\left(\frac{1}{p_1}, r_2\right)}_{\bar p,(s, q_2)} \right\| \leqq c' \|f \mid SB^{\bar r}_{\bar p,\bar q}\| \tag{11}$$

for $s = \min(1, p_1)$. Finally let $f \in S^{\bar r}_{\bar p,\bar q}F$. Then it follows from (5)

$$\|f(x_1, 0) \mid F^{r_1}_{p_1,q_1}(R_1)\|$$
$$\leqq \left(\sum_{k=0}^{\infty} \sum_{n=-1,0,1} \|2^{r_1 j}(F^{-1}\varphi^n_{j,k} Ff)\,(x_1, 0) \mid L_{p_1 \mid x_1}(l_{q_1 \mid j})\|^s \right)^{1/s}$$
$$\leqq c \left(\sum_{k=0}^{\infty} \|2^{r_1 j}(F^{-1}\varphi_{j,k} Ff)\,(x) \,|L_{p_1 \mid x_1}(l_{q_1 \mid j})|\, L_{\infty \mid x_2}\|^s \right)^{1/s}, \tag{12}$$

where $s = \min(1, p_1, q_1)$. As in Step 1 of the proof of Proposition 2.4.1 we prove that

$$\|2^{r_1 j}(F^{-1}\varphi_{j,k}Ff)(x)\,|L_{p_1|x_1}(l_{q_1|j})|\,L_{\infty|x_2}\|$$
$$\leqq c\cdot 2^{\frac{1}{p_2}k}\,\|2^{r_1 j}(F^{-1}\varphi_{j,k}Ff)(x)\,|L_{p_1|x_1}(l_{q_1|j})|\,L_{p_2|x_2}\|\,. \tag{13}$$

We substitute (13) into (12) and obtain for $r_2 > \frac{1}{p_2}$

$$\|f(x_1,0)\mid F^{r_1}_{p_1,q_1}(R_1)\|$$
$$\leqq c\left\|2^{r_1 j+\frac{1}{p_2}k}(F^{-1}\varphi_{j,k}Ff)(x)\,|L_{p_1|x_1}(l_{q_1|j})|\,L_{p_2|x_2}\mid l_{s|k}\right\|$$
$$\leqq c\left(\sum_{k=0}^{\infty}2^{-\left(r_2-\frac{1}{p_2}\right)ks}\right)^{\frac{1}{s}}\|f\mid S^{\bar r}_{\bar p,(q_1,\infty)}F\|\leqq c'\|f\mid S^{\bar r}_{\bar p,\bar q}F\|\,. \tag{14}$$

If we use (6) then we have with $s=\min(1,q_2)$

$$\|2^{r_2 k}[F_1^{-1}\varphi_k F_1 f(0,\cdot)](x_2)\mid l_{q_2|k}\|$$
$$\leqq\left(\sum_{j=0}^{\infty}\sum_{n=-1,0,1}\|2^{r_2 k}(F^{-1}\varphi^n_{j,k}Ff)(0,x_2)\mid l_{q_2|k}\|^s\right)^{1/s}$$
$$\leqq\left(\sum_{j=0}^{\infty}\sum_{n=-1,0,1}\|2^{r_2 k}(F^{-1}\varphi^n_{j,k}Ff)(x)\,|l_{q_2|k}|\,L_{\infty|x_1}\|^s\right)^{1/s}. \tag{15}$$

We claim

$$\|2^{r_2 k}(F^{-1}\varphi^n_{j,k}Ff)(x_1,x_2)\,|l_{q_2|k}|\,L_{\infty|x_1}\|$$
$$\leqq c\cdot 2^{\frac{1}{p_1}j}\,\|2^{r_2 k}(F^{-1}\varphi_{j,k}Ff)(x)\,|l_{q_2|k}|\,L_{p_1|x_1}\|\,, \tag{16}$$

where c is independent of $x_2\in R_1$, and $j=0,1,2,\ldots$ (16) can be derived combining the methods of the proof of (2.4.1/22) and of Step 1 of the proof of Proposition 2.4.1. We substitute (16) into (15) and take the quasi-norm in $L_{p_2|x_2}$. Then we have

$$\|f(0,x_2)\mid F^{r_2}_{p_2,q_2}(R_1)\|$$
$$\leqq c\left\|2^{\frac{1}{p_1}j+r_2 k}(F^{-1}\varphi_{j,k}Ff)(x)\,|l_{q_2|k}|\,L_{p_1|x_1}\,|l_{s|j}|\,L_{p_2|x_2}\right\|$$
$$\leqq c\left(\sum_{j=0}^{\infty}2^{-\left(r_1-\frac{1}{p_1}\right)js}\right)^{\frac{1}{s}}\|f\mid S^{\bar r}_{\bar p,(\infty,q_2)}F\|\leqq c'\|f\mid S^{\bar r}_{\bar p,\bar q}F\|\,. \tag{17}$$

Thus we proved that $\mathfrak{R}_i$ is a linear and bounded operator from $S^{\bar r}_{\bar p,\bar q}B$ $(SB^{\bar r}_{\bar p,\bar q},\ S^{\bar r}_{\bar p,\bar q}F)$ into $B^{r_i}_{p_i,q_i}(R_1)$ $(B^{r_i}_{p_i,q_i}(R_1),\ F^{r_i}_{p_i,q_i}(R_1))$ if $r_i>\frac{1}{p_i}$ $(i=1,2)$.

Step 2. We have to construct corresponding extensions $\mathfrak{S}_i$ $(i=1,2)$ such that $\mathfrak{R}_i$ $(i=1,2)$ becomes a retraction. Let $g(t)\in S(R_1)$ such that $g(0)=1$. We put

$$\left.\begin{aligned}&\mathfrak{S}_1 f=g(x_2)\otimes f\quad\text{and}\\&\mathfrak{S}_2 f=g(x_1)\otimes f,\quad f\in S'(R_1)\end{aligned}\right\}. \tag{18}$$

Here $\otimes$ means the tensor product of distributions. Clearly, $\mathfrak{S}_i f\in S'(R_2)$ $(i=1,2)$. If additionally supp $F_1 f$ is compact we have $f=f(x_i)$,

$$\begin{aligned}&\mathfrak{S}_i f\qquad\quad=(\mathfrak{S}_i f)(x)\quad\text{and}\\&(\mathfrak{R}_i\mathfrak{S}_i f)(x_i)=f(x_i),\quad(i=1,2)\,.\end{aligned} \tag{19}$$

Moreover, let $f \in B^{r_1}_{p_1,q_1}(R_1)$. Then,

$$\|\mathfrak{S}_1 f \mid S^{\bar r}_{\bar p,\bar q}B\| = \|g(x_2) \otimes f \mid S^{\bar r}_{\bar p,\bar q}B\|$$
$$= \|g \mid B^{r_2}_{p_2,q_2}(R_1)\| \, \|f \mid B^{r_1}_{p_1,q_1}(R_1)\| . \tag{20}$$

Thus, together with (19) and Step 1 we have shown that $\mathfrak{R}_1$ is a retraction from $S^{\bar r}_{\bar p,\bar q}B$ onto $B^{r_1}_{p_1,q_1}(R_1)$. Analogously we deal with the other cases. The proof of the theorem is complete.

Remark*. The description of traces on hyperplanes spanned by the coordinate axes for spaces with dominating mixed smoothness properties is relatively simple compared with the analogous problem in isotropic or anisotropic spaces (see also [T, 2.7.2.], or 4.2.3. and the references given there). Direct and inverse embedding theorems for the classical cases $S^{\bar r}_{\bar p,\bar q}B$ (with $\min(p_1, p_2, q_1, q_2) \geqq 1$) and $S^{\bar r}_{\bar p}H$ (with $1 < p_1 < \infty$, $1 < p_2 < \infty$) can be found in T. I. Amanov [1–3], S. M. Nikol'skij [5], A. P. Uninskij [1, 3], P. I. Lizorkin, S. M. Nikol'skij [1], and Ju. A. Bryčkov [1].

2.5. Notes and Comments

2.5.1. Mapping Properties of Integral Operators

Let $\mathscr{K}$ be an integral operator of the type

$$(\mathscr{K}f)(x_2) = \int_a^b K(x_1, x_2) f(x_1) \, \mathrm{d}x_1, \quad x_2 \in (c, d),$$

where (a, b) and (c, d) are intervals. It is a well-known fact that smoothness assumptions for the kernel $K(x_1, x_2)$ ensure mapping properties between certain function spaces, as well as assertions concerning the compactness of $\mathscr{K}$, the distribution of eigenvalues and of several types of approximation numbers and widths etc. We refer to I. Z. Gohberg, M. G. Krein [1, III, § 9, 10] and A. Pietsch [1, Chapter 22]. The aim of this subsection is to describe mapping properties of integral operators where $(a, b) = (c, d) = R_1$ and the corresponding kernels belong to the spaces $S^{\bar r}_{\bar p,\bar q}B$, $SB^{\bar r}_{\bar p,\bar q}$, or $S^{\bar r}_{\bar p,\bar q}F$.

At first let us deal with the concept of an integral operator.

Definition. *Let $K \in S'(R_2)$ and $g \in S(R_1)$. Then the operator $\mathscr{K}$ is given by*

$$(\mathscr{K}f)(g) = K(f(x_1)\, g(x_2)), \quad f \in S(R_1). \tag{1}$$

Because $f(x_1)\, g(x_2) \in S(R_2)$ the definition makes sense and we have $\mathscr{K}f \in S'(R_1)$. If A_1 and A_2 are quasi-Banach spaces with

$$S(R_1) \subset A_1 \subset S'(R_1), \qquad S(R_1) \subset A_2 \subset S'(R_1)$$

then we ask for criterions on K such that $\mathscr{K}$ can be extended to a linear and bounded operator from A_1 into A_2, where A_1 and A_2 are spaces of Besov-Sobolev type on R_1.

Remark 1. Let $K = K(x_1, x_2) \in S'(R_2)$ be a regular distribution. Then,

$$(\mathscr{K}f)(g) = \int_{R_1} \Big[\int_{R_1} K(x_1, x_2) f(x_1) \, \mathrm{d}x_1\Big] g(x_2) \, \mathrm{d}x_2, \quad f \in S(R_1)$$

for all $g \in S(R_1)$. Hence, $\mathscr{K}$ can be identified with the integral operator

$$(\mathscr{K}f)(x_2) = \int_{R_1} K(x_1, x_2) f(x_1) \, \mathrm{d}x_1, \quad f \in S(R_1).$$

Recall that the spaces $B^r_{p,q}(R_1)$ and $F^r_{p,q}(R_1)$ have been defined in 2.2.2/(2), (3) (obvious modifications, cf. also 2.4.2.).

Theorem. *Let* $0 < \nu \leqq 1$, $0 < \mu \leqq 1$, $1 \leqq u \leqq \infty$, $1 \leqq v \leqq \infty$, *and* $0 < q \leqq \infty$. *Let* $\frac{1}{u} + \frac{1}{u'} = 1$, $\frac{1}{v} + \frac{1}{v'} = 1$. *Further let* $-\infty < r < \infty$, $-\infty < t < \infty$, *and* $\tau = \frac{1}{\nu} - 1 - t$.

(i) *If* $0 < p \leqq \infty$ *and* $0 < \mu \leqq \min(1, p)$ *then there exists a positive constant c such that*

$$\|\mathscr{K}f \mid B^r_{p,q}(R_1)\| \leqq c\|K \mid S^{(t,r)}_{(\nu u,p),(\mu v,q)}B\| \; \|f \mid B^\tau_{\nu u',\mu v'}(R_1)\| \tag{2}$$

holds for all $f \in S(R_1)$.

(ii) *If* $0 < p \leqq \infty$ *then there exists a positive constant c such that*

$$\|\mathscr{K}f \mid B^r_{p,q}(R_1)\| \leqq c\|K \mid SB^{(t,r)}_{(\nu u,p),(\mu v,q)}\| \; \|f \mid B^\tau_{\nu u',\mu v'}(R_1)\| \tag{3}$$

holds for all $f \in S(R_1)$.

(iii) *If* $0 < p < \infty$, $1 < u < \infty$ *and* $\nu \leqq \min(1, q)$ *then there exists a positive constant c such that*

$$\|\mathscr{K}f \mid F^r_{p,q}(R_1)\| \leqq c\|K \mid S^{(t,r)}_{(\nu u,p),(\nu v,q)}F\| \; \|f \mid F^\tau_{\nu u',\nu v'}(R_1)\| \tag{4}$$

holds for all $f \in S(R_1)$.

Proof. Step 1. We need some preparations. Let $f \in S(R_1)$, $K \in S'(R_2)$, and let $\varphi = \{\varphi_j\}_{j=0}^\infty \in \Phi(R_1)$. We put $\varphi_{j,l}(x) = \varphi_j(x_1)\varphi_l(x_2)$, $K_{j,l}(x) = (F^{-1}\varphi_{j,l}FK)(x)$; $j, l = 0, 1, 2, \ldots$ As in 2.4.2., F_1, F_1^{-1} denote the one-dimensional Fourier transform and its inverse, respectively. Let $l = 0, 1, 2, \ldots$ Then we claim

$$[F_1^{-1}\varphi_l F_1(\mathscr{K}f)](x_2) = \sum_{j=0}^{\infty} \int_{R_1} K_{j,l}(x_1, x_2) f(x_1)\,dx_1 . \tag{5}$$

By the Paley-Wiener-Schwartz theorem $F_1^{-1}\varphi_l F_1\mathscr{K}f$ and $K_{j,l}$ are infinitely differentiable functions on R_1 and R_2, respectively. Therefore, (5) follows easily from

$$\begin{aligned}(F_1^{-1}\varphi_l F_1\mathscr{K}f)(g) &= K(f \cdot F_1\varphi_l F_1^{-1}g)\\ &= \sum_{j=0}^{\infty} K(F\varphi_{j,l}F^{-1}(f \cdot g)) = \sum_{j=0}^{\infty} K_{j,l}(f \cdot g).\end{aligned}$$

Let us put $f_j(x_1) = (F_1^{-1}\varphi_j F_1 f)(x_1)$; $j = 0, 1, 2, \ldots$, and $f_{-1}(x_1) = 0$, $K_{-1,l} = 0$; $l = 0, 1, 2, \ldots$ We put

$$\begin{aligned}&\int_{R_1} K_{j,l}(x) f(x_1)\,dx_1\\ &= \sum_{w=-1}^{1} \int_{R_1} [F_1^{-1}\varphi_{j+w}(-\xi_1) F_1(K_{j,l}(\cdot, x_2))](x_1) f(x_1)\,dx_1\\ &= \sum_{w=-1}^{1} \int_{R_1} K_{j,l}(x_1, x_2) f_{j+w}(x_1)\,dx_1\end{aligned}$$

into (5) and obtain

$$[F_1^{-1}\varphi_l F_1(\mathscr{K}f)](x_2) = \sum_{w=-1}^{1}\sum_{j=0}^{\infty} \int_{R_1} K_{j,l}(x_1, x_2) f_{j+w}(x_1)\,dx_1 ; \tag{6}$$

$l = 0, 1, 2, \ldots$, $f \in S(R_1)$.

Step 2. We prove (2) and (3) in the case $\nu = 1$. We apply Hölder's inequality and the estimate $(\sum |\ldots|)^\mu \leqq \sum |\ldots|^\mu$. Then the identity (6) yields

$$|F_1^{-1}\varphi_l F_1(\mathscr{K}f)(x_2)|$$
$$\leqq \sum_{w=-1}^{1} \|2^{tj}K_{j,l}(x_1, x_2) | L_{u|x_1}| l_{\mu v|j}\| \, \|2^{-tj}f_{j+w}(x_1) | l_{\mu v'}(L_{u'})\|$$
$$\leqq c\|2^{tj}K_{j,l}(x_1, x_2) | L_{u|x_1}| l_{\mu v|j}\| \, \|f | B^{\tau}_{u',\mu v'}(R_1)\| . \tag{7}$$

Now we apply on both sides of (7) the quasi-norm

$$\|2^{rl}(\ldots) | L_{p|x_2}| l_{q|l}\|$$

and obtain

$$\|\mathscr{K}f | B^r_{p,q}(R_1)\| \leqq c\|K | SB^{(t,r)}_{(u,p),(\mu v,q)}\| \, \|f | B^{\tau}_{u',\mu v'}(R_1)\| . \tag{8}$$

This proves (3) if $\nu = 1$. Now, let $\mu \leqq \min(1, p)$. Then (6) yields similarly as above

$$\|F_1^{-1}\varphi_l F_1(\mathscr{K}f) | L_p\|$$
$$\leqq c \sum_{w=-1}^{1} \left\| \sum_{j=0}^{\infty} (\|K_{j,l}(x) | L_{u|x_1}\|^\mu \, \|f_{j+w} | L_{u'}\|^\mu) | L_{\frac{p}{\mu}|x_2} \right\|^{1/\mu}$$
$$\leqq c\|2^{tj}K_{j,l}(x) | L_{(u,p)}| l_{\mu v|j}\| \, \|f | B^{\tau}_{u',\mu v'}(R_1)\| . \tag{9}$$

This implies (2) in the case $\nu = 1$.

Step 3. Let $0 < \nu < 1$. As a consequence of Theorem 2.4.1(i) we have the embeddings

$$S^{(t,r)}_{(\nu u,p),(\mu v,q)}B \subset S^{(\varkappa,r)}_{(u,p),(\mu v,q)}B , \tag{10}$$
$$SB^{(t,r)}_{(\nu u,p),(\mu v,q)} \subset SB^{(\varkappa,r)}_{(u,p),(\mu v,q)} \tag{11}$$

if $\varkappa = t - \frac{1}{u}\left(\frac{1}{\nu} - 1\right)$. Further, we have the embedding

$$B^{\tau}_{\nu u',\mu v'}(R_1) \subset B^{\tau-\varkappa}_{u',\mu v'}(R_1), \tag{12}$$

where $\varkappa$ has the same meaning as in (10) and (11) (cf. [T, Theorem 2.7.1(i)]). By (11), (12), (13), and the results of Step 2 we obtain (2) and (3) for $0 < \nu \leqq 1$.

Step 4. In order to prove (4) we use the fact that $K_{j,l}(x_1, x_2) f_j(x_1)$ (cf. (6)) is an entire analytic function of exponential type with respect to x_1 ($x_2 \in R_1$ is fixed). Further there exists a constant $c > 0$ such that

$$\operatorname{supp} F_1[K_{j,l}(\cdot, x_2) f_j(\cdot)](y_1) \subset [-c \cdot 2^j, c \cdot 2^j]; \quad j = 0, 1, 2, \ldots,$$

for all $l = 0, 1, 2, \ldots, x_2 \in R_1, f \in S(R_1)$, and $K \in S'(R_2)$. Consequently by Nikol'skij's inequality (cf. [T, Theorem 1.4.1(ii), or Theorem 1.6.2]) we obtain from (6)

$$|F_1^{-1}\varphi_l F_1(\mathscr{K}f)](x_2)|$$
$$\leqq c \sum_{w=-1}^{1} \left(\sum_{j=0}^{\infty} 2^{(1-\nu)j} \int_{R_1} |K_{j,l}(x_1, x_2)|^\nu \, | f_{j+w}(x_1)|^\nu \, dx_1 \right)^{1/\nu}, \tag{13}$$

$0 < \nu \leqq 1$. If $0 < \nu \leqq \min(1, q)$ we proceed as in (9). From (13) follows

$$\|2^{rl}[F_1^{-1}\varphi_l F_1(\mathscr{K}f)](x_2) | l_q\|$$
$$\leqq c \left\| 2^{rl\nu} \int_{R_1} \|2^{tj}K_{j,l}(x) | l_{\nu v|j}\|^\nu \, \|2^{\tau j}f_j(x_1)| l_{\nu v'}\|^\nu \, dx_1 \, | \, l_{\frac{q}{\nu}|l} \right\|^{1/\nu}$$
$$\leqq c\|2^{rl+tj}K_{j,l}(x) | l_{\nu v|j}| l_{q|l} | L_{\nu u|x_1}\| \, \|f | F^{\tau}_{\nu u',\nu v'}(R_1)\| . \tag{14}$$

We take the quasi-norm $\|\cdot | L_{p|x_2}\|$. Then (14) yields (4). The proof is complete.

Remark* **2.** Let $1 \leqq p < \infty$, $1 \leqq u < \infty$, $1 \leqq q < \infty$, $1 \leqq v < \infty$, $0 < t < \infty$, and $0 < r < \infty$. Recall that it has been proved in Theorem 2.3.4/2 and Remark 2.3.4/4 that

$$SB^{(t,r)}_{(u,p),(v,q)} = B^r_{p,q}(R_1, A), \qquad A = B^t_{u,v}(R_1), \tag{15}$$

hold, where $B^r_{p,q}(R_1, A)$ is the vector-valued Besov space from (2.3.4/52). Using this fact, (3) can be reformulated: Given p, u, q, v, t, and r as above. Then,

$$\|\mathscr{K}f \mid B^r_{p,q}(R_1)\| \leqq c\|K \mid B^r_{p,q}(R_1, B^t_{u,v}(R_1))\| \, \|f \mid B^{-t}_{u',v'}(R_1)\|, \tag{16}$$

where $\frac{1}{u} + \frac{1}{u'} = 1$, $\frac{1}{v} + \frac{1}{v'} = 1$, $f \in S(R_1)$. The description of smoothness properties of kernels of integral operators via vector-valued Besov spaces is due to A. Pietsch [2–4]. In particular, he used an assertion of type (16) in the framework of operator ideals in order to investigate qualitative questions such as the distribution of eigenvalues of integral operators. Further results in this direction can be found in A. Pietsch [1–4], B. Carl, T. Kühn [1], and D. Elstner [1]. Note that $B^{-t}_{u',v'}(R_1)$ is the dual space of $B^t_{u,v}(R_1)$ (cf. [T, 2.11.2.]).

Remark 3. Let us discuss another special case of the theorem. In Theorem 2.3.1 we proved that

$$S^{(t,r)}_{(u,p),\bar{2}}F = S^{(t,r)}_{(u,p)}H, \quad \bar{2} = (2, 2), \tag{17}$$

if $1 < u < \infty$, $1 < p < \infty$, $-\infty < r < \infty$, and $-\infty < t < \infty$. Furthermore it is well-known (cf. [T, 2.5.6.]) that

$$F^r_{p,2}(R_1) = H^r_p(R_1), \quad 1 < p < \infty, -\infty < r < \infty, \tag{18}$$

where $H^r_p(R_1)$ are the one-dimensional Bessel-potential spaces. Consequently, (4) yields

$$\|\mathscr{K}f \mid H^r_p(R_1)\| \leqq c\|K \mid S^{(t,r)}_{(u,p)}H\| \, \|f \mid H^{-t}_{u'}(R_1)\|, \tag{19}$$

$\frac{1}{u} + \frac{1}{u'} = 1$. Note again that $H^{-t}_{u'}(R_1)$ is the dual of $H^t_u(R_1)$ (cf. [T, 2.11.2.]). Hence from (19) follows

$$\|\mathscr{K}f \mid W^m_p(R_1)\| \leqq c\|K \mid S^{(l,m)}_{(u,p)}W\| \, \|f \mid (W^l_u(R_1))'\|, \tag{20}$$

where $(W^l_u(R_1))'$ stands for the dual of the one-dimensional Sobolev space $W^l_u(R_1)$. Here, of course, $l = 0, 1, 2, \ldots$ and $m = 0, 1, 2, \ldots$

Remark* **4.** As for the study of mapping properties of integral operators in the framework of Fourier analysis we refer also to H. Triebel [8], where (3) was firstly proved and to H.-J. Schmeisser [2, 5] as far as (2) and (4) are concerned. The theorem shows that spaces with dominating mixed smoothness properties in general can be successfully applied in the theory of integral operators. An analogous assertion should be true in the case of periodic spaces.

2.5.2. Approximation

Spaces with dominating mixed smoothness properties (periodic case) are very useful in approximation theory. In 2.3.2. we proved that $f \in S^{\bar{r}}_{\bar{p},\bar{q}}B$ $(SB^{\bar{r}}_{\bar{p},\bar{q}}, S^{\bar{r}}_{\bar{p},\bar{q}}F)$ can be characterized by approximation with the help of 2-parametric series of entire analytic functions of exponential type (Nikol'skij representations). The periodic counterpart is an approximation with the help of series of trigonometric polynomials. A related problem is the investigation of best approximation of f (periodic distribution) by certain multiple trigonometric polynomials. Mainly two questions lead to spaces with dominating mixed smoothness properties:

(i) Let $1 \leqq p \leqq \infty$. Let $\bar{r} = (r_1, r_2)$, $0 < r_1 < \infty$, $0 < r_2 < \infty$. Let $L_p(T_2)$ be the usual L_p-space on the 2-torus. If $N = 0, 1, 2, \ldots$ and $f \in L_p(T_2)$ then we put

$$E^{\bar{r}}_p(N; f) = \inf_{t_N} \|f - t_N \mid L_p(T_2)\|, \tag{1}$$

where the infimum is taken over all trigonometric polynomials

$$t_N(x) = \sum_{\substack{|k_1|^{r_1}|k_2|^{r_2} \leqq N \\ |k_1|^{r_1} \leqq N, |k_2|^{r_2} \leqq N}} c_{k_1,k_2} \, e^{i(k_1x_1+k_2x_2)}.$$

The behaviour of the numbers $\{E_p^{\bar{r}}(N; f)\}_{N=0}^{\infty}$ and related questions concerning the mixed smoothness of f have been investigated, for example, by K. I. Babenko [1, 2], S. A. Tel'jakovskij [1, 2], Ja. S. Bugrov [1, 3, 5], N. S. Nikol'skaja [1], and V. N. Temljakov [1].

(ii) Let $1 \leqq p \leqq \infty$. Let N, M be natural numbers. If $f \in L_p(T_2)$ then we put

$$\tilde{E}_p(M, N; f) = \inf \|f - (t_{M,1} + t_{N,2}) \mid L_p(T_2)\|, \tag{2}$$

where the infimum is taken over all functions $t_{M,1} \in L_p(T_2)$ and $t_{N,2} \in L_p(T_2)$ such that $t_{M,1} = t_{M,1}(x_1, x_2)$ is a trigonometric polynomial of order M with respect to x_1 and $t_{N,2} = t_{N,2}(x_1, x_2)$ is a trigonometric polynomial of order N with respect to x_2. The behaviour of the sequences $\{\tilde{E}_p(M, N; f)\}_{M,N=0}^{\infty}$ and properties of related spaces with dominating mixed smoothness have been treated, for example, by M. K. Potapov [1–5] and B. Lakovič [1].

We refer also to D. L. Fernandez [6] where approximation in connection with spaces of type $S_{\bar{p},\bar{q}}^{\bar{r}}B(R_2)$ has been studied.

2.5.3. Vector-Valued Besov Spaces

Let $1 \leqq p \leqq \infty$, $1 \leqq q \leqq \infty$ and $0 < r < \infty$. Let A be a Banach space. In Remark 2.3.4/4 we defined the vector-valued Besov space $B_{p,q}^r(R_1, A)$ via differences of functions of the space $L_p(R_1, A)$. Let us remark that the description of spaces with dominating mixed derivatives in the framework of vector-valued spaces (iteration of vector-valued Besov spaces) is due to P. Grisvard [1] (cf. also G. Sparr [1], where this approach is used). It turns out that this method is useful in connection with interpolation theory (cf. 2.5.4.). However, there is no reference where this approach is developed in a satisfactory way. Let us add few remarks. One can try to generalize the Fourier-analysis-approach of the scalar Banach space case (cf. [T, 1.4. and 2.] for $B_{p,q}^r(R_1)$) to the vector-valued case. Then we have to work with A-valued tempered distributions ($S'(R_1, A)$) and with the A-valued Fourier transform ($\mathscr{F}$). A definition of $B_{p,q}^r(R_1, A)$ in this sense reads as follows,

$$B_{p,q}^r(R_1, A) = \{f \mid f \in S'(R_1, A), \|f \mid B_{p,q}^r(R_1, A)\| = \|2^{rj}(\mathscr{F}^{-1}\varphi_j\mathscr{F}f)(x) \mid L_p(R_1, A)| \, l_q\| < \infty\},$$

where $\varphi = \{\varphi_j\} \in \Phi(R_1)$, $1 \leqq p \leqq \infty$, $1 \leqq q \leqq \infty$ and $-\infty < r < \infty$. The main goals should be in analogy with the scalar case a proof of the coincidence of these spaces with the spaces from Remark 2.3.4/4 and a characterization as real interpolation spaces between $L_p(R_1, A)$ and A-valued Sobolev spaces. As mentioned above there is no reference where this theory is worked out in detail. Concerning vector-valued distributions and properties of the vector-valued Fourier transform we refer to L. Schwartz [2, 3], H. Komatsu [1, III], J. Peetre [5], and C. A. McCarthy [1]. Vector-valued Besov spaces are considered also by A. V. Buchvalov [1].

2.5.4. Interpolation

As remarked in the preceding subsection the study of spaces with dominating mixed smoothness properties in connection with interpolation theory is due to P. Grisvard [1] and G. Sparr [1]. In particular, interpolation of several Banach spaces turns out to be a useful and appropriate tool. Two types of questions are of interest: (i) the interpolation of given spaces by some interpolation methods and (ii) the characterization of a given space as an interpolation space of two or several "simpler" spaces. For these topics we refer also to A. Yoshikawa [1], A. Favini [1], C. Foias, J. L. Lions [1], D. L. Fernadez [1–6], J. I. Bertolo, D. L. Fernandez [1], M. Cwikel [1], M. Milman [1], and H. Triebel [4, III–V].

2.5.5. Further Properties

We can ask for further properties of the considered spaces such as duality, bases and discrete representations, convolution algebras, diffeomorphic maps, traces on curves, applications to differential or pseudodifferential operators. Some of these problems can be solved on the basis of the preceding sections by our methods in analogy to the isotropic case treated in [T] (duality, convolution algebras). The other ones seem to require more elaborated techniques. In particular, one should have a practicable interpolation theory (also for $p_i, q_i \leqq 1$) at hand as well as sharp multiplier theorems for $L_{\bar{p}}^{\Omega}(l_{\bar{q}})$ and $Sl_{\bar{q}}(L_{\bar{p}}^{\Omega})$ (cf. Remark 1.10.3/2).

2.5.6. Further Spaces

Spaces on R_n. First of all let us remark that the restriction to R_2 in this chapter is not necessary (cf. H.-J. Schmeisser [1–5] for spaces on $R_{n_1} \times \ldots \times R_{n_d}$).

Periodic Spaces. The theory developed in this chapter has a periodic counterpart (cf. also S. M. Nikol'skij [5], T. I. Amanov [1, 3], Ja. S. Bugrov [1–5] for the case $S_{\bar{p},\bar{q}}^{\bar{r}}B$ with $\min(p_1, p_2, q_1, q_2) \geqq 1$ and N. S. Nikol'skaja [1] for the case $S_{\bar{p}}^{\bar{r}}H$).

Homogeneous Spaces. In contrast to inhomogeneous decompositions of $R_2(R_n)$ as prefered in our definitions, one can deal with homogeneous decompositions of $R_2 - \{x \mid x = (x_1, x_2), x_1 x_2 = 0\}$. We refer to P. I. Lizorkin [2], H. Triebel [4, III–V], B. Stöckert, H. Triebel [1], [F, Chapter 2], and M. L. Gol'dman [4]. We remark that the so-called "dyadic" Hardy spaces are included in this context (cf. [F, 2.5.3.] and H. Triebel [4, III–V]).

The Spaces $S_{\bar{p},\bar{q}}^{\bar{r}^1,\ldots,\bar{r}^N}B$. Given a prescribed set of pairs $\{\bar{r}^1, \ldots, \bar{r}^N\}$, $\bar{p}$, and $\bar{q}$ we put

$$S_{\bar{p},\bar{q}}^{\bar{r}^1,\ldots,\bar{r}^N}B = \bigcap_{i=1}^{N} S_{\bar{p},\bar{q}}^{\bar{r}^i}B .$$

Analogous spaces of type SB or SF can be defined. Problems connected with these spaces are treated, for example, by P. I. Lizorkin, S. M. Nikol'skij [1], S. M. Nikol'skij [3, 4], T. I. Amanov [3], and M. L. Gol'dman [4].

Spaces on Domains. Spaces with dominating mixed smoothness properties on domains are considerd by S. M. Nikol'skij [3, 4] and A. D. Džabrailov [5]. In particular, extensions on R_n and boundary values are treated for a class of domains.

Spaces with Generalized Mixed Smoothness. If we replace the functions $|h_i|^{-r_i}$; ($i = 1, 2$) in the quasi-norms $\|f \mid S_{\bar{p},\bar{q}}^{\bar{r}}B\|^{\Delta}$ (cf. (2.3.4/26)) by more general functions $\varphi_i(|h_i|)$; ($i = 1, 2$) then we obtain also more general spaces. We refer to M. K. Potapov [4, 5] and B. Lakovič [1].

3. Periodic Spaces

3.1. Introduction

In [T, Chapters 1 and 2] the theory of the spaces $B^s_{p,q}(R_n)$ and $F^s_{p,q}(R_n)$ of Besov-Hardy-Sobolev type on the Euclidean n-space R_n has been developed systematically. Here $-\infty < s < \infty$, $0 < p \leqq \infty$ and $0 < q \leqq \infty$ ($p < \infty$ in the case of the spaces $F^s_{p,q}(R_n)$). These two scales contain a lot of classical spaces: Besov spaces (= Lipschitz spaces), Sobolev spaces, Lebesgue spaces (= Bessel-potential spaces), Hölder spaces, Zygmund spaces and non-homogeneous Hardy spaces. In particular, the classical spaces have been extended in a natural way to values p, q with $\min(p, q) < 1$. The basic ingredients are methods of Fourier analysis such as inequalities of Plancheral-Polya-Nikol'skij type, maximal inequalities, and Fourier multipliers for spaces of entire analytic functions of exponential type. As already mentioned in Subsection 2.2.2, this theory can be carried over to other types of spaces on R_n: anisotropic spaces (cf. also Chapter 4), spaces with dominating mixed smoothness properties (cf. Chapters 1, 2), weighted spaces (cf. Chapter 1 and Section 5.1.). The aim of this chapter is to study the periodic counterpart, i.e. we deal with the spaces $B^s_{p,q}(T_n)$ and $F^s_{p,q}(T_n)$ of Besov-Hardy-Sobolev type on the n-dimensional torus T_n, which include the periodic counterparts of the above classical spaces. Periodic Sobolev spaces and Besov spaces $B^s_{p,q}(T_n)$ with $\min(p, q) \geqq 1$ have been studied for a long time, cf. S. M. Nikol'skij [2], and it turned out that they are very useful in approximation theory. It is rather obvious how to deal with spaces on T_n. Roughly speaking entire analytic functions must be replaced by trigonometric polynomials. On the other hand every trigonometric polynomial can be considered as an entire analytic function on R_n. We make use of this fact in order to derive maximal inequalities and Fourier multiplier theorems in Sections 3.3. and 3.4. This is our first goal and it enables us to start a systematic investigation of periodic spaces (Sections 3.5. and 3.6.). However, we do not develop the theory as far as possible. We intend to answer the basic questions such as equivalent quasi-norms, representations, embeddings, interpolation, duality and we describe special cases. Our second goal is to make clear that the methods of Fourier analysis and our approach to periodic spaces can be applied in the field of strong summability (Section 3.7.). Here we shall follow H.-J. Schmeisser, W. Sickel [1], W. Sickel [1, 2], and H. Triebel [17].

3.2. Preliminaries

3.2.1. Distributions on T_n

Let T_n be the n-torus. As usual, we represent T_n in the Euclidean n-space R_n by the cube

$$T_n = \{x \mid x \in R_n, x = (x_1, x_2, \ldots, x_n), |x_j| \leqq \pi \text{ with } j = 1, \ldots, n\}, \tag{1}$$

where opposite points are identified. More precisely, $x \in T_n$ and $y \in T_n$ are identified if and only if $x - y = 2k\pi$, where $k = (k_1, \ldots, k_n) \in Z_n$ and Z_n denotes the set of all lattice-points with integer components (cf. also (1.3.1/5)). Further, let $D(T_n)$ be the collection of all complex-valued infinitely differentiable functions on T_n. In particular, $f(x) = f(y)$ if $x, y \in T_n$, $x - y = 2k\pi$, $k \in Z_n$, and $f \in D(T_n)$. The locally convex topology in $D(T_n)$ is generated by the semi-norms

$$\|f\|_\alpha = \sup_{x \in T_n} |D^\alpha f(x)|, \tag{2}$$

where $\alpha = (\alpha_1, \ldots, \alpha_n)$ is an arbitrary multi-index with non-negative components. Then $D'(T_n)$ is defined to be the topological dual of $D(T_n)$. In other words, $D'(T_n)$ is the set of all linear functionals g on $D(T_n)$ such that

$$|g(f)| \leqq c_N \sum_{|\alpha| \leqq N} \|f\|_\alpha \tag{3}$$

for all $f \in D(T_n)$ and for some natural number N and $c_N > 0$. We use the weak topology on $D'(T_n)$, i.e. $g = \lim_{j\to\infty} g_j$ in $D'(T_n)$ if and only if $g(f) = \lim_{j\to\infty} g_j(f)$ for all $f \in D(T_n)$. We define addition, multiplication by complex numbers, differentiation, and multiplication by functions from $D(T_n)$ in the usual way. Let $0 < p \leqq \infty$. $L_p(T_n)$ has the usual meaning, in particular, we put

$$\left.\begin{aligned} \|g \mid L_p(T_n)\| &= \Big(\int_{T_n} |g(x)|^p \,\mathrm{d}x\Big)^{1/p}, \quad 0 < p < \infty, \\ \|g \mid L_\infty(T_n)\| &= \operatorname*{ess\text{-}sup}_{x \in T_n} |g(x)|. \end{aligned}\right\} \tag{4}$$

If $1 \leqq p \leqq \infty$ then $g \in L_p(T_n)$ can be interpreted in a unique way as an element of $D'(T_n)$ by

$$g(f) = \int_{T_n} g(x)\, f(x)\,\mathrm{d}x, \quad f \in D(T_n). \tag{5}$$

Consequently, we have

$$D(T_n) \subset L_p(T_n) \subset D'(T_n) \tag{6}$$

(topological embeddings) if $1 \leqq p \leqq \infty$. In this sense every trigonometric polynomial

$$t(x) = \sum_{k \in \Lambda} a_k\, \mathrm{e}^{\mathrm{i}kx},$$

where $\Lambda \subset Z_n$ is a finite set and a_k, $k \in \Lambda$, are complex numbers, is an element of $D'(T_n)$. More details about distributions on T_n can be found in R. E. Edwards [1, Chapter 12].

3.2.2. Fourier Coefficients and Fourier Series

Recall that $xy = \sum_{j=1}^{n} x_j y_j$ if $x, y \in R_n$. Let $g \in D'(T_n)$. Then the complex numbers

$$\hat{g}(k) = (2\pi)^{-n}\, g(\mathrm{e}^{-\mathrm{i}kx}), \quad k \in Z_n, \tag{1}$$

are said to be the Fourier coefficients of g. With 3.2.1/(5), (6) it follows

$$\hat{g}(k) = (2\pi)^{-n} \int_{T_n} g(x)\, \mathrm{e}^{-\mathrm{i}kx}\,\mathrm{d}x, \quad k \in Z_n, \tag{2}$$

if $g \in L_p(T_n)$, $1 \leqq p \leqq \infty$. The following facts are well-known (cf. R. E. Edwards [1, Chapter 12]).

(i) Any function $f(x) \in D(T_n)$ can be represented as

$$f(x) = \sum_{k \in Z_n} a_k\, \mathrm{e}^{\mathrm{i}kx} \quad \text{(convergence in } D(T_n)), \tag{3}$$

where $\{a_k\}_{k\in Z_n}$ is a sequence of complex numbers such that

$$|a_k| \leqq c_m(1 + |k|)^{-m}, \quad k \in Z_n, \tag{4}$$

for all $m = 0, 1, 2, \ldots$ Here c_m is an appropriate positive constant. It holds $a_k = \hat{f}(k)$, $k \in Z_n$. Conversely, if $\{a_k\}_{k\in Z_n}$ satisfies (4) then $\sum\limits_{k\in Z_n} a_k\, \mathrm{e}^{ikx}$ converges in $D(T_n)$. If $f(x)$ is its limit function then we have $\hat{f}(k) = a_k$, $k \in Z_n$.

(ii) Any distribution $g \in D'(T_n)$ can be represented as

$$g = \sum_{k\in Z_n} a_k\, \mathrm{e}^{ikx} \quad \text{(convergence in } D'(T_n)), \tag{5}$$

where $\{a_k\}_{k\in Z_n}$ is a sequence of complex numbers such that

$$|a_k| \leqq c_m(1 + |k|)^{m}, \quad k \in Z_n, \tag{6}$$

for some $m = 0, 1, 2, \ldots$ Here c_m is an appropriate positive number. It holds $a_k = \hat{g}(k)$, $k \in Z_n$. Conversely, if $\{a_k\}_{k\in Z_n}$ satisfies (6) for some m, then $\sum\limits_{k\in Z_n} a_k\, \mathrm{e}^{ikx}$ converges in $D'(T_n)$. If g is its limit element then we have $\hat{g}(k) = a_k$, $k \in Z_n$.

Note that

$$g = \sum_{k\in Z_n} \hat{g}(k)\, \mathrm{e}^{ikx}, \quad g \in D'(T_n),$$

is called the Fourier series of g. Further, a sequence $\{a_k\}_{k\in Z_n}$ is said to be of at most polynomial growth if (6) is satisfied.

3.2.3. Periodic Distributions on R_n

Let $S(R_n)$ be the Schwartz space of all complex-valued rapidly decreasing infinitely differentiable functions on R_n and let $S'(R_n)$ be the set of all complex-valued tempered distributions on R_n (cf. 1.2.4.). An element $f \in S'(R_n)$ is said to be a periodic distribution on R_n if

$$f = f(\cdot + 2k\pi) \tag{1}$$

holds for all $k \in Z_n$. This can be reformulated as

$$f(\varphi(x)) = f(\varphi(x + 2k\pi)) \tag{1'}$$

for all $\varphi \in S(R_n)$ and all $k \in Z_n$. The collection of all periodic distributions of R_n is called $S'_\pi(R_n)$. If f is a complex-valued function given on T_n then we extend it periodically on R_n. Let us denote the extended function by the same symbol. In this sense we have, in particular, $\mathrm{e}^{ikx} \in S'_\pi(R_n)$ for all $k \in Z_n$. Moreover, if $\{a_k\}_{k\in Z_n}$ satisfies (3.2.2/6) then $\sum\limits_{k\in Z_n} a_k\, \mathrm{e}^{ikx} \in S'_\pi(R_n)$ (convergence in $S'(R_n)$). Because of (3.2.2/5) any distribution on T_n can be interpreted as a periodic distribution on R_n. The following proposition shows that the converse holds also true. Recall that F denotes the Fourier transform in $S'(R_n)$ (cf. (1.2.4/1)).

Proposition. $f \in S'(R_n)$ *is a periodic distribution if and only if there exists a sequence* $\{a_k\}_{k\in Z_n}$ *of at most polynomial growth such that*

$$f = \sum_{k\in Z_n} a_k\, \mathrm{e}^{ikx} \quad \textit{(convergence in } S'(R_n)). \tag{2}$$

Proof. Let $f \in S'_\pi(R_n)$. Then it follows from (1) that

$$Ff = \mathrm{e}^{i2\pi kx}\, Ff, \quad k \in Z_n.$$

Hence, supp $Ff \subset Z_n$ and

$$Ff = \sum_{k \in Z_n} b_k \delta_k, \tag{3}$$

where δ_k is the δ-distribution with respect to $k \in Z_n$. If $\varphi \in S(R_n)$ with $\varphi(0) = 1$ and supp $\varphi \subset \{x \mid |x| \leqq 1\}$ then

$$b_k = (Ff)(\varphi(\cdot - k)).$$

Well-known properties of $Ff \in S'(R_n)$ yield that $\{b_k\}_{k \in Z_n}$ is of at most polynomial growth. Now,

$$f = F^{-1}(Ff) = \sum_{k \in Z_n} b_k F^{-1} \delta_k = (2\pi)^{-n/2} \sum_{k \in Z_n} b_k \, e^{ikx}$$

implies (2) with $a_k = (2\pi)^{-n/2} b_k$, $k \in Z_n$. The converse direction has been mentioned above.

Remark. The mapping $f \to \{a_k\}_{k \in Z_n}$ in the proposition is one-to-one. This allows to identify $f \in S'_\pi(R_n)$ with $f \in D'(T_n)$ via (3.2.2/5). In this sense we have

$$\|f(x) \mid L_p(T_n)\| = \left(\int_{Q_n} |f(x)|^p \, dx\right)^{1/p}, \quad 1 \leqq p < \infty, \tag{4}$$

where Q_n stands for any cube in R_n with sides of length 2π parallel to the coordinate axes. (Modification if $p = \infty$). The proposition is taken from H. Triebel [17].

3.2.4. Maximal Inequalities

Let $f(x)$ be a complex-valued locally Lebesgue-integrable function on R_n. Then

$$(Mf)(x) = \sup |Q|^{-1} \int_Q |f(y)| \, dy, \quad x \in R_n, \tag{1}$$

is the Hardy-Littlewood maximal function, where the supremum is taken over all cubes Q centered at x (cf. (1.3.3./1)). In 1.3.3., in particular in (1.3.3/3) and (1.3.3/7), we have stated maximal inequalities for the spaces $L_p(R_n)$, $1 < p \leqq \infty$, and $L_p(R_n, l_q)$, $1 < p < \infty$, $1 < q \leqq \infty$. We wish to formulate their periodic counterparts.

Recall that in (1) "Q centered at x" can be replaced by "Q with $x \in Q$". Then the corresponding inequalities (1.3.3/3) and (1.3.3/7) hold true. The spaces $L_p(T_n)$ have been defined in (3.2.1/4). Further we need their l_q-valued versions: If $\{f_j(x)\}_{j=0}^{\infty} \subset L_p(T_n)$, $0 < p < \infty$, we put

$$\|f_j \mid L_p(T_n, l_q)\| = \left(\int_{T_n} \left(\sum_{j=0}^{\infty} |f_j(x)|^q\right)^{p/q} dx\right)^{1/p} \tag{2}$$

if $0 < q < \infty$ and

$$\|f_j \mid L_p(T_n, l_\infty)\| = \left(\int_{T_n} \sup_{j=0,1,\ldots} |f_j(x)|^p \, dx\right)^{1/p}. \tag{3}$$

The corresponding spaces are denoted by $L_p(T_n, l_q)$. Recall that we always assume that all functions and distributions on T_n are 2π-periodically extended on R_n. Then (1) makes sense if $f(x) \in L_p(T_n)$, $1 \leqq p \leqq \infty$ (cf. (3.2.3/4)).

Proposition. *Let* $1 < p < \infty$ *and* $1 < q \leqq \infty$. *Then there exists a positive constant* c *such that*

$$\|Mf_j \mid L_p(T_n, l_q)\| \leqq c \|f_j \mid L_p(T_n, l_q)\| \tag{4}$$

holds for all $\{f_j(x)\}_{j=0}^{\infty} \in L_p(T_n, l_q)$.

Proof. Let Q be an arbitrary cube with sides parallel to the coordinate axes centered at $x \in T_n$. T_n is represented in R_n by (3.2.1/1). Then Q contains N^n, $N = 1, 2, \ldots$, points $x + 2k\pi$, $k \in Z_n$. We decompose Q into N^n mutually disjoint congruent subcubes $Q^{(j)}$, $j = 1, \ldots, N^n$. Then each subcube $Q^{(j)}$ contains exactly one point $x + 2k\pi$. Let $x + 2k^{(j)}\pi \in Q^{(j)}$, $j = 1, \ldots, N^n$. Then by the periodicity of f we obtain

$$\int_Q |f(y)| \, dy = \sum_{j=1}^{N^n} \int_{Q^{(j)} - 2k^{(j)}\pi} |f(y)| \, dy. \tag{5}$$

We have $x \in Q^{(j)} - 2k^{(j)}\pi$ and

$$Q^{(j)} - 2k^{(j)}\pi \subset \{y \mid |y_i| < 3\pi, i = 1, \ldots, n\} = \tilde{T}_n.$$

Hence, $|Q| = N^n |Q^{(j)}|$ and (5) yield

$$\frac{1}{|Q|} \int_Q |f(y)| \, dy \leqq \sup_{\substack{\tilde{Q} \subset \tilde{T}_n \\ x \in \tilde{Q}}} \frac{1}{|\tilde{Q}|} \int_{\tilde{Q}} |f(y)| \, dy. \tag{6}$$

Putting

$$\tilde{f}(y) = \begin{cases} f(y), & y \in \tilde{T}_n \\ 0, & y \notin \tilde{T}_n \end{cases}$$

we obtain from (6)

$$(Mf)(x) \leqq c(M\tilde{f})(x), \quad x \in T_n. \tag{7}$$

Now (4) is an easy consequence of (1.3.3/7) and (3.2.3/4). The proof is complete.

Remark. (4) is the periodic version of the maximal inequality of C. Fefferman, E. M. Stein [1]. Obviously the scalar case ($f_j = 0$ if $j = 1, 2, \ldots$) is valid for $p = \infty$, too.

3.3. Basic Inequalities and Multipliers for Trigonometric Polynomials

3.3.1. Trigonometric Polynomials

Properties of L_p-spaces of entire analytic functions are one of the most powerful tools in the study of function spaces on R_n (cf. [T, Chapters 1 and 2] or Chapters 1, 2, and 5 of this book). We mention inequalities of Plancherel-Polya-Nikol'skij type, maximal inequalities and Fourier multiplier theorems. Now we deal with functions (distributions) on T_n or (what is the same) periodic functions (distributions) on R_n (cf. 3.2.3.). Let F be the Fourier transform in $S'(R_n)$. If $f \in S'_\pi(R_n)$ then we have seen in (3.2.3/3) that supp $Ff \subset Z_n$ holds. Consequently f is a periodic entire analytic function if and only if supp $Ff \subset \Lambda$, where $\Lambda \subset Z_n$ is a finite set of lattice-points. Hence, $f = \sum_{k \in \Lambda} a_k e^{ikx}$ must be a trigonometric polynomial. Let $\Lambda \subset Z_n$ be a finite set. We put

$$T^\Lambda = \left\{ t(x) \mid t(x) = \sum_{k \in \Lambda} a_k e^{ikx}, a_k \text{ complex} \right\}. \tag{1}$$

The spaces T^Λ are the periodic counterparts of the spaces $L_p^\Omega(1, \mu_L)$ from Definition 1.5.1, where μ_L denotes the Lebesgue measure on R_n, or the spaces L_p^Ω of [T, Definition 1.4.1]. In the sense of 3.2.3. we can write

$$\begin{aligned} T^\Lambda &= \{f \mid f \in S'_\pi(R_n), \text{ supp } Ff \subset \Lambda\} \\ &= \{f \mid f \in D'(T_n), \hat{f}(k) = 0 \quad \text{if} \quad k \notin \Lambda\}. \end{aligned} \tag{2}$$

Here $\hat{f}(k)$ has the meaning of (3.2.2/1). Obviously we have

$$T^{\Lambda} \subset L_p(T_n), \quad 0 < p \leqq \infty. \tag{3}$$

On the other hand if we put

$$L_{\infty}^{\Lambda} = \{f \mid f \in S'(R_n), \operatorname{supp} Ff \subset \Lambda, \\ \|f \mid L_{\infty}(R_n)\| = \sup_{x \in R_n} |f(x)| < \infty\}, \tag{4}$$

then we have

$$T^{\Lambda} = L_{\infty}^{\Lambda}. \tag{5}$$

Hence, T^{Λ} coincides with a special L_p^{Ω}-space (or $L_p^{\Omega}(1, \mu_L)$-space). We shall use this fact later on in order to carry over results from the non-periodic spaces to the periodic ones.

The aim of the next subsections and of Section 3.4. is to establish some properties of T^{Λ} which are analogous to Chapter 1 or to [T, Chapter 1]. We restrict ourselves to those assertions which we need in the sequel.

3.3.2. Nikol'skij's Inequality

Let $t = t(x) \in D'(T_n)$ be a trigonometric polynomial. For sake of brevity we put

$$\operatorname{supp} \hat{t} = \{k \mid k \in Z_n, \hat{t}(k) \neq 0\}. \tag{1}$$

Note that $\operatorname{supp} \hat{t} = \operatorname{supp} Ft$ (in the sense of 3.2.3.). Further we denote the number of elements of the finite set $\Lambda, \Lambda \subset Z_n$, by $N(\Lambda)$. Recall that $\|f \mid L_p(T_n)\|$, $0 < p \leqq \infty$, has been defined in (3.2.1/4).

Proposition. *Let $0 < p \leqq q \leqq \infty$ and let p_0 be the smallest integer larger than or equal to $\frac{p}{2}$. Let $t(x)$ be a trigonometric polynomial and let*

$$c_{p_0,t} = (2\pi)^{-n} N(\operatorname{supp} (t^{p_0})^{\wedge}). \tag{2}$$

Then

$$\|t(x) \mid L_q(T_n)\| \leqq (c_{p_0,t})^{1/p-1/q} \|t(x) \mid L_p(T_n)\| \tag{3}$$

holds.

Proof. Step 1. We prove (3) for $q = \infty$. If $t(x)$ is a trigonometric polynomial we denote by

$$D_t(x) = \sum_{k \in \operatorname{supp} \hat{t}} e^{ikx} \tag{4}$$

the corresponding Dirichlet kernel. Let $m = 1, 2, \ldots$ Then $t^m(x)$ is again a trigonometric polynomial and from (3.2.2/2) follows

$$t^m(x) = (2\pi)^{-n} \int_{T_n} t^m(y) D_{t^m}(x-y) \, dy. \tag{5}$$

Choosing $m = p_0$, we obtain from (5)

$$\begin{aligned} |t^{p_0}(x)| &\leqq (2\pi)^{-n} \|t^{p_0} \mid L_2(T_n)\| \|D_{t^{p_0}} \mid L_2(T_n)\| \\ &= (2\pi)^{-n/2} [N(\operatorname{supp} (t^{p_0})^{\wedge})]^{1/2} \|t^{p_0} \mid L_2(T_n)\| \\ &= (c_{p_0,t})^{1/2} \|t^{p_0} \mid L_2(T_n)\|. \end{aligned} \tag{6}$$

We estimate the right-hand side of (6) by

$$\|t^{p_0} \mid L_2(T_n)\| = \left(\int_{T_n} |t(x)|^{2p_0-p+p}\, dx\right)^{1/2}$$
$$\leqq \|t \mid L_\infty(T_n)\|^{p_0-\frac{p}{2}} \|t \mid L_p(T_n)\|^{\frac{p}{2}}. \tag{7}$$

Now, as a consequence of (6) and (7) it follows

$$\|t \mid L_\infty(T_n)\| \leqq (c_{p_0,t})^{1/p} \|t \mid L_p(T_n)\|. \tag{8}$$

This proves (3) in the case $q = \infty$ and $0 < p \leqq \infty$.

Step 2. Let $0 < p < q < \infty$. We have

$$\|t(x) \mid L_q(T_n)\| = \left(\int_{T_n} |t(x)|^{p+q-p}\, dx\right)^{1/q}$$
$$\leqq \|t(x) \mid L_\infty(T_n)\|^{1-\frac{p}{q}} \|t(x) \mid L_p(T_n)\|^{\frac{p}{q}}. \tag{9}$$

Hence (8) and (9) yield immediately the desired inequality (3). Thus, the proof is complete.

The above proposition enables us to formulate a result for the spaces T^Λ from, (3.3.1/1).

Theorem. *Let $\Lambda \subset Z_n$ be a finite set such that*

$$d_\Lambda = \max_{k,l \in \Lambda} |k - l| > 0. \tag{10}$$

If $0 < p \leqq q \leqq \infty$ then there exists a positive constant c depending on p, q, and n such that

$$\|t(x) \mid L_q(T_n)\| \leqq c d_\Lambda^{n\left(\frac{1}{p}-\frac{1}{q}\right)} \|t(x) \mid L_p(T_n)\| \tag{11}$$

holds for all $t(x) \in T^\Lambda$.

Proof. (11) is an easy consequence of (3) and

$$N(\operatorname{supp}(t^{p_0})^\wedge) \leqq c' d_\Lambda^n,$$

where $t \in T^\Lambda$ and p_0 is the number from the proposition.

Remark 1. We are not interested in the best possible constants $c_{p_0,t}$ for all $t \in T^\Lambda$. However, note that $c_{p_0,t} \leqq (2\pi)^{-n} N(\Lambda)$ if $0 < p \leqq 2$. For further considerations we refer to R. J. Nessel, G. Wilmes [1].

Remark 2. Let

$$\Lambda \subset \{k \mid k \in Z_n, |k_j| \leqq N, j = 1, \ldots, n\}, \tag{12}$$

where N is an arbitrary natural number. Then we obtain by (11) that there exists a positive constant c such that

$$\|t(x) \mid L_q(T_n)\| \leqq c N^{n\left(\frac{1}{p}-\frac{1}{q}\right)} \|t(x) \mid L_p(T_n)\| \tag{13}$$

holds for all $t \in T^\Lambda$.

Remark 3. Further, as a consequence of inequality (11) we obtain that in the space T^Λ all quasi-norms $\|t \mid L_p(T_n)\|$, $0 < p \leqq \infty$, are equivalent to each other. Equipped with one of these quasi-norms T^Λ becomes a quasi-Banach space (Banach space if $1 \leqq p \leqq \infty$).

Remark* 4. (3), (11), and (13) are inequalities of Nikol'skij type for trigonometric polynomials, which are due to S. M. Nikol'skij [1]. Inequalities of such a type have been considered by several authors. For references see, for example, S. M. Nikol'skij [2] and R. J. Nessel, G. Wilmes [1, 2, 3] (cf. also Remark* 1.4.3). The proof of the Proposition we have given here (mainly for the convenience of the reader) is due to R. J. Nessel, G. Wilmes [1].

3.3.3. Convolution Inequalities

The inequalities 3.3.2/(3), (11), (13) can be used in order to deduce inequalities for the convolution of trigonometric polynomials. If $t(x)$ and $\tau(x)$ are trigonometric polynomials then their convolution $(t * \tau)(x)$ is well-defined by

$$(t * \tau)(x) = \int_{T_n} t(y)\, \tau(x - y)\, dy. \tag{1}$$

Let $\Lambda \subset Z_n$ be a finite subset. We put

$$\Lambda - \Lambda = \{m \mid m \in Z_n, m = k - l, \text{ where } k \in \Lambda \text{ and } l \in \Lambda\}.$$

Further $N(\Lambda)$ has the same meaning as at the beginning of 3.3.2.

Theorem. *Let $\Lambda \subset Z_n$ be a finite subset and let $0 < p \leqq \infty$. Then there exists a positive constant c depending on p and n such that*

$$\|t * \tau \mid L_p(T_n)\| \leqq c[N(\Lambda - \Lambda)]^{\frac{1}{\min(1,p)} - 1} \|t \mid L_p(T_n)\| \, \|\tau \mid L_p(T_n)\| \tag{2}$$

holds for all $t \in T^\Lambda$ and all $\tau \in T^\Lambda$.

Proof. In order to prove (2) we begin with (1). The case $1 \leqq p \leqq \infty$ is obvious. Let $0 < p < 1$. If $x \in T_n$ is fixed, then $h(x, y) = t(y)\, \tau(x - y)$ is a trigonometric polynomial with respect to y. Moreover, we have

$$\operatorname{supp} (h(x, \cdot)^\wedge \subset \Lambda - \Lambda$$

for all $x \in T_n$. Consequently, (3.3.2/3) yields the existence of a positive constant c independent of $x \in T_n$ such that

$$|(t * \tau)(x)| \leqq c[N(\Lambda - \Lambda)]^{\frac{1}{p} - 1} \left(\int_{T_n} |t(y)\, \tau(x - y)|^p\, dy\right)^{\frac{1}{p}} \tag{3}$$

holds for all $t \in T^\Lambda$ and $\tau \in T^\Lambda$. Raising (3) to the power p and integrating with respect to x we obtain by Fubini's theorem

$$\|(t * \tau)(x) \mid L_p(T_n)\| \leqq c[N(\Lambda - \Lambda)]^{\frac{1}{p} - 1} \left(\int_{T_n} \int_{T_n} |t(y)\, \tau(x - y)|^p\, dx\, dy\right)^{\frac{1}{p}}$$

$$= c[N(\Lambda - \Lambda)]^{\frac{1}{p} - 1} \|t(x) \mid L_p(T_n)\| \, \|\tau(x) \mid L_p(T_n)\| .$$

This proves the theorem.

Remark 1. As a consequence of (2) we obtain the following inequality: Let N be a natural number and let Λ be as in (3.3.2/12). If $0 < p \leqq \infty$ then there exists a positive constant c depending on p and n such that

$$\|t * \tau \mid L_p(T_n)\| \leqq cN^{n\left(\frac{1}{\min(1,p)} - 1\right)} \|t \mid L_p(T_n)\| \, \|\tau \mid L_p(T_n)\| \tag{4}$$

holds for all $t \in T^\Lambda$ and all $\tau \in T^\Lambda$.

Remark* 2. (2) and (4) are the periodic counterparts of [T, Proposition 1.5.3 and (1.5.3/3)] or of Theorem 1.7.3 and (1.7.3/7). In contrast to the convolution inequalities in R_n, now $1 < p \leqq \infty$ is admitted. The proof followed W. Sickel [1, Theorem 2.1/3].

3.3.4. Fourier Multipliers

Let $\Lambda \subset Z_n$ be a finite subset and let $\{M_k\}_{k\in\Lambda}$ be a set of complex numbers. The problem is to find sufficient criterions for $\{M_k\}_{k\in\Lambda}$ such that there exists a positive constant c with

$$\left\| \sum_{k\in\Lambda} M_k \hat{t}(k)\, e^{ikx} \mid L_p(T_n) \right\| \leqq c \| t(x) \mid L_p(T_n) \| \tag{1}$$

for all $t \in T^\Lambda$. Here $0 < p \leqq \infty$. Furthermore, the dependence of the constant c in (1) on Λ and p is of interest. We restrict ourselves to the case where $\{M_k\}_{k\in\Lambda}$ is generated by a continuous function $M(x)$ defined on R_n, i.e. the case $M_k = M(k)$, $k \in \Lambda$.

Recall that F, F^{-1}, and $L_p(R_n)$, $0 < p \leqq \infty$, have been defined in 1.2.4. and 1.3.1., respectively. If $M \in S'(R_n)$ with $F^{-1}M \in L_1(R_n)$ then by well-known properties of the Fourier transform, $M = M(x)$ is a continuous function on R_n. Therefore, $M_k = M(k)$, $k \in \Lambda \subset Z_n$, makes sense. We have

$$\begin{aligned}\sum_{k\in\Lambda} M_k \hat{t}(k)\, e^{ikx} &= \sum_{k\in\Lambda} [F(F^{-1}M)](k)\, \hat{t}(k)\, e^{ikx} \\ &= (2\pi)^{-n/2} \int_{R_n} (F^{-1}M)(y) \left[\sum_{k\in\Lambda} \hat{t}(k)\, e^{-ik(x-y)}\right] dy \\ &= (2\pi)^{-n/2} \int_{R_n} (F^{-1}M)(y)\, t(x-y)\, dy. \end{aligned} \tag{2}$$

This shows the compatibility of (1) with the definition of Fourier multipliers for the spaces L_p^Ω (cf. [T, Definition 1.5.1] or Definition 1.7.1/1 and Remark 1.7.1/1). Now, we use (2) in order to derive criterions on M such that (1) is satisfied. Recall that the spaces $H_2^\varkappa = H_2^\varkappa(R_n)$, $0 \leqq \varkappa < \infty$, have been defined in (1.7.4/1).

Theorem. *Let $\Lambda \subset Z_n$ be a finite set such that*

$$d_\Lambda = \max_{k,l\in\Lambda} |k-l| > 0.$$

Let $0 < p \leqq \infty$ and $0 < \varkappa < \infty$ with

$$\varkappa > \sigma_p = n\left(\frac{1}{\min(1,p)} - \frac{1}{2}\right). \tag{3}$$

Then there exists a positive constant c depending on p and n such that

$$\left\| \sum_{k\in\Lambda} M(k)\, \hat{t}(k)\, e^{ikx} \mid L_p(T_n) \right\| \leqq c \| M(d_\Lambda \cdot) \mid H_2^\varkappa \| \, \| t(x) \mid L_p(T_n) \| \tag{4}$$

holds for all $M \in H_2^\varkappa(R_n)$ and all $t \in T^\Lambda$.

Proof. Without loss of generality we may assume $0 \in \Lambda$ (otherwise consider $e^{-ik^0x}\, t(x)$ instead of $t(x)$). If $0 < p \leqq \infty$ then in any case $\varkappa > \frac{n}{2}$. Hence, by Proposition 1.7.5, $F^{-1}M \in L_1(R_n)$. Then well-known properties of the Fourier transform yield that M must be a continuous function on R_n. Thus, the left-hand side of (4) makes sense. We choose a function $\psi(x) \in S(R_n)$ with $\operatorname{supp}\psi \subset \{x \mid |x| \leqq 2\}$ and $\psi(x) = 1$ if $|x| \leqq 1$. Then $\psi M \in H_2^\varkappa(R_n)$ and

$$\| \psi M \mid H_2^\varkappa(R_n) \| \leqq c_\psi \| M \mid H_2^\varkappa(R_n) \|. \tag{5}$$

Furthermore, Proposition 1.7.5 yields

$$F^{-1}(\psi(d_\Lambda^{-1} \cdot)\, M) \in L_1(R_n).$$

Therefore, formula (2) can be applied and gives

$$\Big|\sum_{k\in\Lambda} M(k)\,\hat{t}(k)\,\mathrm{e}^{ikx}\Big| \leqq c \int_{R_n} |(F^{-1}\psi(d_\Lambda^{-1}\cdot)\,M)\,(y)\,t(x-y)|\,\mathrm{d}y. \tag{6}$$

Let $\tilde{p} = \min(1, p)$. For fixed $x \in T_n$ we have by Nikol'skij's inequality for entire analytic functions (cf. [T, Theorem 1.4.1(ii)] or Theorem 1.5.2(iii))

$$\|F^{-1}(\psi(d_\Lambda^{-1}\cdot)\,M)\,(y)\,t(x-y) \mid L_{\tilde{p}}(R_n)\|$$
$$\leqq \|t \mid L_\infty(R_n)\|\;\|F^{-1}\psi(d_\Lambda^{-1}\cdot)\,M \mid L_{\tilde{p}}(R_n)\| < \infty.$$

Further,

$$\operatorname{supp} F[F^{-1}\psi(d_\Lambda^{-1}\cdot)\,M)\,t(x-\cdot)] \subset \{y \mid |y| \leqq 3d_\Lambda\}.$$

Consequently, the right-hand side of (6) can be estimated again with the help of Nikol'skij's inequality (cf. [T, Theorem 1.4.1(ii) and (1.3.2/5)]). We have

$$\Big|\sum_{k\in\Lambda} M(k)\,\hat{t}(k)\,\mathrm{e}^{ikx}\Big| \leqq c' d_\Lambda^{n\left(\frac{1}{\tilde{p}}-1\right)} \Big(\int_{R_n} |(F^{-1}\psi(d_\Lambda^{-1}\cdot)\,M)\,(y)\,t(x-y)|^{\tilde{p}}\,\mathrm{d}y\Big)^{\frac{1}{\tilde{p}}}. \tag{7}$$

Taking the quasi-norm $\|\cdot \mid L_p(T_n)\|$ on both sides one obtains

$$\Big\|\sum_{k\in\Lambda} M(k)\,\hat{t}(k)\,\mathrm{e}^{ikx} \mid L_p(T_n)\Big\|$$
$$\leqq c' d_\Lambda^{n\left(\frac{1}{\tilde{p}}-1\right)} \|F^{-1}\psi(d_\Lambda^{-1}\cdot)\,M \mid L_{\tilde{p}}(R_n)|\;\|t(x) \mid L_p(T_n)\|$$
$$= c'\|F^{-1}\psi M(d_\Lambda\cdot) \mid L_{\tilde{p}}(R_n)\|\;\|t(x) \mid L_p(T_n)\|$$
$$\leqq c''\|M(d_\Lambda\cdot) \mid H_2^\varkappa\|\;\|t(x) \mid L_p(T_n)\|. \tag{8}$$

In the last estimate we used Proposition 1.7.5 and (5). Thus the theorem is proved.

Remark* 1. We followed W. Sickel [2, Theorem 2]. The above theorem is the periodic counterpart of [T, Theorem 1.5.2]. It tells us that a class of Fourier multipliers for the non-periodic spaces L_p^Ω from [T, Section 1.5.] can be carried over to the periodic case simply by restriction of $M(x)$ on Z_n. Assertions of such a type are well-known in the non-analytic case. For $1 < p < \infty$ we refer to K. de Leeuw [1] and E. M. Stein, G. Weiss [1, Chapter VII.3], and for $0 < p \leqq 1$ (Hardy spaces) to L. Colzani [1]. Criterions for Fourier multipliers in periodic L_p-spaces with $1 < p < \infty$ have been studied extensively. In addition to the literature cited above we refer to R. E. Edwards [1, Chapter 16], R. E. Edwards, G. I. Gaudry [1], G. Gaspar, W. Trebels [1], B. Muckenhoupt, R. L. Wheeden, Wo-Sang Young [2, Chapter 15], H. J. Mertens, R. J. Nessel, G. Wilmes [1], S. M. Nikol'skij [2, 1.5.3], and W. Trebels [1].

As a comparatively simple application of (4) we can deduce an inequality of Bernstein type for trigonometric polynomials, which will be formulated in the following corollary.

Corollary. *Let* $0 < p \leqq \infty$ *and let* $\Lambda \subset Z_n$ *be a finite set with* $\Lambda \subset \{k \mid k \in Z_n, |k_j| \leqq N, j = 1, \ldots, n\}$, *where N is a given natural number. Then there exists a positive constant c such that*

$$\|D^\alpha t(x) \mid L_p(T_n)\| \leqq cN^{|\alpha|}\|t(x) \mid L_p(T_n)\| \tag{9}$$

holds for all $t \in T^\Lambda$ *and all multi-indices* $\alpha = (\alpha_1, \ldots, \alpha_n)$ *with non-negative integers* α_j; $j = 1, \ldots, n$.

Proof. We choose a function $\psi(x) \in S(R_n)$ with a compact support and $\psi(x) = 1$ if $|x_j| \leqq 1$. Then (4) yields

$$\|D^\alpha t(x) \mid L_p(T_n)\| = \left\| \sum_{k \in \Lambda} \psi\left(\frac{k}{N}\right) k_1^{\alpha_1} \dots k_n^{\alpha_n} \hat{t}(k)\, \mathrm{e}^{\mathrm{i}kx} \mid L_p(T_n) \right\|$$
$$\leqq cN^{|\alpha|} \|x_1^{\alpha_1} \dots x_n^{\alpha_n} \psi \mid H_2^\varkappa\| \, \|t(x) \mid L_p(T_n)\|, \tag{10}$$

where $\varkappa > \sigma_p = n\left(\frac{1}{\min(1, p)} - \frac{1}{2}\right)$. This proves (9).

Remark* 2. Inequalities of type (9) have a long history. For references see R. J. Nessel, G. Wilmes [2]. For the extension to values $0 < p \leqq 1$ we refer to P. Oswald [1] and P. G. Nevai [1].

3.3.5. Maximal Inequalities

Let Ω be a compact subset of R_n. We recall

$$L_\infty^\Omega = \{f \mid f \in L_\infty(R_n),\ \operatorname{supp} Ff \subset \Omega\}. \tag{1}$$

Further, Mf denotes the Hardy-Littlewood maximal function of f, (cf. (3.2.4/1)).

Proposition. *Let Ω be a compact subset of R_n. Let $0 < r < \infty$ and let $\alpha = (\alpha_1, \dots, \alpha_n)$ be a multi-index of non-negative integers. Then there exist two positive numbers c_1 and c_2 such that*

$$\sup_{z \in R_n} \frac{|D^\alpha f(x - z)|}{1 + |z|^{n/r}} \leqq c_1 \sup_{z \in R_n} \frac{|f(x - z)|}{1 + |z|^{n/r}}$$
$$\leqq c_2 (M \mid f|^r)^{1/r}(x),\ x \in R_n, \tag{2}$$

holds for all $f \in L_\infty^\Omega$.

Proof. (2) is essentially the unweighted version of Theorem 1.4.2 (cf. also [T, Theorem 1.3.1]). The extension to $f \in L_\infty^\Omega$ causes no problems.

Remark 1. Inequalities of type (2) hold a key position in the theory of (weighted or unweighted) function spaces, in particular for values $p < 1$ (cf. also Remark* 1.4.2/1).

Remark 2. Let Λ be a finite subset of Z_n. Then we have $L_\infty^\Lambda = T^\Lambda$ (cf. (3.3.1/5)). Consequently, (2) can be applied to trigonometric polynomials. Further, in the case $\Omega = \Lambda \subset Z_n$ (finite subset), (2) can be reformulated: One can replace $z \in R_n$ by $z \in T_n$ and $1 + |z|^{n/r}$ can be omitted. Here T_n is represented in R_n as in (3.2.1/1).

Theorem. *Let $0 < r < p \leqq \infty$. Let Λ be a finite subset of Z_n with*

$$d_\Lambda = \max_{k, l \in \Lambda} |k - l| > 0. \tag{3}$$

Then there exists a positive constant c depending on p and n such that

$$\left\| \sup_{z \in R_n} \frac{|t(x - z)|}{1 + |d_\Lambda z|^{n/r}} \mid L_p(T_n) \right\| \leqq c \|t(x) \mid L_p(T_n)\| \tag{4}$$

holds for all sets Λ and all trigonometric polynomials $t \in T^\Lambda$.

Proof. Again we restrict ourselves to subsets $\Lambda \subset Z_n$ with $0 \in \Lambda$. Then consider the function $t(d_\Lambda^{-1} x)$, where $t \in T^\Lambda$. By 3.3.1/(2), (5) it follows $t(d_\Lambda^{-1} x) \in L_\infty^\Omega$ with

$$\Omega = \left\{ y \mid y = \frac{k}{d_\Lambda},\ k \in \Lambda \right\} \subset \{y \mid |y| \leqq 1\}.$$

We put temporarily $\tilde{t}(x) = t(d_A^{-1}x)$. Then (2) yields the estimate

$$\sup_{z\in R_n} \frac{|t(x-z)|}{1+|d_A z|^{n/r}} = \sup_{z\in R_n} \frac{|\tilde{t}(d_A x - z)|}{1+|z|^{n/r}}$$
$$\leqq c(M|\tilde{t}|^r)^{1/r}(d_A x) = c(M|t|^r)^{1/r}(x). \tag{5}$$

Here c is independent of Λ. Now, we apply the scalar version of Proposition 3.2.4 to (5) and obtain (4) by a well-known procedure (cf. (1.5.2/4)). This proves the theorem.

Remark 3. (4) is the periodic counterpart of Theorem 1.5.2(i) (cf. also [T, Theorem 1.4.1(i)] for the unweighted non-periodic case).

3.4. Vector-Valued L_p-Spaces of Periodic Functions

3.4.1. Maximal Inequalities and Fourier Multipliers

The aim of this section is twofold. On the one hand we extend the results of 3.3.4. and 3.3.5. to the vector-valued case or, more precisely, to spaces of type $L_p(T_n, l_q)$ from 3.2.4/(2), (3) of trigonometric polynomials. On the other hand we deal with Fourier multipliers for $L_p(T_n, l_q)$, $1 < p < \infty$, $1 < q < \infty$ (non-analytic case). Thus, the following can be considered as the counterpart of Section 1.9., where weighted spaces have been treated.

Let $\Lambda = \{\Lambda_j\}_{j=0}^{\infty}$ be a sequence of finite subsets of Z_n. Further let

$$d_j = \max_{k,l\in\Lambda_j} |k-l| > 0; \quad j = 0, 1, \ldots$$

We recall that $L_p(T_n, l_q)$ and $T^{\Omega}, \Omega \subset Z_n$, have been defined in 3.2.4/(2), (3) and (3.3.1/1), respectively. If $0 < p < \infty$, $0 < q \leqq \infty$ and if Λ is given as above then we introduce

$$L_p^{\Lambda}(T_n, l_q) = \{t \mid t = \{t_j(x)\}_{j=0}^{\infty},\ t_j \in T^{\Lambda_j} \text{ if } j = 0, 1, \ldots,$$
$$\|t_j \mid L_p(T_n, l_q)\| < \infty\}. \tag{1}$$

The spaces $L_p^{\Lambda}(T_n, l_q)$ are quasi-Banach spaces (Banach spaces if $\min(p, q) \geqq 1$). They are the periodic counterparts of the spaces $L_p^{\Omega}(R_n, l_q)$ from [T, Definition 1.6.1].

Theorem 1. *Let $0 < p < \infty$ and $0 < q \leqq \infty$. Let $\Lambda = \{\Lambda_j\}_{j=0}^{\infty}$ be a sequence of finite subsets of Z_n with $d_j > 0$; $j = 0, 1, 2, \ldots$ If $0 < r < \min(p, q)$ then there exists a positive constant c depending on p, q, r, and n such that*

$$\left\| \sup_{z\in R_n} \frac{|t_j(\cdot - z)|}{1+|d_j z|^{n/r}} \,\Big|\, L_p(T_n, l_q) \right\| \leqq c\|t_j \mid L_p(T_n, l_q)\| \tag{2}$$

holds for all $t = \{t_j\}_{j=0}^{\infty} \in L_p^{\Lambda}(T_n, l_q)$.

Proof. Let $t = \{t_j\}_{j=0}^{\infty} \in L_p^{\Lambda}(T_n, l_q)$. It is easy to see that we may restrict ourselves to sequences $\Lambda = \{\Lambda_j\}_{j=0}^{\infty}$ with $0 \in \Lambda_j$. Now we use (3.3.5/5) with t_j and Λ_j instead of t and Λ. Consequently,

$$\sup_{z\in R_n} \frac{|t_j(x-z)|}{1+|d_j z|^{n/r}} \leqq c(M|t_j|^r)^{1/r}(x),\ x \in R_n; \quad j = 0, 1, 2, \ldots, \tag{3}$$

where the constant c is independent of $\{t_j\}$ and $\{\Lambda_j\}$. Then (2) follows from (3) and Proposition 3.2.4 by standard arguments (cf. 1.9.1/(11), (12)).

Remark 1. The Theorem is the vector-valued counterpart of Theorem 3.3.5. In contrast to the latter theorem we have now supposed $p < \infty$. This is necessary because Proposition 3.2.4 does not hold true in the cases $p = \infty$ and $1 < q < \infty$. (Cf. also [T, Remark 1.6.2] and C. Fefferman, E. M. Stein [1].)

Recall that the spaces $H_2^\varkappa = H_2^\varkappa(R_n)$ have been defined in (1.7.4/1).

Theorem 2. *Let $0 < p < \infty$ and $0 < q \leqq \infty$. Let $\Lambda = \{\Lambda_j\}_{j=0}^\infty$ be a sequence of finite subsets of Z_n with $d_j > 0; j = 0, 1, 2, \ldots$ If*

$$\varkappa > n\left(\frac{1}{\min(p, q)} + \frac{1}{2}\right), \tag{4}$$

then there exists a positive constant c such that

$$\Big\|\sum_{k\in Z_n} M_j(k)\, \hat{t}_j(k)\, \mathrm{e}^{ikx} \mid L_p(T_n, l_q)\Big\| \leqq c \sup_j \|M_j(d_j\cdot) \mid H_2^\varkappa\|\, \|t_j \mid L_p(T_n, l_q)\| \tag{5}$$

holds for all systems $t = \{t_j(x)\}_{j=0}^\infty \in L_p^\Lambda(T_n, l_q)$ and all sequences $\{M_j(x)\}_{j=0}^\infty \subset H_2^\varkappa(R_n)$.

Proof. Because of (4) and Proposition 1.7.5 we have $F^{-1}M_j \in L_1(R_n)$. Consequently, by well-known properties of the Fourier transform M_j is a continuous function on R_n. Therefore, the functions M_j are well-defined for all $k \in Z_n$ and we obtain by (3.3.4/2)

$$\Big|\sum_{k\in Z_n} M_j(k)\, \hat{t}_j(k)\, \mathrm{e}^{ikx}\Big| \leqq c \int_{R_n} |(F^{-1}M_j)(y)\, t_j(x-y)|\, \mathrm{d}y, \tag{6}$$

$x \in R_n, j = 0, 1, 2, \ldots$ For sake of brevity we put

$$t_j^*(x) = \sup_{z\in R_n} \frac{|t_j(x-z)|}{1 + |d_j z|^{n/r}}, \tag{7}$$

where we choose r later on. Then we have

$$\begin{aligned}
\Big|\sum_{k\in Z_n} M_j(k)\, \hat{t}_j(k)\, \mathrm{e}^{ik(x-z)}\Big| &\leqq c \int_{R_n} |(F^{-1}M_j)(x-z-y)|\, |t_j(y)|\, \mathrm{d}y \\
&\leqq c t_j^*(x) \int_{R_n} |(F^{-1}M_j)(x-z-y)|\, (1 + |d_j(x-y)|^{n/r})\, \mathrm{d}y \\
&\leqq c' t_j^*(x)\, (1 + |d_j z|^{n/r})\, d_j^{-n} \int_{R_n} |(F^{-1}M_j)(d_j^{-1}y)|\, (1 + |y|^{n/r})\, \mathrm{d}y \\
&= c' t_j^*(y)\, (1 + |d_j z|^{n/r})\, \|(1 + |y|^{n/r})\, F^{-1}M_j(d_j\cdot) \mid L_1(R_n)\|\,.
\end{aligned} \tag{8}$$

By Proposition 1.7.5, $0 < r < \min(p, q)$, and $\varkappa > \dfrac{n}{r} + \dfrac{n}{2}$ we obtain

$$\sup_{z\in R_n} \frac{\Big|\sum_{k\in Z_n} M_j(k)\, \hat{t}_j(k)\, \mathrm{e}^{ik(x-z)}\Big|}{1 + |d_j z|^{n/r}} \leqq c \|M_j(d_j\cdot)|\, H_2^\varkappa\|\, t_j^*(x), \tag{9}$$

where c is independent of x and j_x. Clearly,

$$\Big|\sum_{k\in Z_n} M_j(k)\, \hat{t}_j(k)\, \mathrm{e}^{ikx}\Big| \leqq \sup_{z\in R_n} \frac{\Big|\sum_{k\in Z_n} M_j(k)\, \hat{t}_j(k)\, \mathrm{e}^{ik(x-z)}\Big|}{1 + |d_j z|^{n/r}}\,. \tag{10}$$

Now, (5) is an immediate consequence of (9), (10), Theorem 1, and $0 < r < \min(p, q)$. Thus, the theorem is proved.

Remark 2. (5) is the vector-valued counterpart of (3.3.4/4) and the periodic counterpart of [T, Theorem 1.6.3]. Condition (4) concerning the value of $\varkappa$ depends on the maximal technique. As we shall see later on (cf. Theorem 3.6.4) it can be improved essentially.

Remark* 3. In H. Triebel [17] an inequality of type (5) has been proved (with a somewhat larger $\varkappa$) via weighted spaces. Our proof followed W. Sickel [1, 2].

3.4.2. Fourier Multipliers for $L_p(T_n, l_q)$, $1 < p < \infty$, $1 < q < \infty$

Our approach to Fourier multipliers for the spaces $L_p(T_n, l_q)$, $1 < p < \infty$, $1 < q < \infty$, can be described as follows: We suppose that the Fourier multiplier theorems for $L_p(R_n, l_q)$, $1 < p < \infty$, $1 < q < \infty$, are known and show that every such a multiplier implies a Fourier multiplier for $L_p(T_n, l_q)$, $1 < p < \infty$, $1 < q < \infty$ (if it makes sense). This method is due to K. de Leeuw [1] (cf. E. M. Stein, G. Weiss [1, Chapter 7, Section 3]). It turns out that all multipliers which we need in the sequel can be carried over from the non-periodic case to the periodic one in this way.

Recall that we have defined the spaces $L_p(R_n, l_q)$ and $L_p(T_n, l_q)$ in (1.3.1/8) and (3.2.4/2), respectively.

Definition 1. *Let $1 < p < \infty$ and $1 < q < \infty$. Let $M = \{M_{\mu,\nu}(x)\}_{\mu,\nu=0}^{\infty} \subset L_\infty(R_n)$. Then M is said to be a Fourier multiplier for $L_p(R_n, l_q)$ if there exists a positive constant c_M such that*

$$\left\| \left\{ \sum_{\nu=0}^{\infty} F^{-1} M_{\mu,\nu} F f_\nu \right\}_\mu \mid L_p(R_n, l_q) \right\| \leqq c_M \| \{f_\nu\}_\nu \mid L_p(R_n, l_q) \| \tag{1}$$

holds for all finite sequences $\{f_\nu(x)\}_\nu \subset S(R_n)$.

Remark 1. In (1) we restricted ourselves to finite sequences $\{f_\nu\} \subset S(R_n)$. However, the collection of all these sequences is dense in $L_p(R_n, l_q)$ because $1 < p < \infty$ and $1 < q < \infty$. Consequently, if M is a Fourier multiplier then (1) holds for all elements $\{f_\nu\}_\nu \in L_p(R_n, l_q)$, where the left-hand side is defined via the limiting process. The above definition coincides essentially with the notation of a Fourier multiplier as it had been used in [I, 2.2.4.]. The restriction $M \subset L_\infty(R_n)$ is quite natural because we can easily see that each $M_{\mu,\nu}$, $\mu = 0, 1, \ldots, \nu = 0, 1, \ldots$, is a Fourier multiplier for $L_p(R_n)$ and thus an essentially bounded function on R_n (cf. E. M. Stein, G. Weiss [1, Chapter 1]).

Let us denote by $l_\infty(Z_n)$ the space of all complex-valued sequences $\{a_k\}_{k \in Z_n}$ with $\sup_{k \in Z_n} |a_k| < \infty$.

Definition 2. *Let $1 < p < \infty$ and $1 < q < \infty$. Let $\tilde{M} = \{\{\tilde{M}_{\mu,\nu}(k)\}_k\}_{\mu,\nu=0}^{\infty} \subset l_\infty(Z_n)$. Then $\tilde{M}$ is said to be a Fourier multiplier for $L_p(T_n, l_q)$ if there exists a positive constant $\tilde{c}_M$ such that*

$$\left\| \left\{ \sum_{\nu=0}^{\infty} \left[\sum_{k \in Z_n} \tilde{M}_{\mu,\nu}(k) \hat{f}_\nu(k)\, \mathrm{e}^{\mathrm{i}kx} \right] \right\}_\mu \mid L_p(T_n, l_q) \right\| \leqq \tilde{c}_M \| \{f_\nu\}_\nu \mid L_p(T_n, l_q) \| \tag{2}$$

holds for all finite systems $f = \{f_\nu\}_\nu \in L_p(T_n, l_q)$.

Remark 2. 3.2.2(ii) shows that the left-hand side of (2) is meaningful. In particular, each element

$$\sum_{k \in Z_n} \tilde{M}_{\mu,\nu}(k) \hat{f}_\nu(k)\, \mathrm{e}^{\mathrm{i}kx} \in D'(T_n) \quad (\text{or } S'_\pi(R_n))$$

belongs to $L_p(T_n)$. Moreover, each sequence $\{\tilde{M}_{\mu,\nu}(k)\}_{k \in Z_n}$ is a Fourier multiplier for $L_p(T_n)$. Again, because of density arguments it is sufficient to consider (2) for all finite sequences of trigonometric polynomials.

Remark 3. In Definition 1 and Definition 2 we have implicitly defined Fourier multipliers for $L_p(R_n)$ $(L_p(T_n))$, $1 < p < \infty$, namely as the special cases $M_{\mu,\nu}(x) = 0$ $(\widetilde{M}_{\mu,\nu} = 0)$ for $\mu + \nu > 0$, $M_{0,0}(x) = M(x)$ $(\widetilde{M}_{0,0} = \widetilde{M})$, $f_\nu(x) = 0$ for $\nu = 1, 2, \ldots$ and $f_0(x) = f(x)$. This coincides with the corresponding definitions in E. M. Stein, G. Weiss [1, Chapters 1 and 7].

Theorem. *Let $1 < p < \infty$ and $1 < q < \infty$. Let $M = \{M_{\mu,\nu}(x)\}_{\mu,\nu=0}^{\infty} \subset L_\infty(R_n)$ be a Fourier multiplier for $L_p(R_n, l_q)$. Suppose additionally that $M_{\mu,\nu}(x)$ is continuous at all points $k \in Z_n$ for all $\mu = 0, 1, \ldots,$ and $\nu = 0, 1, \ldots$ Then $\widetilde{M} = \{\{M_{\mu,\nu}(k)\}_k\}_{\mu,\nu=0}^{\infty}$ is a Fourier multiplier for $L_p(T_n, l_q)$ with $\tilde{c}_M \leqq c_M$ for the constants from* (1) *and* (2).

Proof. Step 1. Let $f(x)$ be a continuous periodic function defined on R_n. Then we have

$$\int_{T_n} f(x)\,\mathrm{d}x = \lim_{\varepsilon \downarrow 0} (2\pi\varepsilon)^{\frac{n}{2}} \int_{R_n} f(x)\,\mathrm{e}^{-\frac{\varepsilon}{2}|x|^2}\,\mathrm{d}x. \tag{3}$$

This is a well-known assertion and not difficult to prove. It can be found in E. M. Stein, G. Weiss [1, Chapter 7, Lemma 3.9].

Step 2. We need a further preparation. Let $t(x)$ and $\tau(x)$ be trigonometric polynomials. Let $M(x) \in L_\infty(R_n)$ be continuous at every point $k \in Z_n$. We put

$$\omega_\delta(x) = \mathrm{e}^{-\frac{\delta}{2}|x|^2}, \quad \delta > 0. \tag{4}$$

Then it follows

$$\int_{T_n} \Big(\sum_{k \in Z_n} M(k)\,\hat{t}(k)\,\mathrm{e}^{\mathrm{i}kx}\Big)\,\overline{\tau(x)}\,\mathrm{d}x$$
$$= \lim_{\varepsilon \to 0} (2\pi\varepsilon)^{n/2} \int_{R_n} [F^{-1}MF(t\omega_{\varepsilon\alpha})](x)\,\overline{\tau(x)}\,\omega_{\varepsilon\beta}(x)\,\mathrm{d}x \tag{5}$$

for all positive numbers α, β with $\alpha + \beta = 1$. (5) is a reformulation of Lemma 3.11 in E. M. Stein, G. Weiss [1, Chapter 7].

Step 3. To prove the theorem we use duality arguments. Let $\{t_\nu(x)\}_\nu$ and $\{\tau_\nu(x)\}_\nu$ be finite sequences of trigonometric polynomials. We put $f_{\nu,\varepsilon}(x) = t_\nu(x)\,\omega_{\varepsilon\alpha}(x)$ and $g_{\nu,\varepsilon}(x) = \tau_\nu(x)\,\omega_{\varepsilon\beta}(x)$, $\nu = 0, 1, \ldots, \varepsilon > 0$, where α and β are positive numbers with $\alpha + \beta = 1$ and where ω_δ, $\delta > 0$, has the meaning of (4). Obviously, all functions $f_{\nu,\varepsilon}$ and $g_{\nu,\varepsilon}$ belong to $S(R_n)$. Then by Hölder's inequality we obtain

$$\left|\int_{R_n} \sum_{\mu=0}^{\infty} \Big(\sum_{\nu=0}^{\infty} (F^{-1}M_{\mu,\nu}Ff_{\nu,\varepsilon})(x)\Big)\,\overline{g_{\mu,\varepsilon}(x)}\,\mathrm{d}x\right|$$
$$\leqq \left\|\Big\{\sum_{\nu=0}^{\infty} (F^{-1}M_{\mu,\nu}Ff_{\nu,\varepsilon})(x)\Big\}_\mu \mid L_p(R_n, l_q)\right\| \, \|\{g_{\mu,\varepsilon}\}_\mu \mid L_{p'}(R_n, l_{q'})\|. \tag{6}$$

Here $\dfrac{1}{p} + \dfrac{1}{p'} = \dfrac{1}{q} + \dfrac{1}{q'} = 1$. Because M is assumed to be a Fourier multiplier in $L_p(R_n, l_q)$ it follows from (6) and the fact that all sums are finite

$$\left|\sum_{\mu=0}^{\infty} \sum_{\nu=0}^{\infty} \int_{R_n} [F^{-1}M_{\mu,\nu}Ff_{\nu,\varepsilon}](x)\,\overline{g_{\mu,\varepsilon}(x)}\,\mathrm{d}x\right|$$
$$\leqq c_M \Big(\int_{R_n} \|t_\nu(x) \mid l_q\|^p\,\omega_{\varepsilon\alpha p}(x)\,\mathrm{d}x\Big)^{1/p} \Big(\int_{R_n} \|\tau_\nu(x) \mid l_{q'}\|^{p'}\,\omega_{\varepsilon\beta p'}(x)\,\mathrm{d}x\Big)^{1/p'}, \tag{7}$$

where c_M denotes the constant from (1). We multiply (7) with $(2\pi\varepsilon)^{n/2}$, choose $\alpha = p^{-1}$ and $\beta = p'^{-1}$ and consider the limit process $\varepsilon \to 0$. Then we obtain by means of (3) and (5)

$$\left| \int_{T_n} \sum_{\mu=0}^{\infty} \left[\sum_{\nu=0}^{\infty} \left(\sum_{k \in Z_n} M_{\mu,\nu}(k)\, \hat{t}_\nu(k)\, \mathrm{e}^{ikx} \right) \right] \overline{\tau_\mu(x)}\, \mathrm{d}x \right|$$
$$\leqq c_M \| t_\nu \mid L_p(T_n, l_q) \| \, \| \tau_\nu \mid L_{p'}(T_n, l_{q'}) \| \,. \tag{8}$$

The sequence

$$\left\{ \sum_{\nu=0}^{\infty} \left(\sum_{k \in Z_n} M_{\mu,\nu}(k)\, \hat{t}_\nu(k)\, \mathrm{e}^{ikx} \right) \right\}_{\mu=0}^{\infty}$$

belongs to $L_p(T_n, l_q)$ because the sum $\sum_\nu (\ldots)$ is finite. Hence it can be interpreted as a linear functional on $L_{p'}(T_n, l_{q'})$ (see for example R. E. Edwards [2, Theorem 8.20.5]). This shows that (8) remains true if we replace the finite sequence $\{\tau_\mu(x)\}_\mu$ by an arbitrary sequence $\{\varphi_\mu(x)\}_{\mu=0}^{\infty} \in L_{p'}(T_n, l_{q'})$. Taking on both sides of the modified inequality (8) the supremum over all sequences $\{\varphi_\nu\}_{\nu=0}^{\infty}$ with $\|\varphi_\nu \mid L_{p'}(T_n, l_{q'})\| \leqq 1$ we obtain by well-known facts of functional analysis that

$$\left\| \left\{ \sum_{\nu=0}^{\infty} \left(\sum_{k \in Z_n} M_{\mu,\nu}(k)\, \hat{t}_\nu(k)\, \mathrm{e}^{ikx} \right) \right\}_\mu \Big| \, L_p(T_n, l_q) \right\|$$
$$= \left\| \left\{ \sum_{\nu=0}^{\infty} \left(\sum_{k \in Z_n} M_{\mu,\nu}(k)\, \hat{t}_\nu(k)\, \mathrm{e}^{ikx} \right) \right\}_\mu \Big| \, [L_{p'}(T_n, l_{q'})]' \right\|$$
$$\leqq c_M \| t_\nu(x) \mid L_p(T_n, l_q) \| \,. \tag{9}$$

Here $[L_{p'}(T_n, l_{q'})]'$ denotes the dual space of $L_{p'}(T_n, l_{q'})$. This proves the theorem.

Remark 4. The formulas (3) and (5) are the main ingredients of the proof. Let us note that the assumptions concerning the continuity of $M_{\mu,\nu}(x)$ at $k \in Z_n$ can be weakened. It suffices to suppose that all lattice-points k are Lebesgue points of $M_{\mu,\nu}(x)$, $\mu = 0, 1, \ldots, \nu = 0, 1, \ldots$ This follows from the fact that (5) is also true in this case (cf. E. M. Stein, G. Weiss [1, Chapter 7, Corollary 3.16]).

Remark* 5. The above proof of the theorem is the vector-valued version of E. M. Stein, G. Weiss [1, Chapter 7, Theorem 3.8]. We followed H.-J. Schmeisser, W. Sickel [1, Section 2.1.]. As already mentioned the method is due to K. de Leeuw [1]. An analogous assertion concerning Fourier multipliers for non-periodic and periodic Hardy spaces $H_p(R_n)$ and $H_p(T_n)$, $0 < p \leqq 1$, respectively, can be found in L. Colzani [1, Theorem 3].

3.4.3. Special Multipliers for $L_p(T_n, l_q)$, $1 < p < \infty$, $1 < q < \infty$

In this subsection and in the next one we give explicit criterions for multipliers in $L_p(T_n, l_q)$. In particular, by Theorem 3.4.2 we are now in a position to state the periodic counterparts of multiplier theorems in $L_p(R_n, l_q)$ proved by H. Triebel [18] (cf. also [I, 2.2.4.]) and P. I. Lizorkin [1]. However, we are not interested in most general formulation, on the contrary, we restrict ourselves to assertions needed in the sequel. At first let us deal with the case of diagonal Fourier multipliers. That means we restrict ourselves to matrices $M = \{\{M_{\mu,\nu}(k)\}_k\}_{\mu,\nu=0}^{\infty}$ with

$$M_{\mu,\nu}(k) = \begin{cases} M_\mu(k) & \text{if} \quad \mu = \nu, \\ 0 & \text{if} \quad \mu \neq \nu \end{cases} \qquad k \in Z_n, \tag{1}$$

for all $\mu = 0, 1, 2, \ldots$ and $\nu = 0, 1, 2, \ldots$ (cf. also (1.9.2/1)).

Recall that the spaces $H_2^\varkappa = H_2^\varkappa(R_n)$, $\varkappa > 0$, have been defined in (1.7.4/1). $L_p(T_n, l_q)$ and $L_p(R_n, l_q)$ have the meaning of the preceding subsection.

Theorem 1. *Let* $1 < p < \infty$ *and* $1 < q < \infty$. *Let* $\psi(x) \in S(R_n)$ *with*

$$0 \leqq \psi(x) \leqq 1, \operatorname{supp} \psi \subset \{y \mid \tfrac{1}{4} \leqq |y| \leqq 4\}, \psi(x) = 1 \quad \textit{if} \ \tfrac{1}{2} \leqq |x| \leqq 2. \quad (2)$$

If $\varkappa > \frac{n}{2}$ *then there exists a positive constant* c *such that*

$$\begin{aligned} &\Big\| \sum_{k \in Z_n} M_\mu(k) \hat{f}_\mu(k)\, \mathrm{e}^{ikx} \mid L_p(T_n, l_q) \Big\| \\ &\leqq c \sup_{j=0, \pm 1, \pm 2, \dots} \Big(\sum_{\nu=0}^{\infty} \| \psi(\cdot)\, M_\nu(2^j \cdot) \mid H_2^\varkappa \|^2 \Big)^{1/2} \| f_\mu \mid L_p(T_n, l_q) \| \end{aligned} \quad (3)$$

holds for all systems $\{M_\mu(x)\}_{\mu=0}^\infty \subset H_2^\varkappa(R_n)$ *and all systems* $\{f_\mu(x)\}_{\mu=0}^\infty \in L_p(T_n, l_q)$.

Proof. We recall a version of the Fourier multiplier theorem of Michlin-Hörmander type for $L_p(R_n, l_q)$ as it is stated in [T, (2.4.8/8)] (cf. also (1.9.2/14)). For a detailed proof see [I, pp. 161–165]. It holds

$$\begin{aligned} &\| F^{-1} M_\mu F f_\mu \mid L_p(R_n, l_q) \| \\ &\leqq c \sup_{j=0, \pm 1, \pm 2, \dots} \Big(\sum_{\nu=0}^{\infty} \| \psi(\cdot)\, M_\nu(2^j \cdot) \mid H_2^\varkappa \|^2 \Big)^{1/2} \| f_\mu \mid L_p(R_n, l_q) \| \end{aligned} \quad (4)$$

if $\varkappa > \frac{n}{2}$ and $1 < p < \infty$, $1 < q < \infty$. We put $M_{\mu,\nu} = M_\mu$ if $\mu = \nu$ and $M_{\mu,\nu} = 0$ if $\mu \neq \nu$ $(\mu = 0, 1, 2, \dots, \nu = 0, 1, 2, \dots)$. Then $M = \{M_{\mu,\nu}\}$ satisfies the assumptions of Theorem 3.4.2. As an immediate consequence we obtain (3).

Remark 1. Theorem 1 says that every Fourier multiplier for $L_p(R_n, l_q)$ of Michlin-Hörmander type is also a Fourier multiplier for $L_p(T_n, l_q)$ if it is restricted to Z_n. Clearly, the scalar case (i.e. Fourier multipliers of this type for $L_p(T_n)$, $1 < p < \infty$) is contained in our theorem as a special case.

As a corollary of Theorem 1 we can improve the criterion on Fourier multipliers for the spaces $L_p^A(T_n, l_q)$ with $1 < p < \infty$ and $1 < q < \infty$ of Theorem 3.4.1/2.

Theorem 2. *Let* $1 < p < \infty$ *and* $1 < q < \infty$. *Let* $A = \{A_j\}_{j=0}^\infty$ *be a sequence of subsets of* Z_n *such that*

$$A_j \subset \{k \mid k \in Z_n, |k| \leqq 2^j\}, \quad j = 0, 1, \dots \quad (5)$$

If $\varkappa > \frac{n}{2}$ *then there exists a positive constant* c *such that*

$$\begin{aligned} &\Big\| \sum_{k \in Z_n} M_j(k)\, \hat{t}_j(k)\, \mathrm{e}^{ikx} \mid L_p(T_n, l_q) \Big\| \\ &\leqq c \sup_{j=0,1,2,\dots} \| M_j(2^j \cdot) \mid H_2^\varkappa \| \, \| t_j \mid L_p(T_n, l_q) \| \end{aligned} \quad (6)$$

holds for all systems $\{M_j(x)\}_{j=0}^\infty \subset H_2^\varkappa(R_n)$ *and all systems* $\{t_j(x)\}_{j=0}^\infty \in L_p^A(T_n, l_q)$.

Proof. At first we claim that it is sufficient to prove (6) in the case that

$$A_j \subset \{k \mid k \in Z_n, 2^{j+1} \leqq |k| \leqq 2^{j+2}\}, \quad j = 0, 1, 2, \dots \quad (7)$$

This follows from a simple translation argument if one considers $\{\mathrm{e}^{ik^{(j)}x} t_j(x)\}_{j=0}^\infty$ with an appropriate sequence $\{k^{(j)}\}_{j=0}^\infty \subset Z_n$ instead of $\{t_j(x)\}_{j=0}^\infty$. Now let

$\Lambda = \{\Lambda_j\}_{j=0}^{\infty}$ be a sequence of subsets of Z_n with the property (7). Further let ψ be the function from (2). Let $\{t_j(x)\}_{j=0}^{\infty} \in L_p^{\Lambda}(T_n, l_q)$. Then it holds

$$\sum_{k \in Z_n} M_\mu(k)\, \hat{t}_\mu(k)\, e^{ikx} = \sum_{k \in Z_n} M_\mu(k)\, \psi(2^{-\mu-1}k)\, \hat{t}_\mu(k)\, e^{ikx}.$$

By Theorem 1 we have

$$\Big\| \sum_{k \in Z_n} M_\mu(k)\, \hat{t}_\mu(k)\, e^{ikx} \mid L_p(T_n, l_q) \Big\|$$
$$\leqq c \sup_{j=0, \pm 1, \pm 2, \dots} \left(\sum_{\nu=0}^{\infty} \| \psi(\cdot)\, \psi(2^{j-\nu-1}\cdot)\, M_\nu(2^j \cdot) \mid H_2^{\varkappa} \|^2 \right)^{1/2} \| t_\mu \mid L_p(T_n, l_q) \|. \tag{8}$$

If j is an integer and if $\nu = 0, 1, 2, \dots$ then the function $\psi(x)\, \psi(2^{j-\nu-1}x)$ vanishes if $\nu < j - 5$ or $\nu > j + 3$. Hence we find constants c, c' such that

$$\| \psi(\cdot)\, \psi(2^{j-\nu-1}\cdot)\, M_\nu(2^j\cdot) \mid H_2^{\varkappa} \| \leqq c \| M_\nu(2^j\cdot) \mid H_2^{\varkappa} \|$$
$$\leqq c' \| M_\nu(2^\nu\cdot) \mid H_2^{\varkappa} \|.$$

Hence it follows

$$\sup_{j=0, \pm 1, \pm 2, \dots} \left(\sum_{\nu=0}^{\infty} \| \psi(\cdot)\, \psi(2^{j-\nu-1}\cdot)\, M_\nu(2^j\cdot) \mid H_2^{\varkappa} \|^2 \right)^{1/2}$$
$$\leqq c \sup_{j=0,1,2,\dots} \| M_j(2^j\cdot) \mid H_2^{\varkappa} \|, \tag{9}$$

with an appropriate constant c independent of j. (8) and (9) prove (6).

Remark 2. (6) is an improvement of (3.4.1/5) concerning the number $\varkappa$ provided that $1 < p < \infty$ and $1 < q < \infty$. In this case $\varkappa > \frac{n}{2}$ is a natural condition (also in the scalar case). In addition to (3.4.1/5) we specialized the sequence $\Lambda = \{\Lambda_j\}_{j=0}^{\infty} \subset Z_n$ by (5). The numbers 2^j, $j = 0, 1, 2, \dots$, come from Theorem 1. We remark that the proof of (6) goes also through if

$$0 < d_j = \max_{k, l \in \Lambda_j} |k - l| \leqq 2^j$$

is assumed. Later on we show by means of complex interpolation that also in the case $\min(p, q) \leqq 1$, (3.4.1/5) can be essentially strengthened.

Theorem 3. *Let $1 < p < \infty$ and $1 < q < \infty$. Let $M = \{M_j(x)\}_{j=0}^{\infty} \subset L_\infty(R_n)$ be a sequence of functions which can be represented as*

$$M_j(x) = \int_{-\infty}^{x_1} \dots \int_{-\infty}^{x_n} d\mu_j(y); \quad j = 0, 1, 2, \dots, \tag{10}$$

where the μ_j's; $j = 0, 1, 2, \dots$; are finite measures with uniformly bounded variation, i.e.

$$\operatorname{var} \mu_j = \int_{R_n} |d\mu_j| \leqq c_M \tag{11}$$

for all $j = 0, 1, 2, \dots$ Let all functions M_j; $j = 0, 1, 2, \dots$; be continuous at all points $k \in Z_n$. Then there exists a positive constant c depending on p, q, n such that

$$\Big\| \sum_{k \in Z_n} M_j(k)\, \hat{f}_j(k)\, e^{ikx} \mid L_p(T_n, l_q) \Big\| \leqq c c_M \| f_j \mid L_p(T_n, l_q) \| \tag{12}$$

holds for all $\{f_j\}_{j=0}^{\infty} \in L_p(T_n, l_q)$.

Proof. If M satisfies the hypotheses of the theorem then it is a Fourier multiplier for $L_p(R_n, l_q)$. This is a result by P. I. Lizorkin [1, Theorem 5, p. 241]. Consequently (12) follows from Theorem 3.4.2.

Remark 3. Because of $\varkappa > \frac{n}{2}$ in (3) the multipliers of Theorem 1 are generated by sequences of continuous functions on $R_n \setminus \{0\}$. But this is not necessary as Theorem 3 shows. It yields sharper results than Theorem 1. Of peculiar interest is the case where the μ_j's are Dirac measures. Let $\{a^{(j)}\}_{j=0}^{\infty} \subset R_n$ be a sequence of points $a^{(j)} = (a_1^{(j)}, \ldots, a_n^{(j)})$ with $a_i^{(j)} \neq$ integer for all $i = 1, \ldots, n$ and all $j = 0, 1, 2, \ldots$ Then by (10) and (11) we have

$$M_j(x) = \begin{cases} 1 & \text{if} \quad x_i \geqq a_i^{(j)} \quad \text{for all} \quad i = 1, \ldots, n \\ 0 & \text{otherwise} \end{cases} \tag{13}$$

for $j = 0, 1, 2, \ldots$ and $c_M = 1$. Furthermore, $M = \{\{M_j(k)\}_k\}_{j=0}^{\infty}$ is a multiplier for $L_p(T_n, l_q)$ in the sense of (12). By simple linear combinations of the above multipliers one obtains new multipliers. In particular if

$$\Lambda_j = \{k \mid k \in Z_n, a_i^{(j)} \leqq k_i \leqq b_i^{(j)}; \quad i = 1, \ldots, n\}, \tag{14}$$

where $\{a^{(j)}\}_{j=0}^{\infty} \subset Z_n$ and $\{b^{(j)}\}_{j=0}^{\infty} \subset Z_n$ are sequences with $a_i^{(j)} < b_i^{(j)}$ ($j = 0, 1, 2, \ldots; i = 1, \ldots, n$), then we have

$$\left\| \sum_{k \in \Lambda_j} \hat{f}_j(k)\, \mathrm{e}^{ikx} \mid L_p(T_n, l_q) \right\| \leqq c \| f_j \mid L_p(T_n, l_q) \| . \tag{15}$$

Here the constant c is independent of $\Lambda = \{\Lambda_j\}_{j=0}^{\infty}$.

3.4.4. The Spaces $L_p(T_n)$, $1 < p < \infty$

The aim of this subsection is to shed some light on Littlewood-Paley theorems for the spaces $L_p(T_n)$, $1 < p < \infty$, by means of our approach to periodic spaces. We follow the principle of [I, Theorem 2.3.3] (non-periodic case) or of Section 1.9.4 (weighted case). The idea is to use vertical and horizontal multipliers for the spaces $L_p(T_n, l_2)$, $1 < p < \infty$. First we recall the concepts of vertical and horizontal multipliers. Let $M = \{\{M_{\mu,\nu}(k)\}_k\}_{\mu,\nu=0}^{\infty} \subset l_\infty(Z_n)$. Then M is called a vertical matrix if

$$M_{\mu,0}(k) = M_\mu(k) \quad \text{and} \quad M_{\mu,\nu}(k) = 0 \tag{1}$$

for $\mu = 0, 1, 2, \ldots$ and $\nu = 1, 2, \ldots$ On the other hand, if

$$M_{0,\nu}(k) = M_\nu(k) \quad \text{and} \quad M_{\mu,\nu}(k) = 0 \tag{2}$$

for $\mu = 1, 2, \ldots$ and $\nu = 0, 1, 2, \ldots$ then M is said to be a horizontal matrix. If additionally M is a Fourier multiplier we shall denote it as vertical or horizontal multiplier, respectively. As in the preceding subsection $H_2^\varkappa = H_2^\varkappa(R_n)$ has the meaning of (1.7.4/1).

Proposition. *Let $1 < p < \infty$ and let $\varphi = \{\varphi_j(x)\}_{j=0}^{\infty} \subset S(R_n)$ with*

$$\operatorname{supp} \varphi_0 \subset \{y \mid |y| \leqq 2\}, \operatorname{supp} \varphi_j \subset \{y \mid 2^{j-1} \leqq |y| \leqq 2^{j+1}\}; \quad j = 1, 2, \ldots \tag{3}$$

Let $\varphi = \{\{\varphi_{\mu,\nu}(k)\}_k\}_{\mu,\nu=0}^{\infty}$ be either a vertical or a horizontal matrix (in the sense of (1) and (2) with φ_μ instead of M_μ). Let $\varkappa > \frac{n}{2}$. Then there exists a positive constant c such that

$$\begin{aligned} &\left\| \sum_{\nu=0}^{\infty} \left[\sum_{k \in Z_n} \varphi_{\mu,\nu}(k) \hat{f}_\nu(k)\, \mathrm{e}^{ikx} \right] \mid L_p(T_n, l_2) \right\| \\ &\leqq c \sup_{j=0,1,2,\ldots} \| \varphi_j(2^j \cdot) \mid H_2^\varkappa \| \, \| f_\mu \mid L_p(T_n, l_2) \| \end{aligned} \tag{4}$$

holds for all systems $\{f_\mu\}_{\mu=0}^{\infty} \in L_p(T_n, l_2)$.

Proof. Analogously to the proof of Theorem 3.4.3/1 we use the corresponding assertion for the spaces $L_p(R_n, l_2)$ from [I, Theorem 2.2.4]. If $\psi \in S(R_n)$ is the function from (3.4.3/2) then

$$\left\| \sum_{\nu=0}^{\infty} F^{-1}\varphi_{\mu,\nu} F f_\nu \mid L_p(R_n, l_2) \right\|$$
$$\leqq c \sup_{j=0,\pm 1,\pm 2,\ldots} \left(\sum_{\nu=0}^{\infty} \|\psi(\cdot)\, \varphi_\nu(2^j \cdot) \mid H_2^\varkappa\|^2 \right)^{1/2} \|f_\mu \mid L_p(R_n, l_2)\| \tag{5}$$

holds (cf. [I, Theorem 2.2.4(a)] and the remarks in Step 2 of the proof of [T, Proposition 2.4.8] or (1.9.2/14)). In the same way as in the proof of Theorem 3.4.3/2 we see that

$$\sup_{j=0,\pm 1,\pm 2,\ldots} \left(\sum_{\nu=0}^{\infty} \|\psi(\cdot)\, \varphi_\nu(2^j \cdot) \mid H_2^\varkappa\|^2 \right)^{1/2} \leqq c \sup_{j=0,1,\ldots} \|\varphi_j(2^j \cdot) \mid H_2^\varkappa\| \tag{6}$$

holds. Now, (5), (6), and Theorem 3.4.2 lead to (4). Thus, the proposition is proved.

Remark 1. In contrast to Theorem 3.4.3/1 (diagonal case) we restrict ourselves to matrices with property (3). This is not necessary but convenient (and sufficient for our later purposes). However, the restriction to $q = 2$ is natural.

Theorem. *Let* $1 < p < \infty$.

(i) *Let* $\varphi = \{\varphi_j(x)\}_{j=0}^{\infty} \subset S(R_n)$ *with* (3),

$$\sum_{j=0}^{\infty} \varphi_j(x) = 1, \quad x \in R_n, \tag{7}$$

and

$$\sup_{j=0,1,2,\ldots} \|\varphi_j(2^j \cdot) \mid H_2^\varkappa\| < \infty \tag{8}$$

for some $\varkappa > \frac{n}{2}$. *Then there exist two positive constants* c_1 *and* c_2 *such that*

$$c_1 \|f \mid L_p(T_n)\| \leqq \left\| \left\{ \sum_{k \in Z_n} \varphi_j(k) \hat{f}(k)\, \mathrm{e}^{\mathrm{i}kx} \right\}_j \mid L_p(T_n, l_2) \right\|$$
$$\leqq c_2 \|f \mid L_p(T_n)\| \tag{9}$$

holds for all $f \in L_p(T_n)$.

(ii) *Let* $\Lambda = \{\Lambda_j\}_{j=0}^{\infty} \subset Z_n$ *with*

$$\Lambda_0 = \{k \mid k \in Z_n, |k_l| \leqq 1;\ l = 1, \ldots, n\},$$
$$\Lambda_j = \{k \mid k \in Z_n, |k_l| \leqq 2^j;\ l = 1, \ldots, n\} \setminus \{k \mid k \in Z_n, |k_l| \leqq 2^{j-1};\ l = 1, \ldots, n\}; \tag{10}$$

$j = 1, 2, \ldots$ *Then there exist two positive constants* c_1 *and* c_2 *such that*

$$c_1 \|f \mid L_p(T_n)\| \leqq \left\| \left\{ \sum_{k \in \Lambda_j} \hat{f}(k)\, \mathrm{e}^{\mathrm{i}kx} \right\}_j \mid L_p(T_n, l_2) \right\| \leqq c_2 \|f \mid L_p(T_n)\| \tag{11}$$

holds for all $f \in L_p(T_n)$.

Proof. Step 1. We prove (9). The right-hand side is an immediate consequence of the vertical case of the preceding Proposition. In order to show the left-hand side we use (7) and (3) and obtain

$$f = \sum_{j=0}^{\infty} \left[\sum_{k \in Z_n} \varphi_j(k) \hat{f}(k)\, \mathrm{e}^{\mathrm{i}kx} \right] = \sum_{r=-1}^{1} \left(\sum_{j=0}^{\infty} \left[\sum_{k \in Z_n} \varphi_{j+r}(k) \hat{f}_j(k)\, \mathrm{e}^{\mathrm{i}kx} \right] \right), \tag{12}$$

where

$$f_j(x) = \sum_{k \in Z_n} \varphi_j(k) \hat{f}(k) e^{ikx}; \quad j = 0, 1, 2, \ldots,$$

and $\varphi_{-1}(x) = 0$. Now, (12) and the horizontal case of the Proposition imply the left-hand side of (9).

Step 2. We prove (11). According to (9) it is sufficient to show that

$$\left\| \left\{ \sum_{k \in Z_n} \varphi_j(k) \hat{f}(k) e^{ikx} \right\}_j \mid L_p(T_n, l_2) \right\|$$

can be estimated from below and from above by

$$\left\| \left\{ \sum_{k \in \Lambda_j} \hat{f}(k) e^{ikx} \right\}_j \mid L_p(T_n, l_2) \right\|.$$

Let χ_j be the characteristic function of Λ_j; $j = 0, 1, 2, \ldots$; and let $\{\varphi_j\}_{j=0}^{\infty}$ be the system from (9). Then we observe

$$\chi_j = \sum_{r=-N}^{N} \chi_j \varphi_{j+r} \quad \text{and} \quad \varphi_j = \sum_{r=-\tilde{N}}^{\tilde{N}} \varphi_j \chi_{j+r}; \quad j = 0, 1, 2, \ldots,$$

where the numbers N and $\tilde{N}$ are independent of $j = 0, 1, 2, \ldots$ (with $\varphi_j = 0$ and $\chi_j = 0$ if $j < 0$). Now we argue in the same way as in (12) and use that $\{\varphi_j\}_{j=0}^{\infty}$ and $\{\chi_j\}_{j=0}^{\infty}$ are (diagonal) Fourier multipliers for $L_p(T_n, l_2)$ (cf. Theorem 3.4.3/2, Theorem 3.4.3/3, and (3.4.3/15)). This leads to the desired estimates. Thus, the proof of the theorem is complete.

Remark* 2. (9) and (11) are Littlewood-Paley theorems for $L_p(T_n)$. (9) uses smooth partitions of unity whereas (11) is related to characteristic functions of dyadic corridors. Theorems of this type are well-known. We refer to S. M. Nikol'skij [2, (1.5.2/13)] and R. E. Edwards, G. I. Gaudry [1, Chapter 7]. For mixed $L_{\bar{p}}(T_n)$-spaces we refer to O. V. Besov, V. P. Il'in, S. M. Nikol'skij [1, Lemma 15.2]. For the non-periodic case see, for example, P. I. Lizorkin [3], E. M. Stein [1], S. M. Nikol'skij [2], [I, 2.3.3], and R. E. Edwards, G. I. Gaudry [1] (cf. also 1.9.4. for weighted spaces and 2.3.1. for mixed spaces).

3.5. The Spaces $B^s_{p,q}(T_n)$ and $F^s_{p,q}(T_n)$

3.5.1. Definitions and Basic Properties

In Section 3.5. we give the definitions of the spaces $B^s_{p,q}(T_n)$ and $F^s_{p,q}(T_n)$. Further we collect the fundamental properties of these spaces and investigate their relations to classical periodic spaces of Sobolev-Besov type. The proofs are based on the results of Sections 3.3. and 3.4. and can be carried out in complete analogy to the non-periodic case as it has been treated in [T, Chapter 2], for example. Therefore we allow us to omit any detailed proofs of the assertions stated in this section. We refer also to H.-J. Schmeisser, W. Sickel [1], W. Sickel [1, 2], and S. Mundlos [1].

We use the notations from 3.2., 3.3., and 3.4. Further we recall that the class $\Phi(R_n)$ of systems of test functions has been introduced in Definition 2.1/1.

Definition. *Let* $-\infty < s < \infty$ *and* $0 < q \leqq \infty$. *Let* $\varphi = \{\varphi_j(x)\}_{j=0}^{\infty} \in \Phi(R_n)$.

(i) *If* $0 < p \leqq \infty$, *then*

$$B^s_{p,q}(T_n) = \left\{ f \mid f \in D'(T_n),\ \|f \mid B^s_{p,q}(T_n)\|^{\varphi} = \left\| 2^{sj} \sum_{k \in Z_n} \varphi_j(k) \hat{f}(k) e^{ikx} \mid l_q(L_p(T_n)) \right\| < \infty \right\}. \tag{1}$$

(ii) *If* $0 < p < \infty$, *then*

$$F^s_{p,q}(T_n) = \left\{ f \mid f \in D'(T_n), \; \|f \mid F^s_{p,q}(T_n)\|^{\varphi} = \left\| 2^{sj} \sum_{k \in Z_n} \varphi_j(k) \hat{f}(k)\, e^{ikx} \mid L_p(T_n, l_q) \right\| < \infty \right\}. \tag{2}$$

Remark 1. This is the periodic counterpart of [T, Definition 2.3.1/2]. $D'(T_n)$ in (1) and (2) can be replaced by $S'_\pi(R_n)$, cf. 3.2.3. The sums

$$\sum_{k \in Z_n} \varphi_j(k) \hat{f}(k)\, e^{ikx}$$

are finite for every $j = 0, 1, 2, \ldots$ In particular, we have

$$\sum_{k \in Z_n} \varphi_j(k) \hat{f}(k)\, e^{ikx} \in T^{\Lambda_j}$$

with

$$\Lambda_0 = \{k \mid k \in Z_n, |k| < 2\},$$

$$\Lambda_j = \{k \mid k \in Z_n, 2^{j-1} < |k| < 2^{j+1}\}; \quad j = 1, 2, \ldots, \tag{3}$$

cf. (3.3.1/2). Thus it is clear that the results from 3.3. and 3.4. can be applied in order to study the spaces $B^s_{p,q}(T_n)$ and $F^s_{p,q}(T_n)$. Further we find that

$$F^{-1}\varphi_j F f = \sum_{k \in Z_n} \varphi_j(k) \hat{f}(k)\, e^{ikx} \tag{4}$$

if we identify $D'(T_n)$ and $S'_\pi(R_n)$ (cf. 3.2.3. and (3.3.4/2)). This shows that the construction of the spaces in (1) and (2) is the same as in [T, Definition 2.3.1/2] (cf. also 2.2.2/(2), (3), where the case $n = 2$ is described).

Theorem 1. *Let* $-\infty < s < \infty$ *and* $0 < q \leqq \infty$.

(i) *Let* $0 < p \leqq \infty$. *Then* $B^s_{p,q}(T_n)$ *is a quasi-Banach space (Banach space if* $\min(p, q) \geqq 1$). *The quasi-norms* $\|f \mid B^s_{p,q}(T_n)\|^{\varphi}$ *and* $\|f \mid B^s_{p,q}(T_n)\|^{\psi}$ *with* $\varphi \in \Phi(R_n)$ *and* $\psi \in \Phi(R_n)$ *are equivalent. Furthermore,*

$$D(T_n) \subset B^s_{p,q}(T_n) \subset D'(T_n). \tag{5}$$

If $\max(p, q) < \infty$ *then the set of all trigonometric polynomials is dense in* $B^s_{p,q}(T_n)$.

(ii) *Let* $0 < p < \infty$. *Then* $F^s_{p,q}(T_n)$ *is a quasi-Banach space (Banach space if* $\min(p, q) \geqq 1$). *The quasi-norms* $\|f \mid F^s_{p,q}(T_n)\|^{\varphi}$ *and* $\|f \mid F^s_{p,q}(T_n)\|^{\psi}$ *with* $\varphi \in \Phi(R_n)$ *and* $\psi \in \Phi(R_n)$ *are equivalent. Furthermore,*

$$D(T_n) \subset F^s_{p,q}(T_n) \subset D'(T_n). \tag{6}$$

If $q < \infty$ *then the set of all trigonometric polynomials is dense in* $F^s_{p,q}(T_n)$.

(iii) *If* $s > 0$ *and* $1 \leqq p \leqq \infty$ *then*

$$B^s_{p,q}(T_n) \subset L_p(T_n). \tag{7}$$

If $s > n\left(\frac{1}{p} - 1\right)$ *and* $0 < p < 1$ *then*

$$B^s_{p,q}(T_n) \subset L_1(T_n). \tag{8}$$

(iv) *If* $s > 0$ *and* $1 \leqq p < \infty$ *then*

$$F^s_{p,q}(T_n) \subset L_p(T_n). \tag{9}$$

If $s > n\left(\frac{1}{p} - 1\right)$ *and* $0 < p < 1$ *then*

$$F^s_{p,q}(T_n) \subset L_1(T_n). \tag{10}$$

Remark 2. The equivalence of the quasi-norms $\|\cdot\|^{\varphi}$ for different $\varphi \in \Phi(R_n)$ and hence the independence of the spaces $B^s_{p,q}(T_n)$ and $F^s_{p,q}(T_n)$ of φ is a consequence of the Fourier multiplier theorems 3.3.4 and 3.4.1/2. If we have in mind (4) the proof follows the proof of [T, Proposition 2.3.2/1]. Similarly the remaining assertions in (i) and (ii) can be carried over from [T, Theorem 2.3.3]. The embeddings (7)–(10) can be derived as in [T, Remark 2.5.3/1] using Nikol'skij's inequality for trigonometric polynomials (3.3.2/11).

Remark 3. Part (iii) and part (iv) of the theorem show that $f \in B^s_{p,q}(T_n)$ (or $f \in F^s_{p,q}(T_n)$) is a regular distribution if

$$s > \max\left(n\left(\frac{1}{p} - 1\right), 0\right). \tag{11}$$

Because of $L_1(T_n) \subset L_p(T_n)$ if $0 < p < 1$, f belongs to $L_p(T_n)$, $0 < p \leqq \infty$, provided that (11) is satisfied. On the other hand

$$\delta \in B^s_{p,q}(T_n) \quad \text{if} \quad s < n\left(\frac{1}{p} - 1\right), \qquad 0 < p \leqq \infty, \tag{12}$$

(cf. [T, Remark 2.5.3/3] and W. Sickel [1, Remark 9]). Here δ stands for the Delta-distribution from $D'(T_n)$.

Remark 4. Obviously we have

$$B^s_{p,p}(T_n) = F^s_{p,p}(T_n), \quad -\infty < s < \infty, \quad 0 < p < \infty. \tag{13}$$

With this trivial exception the spaces $B^{s_0}_{p_0,q_0}(T_n)$ and $F^{s_1}_{p_1,q_1}(T_n)$ are always different. Furthermore, for different triplets (s_0, p_0, q_0) and (s_1, p_1, q_1) the spaces $B^{s_0}_{p_0,q_0}(T_n)$ and $B^{s_1}_{p_1,q_1}(T_n)$ are different, and also the spaces $F^{s_0}_{p_0,q_0}(T_n)$ and $F^{s_1}_{p_1,q_1}(T_n)$ are different. This can be proved analogously to [T, 2.3.9.]. However we have some elementary embeddings between these spaces. In particular, it holds

$$F^s_{p_1,q}(T_n) \subset B^s_{p_0,q}(T_n) \quad \text{if} \quad 0 < p_0 \leqq p_1 \leqq \infty, 0 < q \leqq \infty, -\infty < s < \infty, \tag{14}$$

$$F^s_{p_1,q}(T_n) \subset F^s_{p_0,q}(T_n) \quad \text{if} \quad 0 < p_0 \leqq p_1 < \infty, 0 < q \leqq \infty, -\infty < s < \infty, \tag{15}$$

$$B^s_{p,q_0}(T_n) \subset B^s_{p,q_1}(T_n) \quad \text{if} \quad 0 < p \leqq \infty, 0 < q_0 \leqq q_1 \leqq \infty, -\infty < s < \infty, \tag{16}$$

$$F^s_{p,q_0}(T_n) \subset F^s_{p,q_1}(T_n) \quad \text{if} \quad 0 < p < \infty, 0 < q_0 \leqq q_1 \leqq \infty, -\infty < s < \infty, \tag{17}$$

$$B^{s+\varepsilon}_{p,q_0}(T_n) \subset B^s_{p,q_1}(T_n) \quad \text{if} \quad \varepsilon > 0, 0 < p \leqq \infty, 0 < q_0, q_1 \leqq \infty, -\infty < s < \infty, \tag{18}$$

$$F^{s+\varepsilon}_{p,q_0}(T_n) \subset F^s_{p,q_1}(T_n) \quad \text{if} \quad \varepsilon > 0, 0 < p < \infty, 0 < q_0, q_1 \leqq \infty, -\infty < s < \infty, \tag{19}$$

$$B^s_{p,\min(p,q)}(T_n) \subset F^s_{p,q}(T_n) \subset B^s_{p,\max(p,q)}(T_n) \tag{20}$$

if $0 < p < \infty$, $0 < q \leqq \infty$, $-\infty < s < \infty$.

Theorem 2. *Let $0 < q \leqq \infty$ and $-\infty < s < \infty$. Let $\{\Lambda_j\}_{j=0}^{\infty}$ be the sequence of finite subsets of Z_n defined in* (3).

(i) *If $0 < p \leqq \infty$, then*

$$B^s_{p,q}(T_n) = \{f \mid f \in D'(T_n), \exists \{t_j(x)\}_{j=0}^{\infty} \subset D(T_n) \text{ with } t_j(x) \in T^{\Lambda_j};$$

$$j = 0, 1, 2, \ldots, \text{ such that } f = \sum_{j=0}^{\infty} t_j(x) \text{ in } D'(T_n) \text{ and}$$

$$\|2^{sj} t_j(x) \mid l_q(L_p(T_n))\| < \infty\}. \tag{21}$$

Furthermore,

$$\inf \|2^{sj} t_j(x) \mid l_q(L_p(T_n))\|$$

is an equivalent quasi-norm in $B^s_{p,q}(T_n)$, where the infimum is taken over all admissible representations.

(ii) *If* $0 < p < \infty$, *then*

$$F^s_{p,q}(T_n) = \{f \mid f \in D'(T_n),\ \exists \{t_j(x)\}_{j=0}^{\infty} \subset D(T_n) \quad \text{with} \quad t_j(x) \in T^{\Delta_j};$$
$$j = 0, 1, 2, \ldots, \text{ such that } f = \sum_{j=0}^{\infty} t_j(x) \quad \text{in} \quad D'(T_n) \quad \text{and}$$
$$\|2^{sj} t_j(x) \mid L_p(T_n, l_q)\| < \infty\}. \tag{22}$$

Furthermore,

$$\inf \|2^{sj} t_j(x) \mid L_p(T_n, l_q)\|$$

is an equivalent quasi-norm in $F^s_{p,q}(T_n)$, *where the infimum is taken over all admissible representations.*

Remark 5. The theorem gives an equivalent definition of the spaces $B^s_{p,q}(T_n)$ and $F^s_{p,q}(T_n)$. Again the proof is a consequence of Theorem 3.3.4 and Theorem 3.4.1/2 (cf. [T, Theorem 2.5.2]). There are other equivalent definitions of our spaces: Instead of coverings of R_n in the definition of the class $\Phi(R_n)$ (or of Z_n in (3)) by differences of dyadic balls one can use differences of cubes or appropriate dyadic cubes. Further, the number "2" appearing in the dyadic converings can be replaced by an arbitrary number $a > 1$. For more details we refer to the non-periodic case as it is treated in [S, 2.2.1.]. If $p > 1$ then we have also so-called Lizorkin representations (cf. 3.5.3.). If s is sufficiently large (depending on p, q) then $B^s_{p,q}(T_n)$ and $F^s_{p,q}(T_n)$ can be characterized by means of differences and derivatives (cf. 3.5.4/(16), (18) and Remark 3.5.4/4).

Remark *6. As we shall see in the next subsections the spaces $F^s_{p,2}(T_n)$, $1 < p < \infty$, coincide with the periodic Sobolev-Lebesgue spaces and the spaces $B^s_{p,q}(T_n)$ with $s > 0$ and $\min(p, q) \geqq 1$ are the well-known periodic Besov (or Lipschitz) spaces. These spaces have been studied for a long time, cf. S. M. Nikol'skij [2], where also many references can be found. We refer also to P. I. Lizorkin [6]. P. Oswald [2, 3] has given a different definition of the spaces of type $B^s_{p,q}(T_1)$ and $F^s_{p,q}(T_1)$. In particular, the classical periodic Hardy spaces are contained in his scale $F^s_{p,q}(T_1)$, $0 < p \leqq 1$.

3.5.2. The Spaces $F^s_{\infty,q}(T_n)$

It makes no sense to admit $p = \infty$ in the definition of the spaces $F^s_{p,q}(T_n)$ (cf. Definition 3.5.1(ii)). In particular, the corresponding spaces would not be independent of the chosen system $\varphi \in \Phi(R_n)$. The reason is that the Fourier multiplier theorem for $L^{\Lambda}_p(T_n, l_q)$ (cf. Theorem 3.4.3/2) does not hold true if $p = \infty$ and $1 < q < \infty$. Thus, the problem arises whether there is a natural extension of the scale of spaces $F^s_{p,q}(T_n)$, $0 < p < \infty$, $0 < q \leqq \infty$, $-\infty < s < \infty$, to the case $p = \infty$, $0 < q < \infty$, and $-\infty < s < \infty$. In order to give an answer of this question we use the following observation.

Proposition. *Let* $1 < p < \infty$, $1 < q < \infty$, *and* $-\infty < s < \infty$. *Let* $\varphi \in \Phi(R_n)$. *Then*

$$F^s_{p,q}(T_n) = \{f \mid f \in D'(T_n),\ \exists \{f_j(x)\}_{j=0}^{\infty} \subset L_p(T_n) \quad \text{such that}$$
$$f = \sum_{j=0}^{\infty} \Big(\sum_{k \in Z_n} \hat{f}_j(k)\, \varphi_j(k)\, e^{ikx} \Big) \quad \text{in} \quad D'(T_n) \quad \text{and}$$
$$\|2^{sj} f_j(x) \mid L_p(T_n, l_q)\| < \infty\} \tag{1}$$

holds. Furthermore,

$$\inf \|2^{sj} f_j(x) \mid L_p(T_n, l_q)\|,$$

where the infimum is taken over all representations of f in the sense of (1), *is an equivalent norm in* $F^s_{p,q}(T_n)$.

Remark 1. The proposition is the periodic analogon of [T, Proposition 2.3.4/1]. It is clear that $F^s_{p,q}(T_n)$ is a subspace of the space described on the right-hand side of (1). The converse direction follows from the Fourier multiplier theorem 3.4.3/3 as in the non-periodic case.

Recall that the spaces $L_p(T_n, l_q)$, $0 < p < \infty$, $0 < q \leqq \infty$, have been defined in 3.2.4/(2), (3). We introduce $L_\infty(T_n, l_q)$, $0 < q < \infty$, by replacing the L_p-quasi-norm ($0 < p < \infty$) in (3.2.4/2) by the norm in $L_\infty(T_n)$.

Definition. *Let* $-\infty < s < \infty$, $1 < q < \infty$. *Let* $\varphi \in \Phi(R_n)$. *Then*

$$F^s_{\infty,q}(T_n) = \{f \mid f \in D'(T_n),\ \exists \{f_j(x)\}_{j=0}^\infty \subset L_\infty(T_n) \quad \textit{such that}$$

$$f = \sum_{j=0}^\infty \Big(\sum_{k \in Z_n} \varphi_j(k) \hat{f}_j(k)\, e^{ikx} \Big) \quad \textit{in} \quad D'(T_n) \quad \textit{and}$$

$$\| 2^{sj} f_j(x) \mid L_\infty(T_n, l_q) \| < \infty\} \tag{2}$$

and

$$\| f \mid F^s_{\infty,q}(T_n) \|^\varphi = \inf \| 2^{sj} f_j(x) \mid L_\infty(T_n, l_q) \|, \tag{3}$$

where the infimum is taken over all representations of f in the sense of (2).

Remark 2. Of course, the definition fits very well with the above proposition. However, the most interesting fact is the relation

$$(F^s_{p,q}(T_n))' = F^{-s}_{p',q'}(T_n), \tag{4}$$

where $1 \leqq p < \infty$, $1 < q < \infty$, and $-\infty < s < \infty$. Here $(A)'$ stands for the dual space of the Banach space A, $\frac{1}{p} + \frac{1}{p'} = 1$, $\frac{1}{q} + \frac{1}{q'} = 1$ (with $p' = \infty$ if $p = 1$). We refer to 3.5.6. In particular, it holds

$$(F^s_{1,q}(T_n))' = F^{-s}_{\infty,q'}(T_n), \tag{5}$$

$1 < q < \infty$, $-\infty < s < \infty$. Hence, by duality arguments properties of $F^s_{1,q}(T_n)$ can be carried over to $F^s_{\infty,q}(T_n)$. We have the following theorem.

Theorem. *Let* $1 < q < \infty$ *and* $-\infty < s < \infty$. *The spaces* $F^s_{\infty,q}(T_n)$ *are Banach spaces. They are independent of* $\varphi \in \Phi(R_n)$. *There hold the embeddings*

$$D(T_n) \subset F^s_{\infty,q}(T_n) \subset D'(T_n). \tag{6}$$

If $\varphi \in \Phi(R_n)_1$ *and* $\psi \in \Phi(R_n)$ *then the corresponding norms* $\| f \mid F^s_{\infty,q}(T_n)\|^\varphi$ *and* $\| f \mid F^s_{\infty,q}(T_n)\|^\psi$ *are equivalent to each other.*

Remark 3. It remains a gap. Namely, we can not deal with the case $0 < q \leqq 1, p = \infty, -\infty < s < \infty$. Of course (2) makes sense in this case. However, the duality procedure (5) does not work. Note that the case $p = \infty$, $q = \infty$ is covered by the spaces $B^s_{\infty,\infty}(T_n)$ (cf. Definition 3.5.1(i)).

3.5.3. Lizorkin Representations

In the definition of the spaces $B^s_{p,q}(T_n)$ and $F^s_{p,q}(T_n)$ we used smooth partitions of unity. Theorem 3.4.3/3 and Remark 3.4.3/3 imply that in the case $1 < p < \infty$ and $1 < q < \infty$ this assumption can be essentially weakened. Roughly speaking the functions $\varphi_j(x)$ from $\varphi \in \Phi(R_n)$ can be replaced by characteristic functions of dyadic corridors or dyadic cubes. To this end we introduce

$$\left.\begin{aligned} K_0 &= \{0 \mid 0 = (0, \dots, 0) \in Z_n\}, \\ K_j &= \{k \mid k \in Z_n, |k_m| < 2^j; m = 1, \dots, n\} \setminus \{k \mid k \in Z_n, |k_m| < 2^{j-1}; \\ &\qquad m = 1, \dots, n\}, \end{aligned}\right\} \tag{1}$$

$j = 1, 2, \dots$ Then $Z_n = \bigcup_{j=0}^{\infty} K_j$ is a decomposition of Z_n in dyadic "corridors".

Theorem. *Let* $-\infty < s < \infty$ *and* $1 < p < \infty$.

(i) *If* $0 < q \leqq \infty$, *then*

$$B^s_{p,q}(T_n) = \Big\{f \mid f \in D'(T_n),\ \|f \mid B^s_{p,q}(T_n)\|^* = \Big\| 2^{sj} \sum_{k \in K_j} \hat{f}(k)\, \mathrm{e}^{ikx} \mid l_q(L_p(T_n)) \Big\| < \infty \Big\}, \tag{2}$$

where $\|f \mid B^s_{p,q}(T_n)\|^*$ *is an equivalent quasi-norm in* $B^s_{p,q}(T_n)$ *(norm if* $q \geqq 1$).

(ii) *If* $1 < q < \infty$, *then*

$$F^s_{p,q}(T_n) = \Big\{f \mid f \in D'(T_n),\ \|f \mid F^s_{p,q}(T_n)\|^* = \Big\| 2^{sj} \sum_{k \in K_j} \hat{f}(k)\, \mathrm{e}^{ikx} \mid L_p(T_n, l_q) \Big\| < \infty \Big\}, \tag{3}$$

where $\|f \mid F^s_{p,q}(T_n)\|^*$ *is an equivalent norm in* $F^s_{p,q}(T_n)$.

Remark 1. The proof of (2) and (3) follows the non-periodic case taking into consideration Theorem 3.4.3/3 and Remark 3.4.3/3. Representations of such a type are known as Lizorkin representations. We refer to P. I. Lizorkin [6] for the spaces $B^s_{p,q}(T_1)$ or to P. I. Lizorkin [1, 4], and S. M. Nikol'skij [2, 8.10.1.] for spaces on R_n (cf. also the remarks in [T, 2.5.4.]).

Remark 2. The assertions of the theorem can be modified subdividing the "corridors" K_j with $j = 1, 2, \dots$ in congruent "cubes". We refer to [T, 2.5.4/(3), (4)], where the non-periodic case has been treated.

3.5.4. Periodic Besov and Sobolev-Lebesgue Spaces

As we have already mentioned in Remark 3.5.1/6 the spaces $B^s_{p,q}(T_n)$ and $F^s_{p,q}(T_n)$ contain a lot of classical periodic spaces as special cases. It is the aim of this subsection to clarify these relations. At first we explain what is meant by "classical" spaces. We put

$$C(T_n) = \{f \mid f = f(x) \text{ uniformly continuous on } T_n\}, \tag{1}$$

$$\|f \mid C(T_n)\| = \max_{x \in T_n} |f(x)|, \tag{2}$$

$$C^m(T_n) = \{f \mid D^\alpha f \in C(T_n) \quad \text{for all } \alpha, |\alpha| \leqq m\}, \tag{3}$$

$$\|f \mid C^m(T_n)\| = \sum_{|\alpha| \leqq m} \|D^\alpha f \mid C(T_n)\|. \tag{4}$$

Here $m = 1, 2, \dots$ In the sense of 3.2.3., $C(T_n)$ consists of the set of all 2π-periodic uniformly continuous functions on R_n. Of course,

$$D(T_n) \subset C(T_n) \subset D'(T_n). \tag{5}$$

The Periodic Hölder or Lipschitz Spaces $C^s(T_n)$. If s is a real number, then

$$s = [s] + \{s\} \quad \text{with } [s] \text{ integer and} \quad 0 \leqq \{s\} < 1. \tag{6}$$

Let $0 < s \neq$ integer. Then,

$$C^s(T_n) = \Big\{ f \mid f \in C^{[s]}(T_n),\ \|f \mid C^s(T_n)\| = \|f \mid C^{[s]}(T_n)\| + \sum_{|\alpha| = [s]} \sup_{\substack{x \neq y \\ x, y \in T_n}} \frac{|D^\alpha f(x) - D^\alpha f(y)|}{|x - y|^{\{s\}}} < \infty \Big\}. \tag{7}$$

The Periodic Zygmund Spaces $\mathscr{C}^s(T_n)$. If s is a real number, then

$$s = [s]^- + \{s\}^+ \text{ with } [s]^- \text{ integer and } 0 < \{s\}^+ \leqq 1. \tag{8}$$

Further, if $f(x)$ is defined on T_n and extended 2π-periodically on R_n, then we put

$$\left.\begin{aligned} \Delta_h^1 f(x) &= f(x + h) - f(x), \quad h \in T_n, x \in T_n, \\ \Delta_h^m f(x) &= \Delta_h^1 (\Delta_h^{m-1} f)(x), \quad m = 2, 3, \ldots, h \in T_n, x \in T_n. \end{aligned}\right\} \tag{9}$$

Let $s > 0$ be an arbitrary real number. Then,

$$\mathscr{C}^s(T_n) = \Big\{ f \mid f \in C^{[s]^-}(T_n),\ \|f \mid \mathscr{C}^s(T_n)\| = \|f \mid C^{[s]^-}(T_n)\| + \sum_{|\alpha| = [s]^-} \sup_{0 \neq h \in T_n} |h|^{-\{s\}^+} \|\Delta_h^2 f(x) \mid C(T_n)\| < \infty \Big\}. \tag{10}$$

The Periodic Sobolev Spaces $W_p^m(T_n)$. Let $1 < p < \infty$ and $m = 1, 2, \ldots$ Then,

$$W_p^m(T_n) = \Big\{ f \mid f \in L_p(T_n),\ \|f \mid W_p^m(T_n)\| = \sum_{|\alpha| \leqq m} \|D^\alpha f \mid L_p(T_n)\| < \infty \Big\}. \tag{11}$$

Further we put $W_p^0(T_n) = L_p(T_n)$, $1 < p < \infty$.

The Periodic Besov (or Lipschitz) Spaces $\Lambda_{p,q}^s(T_n)$. Let $0 < s < \infty$, $1 \leqq p < \infty$, and $1 \leqq q < \infty$. Then,

$$\Lambda_{p,q}^s(T_n) = \Big\{ f \mid f \in W_p^{[s]^-}(T_n),\ \|f \mid \Lambda_{p,q}^s(T_n)\| = \|f \mid W_p^{[s]^-}(T_n)\| + \sum_{|\alpha| = [s]^-} \Big(\int_{T_n} |h|^{-\{s\}^+ q} \|\Delta_h^2 D^\alpha f(x) \mid L_p(T_n)\|^q \frac{dh}{|h|^n} \Big)^{1/q} < \infty \Big\}. \tag{12}$$

If $0 < s < \infty$, $1 \leqq p < \infty$, then

$$\Lambda_{p,\infty}^s(T_n) = \Big\{ f \mid f \in W_p^{[s]^-}(T_n),\ \|f \mid \Lambda_{p,\infty}^s(T_n)\| = \|f \mid W_p^{[s]^-}(T_n)\| + \sum_{|\alpha| = [s]^-} \sup_{0 \neq h \in T_n} |h|^{-\{s\}^+} \|\Delta_h^2 D^\alpha f(x) \mid L_p(T_n)\| < \infty \| \Big\}. \tag{13}$$

The Periodic Bessel-Potential (or Lebesgue or Liouville) Spaces $H_p^s(T_n)$. Let $1 < p < \infty$ and $-\infty < s < \infty$. Then,

$$H_p^s(T_n) = \Big\{ f \mid f \in D'(T_n),\ \|f \mid H_p^s(T_n)\| = \Big\| \sum_{k \in Z_n} (1 + |k|^2)^{s/2} \hat{f}(k)\, e^{ikx} \mid L_p(T_n) \Big\| < \infty \Big\}. \tag{14}$$

Remark 1. The spaces defined in (7), (10)–(14) are the so-called classical spaces. They are widely used in approximation theory (at least as far as the one-dimensional case is concerned). For the literature and historical remarks we refer to S. M. Nikol'skij [2]. Sometimes the spaces $\Lambda_{p,\infty}^s(T_n)$ are denoted as Nikol'skij spaces.

Theorem. (i) *If $s > 0$ is not an integer, then*

$$B_{\infty,\infty}^s(T_n) = C^s(T_n). \tag{15}$$

(ii) *If* $0 < s < \infty$, *then*

$$B^s_{\infty,\infty}(T_n) = \mathscr{C}^s(T_n). \tag{16}$$

(iii) *If* $1 < p < \infty$ *and if* $m = 0, 1, 2, \ldots$, *then*

$$F^m_{p,2}(T_n) = W^m_p(T_n). \tag{17}$$

(iv) *If* $1 \leqq p < \infty$, $1 \leqq q \leqq \infty$, *and* $0 < s < \infty$, *then*

$$B^s_{p,q}(T_n) = \Lambda^s_{p,q}(T_n). \tag{18}$$

(v) *If* $1 < p < \infty$ *and* $-\infty < s < \infty$, *then*

$$F^s_{p,2}(T_n) = H^s_p(T_n). \tag{19}$$

Remark 2. The theorem is the periodic counterpart of [T, Theorem 2.5.6 and Theorem 2.5.7]. The proof is similar. The main ingredients are the embeddings

$$B^0_{\infty,1}(T_n) \subset C(T_n) \subset B^0_{\infty,\infty}(T_n) \tag{20}$$

and

$$B^0_{p,1}(T_n) \subset L_p(T_n) \subset B^0_{p,\infty}(T_n), \quad 1 \leqq p < \infty, \tag{21}$$

as well as the Littlewood-Paley theorem

$$F^0_{p,2}(T_n) = L_p(T_n), \quad 1 < p < \infty. \tag{22}$$

(20) and (21) follow from the definition of $B^0_{p,q}(T_n)$. (22) has been proved in Theorem 3.4.4(i). Now we can use interpolation theory (cf. also 3.6.1.) and Fourier multipliers. For the details we refer to [T, 2.5.6. and 2.5.7.].

Remark 3. Immediate corollaries are the well-known formulas

$$C^s(T_n) = \mathscr{C}^s(T_n) \tag{23}$$

if s is not an integer and

$$W^m_p(T_n) = H^m_p(T_n), \tag{24}$$

$1 < p < \infty$, $m = 0, 1, 2, \ldots$ Note, that the spaces $C^m(T_n)$, $m = 0, 1, 2, \ldots, L_1(T_n)$, and $L_\infty(T_n)$ are not contained in the scales $B^s_{p,q}(T_n)$ and $F^s_{p,q}(T_n)$ (cf. [S, Proposition 1.3] for the non-periodic case).

Remark 4. Let us emphasize that also the spaces $B^s_{p,q}(T_n)$ with $0 < p < 1$ and, moreover, the spaces $F^s_{p,q}(T_n)$ with $0 < p < \infty$ and $0 < q \leqq \infty$ can be characterized by means of differences and derivatives provided that s is large enough. Again we can modify the methods from the non-periodic case on the basis of the results from Sections 3.3. and 3.4. However it is not our aim to point it out here in detail (cf. [T, 2.5.9.–2.5.12.]).

Remark 5. The Hardy spaces on the n-torus $H_p(T_n)$ are not included in our scales of spaces. These spaces are covered by the periodic spaces of Besov-Sobolev-Hardy type (one-dimensional case) of P. Oswald [2, 3]. In contrast to our approach he deals with periodic distributions of power series type, i.e. $f \in D'(T_1)$ with $\hat{f}(k) = 0$ if $k < 0$, $k \in Z_1$. It would be of interest to give explicit characterizations of $F^0_{p,2}(T_n)$, $0 < p \leqq 1$, in the sense of [T, 2.5.8.] (local Hardy spaces).

3.5.5. Embedding Theorems for Different Metrics

From the point of view of our applications to the problem of strong summability in Section 3.7. and also from the historical point of view the embedding theorems for periodic function spaces which we establish in the following theorem are of peculiar interest.

Theorem. (i) *Let* $0 < p_0 \leqq p_1 \leqq \infty$, $0 < q \leqq \infty$, *and* $-\infty < s_1 \leqq s_0 < \infty$. *Then,*

$$B^{s_0}_{p_0,q}(T_n) \subset B^{s_1}_{p_1,q}(T_n) \quad \text{if} \quad s_0 - \frac{n}{p_0} = s_1 - \frac{n}{p_1}. \tag{1}$$

(ii) *Let* $0 < p_0 < p_1 < \infty$, $0 < q \leqq \infty$, $0 < r \leqq \infty$, *and* $-\infty < s_1 < s_0 < \infty$. *Then,*

$$F^{s_0}_{p_0,q}(T_n) \subset F^{s_1}_{p_1,r}(T_n) \quad \text{if} \quad s_0 - \frac{n}{p_0} = s_1 - \frac{n}{p_1}. \tag{2}$$

Remark 1. This is the counterpart of [T, Theorem 2.7.1]. (1) is an immediate consequence of the definition of the spaces $B^s_{p,q}(T_n)$ and Nikol'skij's inequality for trigonometric polynomials (3.3.2/11). In order to prove (2) one uses the method of B. Jawerth [1] and proceeds in analogy to the R_n-case (cf. [T, Step 2 of the proof of Theorem 2.7.1]). If $\min(p_0, q, p_1) \geqq 1$ and $s_1 \geqq 0$ then (1) is a classical assertion (cf. S. M. Nikol'skij [2, 5.3. and 6.10.1.]).

Corollary. *Let* $0 < s < \infty$.

(i) *If* $0 < p \leqq \infty$ *and* $0 < q \leqq \infty$, *then*

$$B^{s+\frac{n}{p}}_{p,q}(T_n) \subset \mathscr{C}^s(T_n). \tag{3}$$

(ii) *If* $0 < p < \infty$ *and* $0 < q \leqq \infty$, *then*

$$F^{s+\frac{n}{p}}_{p,q}(T_n) \subset \mathscr{C}^s(T_n). \tag{4}$$

(iii) *If* $1 < q < \infty$, *then*

$$F^s_{\infty,q} \subset \mathscr{C}^s(T_n). \tag{5}$$

Remark 2. (3) follows from (1) with $p_1 = \infty$, (3.5.1/16), and (3.5.4/16). Further, (4) is a consequence of (3.5.1/20) and part (i) of the theorem. Finally, (5) follows from

$$B^{-s}_{1,1}(T_n) \subset F^{-s}_{1,q'}(T_n),$$

which is due to (3.5.1/20), and from the duality assertions (3.5.6/3) and (3.5.6/4).

Remark 3. If s is not an integer then $\mathscr{C}^s(T_n)$ in the Corollary can be replaced by $C^s(T_n)$ (cf. (3.5.4/23)). Let $m = 0, 1, 2, \ldots$ be a non-negative integer. Then we have

$$B^{m+\frac{n}{p}}_{p,q}(T_n) \subset C^m(T_n), \tag{6}$$

provided that $0 < p \leqq \infty$, $0 < q \leqq 1$. Here $C^m(T_n)$ has the meaning of (3.5.4/3). (6) is a consequence of (3.5.4/20) and the estimate

$$\|D^\alpha f \mid B^0_{\infty,q}(T_n)\| \leqq c\|f \mid B^m_{\infty,q}(T_n)\| \quad \text{for} \quad |\alpha| \leqq m. \tag{7}$$

We obtain (7) with the help of Bernstein's inequality (3.3.4/9) and the definition of $B^s_{\infty,q}(T_n)$.

Remark 4. Theorem 3.5.4 enables us to establish further special embedding theorems in connection with the above theorem and the corollary. For example, embeddings of periodic Sobolev and Lebesgue (or Liouville) spaces can be derived. Further, note that part (ii) of the theorem can be complemented by

$$B^{s_2}_{p_2,p_0}(T_n) \subset F^{s_0}_{p_0,q}(T_n) \subset B^{s_1}_{p_1,p_0}(T_n), \tag{8}$$

where $0 < p_2 < p_0 < p_1 < \infty$, $0 < q \leqq \infty$, and $-\infty < s_1 < s_0 < s_2 < \infty$ with

$$s_2 - \frac{n}{p_2} = s_0 - \frac{n}{p_0} = s_1 - \frac{n}{p_1}$$

(cf. B. Jawerth [1] and J. Franke [2]).

3.5.6. Dual Spaces

Recall that we have introduced the spaces $F^s_{\infty,q}(T_n)$, $1 < q < \infty$, $-\infty < s < \infty$, in Subsection 3.5.2. As mentioned in Remark 3.5.2/2 the main tool for the study of $F^s_{\infty,q}(T_n)$ is duality theory. For the sake of completeness we shall describe the dual spaces of $B^s_{p,q}(T_n)$ and $F^s_{p,q}(T_n)$, where $0 < p < \infty$, $0 < q < \infty$ and $-\infty < s < \infty$. In these cases $D(T_n)$ is dense in $B^s_{p,q}(T_n)$ and $F^s_{p,q}(T_n)$ (cf. Theorem 3.5.1/(i, ii)). Further, the embeddings (3.5.1/5) and (3.5.1/6) hold. Therefore, we have with the usual interpretation

$$\left.\begin{aligned} &D(T_n) \subset (B^s_{p,q}(T_n))' \subset D'(T_n) \\ \text{and}\quad &D(T_n) \subset (F^s_{p,q}(T_n))' \subset D'(T_n) \end{aligned}\right\} \tag{1}$$

if $0 < p < \infty$, $0 < q < \infty$, and $-\infty < s < \infty$. Here $(A)'$ denotes the dual space of the quasi-Banach space A. Moreover, $g \in (B^s_{p,q}(T_n))'$ with $0 < p < \infty$, $0 < q < \infty$, and $-\infty < s < \infty$ if and only if $g \in D'(T_n)$ and

$$|g(\varphi)| \leqq c\|\varphi \mid B^s_{p,q}(T_n)\| \tag{2}$$

for all $\varphi \in D(T_n)$, where c is an appropriate positive constant. Analogously for $F^s_{p,q}(T_n)$ with $0 < p < \infty$, $0 < q < \infty$, and $-\infty < s < \infty$. In this sense we have to understand the following theorem.

Recall that $\frac{1}{p} + \frac{1}{p'} = 1$ if $1 < p < \infty$ and $p' = \infty$ if $0 < p \leqq 1$.

Theorem. *Let* $-\infty < s < \infty$.

(i) *If* $1 \leqq p < \infty$ *and* $0 < q < \infty$, then

$$(B^s_{p,q}(T_n))' = B^{-s}_{p',q'}(T_n). \tag{3}$$

(ii) *If* $1 \leqq p < \infty$ *and* $1 < q < \infty$, *then*

$$(F^s_{p,q}(T_n))' = F^{-s}_{p',q'}(T_n). \tag{4}$$

(iii) *If* $0 < p < 1$ *and* $0 < q < \infty$, *then*

$$(B^s_{p,q}(T_n))' = B^{-s+n\left(\frac{1}{p}-1\right)}_{\infty,q'}(T_n). \tag{5}$$

(iv) *If* $0 < p < 1$ *and* $0 < q < \infty$, *then*

$$(F^s_{p,q}(T_n))' = B^{-s+n\left(\frac{1}{p}-1\right)}_{\infty,\infty}(T_n). \tag{6}$$

Remark 1. The proof of the theorem can be obtained on the same line as in [T, 2.11.2. and 2.11.3.] for spaces on R_n with more or less obvious modifications. For example, in order to show (iii) and (iv) we have to use Nikol'skij's inequality for trigonometric polynomials (cf. (3.3.2/11)) and the embedding theorems for different metrics established in the preceding subsection. For historical remarks we refer to [T, Remark 2.11.2/3 and 2.11.3/2].

Remark 2. As in the non-periodic case the restrictions to $p < \infty$ and $q < \infty$ in our theorem are natural. They can be removed if one turns to the spaces $\tilde{B}^s_{p,q}(T_n)$ and $\tilde{F}^s_{p,q}(T_n)$ which are defined as the completions of $D(T_n)$ in the spaces $B^s_{p,q}(T_n)$ and $F^s_{p,q}(T_n)$, respectively (cf. [T, Remark 2.11.2/2] and [S, Remarks 2.5.1/7 and 2.5.2/3]). We omit the details.

3.5.7. Remarks on Further Properties

There is no doubt that further properties of spaces on R_n can be carried over to the periodic case. This concerns further equivalent characterizations (such as via maximal functions, or via approximation), interpolation results, Fourier multiplier assertions, traces, convolution algebras, pointwise multipliers, or bases. However, it is not our aim to establish all these properties in detail here. For the classical spaces $\Lambda^s_{p,q}(T_n)$ (Besov (or Lipschitz) spaces) and $H^s_p(T_n)$ (Sobolov-Lebesgue (or Bessel-potential, or Liouville) spaces) some of them can be found in the literature. We refer to the book by S. M. Nikol'skij [2, Chapter 2, 5.3., 5.4.3., 5.5.8., 6.10.1., and the remarks to Chapter 5, Chapter 8, and Chapter 9] (equivalent norms, theorems of Jackson and Bernstein type, embeddings and traces). There are given a lot of further references on the periodic case. P. I. Lizorkin [6] constructs an unconditional Schauder basis in $B^s_{p,q}(T_1)$, $0 < s < \infty$, $1 < p < \infty$ and $1 < q < \infty$, and proves Fourier multiplier theorems for $B^s_{p,q}(T_n)$, $1 < p < \infty$, $1 < q < \infty$, $0 < s < \infty$. Generalizations to the n-dimensional case can be found in P. I. Lizorkin, D. G. Orlovskij [1] and D. G. Orlovskij [1].

V. I. Burenkov, M. L. Gol'dman [1, 2] investigated the question how to translate properties of Banach spaces of functions defined on R_n to Banach spaces of functions on T_n and vice versa. Their method can be applied to the spaces $B^s_{p,q}(T_n)$ and $F^s_{p,q}(T_n)$ with $\min(p, q) \geqq 1$ and $s > 0$. In particular, we obtain on this way direct and inverse embedding theorems for different metrics and different dimensions (cf. V. I. Burenkov, M. L. Gol'dman [2, Section 2]) as well as mapping properties of linear operators (cf. V. I. Burenkov, M. L. Gol'dman [2, Section 1]).

For investigations of periodic Hardy spaces $H_p(T_1)$, $0 < p < \infty$, and related spaces of Besov-Sobolev type we refer to E. A. Storoženko [1, 2], P. Oswald [2, 3]. See also E. A. Storoženko, P. Oswald [1, 2], E. A. Storoženko, V. G. Krotov, P. Oswald [1], and L. Colzani [1] as far as periodic L_p and H_p-spaces with $0 < p \leqq 1$ are concerned.

It turns out that the basic tools established in 3.3. and 3.4. are not sufficient to give satisfactory answers of the above mentioned questions if one is interested in the spaces $B^s_{p,q}(T_n)$ and $F^s_{p,q}(T_n)$ with $\min(p, q) < 1$. In particular, we need an improvement of the Fourier multiplier theorem for the spaces $L^A_p(T_n, l_q)$ (Theorem 3.4.1/2) as well as a complex interpolation method for these spaces. Then, again the results from the non-periodic case can be carried over to the periodic one. We shall restrict ourselves to describe the interpolation results and their use in order to achieve Fourier multiplier criterions. This will be done in the next section. It is also sufficient for our later considerations of approximation and strong summability.

3.6. Complex Interpolation and Fourier Multipliers

3.6.1. Real and Complex Interpolation

Interpolation methods are frequently and successfully used in the theory of function spaces (cf. [I], [F], [S], [T], J. Bergh, J. Löfström [1], and J. Peetre [4]). The main aim of this section is to modify a new complex interpolation method for quasi-Banach spaces which was developed by H. Triebel [10, 11] and L. Päivärinta [1] for spaces on R_n in order to give criterions for Fourier multipliers in $B^s_{p,q}(T_n)$ and $F^s_{p,q}(T_n)$ as well as to improve the multiplier theorem for the spaces $L^A_p(T_n, l_q)$ with $\min(p, q) \leqq 1$ (cf. Theorem 3.4.1/2). The latter is of interest if one wishes to obtain reasonable results in the case of the spaces $F^s_{p,q}(T_n)$ with $\min(p, q) \leqq 1$. It turns out that the proofs of our assertions can be carried out in analogy to the non-periodic case as it is stated in [T, 2.4.] if one uses the preparations in Sections 3.3. and 3.4. Therefore, we restrict ourselves to the formulation of the main definitions and theorems and to some remarks concerning the proofs as in the preceding section.

For completeness we shall state some well-known interpolation results of the periodic spaces $B^s_{p,q}(T_n)$ and $F^s_{p,q}(T_n)$. As usual, $(A_0, A_1)_{\theta,q}$, $0 < \theta < 1$, $0 < q \leqq \infty$, denotes the real interpolation space of two quasi-Banach spaces which are continuously embedded in the same linear Hausdorff space. Further, $[A_0, A_1]_\theta$, $0 < \theta < 1$, means the complex interpolation space of J. L. Lions, A. P. Calderón, and S. G. Krejn. Here A_0 and A_1 are supposed to be Banach spaces embedded in a linear Hausdorff space. Definitions and properties of these interpolation spaces can be found in [I] or in J. Bergh, J. Löfström [1]. Note that $[A_0, A_1]_\theta$ makes no sense if A_0 and A_1 are quasi-Banach spaces.

Theorem 1. *Let* $0 < \theta < 1$, $0 < q_0 \leqq \infty$, $0 < q_1 \leqq \infty$, *and* $0 < q \leqq \infty$. *Furthermore, let* $-\infty < s_0 < \infty$, $-\infty < s_1 < \infty$, $s_0 \neq s_1$ *and* $s = (1 - \theta)\, s_0 + \theta s_1$.

(i) *If* $0 < p \leqq \infty$, *then*

$$(B^{s_0}_{p,q_0}(T_n), B^{s_1}_{p,q_1}(T_n))_{\theta,q} = B^s_{p,q}(T_n). \tag{1}$$

(ii) *If* $0 < p < \infty$, *then,*

$$(F^{s_0}_{p,q_0}(T_n), F^{s_1}_{p,q_1}(T_n))_{\theta,q} = B^s_{p,q}(T_n). \tag{2}$$

Remark 1. The proof is the same as in [T, Theorem 2.4.2] (with the usual modifications).

Remark 2. If $1 \leqq p \leqq \infty$, $1 \leqq q_0 \leqq \infty$, $1 \leqq q_1 \leqq \infty$, $0 \leqq s_0 < \infty$, and $0 \leqq s_1 < \infty$, then (1) and (2) lead together with Theorem 3.5.4 to interpolation results for the classical periodic spaces of Sobolev and Besov type. In particular, we have

$$(L_p(T_n), W^m_p(T_n))_{\theta,q} = \Lambda^{\theta m}_{p,q}(T_n), \tag{3}$$

where $1 < p < \infty$, $m = 1, 2, \ldots$, $0 < q \leqq \infty$, and $0 < \theta < 1$. This is a very classical formula. We refer also to S. Mundlos [1]. Further, we observe that

$$(C(T_n), C^m(T_n))_{\theta,\infty} = \mathscr{C}^{\theta m}(T_n), \tag{4}$$

$0 < \theta < 1$, $m = 1, 2, \ldots$ This is a consequence of (1), the embeddings

$$B^m_{\infty,1}(T_n) \subset C^m(T_n) \subset B^m_{\infty,\infty}(T_n), \tag{5}$$

and Theorem 3.5.4 (ii).

Theorem 2. *Let* $0 < \theta < 1$, $1 < q_0 < \infty$, $1 < q_1 < \infty$, $1 < p_0 < \infty$, *and* $1 < p_1 < \infty$. *Furthermore, let* $-\infty < s_0 < \infty$, $-\infty < s_1 < \infty$, *and*

$$s = (1 - \theta)\, s_0 + \theta s_1, \qquad \frac{1}{p} = \frac{1-\theta}{p_0} + \frac{\theta}{p_1}, \qquad \frac{1}{q} = \frac{1-\theta}{q_0} + \frac{\theta}{q_1}. \tag{6}$$

Then we have

$$[B^{s_0}_{p_0,q_0}(T_n), B^{s_1}_{p_1,q_1}(T_n)]_\theta = B^s_{p,q}(T_n), \tag{7}$$

and

$$[F^{s_0}_{p_0,q_0}(T_n), F^{s_1}_{p_1,q_1}(T_n)]_\theta = F^s_{p,q}(T_n). \tag{8}$$

Remark 3. The proof is the same as in [I, Theorem 2.4.1(d) and Theorem 2.4.2(d)] (obvious modifications).

Remark 4. As special cases we obtain from Theorem 3.5.4(iii, v) and (8)

$$[W^{m_0}_{p_0}(T_n), W^{m_1}_{p_1}(T_n)]_\theta = H^s_p(T_n) \tag{9}$$

and

$$[L_{p_0}(T_n), L_{p_1}(T_n)]_\theta = L_p(T_n) \tag{10}$$

if $1 < p_0 < \infty$, $1 < p_1 < \infty$, $0 < \theta < 1$, $m_0 = 0, 1, 2, \ldots$, $m_1 = 0, 1, 2, \ldots$ and

$$s = (1 - \theta)\, m_0 + \theta m_1, \qquad \frac{1}{p} = \frac{1-\theta}{p_0} + \frac{\theta}{p_1}.$$

Remark 5. The real and complex interpolation method in the above theorems are general interpolation methods applicable to arbitrary quasi-Banach and Banach spaces, respectively. However, the above complex method cannot be extended to general quasi-Banach spaces. But this would be desirable in order to interpolate spaces $B^{s_0}_{p_0,q_0}(T_n)$ $(F^{s_0}_{p_0,q_0}(T_n))$ and $B^{s_1}_{p_1,q_1}(T_n)$ $(F^{s_1}_{p_1,q_1}(T_n))$ with $\min(p_0, p_1, q_0, q_1) < 1$ and $p_0 \neq p_1$ (cf. also [T, 2.4.3.] or [I, 2.4.] for the real method). This will be done in the next subsection.

3.6.2. Complex Interpolation for the Spaces $B^s_{p,q}(T_n)$ and $F^s_{p,q}(T_n)$

Let $A = \{z \mid 0 < \operatorname{Re} z < 1\}$ be a strip in the complex plane. Its closure $\{z \mid 0 \leqq \operatorname{Re} z \leqq 1\}$ is denoted by $\bar{A}$. A function $f = f(z)$ defined on $\bar{A}$ with values in $D'(T_n)$ is said to be a $D'(T_n)$-analytic function in A if the following properties are satisfied:

(i) If $f(z)$ is represented in $D'(T_n)$ by

$$f(z) = \sum_{k \in Z_n} a_k(z)\, e^{ikx}, \quad z \in \bar{A},$$

(cf. 3.2.2) then

$$\sum_{k \in Z_n} \varphi(k)\, a_k(z)\, e^{ikx} \tag{1}$$

is a uniformly continuous and bounded function on $T_n \times \bar{A}$ for any $\varphi \in S(R_n)$ with compact support in R_n.

(ii) For any fixed $x \in T_n$ and any $\varphi \in S(R_n)$ with compact support the function from (1) is analytic in A.

Definition 1. *Let* $-\infty < s_0 < \infty$, $-\infty < s_1 < \infty$, $0 < q_0 \leqq \infty$, *and* $0 < q_1 \leqq \infty$.

(i) *If* $0 < p_0 \leqq \infty$ *and* $0 < p_1 \leqq \infty$ *then*

$$\begin{aligned} &F(B^{s_0}_{p_0,q_0}(T_n), B^{s_1}_{p_1,q_1}(T_n)) \\ &= \Big\{ f(z) \mid f(z) \ \ D'(T_n)\text{-analytic in } A, \\ &\quad f(it) \in B^{s_0}_{p_0,q_0}(T_n),\ f(1+it) \in B^{s_1}_{p_1,q_1}(T_n) \text{ for every } t \in R_1, \\ &\quad \| f(z) \mid F(B^{s_0}_{p_0,q_0}(T_n), B^{s_1}_{p_1,q_1}(T_n)) \| \\ &\quad = \max_{\mu=0,1} \sup_{t \in R_1} \| f(\mu + it) \mid B^{s_\mu}_{p_\mu,q_\mu}(T_n) \| < \infty \Big\}. \end{aligned} \tag{2}$$

(ii) *If* $0 < p_0 < \infty$ *and* $0 < p_1 < \infty$, *then*

$$\begin{aligned} &F(F^{s_0}_{p_0,q_0}(T_n), F^{s_1}_{p_1,q_1}(T_n)) \\ &= \Big\{ f(z) \mid f(z) \ \ D'(T_n)\text{-analytic in } A, \\ &\quad f(it) \in F^{s_0}_{p_0,q_0}(T_n),\ f(1+it) \in F^{s_1}_{p_1,q_1}(T_n) \text{ for every } t \in R_1, \\ &\quad \| f(z) \mid F(F^{s_0}_{p_0,q_0}(T_n), F^{s_1}_{p_1,q_1}(T_n)) \| \\ &\quad = \max_{\mu=0,1} \sup_{t \in R_1} \| f(\mu + it) \mid F^{s_\mu}_{p_\mu,q_\mu}(T_n) \| < \infty \Big\}. \end{aligned} \tag{3}$$

Definition 2. *Let* $-\infty < s_0 < \infty$, $-\infty < s_1 < \infty$, $0 < q_0 \leqq \infty$, $0 < q_1 \leqq \infty$ *and* $0 < \theta < 1$.

(i) *If* $0 < p_0 \leqq \infty$ *and* $0 < p_1 \leqq \infty$, *then*

$$(B^{s_0}_{p_0,q_0}(T_n), B^{s_1}_{p_1,q_1}(T_n))_\theta = \{g \mid \exists f(z) \in F(B^{s_0}_{p_0,q_0}(T_n), B^{s_1}_{p_1,q_1}(T_n)) \text{ with } g = f(\theta)\}, \quad (4)$$

$$\|g \mid (B^{s_0}_{p_0,q_0}(T_n), B^{s_1}_{p_1,q_1}(T_n))_\theta\| = \inf \|f(z) \mid F(B^{s_0}_{p_0,q_0}(T_n), B^{s_1}_{p_1,q_1}(T_n))\|, \quad (5)$$

where the infimum is taken over all admissible functions $f(z)$ *in the sense of* (4).

(ii) *If* $0 < p_0 < \infty$ *and* $0 < p_1 < \infty$, *then*

$$(F^{s_0}_{p_0,q_0}(T_n), F^{s_1}_{p_1,q_1}(T_n))_\theta = \{g \mid \exists f(z) \in F(F^{s_0}_{p_0,q_0}(T_n), F^{s_1}_{p_1,q_1}(T_n)) \text{ with } g = f(\theta)\}, \quad (6)$$

$$\|g \mid (F^{s_0}_{p_0,q_0}(T_n), F^{s_1}_{p_1,q_1}(T_n))_\theta\| = \inf \|f(z) \mid F(F^{s_0}_{p_0,q_0}(T_n), F^{s_1}_{p_1,q_1}(T_n))\|, \quad (7)$$

where the infimum is taken over all admissible functions $f(z)$ *in the sense of* (6).

Remark 1. Definition 1 and Definition 2 are the periodic counterparts of Definition 1 and Definition 2 in [T, 2.4.4.] (or H. Triebel [11]). In contrast to the complex interpolation $[\cdot, \cdot]_\theta$ used in the preceding subsection we denote the above complex method with $(\cdot, \cdot)_\theta$. The idea to develop a complex interpolation method for certain quasi-Banach spaces is due to A. P. Calderón, A. Torchinsky [1, II]. We refer also to H. Triebel [10, 11] and L. Päivärinta [1].

Remark 2. All spaces defined in (2), (3), (4), and (6) are quasi-Banach spaces (Banach spaces if $\min(p_0, p_1, q_0, q_1) \geqq 1$). This can be proved on the same line as in H. Triebel [10].

Theorem. *Let* $-\infty < s_0 < \infty$, $-\infty < s_1 < \infty$, $0 < q_0 \leqq \infty$, $0 < q_1 \leqq \infty$, *and* $0 < \theta < 1$.

(i) *If* $0 < p_0 \leqq \infty$, $0 < p_1 \leqq \infty$, *and*

$$s = (1-\theta)\, s_0 + \theta s_1, \qquad \frac{1}{p} = \frac{1-\theta}{p_0} + \frac{\theta}{p_1}, \qquad \frac{1}{q} = \frac{1-\theta}{q_0} + \frac{\theta}{q_1}, \quad (8)$$

then

$$(B^{s_0}_{p_0,q_0}(T_n), B^{s_1}_{p_1,q_1}(T_n))_\theta = B^s_{p,q}(T_n). \quad (9)$$

(ii) *If* $0 < p_0 < \infty$, $0 < p_1 < \infty$, *and if* (8) *is satisfied, then*

$$(F^{s_0}_{p_0,q_0}(T_n), F^{s_1}_{p_1,q_1}(T_n))_\theta = F^s_{p,q}(T_n). \quad (10)$$

Remark 3. For the proof of the theorem we refer to [T, 2.4.7.] or H. Triebel [11] and the usual modifications which must be done if one turns from the non-periodic to the periodic case (cf. also Sections 3.3. and 3.4.).

Remark 4. As we shall see in the next subsection formulas (9) and (10) are useful instruments in the theory of our spaces. Theorem 3.6.1/2 shows that in the case $1 < p_0 < \infty$, $1 < p_1 < \infty$, $1 < q_0 < \infty$, $1 < q_1 < \infty$ both the classical complex method and the method described in this subsection yield the same results. In general this is not true (cf. [T, Remark 2.4.7/2]).

Remark 5. Let us mention that it is not obvious whether the complex method $(\cdot, \cdot)_\theta$ has the interpolation property as it is known for the real method or for the classical method $[\cdot, \cdot]_\theta$ (cf., for example, [T, Proposition 2.4.1] and [I, Theorem 1.9.3(a)]). We refer to H. Triebel [10, Theorem 2], where this question is discussed in the framework of $L_p(l_q)$-spaces of entire analytic functions. One needs additional assumptions for the linear bounded operator being under consideration. On the other hand this method has the advantage that families of operators can be considered (cf. [T, 2.4.8.] for the example of Fourier multipliers). Thus, we have to show the interpolation property for each concrete operator we are interested in. However J. Franke [3] proved recently that the interpolation property for $(\cdot, \cdot)_\theta$ holds under very mild restrictions for the underlying operator (non-periodic case, but there is no doubt that this result has a periodic counterpart).

3.6.3. Fourier Multipliers for $B^s_{p,q}(T_n)$ and $F^s_{p,q}(T_n)$

We formulate a theorem on Fourier multipliers for the spaces $B^s_{p,q}(T_n)$ and $F^s_{p,q}(T_n)$. Let $\psi(x) \in S(R_n)$ and $\varphi(x) \in S(R_n)$ with

$$0 \leqq \psi(x) \leqq 1, \operatorname{supp} \psi \subset \{y \mid |y| \leqq 4\}, \psi(x) = 1 \text{ if } |x| \leqq 2, \tag{1}$$

$$0 \leqq \varphi(x) \leqq 1, \operatorname{supp} \varphi \subset \{y \mid \tfrac{1}{4} \leqq |y| \leqq 4\}, \varphi(x) = 1 \text{ if } \tfrac{1}{2} \leqq |x| \leqq 2. \tag{2}$$

For the sake of brevity we put

$$\|m \mid h^{\varkappa}_2(R_n)\| = \|\psi m \mid H^{\varkappa}_2(R_n)\| + \sup_{j=0,1,\dots} \|\varphi(\cdot)\, m(2^j \cdot) \mid H^{\varkappa}_2(R_n)\|. \tag{3}$$

Here $H^{\varkappa}_2(R_n)$ has the meaning of (1.7.4/1).

Theorem. (i) *Let* $0 < p \leqq \infty$, $0 < q \leqq \infty$, $-\infty < s < \infty$, *and*

$$\varkappa > \sigma_p = n\left(\frac{1}{\min(1,p)} - \frac{1}{2}\right). \tag{4}$$

Then there exists a positive constant c such that

$$\left\| \sum_{k \in Z_n} m(k) \hat{f}(k)\, e^{ikx} \mid B^s_{p,q}(T_n) \right\| \leqq c \|m \mid h^{\varkappa}_2(R_n)\| \, \|f \mid B^s_{p,q}(T_n)\| \tag{5}$$

holds for all functions $m(x) \in L_\infty(R_n)$ *and all* $f \in B^s_{p,q}(T_n)$.

(ii) *Let* $0 < p < \infty$, $0 < q < \infty$, $-\infty < s < \infty$, *and*

$$\varkappa > \sigma_{p,q} = n\left(\frac{1}{\min(1,p,q)} - \frac{1}{2}\right). \tag{6}$$

Then there exists a positive constant c such that

$$\left\| \sum_{k \in Z_n} m(k) \hat{f}(k)\, e^{ikx} \mid F^s_{p,q}(T_n) \right\| \leqq c \|m \mid h^{\varkappa}_2(R_n)\| \, \|f \mid F^s_{p,q}(T_n)\| \tag{7}$$

holds for all $m \in L_\infty(R_n)$ *and all* $f \in F^s_{p,q}(T_n)$.

Remark 1. The theorem is the periodic counterpart of [T, (2.6.1/2)] and [T, Theorem 2.4.8]. Part (i) is an easy consequence of Theorem 3.3.4 and the definition of the spaces $B^s_{p,q}(T_n)$. By the same arguments we obtain (7) in the case $1 < p < \infty$ and $1 < q < \infty$ from Theorem 3.4.3/2. On the other hand Theorem 3.4.1/2 leads to (7) in the case $0 < p < \infty$ and $0 < q < \infty$, provided that

$$\varkappa > \frac{n}{2} + \frac{n}{\min(p,q)}. \tag{8}$$

Now one can use the complex interpolation method of the preceding subsection, in particular, Theorem 3.6.2 (ii). To this end we have to investigate the operator

$$T: f \to \sum_{k \in Z_n} m(k) \hat{f}(k)\, e^{ikx}. \tag{9}$$

This can be done (with the necessary and obvious modifications) step by step in the same way as in the non-periodic case (cf. [T, Theorem 2.4.8]). Then, by means of Theorem 3.6.2(ii), (8) can be improved and we obtain (6).

Remark 2. The interpolation method cannot be applied to the spaces $F^s_{p,\infty}(T_n)$, $0 < p < \infty$, $-\infty < s < \infty$. However, in Theorem 3.4.1/2 the case $q = \infty$ is admitted. We obtain (7) for

$0 < p < \infty$, $q = \infty$, $-\infty < s < \infty$ with

$$\varkappa > \frac{n}{2} + \frac{n}{p}. \tag{10}$$

Remark 3. (5) and (7) are multiplier assertions of Michlin-Hörmander type. To see this we put

$$\|m\|_N = \sup_{|\alpha| \leqq N} \sup_{x \in R_n} (1 + |x|^2)^{\frac{|\alpha|}{2}} |D^\alpha m(x)|, \tag{11}$$

where N is a natural number. Then, (5) yields

$$\left\| \sum_{k \in Z_n} m(k) \hat{f}(k)\, e^{ikx} \mid B^s_{p,q}(T_n) \right\| \leqq c \|m\|_{[\sigma_p]+1} \|f \mid B^s_{p,q}(T_n)\| \tag{12}$$

if $0 < p \leqq \infty$, $0 < q \leqq \infty$, and $-\infty < s < \infty$. Finally, (7) yields

$$\left\| \sum_{k \in Z_n} m(k) \hat{f}(k)\, e^{ikx} \mid F^s_{p,q}(T_n) \right\| \leqq c \|m\|_{[\sigma_{p,q}]+1} \|f \mid F^s_{p,q}(T_n)\| \tag{13}$$

if $0 < p < \infty$, $0 < q < \infty$, and $-\infty < s < \infty$. Here σ_p and $\sigma_{p,q}$ have the meaning of (4) and (6), respectively.

Remark* 4. Fourier multipliers from $B^{s_0}_{p_0, q_0}(T_n)$ into $B^{s_1}_{p_1, q_1}(T_n)$ with $1 < p_i < \infty$, $1 < q_i < \infty$, $0 < s_i < \infty$ $(i = 0, 1)$ are studied by P. I. Lizorkin [6] and D. G. Orlovskij [1] via discrete representations of the considered spaces.

3.6.4. Fourier Multipliers for $L^A_p(T_n, l_q)$

Theorem 3.6.3 has a counterpart within the framework of spaces $L^A_p(T_n)$ (or T^A), $0 < p \leqq \infty$, and $L^A_p(T_n, l_q)$, $0 < p < \infty$ and $0 < q \leqq \infty$. Recall that these spaces have been defined in (3.4.1/1). The scalar case is settled by Theorem 3.3.4. Theorem 3.4.3/2 covers the case $L^A_p(T_n, l_q)$, provided that $1 < p < \infty$, $1 < q < \infty$, and

$$A = \{A_j\}_{j=0}^\infty \quad \text{with} \quad A_j \subset \{k \mid k \in Z_n, |k| \leqq 2^j\}; \quad j = 0, 1, 2, \dots \tag{1}$$

However, Theorem 3.4.1/2 does not give a final result for $L^A_p(T_n, l_q)$ in the case $\min(p, q) \leqq 1$. It holds the following improvement.

Theorem. *Let* $0 < p < \infty$ *and* $0 < q < \infty$. *Let* A *be defined as in* (1). *If*

$$\varkappa > \sigma_{p,q} = n\left(\frac{1}{\min(1, p, q)} - \frac{1}{2}\right), \tag{2}$$

then there exists a positive contant c such that

$$\left\| \sum_{k \in Z_n} M_j(k)\, \hat{t}_j(k)\, e^{ikx} \mid L_p(T_n, l_q) \right\|$$
$$\leqq c \sup_{j=0,1,2,\dots} \|M_j(2^j \cdot) \mid H^\varkappa_2(R_n)\| \, \|t_j \mid L_p(T_n, l_q)\| \tag{3}$$

holds for all systems $\{M_j(x)\}_{j=0}^\infty \subset H^\varkappa_2(R_n)$ *and all systems* $\{t_j(x)\}_{j=0}^\infty \in L^A_p(T_n, l_q)$.

Remark. (3) can be derived from Theorem 3.6.3(ii). This is a technical matter and shown in W. Sickel [2, 3.3.]. On the other hand we can develop an analog of the complex interpolation method for $F^s_{p,q}(T_n)$ in the framework of the $L^A_p(T_n, l_q)$-spaces (A as in (1)). Then, this interpolation method, Theorem 3.4.1/2, and Theorem 3.4.3/2 yield the above theorem. For the non-periodic counterpart we refer to H. Triebel [10, 19], L. Päivärinta [1], and to [T, 2.4.9.].

3.7. Approximation and Strong Summability

3.7.1. Characterizations of $B^s_{p,q}(T_n)$ and $F^s_{p,q}(T_n)$ by Approximation

It is well-known that the classical periodic Besov (or Lipschitz) spaces can be characterized by means of approximation by trigonometric polynomials (cf. S. M. Nikol'skij [2, 5.5. and 8.9.] and the literature mentioned there). On the other hand [T, 2.5.3.] shows that in the non-periodic case this is also possible if $\min(p, q) < 1$. Furthermore, the considerations can be extended to spaces of type $F^s_{p,q}$. In this subsection this question is studied in the case of the spaces $B^s_{p,q}(T_n)$ and $F^s_{p,q}(T_n)$, where $0 < p \leqq \infty$, $0 < q \leqq \infty$, and $0 < s < \infty$, ($p < \infty$ in the case of $F^s_{p,q}(T_n)$).

Let $0 < p \leqq \infty$ and $0 < q \leqq \infty$. Recall that

$$\sigma_p = n\left(\frac{1}{\min(1, p)} - \frac{1}{2}\right) \quad \text{and} \quad \sigma_{p,q} = n\left(\frac{1}{\min(1, p, q)} - \frac{1}{2}\right). \tag{1}$$

We put

$$\tilde{\sigma}_p = n\left(\frac{1}{\min(1, p)} - 1\right) \quad \text{and} \quad \tilde{\sigma}_{p,q} = n\left(\frac{1}{\min(1, p, q)} - 1\right). \tag{2}$$

Moreover, we introduce

$$B_j = \{k \mid k \in Z_n, |k| < 2^{j+1}\}, \quad j = 0, 1, 2, \ldots \tag{3}$$

Theorem. (i) *If $0 < p \leqq \infty$, $0 < q \leqq \infty$, and $s > \tilde{\sigma}_p$, then*

$$\begin{aligned} B^s_{p,q}(T_n) = \{f \mid f \in D'(T_n), \exists t = \{t_{2^j}(x)\}_{j=0}^\infty \subset D(T_n) \text{ with } \operatorname{supp} \hat{t}_{2^j} \subset B_j \\ (j = 0, 1, 2, \ldots), \text{ such that } f = \lim_{j\to\infty} t_{2^j}(x) \text{ in } D'(T_n) \text{ and} \\ \|2^{sj}(f - t_{2^j}(x)) \mid l_q(L_p(T_n))\| < \infty\}. \end{aligned} \tag{4}$$

Furthermore,

$$\|f \mid B^s_{p,q}(T_n)\|_t = \inf\left(\|t_{2^0}(x) \mid L_p(T_n)\| + \|2^{sj}(f - t_{2^j}(x)) \mid l_q(L_p(T_n))\|\right)$$

is an equivalent quasi-norm in $B^s_{p,q}(T_n)$, where the infimum is taken over all admissible representations.

(ii) *If $0 < p < \infty$, $0 < q < \infty$, and $s > \tilde{\sigma}_{p,q}$, then*

$$\begin{aligned} F^s_{p,q}(T_n) = \{f \mid f \in D'(T_n), \exists t = \{t_{2^j}(x)\}_{j=0}^\infty \subset D(T_n) \text{ with } \operatorname{supp} \hat{t}_{2^j} \subset B_j \\ (j = 0, 1, 2, \ldots), \text{ such that } f = \lim_{j\to\infty} t_{2^j}(x) \text{ in } D'(T_n) \text{ and} \\ \|2^{sj}(f - t_{2^j}(x)) \mid L_p(T_n, l_q)\| < \infty\}. \end{aligned} \tag{5}$$

Furthermore,

$$\|f \mid F^s_{p,q}(T_n)\|_t = \inf\left(\|t_{2^0}(x) \mid L_p(T_n)\| + \|2^{sj}(f - t_{2^j}(x)) \mid L_p(T_n, l_q)\|\right)$$

is an equivalent quasi-norm in $F^s_{p,q}(T_n)$, where the infimum is taken over all admissible representations.

Proof. Step 1. Let $f \in F^s_{p,q}(T_n)$, $0 < p < \infty$, $0 < q < \infty$, and $s > \tilde{\sigma}_{p,q}$. Then f belongs to $L_p(T_n)$, cf. Theorem 3.5.1/1(iv) and also Remark 3.5.1/3. Let $\varphi = \{\varphi_j(x)\}_{j=0}^\infty \in \Phi(R_n)$. If

$$t_{2^j}(x) = \sum_{l=0}^{j} \left(\sum_{k \in Z_n} \varphi_l(k) \hat{f}(k)\, e^{ikx}\right),$$

then $f = \lim_{j\to\infty} t_{2^j}(x)$ in $D'(T_n)$ and $\operatorname{supp} t_{2^j}(x) \subset B_j$ $(j = 0, 1, 2, \ldots)$. Further we have the estimates

$$\begin{aligned}
&\|2^{sj}(f - t_{2^j}(x)) \mid L_p(T_n, l_q)\| \\
&= \left\| 2^{sj} \sum_{l=1}^{\infty} \left(\sum_{k\in Z_n} \varphi_{l+j}(k) \hat{f}(k)\, \mathrm{e}^{ikx} \right) \mid L_p(T_n, l_q) \right\| \\
&\leqq \left\| \sum_{l=1}^{\infty} 2^{-slr} \cdot 2^{s(j+l)r} \left| \sum_{k\in Z_n} \varphi_{l+j}(k) \hat{f}(k)\, \mathrm{e}^{ikx} \right|^r \mid L_{p/r}(T_n, l_{q/r}) \right\|^{1/r} \\
&\leqq \left(\sum_{l=1}^{\infty} 2^{-slr} \right)^{1/r} \|f \mid F^s_{p,q}(T_n)\| = c\|f \mid F^s_{p,q}(T_n)\| .
\end{aligned} \tag{6}$$

Here $r = \min(1, p, q)$. Consequently, f belongs to the space on the right-hand side of (5) and $\|f \mid F^s_{p,q}(T_n)\|_t \leqq c'\|f \mid F^s_{p,q}(T_n)\|$.

Step 2. We prove the converse direction of (5). Let $t = \{t_{2^j}(x)\}_{j=0}^{\infty}$ be an admissible sequence in the sense of (5). Then,

$$f = t_{2^0}(x) + \sum_{l=1}^{\infty} (t_{2^l}(x) - t_{2^{l-1}}(x)) = \sum_{l=0}^{\infty} \tau_{2^l}(x) \text{ in } D'(T_n). \tag{7}$$

If $\varphi = \{\varphi_j(x)\}_{j=0}^{\infty} \in \Phi(R_n)$ and $r = \min(1, p, q)$, then

$$\begin{aligned}
\|f \mid F^s_{p,q}(T_n)\| &= \left\| 2^{sj} \sum_{k\in Z_n} \varphi_j(k) \hat{f}(k)\, \mathrm{e}^{ikx} \mid L_p(T_n, l_q) \right\| \\
&= \left\| 2^{sj} \sum_{l=-1}^{\infty} \left(\sum_{k\in Z_n} \varphi_j(k)\, \hat{\tau}_{2^{j+l}}(k)\, \mathrm{e}^{ikx} \right) \mid L_p(T_n, l_q) \right\| \\
&\leqq \left(\sum_{l=-1}^{\infty} 2^{-slr} \left\| 2^{s(j+l)} \sum_{k\in Z_n} \varphi_j(k)\, \hat{\tau}_{2^{j+l}}(k)\, \mathrm{e}^{ikx} \mid L_p(T_n, l_q) \right\|^r \right)^{1/r} .
\end{aligned} \tag{8}$$

Of course, $\tau_{2^{-1}}(x) = 0$. Now, we estimate the right-hand side of (8) with the help of Theorem 3.6.4. We obtain

$$\begin{aligned}
&\|f \mid F^s_{p,q}(T_n)\| \\
&\leqq c \left(\sum_{l=-1}^{\infty} 2^{-slr} \left(\sup_{j=0,1,\ldots} \|\varphi_j(2^{j+l+1}\cdot) \mid H^{\varkappa}_2\|^r \right)^{1/r} \times \|2^{sj}\tau_{2^j}(x) \mid L_p(T_n, l_q)\| \right)
\end{aligned} \tag{9}$$

with $\varkappa > \sigma_{p,q}$. If we choose $\varkappa = \sigma_{p,q} + \varepsilon$ with $\tilde{\sigma}_{p,q} < \tilde{\sigma}_{p,q} + \varepsilon < s$, then we have

$$\sup_{j=0,1,2,\ldots} \|\varphi_j(2^{j+l+1}\cdot) \mid H^{\varkappa}_2\| \leqq c \cdot 2^{l(\tilde{\sigma}_{p,q}+\varepsilon)} \tag{10}$$

and consequently by (9) and (7)

$$\begin{aligned}
&\|f \mid F^s_{p,q}(T_n)\| \\
&\leqq c \left(\sum_{l=-1}^{\infty} 2^{-(s-\tilde{\sigma}_{p,q}-\varepsilon)lr} \right)^{1/r} \|2^{sj}\tau_{2^j}(x) \mid L_p(T_n, l_q)\| \\
&\leqq c' \left(\|t_{2^0}(x) \mid L_p(T_n)\| + \left\| \left(\sum_{j=1}^{\infty} 2^{sjq} \mid t_{2^j}(x) - f + f - t_{2^{j-1}}(x)|^q \right)^{1/q} \mid L_p(T_n) \right\| \right) \\
&\leqq c''(\|t_{2^0}(x) \mid L_p(T_n)\| + \|2^{sj}(f - t_{2^j}(x)) \mid L_p(T_n, l_q)\|).
\end{aligned} \tag{11}$$

Taking the infimum over all admissible sequences $\{t_{2^j}(x)\}_{j=0}^{\infty}$, (11) yields that $\|f \mid F^s_{p,q}(T_n)\| \leqq c\|f \mid F^s_{p,q}(T_n)\|_t$. This proves part (ii) of the theorem.

Step 3. The proof of part (i) follows by Step 1 and Step 2 with obvious modifications. Instead of Theorem 3.6.4 we use Theorem 3.3.4.

Remark 1. Step 2 of the proof and Theorem 3.4.1/2 show that part (ii) of the theorem can be extended to $q = \infty$, $0 < p < \infty$, provided that $s > \frac{n}{p}$. Note that $\frac{n}{p} > \tilde{\sigma}_{p,\infty} = n\left(\frac{1}{\min(1,p)} - 1\right)$.

The above theorem enables us to derive the classical characterization of periodic Besov spaces by means of best approximation by trigonometric polynomials. Let $0 < p \leqq \infty$ and let $f \in L_p(T_n)$. Then we put

$$E_p(\nu, f) = \inf \|f(x) - t(x) \mid L_p(T_n)\|; \quad \nu = 0, 1, 2, \ldots, \tag{12}$$

where the infimum is taken over all trigonometric polynomials $t(x)$ with

$$\operatorname{supp} \hat{t} \subset \{k \mid k \in Z_n, |k| \leqq \nu\}; \quad \nu = 0, 1, 2, \ldots \tag{13}$$

We have the following corollary.

Corollary. *Let* $0 < p \leqq \infty$, $0 < q \leqq \infty$, *and* $s > \tilde{\sigma}_p$.

(i) *It holds*

$$B^s_{p,q}(T_n) = \Big\{f \mid f \in D'(T_n) \cap L_p(T_n), \|f \mid B^s_{p,q}(T_n)\|_E$$
$$= \|f \mid L_p(T_n)\| + \left(\sum_{j=0}^{\infty} 2^{sjq} E_p(2^j, f)^q\right)^{1/q} < \infty\Big\} \tag{14}$$

(*modification if* $q = \infty$). *Furthermore*, $\|f \mid B^s_{p,q}(T_n)\|_E$ *is an equivalent quasi-norm in* $B^s_{p,q}(T_n)$.

(ii) *It holds*

$$B^s_{p,q}(T_n) = \Big\{f \mid f \in D'(T_n) \cap L_p(T_n), \|f \mid B^s_{p,q}(T_n)\|^+_E$$
$$= \|f \mid L_p(T_n)\| + \left(\sum_{\nu=1}^{\infty} \nu^{\left(s - \frac{1}{q}\right)q} E_p(\nu - 1, f)^q\right)^{\frac{1}{q}} < \infty\Big\} \tag{15}$$

(*modification if* $q = \infty$). *Furthermore*, $\|f \mid B^s_{p,q}(T_n)\|^+_E$ *is an equivalent quasi-norm in* $B^s_{p,q}(T_n)$.

Proof. Step 1. (14) is essentially a reformulation of part (i) of the Theorem. Obviously, by (4) and formulas 3.5.1/(7), (8) we have

$$\|f \mid B^s_{p,q}(T_n)\|_E \leqq c\|f \mid B^s_{p,q}(T_n)\| + \|f \mid B^s_{p,q}(T_n)\|_t \leqq c'\|f \mid B^s_{p,q}(T_n)\|.$$

On the other hand if f belongs to the right-hand side of (14) then we choose a sequence $\{t_{2^j}(x)\}_{j=0}^{\infty} \subset D(T_n)$ with $\operatorname{supp} \hat{t}_{2^j} \subset B_j$ (cf. (3)), such that

$$\|f - t_{2^j}(x) \mid L_p(T_n)\| \leqq 2E_p(2^{j+1} - 1, f); \quad j = 0, 1, 2, \ldots \tag{16}$$

Consequently,

$$\|t_{2^0}(x) \mid L_p(T_n)\| + \left(\sum_{j=0}^{\infty} 2^{sjq}\|f - t_{2^j}(x) \mid L_p(T_n)\|^q\right)^{1/q}$$
$$\leqq c\left(\|f \mid L_p(T_n)\| + \left(\sum_{j=0}^{\infty} 2^{sjq} E_p(2^{j+1} - 1, f)^q\right)^{1/q}\right) \leqq c'\|f \mid B^s_{p,q}(T_n)\|_E. \tag{17}$$

Moreover, it is not difficult to see that $\lim\limits_{j\to\infty} t_{2^j}(x) = f$ in $D'(T_n)$. Here we need

$s > \tilde{\sigma}_p$. Hence, by (17) and part (i) of the Theorem we have

$$\|f \mid B^s_{p,q}(T_n)\| \leqq c\|f \mid B^s_{p,q}(T_n)\|_E.$$

Step 2. In order to prove (15) we observe

$$\sum_{\nu=1}^{\infty} \nu^{\left(s-\frac{1}{q}\right)q} E_p(\nu-1, f)^q = \sum_{j=0}^{\infty} \left(\sum_{\nu=2^j}^{2^{j+1}-1} \nu^{\left(s-\frac{1}{q}\right)q} E_p(\nu-1, f)^q\right)$$
$$\leqq \sum_{j=0}^{\infty} \left(\sum_{\nu=2^j}^{2^{j+1}-1} 2^{(j+1)\left(s-\frac{1}{q}\right)q} E_p(2^j-1, f)^q\right)$$
$$\leqq c \sum_{j=0}^{\infty} 2^{sjq} E_p(2^j-1, f)^q, \tag{18}$$

and

$$\sum_{j=1}^{\infty} 2^{sjq} E_p(2^j, f)^q = \sum_{j=1}^{\infty} 2^{-j} \sum_{\nu=2^{j-1}}^{2^j-1} 2^{sjq} E_p(2^j, f)^q \leqq c \sum_{\nu=1}^{\infty} \nu^{\left(s-\frac{1}{q}\right)q} E_p(\nu-1, f)^q. \tag{19}$$

Now, (15) is a consequence of (18), (19), and (14). This proves the corollary.

Remark* 2. If $\min(p, q) \geqq 1$ then (14) and Theorem 3.5.4 yield the classical characterizations of the periodic Besov (or Lipschitz) spaces $\Lambda^s_{p,q}(T_n)$ and Hölder-Zygmund spaces $\mathscr{C}^s(T_n)$ by means of best approximation (cf. S. M. Nikol'skij [2, 5.5.]). Part (ii) of the Theorem and Theorem 3.5.4 yield characterizations of the periodic Sobolev-Lebesgue spaces $H^s_p(T_n)$, $1 < p < \infty$, $s > 0$, by means of approximation.

3.7.2. Approximation by Partial Sums and Smooth Means

Let the assumptions of Theorem 3.7.1 be satisfied. The question arises whether one can find an appropriate approximation process $f = \lim_{j\to\infty} t_{2^j}(x)$ such that

$$\|t_{2^0}(x) \mid L_p(T_n)\| + \|2^{sj}(f - t_{2^j}(x) \mid l_q(L_p(T_n))\|$$

and

$$\|t_{2^0}(x) \mid L_p(T_n)\| + \|2^{sj}(f - t_{2^j}(x)) \mid L_p(T_n, l_q)\|$$

are equivalent quasi-norms in $B^s_{p,q}(T_n)$ and $F^s_{p,q}(T_n)$, respectively. We discuss this problem in the following.

Let $\psi(x)$ be a real continuous function on R_n with

$$\operatorname{supp} \psi(x) \subset \{x \mid |x| \leqq 2\} \quad \text{and} \quad \psi(0) = 1. \tag{1}$$

Then we have $\lim_{\nu\to\infty} \psi(\nu^{-1}x) = 1$ for all $x \in R_n$. If $f \in D'(T_n)$ and if $\nu = 1, 2, \ldots$, then we put

$$M^\psi_\nu f(x) = \sum_{k \in Z_n} \psi(\nu^{-1}k) \hat{f}(k)\, e^{ikx}. \tag{2}$$

Then,

$$f = \lim_{\nu\to\infty} M^\psi_\nu f(x) = \lim_{j\to\infty} M^\psi_{2^j} f(x) \tag{3}$$

in $D'(T_n)$. Recall that σ_p, $\tilde{\sigma}_p$, $\sigma_{p,q}$, and $\tilde{\sigma}_{p,q}$ have the meaning of (3.7.1/(1), (2).

Theorem 1. *Let* ψ *and* M^ψ_ν, $\nu = 1, 2, \ldots$, *be given by* (1) *and* (2). *Let*

$$\||x|^{-\sigma} (\psi(2x) - \psi(x)) \mid H^\lambda_2\| < \infty \tag{4}$$

for some numbers $0 < \sigma < \infty$ *and* $0 < \lambda < \infty$.

(i) *If* $0 < p \leqq \infty$, $0 < q \leqq \infty$, $\lambda > \sigma_p$, *and* $\tilde{\sigma}_p < s < \sigma - \frac{n}{2}$, *then*

$$B^s_{p,q}(T_n) = \{f \mid f \in D'(T_n),\ \|f \mid B^s_{p,q}(T_n)\|_{M^\psi} = \|M^\psi_1 f \mid L_p(T_n)\| + \|2^{sj}(f - M^\psi_{2^j} f(x)) \mid l_q(L_p(T_n))\| < \infty\}. \quad (5)$$

Furthermore, $\|f \mid B^s_{p,q}(T_n)\|_{M^\psi}$ *is an equivalent quasi-norm in* $B^s_{p,q}(T_n)$.

(ii) *If* $0 < p < \infty$, $0 < q < \infty$, $\lambda > \sigma_{p,q}$, *and* $\tilde{\sigma}_{p,q} < s < \sigma - \frac{n}{2}$, *then*

$$F^s_{p,q}(T_n) = \{f \mid f \in D'(T_n),\ \|f \mid F^s_{p,q}(T_n)\|_{M^\psi} = \|M^\psi_1 f \mid L_p(T_n)\| + \|2^{sj}(f - M^\psi_{2^j} f(x)) \mid L_p(T_n, l_q)\| < \infty\}. \quad (6)$$

Furthermore, $\|f \mid F^s_{p,q}(T_n)\|_{M^\psi}$ *is an equivalent quasi-norm in* $F^s_{p,q}(T_n)$.

Proof. Step 1. Because $s > \tilde{\sigma}_{p,q}$ we obtain as an immediate consequence of Theorem 3.7.1(ii) and (1)

$$\|f \mid F^s_{p,q}(T_n)\| \leqq c\|f \mid F^s_{p,q}(T_n)\|_t \leqq c\|f \mid F^s_{p,q}(T_n)\|_{M^\psi}. \quad (7)$$

Step 2. We prove

$$\|f \mid F^s_{p,q}(T_n)\|_{M^\psi} \leqq c\|f \mid F^s_{p,q}(T_n)\|. \quad (8)$$

Let $j = 0, 1, 2, \ldots$ and let $f \in F^s_{p,q}(T_n)$. Then

$$f - M^\psi_{2^j} f(x) = \sum_{l=j}^{\infty} [M^\psi_{2^{l+1}} f(x) - M^\psi_{2^l} f(x)] = \sum_{l=j}^{\infty} \Big(\sum_{k \in Z_n} \varrho(2^{-l}k) \hat{f}(k)\, e^{ikx} \Big) \quad (9)$$

holds, where $\varrho(x) = \psi\left(\frac{x}{2}\right) - \psi(x)$ (convergence in $D'(T_n)$). Starting with (9) we obtain in the same way as in (3.7.1/6)

$$\|2^{sj}(f - M^\psi_{2^j} f(x)) \mid L_p(T_n, l_q)\| \leqq c \left\| 2^{sj} \sum_{k \in Z_n} \varrho(2^{-j}k) \hat{f}(k)\, e^{ikx} \mid L_p(T_n, l_q) \right\|. \quad (10)$$

Now, let $\varphi = \{\varphi_l(x)\}_{l=0}^\infty \in \Phi(R_n)$. We put

$$f_l(x) = \sum_{k \in Z_n} |2^{-l}k|^\sigma \varphi_l(k) \hat{f}(k)\, e^{ikx}, \quad \text{if} \quad l = 0, 1, 2, \ldots,$$

and $f_l(x) = 0$ if $l = -1, -2, \ldots$ Then we have for $j = 0, 1, 2, \ldots$

$$\sum_{k \in Z_n} \varrho(2^{-j}k) \hat{f}(k)\, e^{ikx} = \sum_{l=-\infty}^{j+2} 2^{\sigma(l-j)} \sum_{k \in Z_n} |2^{-j}k|^{-\sigma} \varrho(2^{-j}k) \hat{f}_l(k)\, e^{ikx}$$
$$= c \sum_{l=-\infty}^{0} 2^{\sigma l} \sum_{k \in Z_n} |2^{-j}k|^{-\sigma} \varrho(2^{-j}k) \hat{f}_{l+j+2}(k)\, e^{ikx}. \quad (11)$$

We choose $r = \min(1, p, q)$. Then (10) and (11) lead to

$$\|2^{sj}(f - M^\psi_{2^j} f(x)) \mid L_p(T_n, l_q)\|$$
$$\leqq c \left(\sum_{l=-\infty}^{0} 2^{(\sigma-s)lr} \left\| 2^{s(l+j)} \sum_{k \in Z_n} |2^{-j}k|^{-\sigma} \varrho(2^{-j}k) \hat{f}_{l+j+2}(k)\, e^{ikx} \mid L_p(T_n, l_q) \right\|^r \right)^{1/r}. \quad (12)$$

Assumption (4) yields

$$\| |2^l x|^{-\sigma} \varrho(2^l x) \mid H_2^\lambda \| \leqq c \cdot 2^{-\frac{n}{2}l}; \quad l = 0, -1, -2, \ldots \tag{13}$$

By Theorem 3.6.4, (13), and $\sigma - \frac{n}{2} > s$ the right-hand side of (12) can be estimated by

$$\begin{aligned} &\|2^{sj}(f - M_{2^j}^{\psi} f(x)) \mid L_p(T_n, l_q)\| \\ &\leqq c \left(\sum_{l=-\infty}^{0} 2^{\left(\sigma - \frac{n}{2} - s\right) lr} \right)^{\frac{1}{r}} \|2^{sj} f_j(x) \mid L_p(T_n, l_q)\| \\ &= c' \, \| 2^{sj} f_j(x) \mid L_p(T_n, l_q)\| . \end{aligned} \tag{14}$$

Applying Theorem 3.6.4 again, the right-hand side of (14) can be estimated from above by $c\|f \mid F_{p,q}^s(T_n)\|$. This yields

$$\|2^{sj}(f - M_{2^j}^{\psi} f(x)) \mid L_p(T_n, l_q)\| \leqq c\|f \mid F_{p,q}^s(T_n)\| . \tag{15}$$

(15) and the obvious inequality

$$\|M_1^{\psi} f(x) \mid L_p(T_n)\| \leqq c(\|f \mid L_p(T_n)\| + \|f - M_1^{\psi} f(x) \mid L_p(T_n)\|) \tag{16}$$

yield (8).

Step 3. The proof of part (i) proceeds analogously with the usual modifications. Instead of Theorem 3.6.4 we use Theorem 3.3.4. The proof is complete.

Remark 1. The theorem shows that the spaces $B_{p,q}^s(T_n)$ and $F_{p,q}^s(T_n)$ can be characterized with the help of the smooth means $M_{2^j}^{\psi} f$ $(j = 0, 1, \ldots)$. If $p < 1$ and $q < 1$ then we need a higher smoothness of the considered means. This follows from the conditions $\lambda > \sigma_p$ and $\lambda > \sigma_{p,q}$, respectively. If $\min(p, q) \geqq 1$ then $\lambda > \frac{n}{2}$ is sufficient.

As an example let us consider the so-called generalized de la Vallée-Poussin means. To this end we choose a real continuous function $\psi(x)$ defined on R_n such that

$$0 \leqq \psi(x) \leqq 1, \qquad \psi(x) = 1 \text{ if } |x| \leqq 1, \qquad \psi(x) = 0 \text{ if } |x| \geqq 2. \tag{17}$$

We put

$$V_\nu^{\psi} f(x) = \sum_{k \in Z_n} \psi(\nu^{-1} k) \hat{f}(k) \, e^{ikx}, \; f \in D'(T_n); \quad \nu = 1, 2, \ldots \tag{18}$$

Obviously, $V_\nu^{\psi} f$ $(\nu = 1, 2, \ldots)$ is a special case of (2). As a concequence of Theorem 1 we obtain the following corollary.

Corollary. *Let $\psi(x) \in H_2^\lambda(R_n)$, $0 < \lambda < \infty$, and let V_ν^{ψ} $(\nu = 1, 2, \ldots)$ be given by (18).*
(i) *If $0 < p \leqq \infty$, $0 < q \leqq \infty$, $\tilde{\sigma}_p < s < \infty$, and $\lambda > \sigma_p$, then*

$$\begin{aligned} B_{p,q}^s(T_n) = \{f \mid f \in D'(T_n), \|f \mid B_{p,q}^s(T_n)\|_{V^\psi} \\ = \|V_1^{\psi} f \mid L_p(T_n)\| + \|2^{sj}(f - V_{2^j}^{\psi} f(x)) \mid l_q(L_p(T_n))\| < \infty\}. \end{aligned} \tag{19}$$

Furthermore, $\|f \mid B_{p,q}^s(T_n)\|_{V^\psi}$ is an equivalent quasi-norm in $B_{p,q}^s(T_n)$.

(ii) *If $0 < p < \infty$, $0 < q < \infty$, $\tilde{\sigma}_{p,q} < s < \infty$, and $\lambda > \sigma_{p,q}$, then*

$$\begin{aligned} F_{p,q}^s(T_n) = \{f \mid f \in D'(T_n), \|f \mid F_{p,q}^s(T_n)\|_{V^\psi} \\ = \|V_1^{\psi} f \mid L_p(T_n)\| + \|2^{sj}(f - V_{2^j}^{\psi} f(x)) \mid L_p(T_n, l_q)\| < \infty\}. \end{aligned} \tag{20}$$

Furthermore, $\|f \mid F_{p,q}^s(T_n)\|_{V^\psi}$ is an equivalent quasi-norm in $F_{p,q}^s(T_n)$.

Remark 2. Theorem 1 and the Corollary coincide essentially with W. Sickel [1, Theorem 5 and Section 5.1.]. There one can find characterizations of type (19) and (20) for further special means (cf. also Remark 3.7.4/4).

Remark 3. The representations (6) and (20) can be extended to the spaces $F^s_{p,\infty}(T_n)$ using Theorem 3.4.1/2 instead of Theorem 3.6.4 and Remark 3.7.1/1. In particular, (6) holds true if $0 < p < \infty$, $q = \infty$, $\frac{n}{p} < s < \sigma - \frac{n}{2}$, and $\lambda > n\left(\frac{1}{p} + \frac{1}{2}\right)$. (20) is also valid if $0 < p < \infty$, $q = \infty$, $\frac{n}{p} < s < \infty$, and $\lambda > n\left(\frac{1}{p} + \frac{1}{2}\right)$.

Let us conclude this subsection with a discussion of approximation by partial sums of Fourier series. This is suggested by the above theorem on the one hand and by Lizorkin's representation (cf. Theorem 3.5.3) on the other hand. Let $f \in D'(T_n)$ and let $\nu = 1, 2, \ldots$ Then we put

$$S_\nu f(x) = \sum_{|k_1| < \nu} \cdots \sum_{|k_n| < \nu} \hat{f}(k)\, e^{ikx}. \tag{21}$$

Theorem 2. *Let $1 < p < \infty$ and let $0 < s < \infty$.*

(i) *If $0 < q \leqq \infty$, then*

$$\begin{aligned} B^s_{p,q}(T_n) = \{f \mid f \in D'(T_n),\ &\|f \mid B^s_{p,q}(T_n)\|_S \\ &= \|S_1 f(x) \mid L_p(T_n)\| + \|2^{sj}(f - S_{2^j} f(x)) \mid l_q(L_p(T_n))\| < \infty\}. \end{aligned} \tag{22}$$

Furthermore, $\|f \mid B^s_{p,q}(T_n)\|_S$ is an equivalent quasi-norm (norm if $q \geqq 1$) in $B^s_{p,q}(T_n)$.

(ii) *If $1 < q < \infty$, then*

$$\begin{aligned} F^s_{p,q}(T_n) = \{f \mid f \in D'(T_n),\ &\|f \mid F^s_{p,q}(T_n)\|_S \\ &= \|S_1 f(x) \mid L_p(T_n)\| + \|2^{sj}(f - S_{2^j} f(x)) \mid L_p(T_n, l_q)\| < \infty\}. \end{aligned} \tag{23}$$

Furthermore, $\|f \mid F^s_{p,q}(T_n)\|_S$ is an equivalent norm in $F^s_{p,q}(T_n)$

Proof. Let $\chi_j(k)$, $k \in Z_n$, be the characteristic function of K_j, $j = 0, 1, \ldots$, where K_j has the meaning of (3.5.3/1). Further, let us denote

$$g_\nu(x) = (f - S_\nu f)(x), \quad \nu = 1, 2, \ldots, f \in L_p(T_n), \quad 1 < p < \infty. \tag{24}$$

Then it follows from Theorem 3.4.3/3 (cf. also Remark 3.4.3/3)

$$\begin{aligned} \|f \mid F^s_{p,q}(T_n)\|^* &= \|2^{sj} \sum_{k \in K_j} \hat{f}(k)\, e^{ikx} \mid L_p(T_n, l_q)\| \\ &= \|S_1 f(x) \mid L_p(T_n)\| + \left\| 2^{sj} \sum_{k \in Z_n} \chi_{j+1}(k)\, \hat{g}_{2^j}(k)\, e^{ikx} \mid L_p(T_n, l_q) \right\| \\ &\leqq \|S_1 f(x) \mid L_p(T_n)\| + c \left\| 2^{sj} \sum_{k \in Z_n} \hat{g}_{2^j}(k)\, e^{ikx} \mid L_p(T_n, l_q) \right\| \\ &\leqq c' \|f \mid F^s_{p,q}(T_n)\|_S. \end{aligned} \tag{25}$$

Consequently, by Theorem 3.5.3(ii) we have

$$\|f \mid F^s_{p,q}(T_n)\| \leqq c \|f \mid F^s_{p,q}(T_n)\|_S.$$

On the other hand we obtain the converse direction as a consequence of

$$\begin{aligned} &\|2^{sj}(f - S_{2^j} f)(x) \mid L_p(T_n, l_q)\| \\ &= \left\| \sum_{l=1}^{\infty} 2^{-sl} \left(2^{s(l+j)} \sum_{k \in K_{l+j}} \hat{f}(k)\, e^{ikx} \right) \mid L_p(T_n, l_q) \right\| \\ &\leqq \sum_{l=1}^{\infty} 2^{-sl} \left\| 2^{s(l+j)} \sum_{k \in K_{l+j}} \hat{f}(k)\, e^{ikx} \mid L_p(T_n, l_q) \right\| \\ &\leqq c \|f \mid F^s_{p,q}(T_n)\|^* \end{aligned} \tag{26}$$

and

$$\|S_1 f(x) \mid L_p(T_n)\| \leqq \|f \mid L_p(T_n)\| + \|f - S_1 f(x) \mid L_p(T_n)\|, \tag{27}$$

as well as Theorem 3.5.3(ii). Similarly we prove part (i) of the theorem.

Remark 4. We cannot expect to obtain equivalent characterizations as in Theorem 1, Theorem 2 or in the Corollary if $s < \tilde{\sigma}_p = n\left(\frac{1}{\min(1, p)} - 1\right)$. If $0 < p < 1$ and $s < n\left(\frac{1}{p} - 1\right)$, then $\delta \in B^s_{p,q}(T_n)$, $0 < q \leqq \infty$ (cf. Remark 3.5.1/3). On the other hand if $1 \leqq p \leqq \infty$, $0 < q \leqq \infty$, and $s < 0$ then it is not difficult to prove that $\|f \mid B^s_{p,q}(T_n)\|_{M^\psi} \sim \|f \mid L_p(T_n)\|$. Further, $\|f \mid B^s_{p,q}(T_n)\|_S \sim \|f \mid L_p(T_n)\|$ if $1 < p < \infty$, $1 < q < \infty$, and $s < 0$. Similarly, in the case of $F^s_{p,q}(T_n)$.

3.7.3. Strong Summability of Partial Sums

The problem of strong summability can be described as follows. Let $S_\nu f$, $\nu = 1, 2, \ldots$, be the ν-th (cubic) partial sum of $f \in D'(T_n)$ as defined in (3.7.2/21). Then we have $\lim_{\nu \to \infty} S_\nu f = f$ at least in $D'(T_n)$. If $f = f(x)$ is a regular distribution then one can try to measure the degree of approximation of $|f(x) - S_\nu f(x)|$. In particular, one can ask whether

$$\left\| \left(\sum_{\nu=1}^{\infty} \nu^{sq} |f(x) - S_\nu f(x)|^q \right)^{1/q} \mid L_p(T_n) \right\| < \infty \tag{1}$$

holds, where $0 < p \leqq \infty$, $0 < q \leqq \infty$, and s are given real numbers. This is the problem of strong summability. Conversely, if (1) holds for some f what can be said about properties of the function f? The study of these problems (with $p = \infty$ and $n = 1$ in (1)) goes back to G. Alexits [1] and G. Freud [2]. The theory has been developed mainly by L. Leindler in a series of papers. We refer to L. Leindler [10, 11], where historical remarks and references can be found. Further, let us mention the papers by V. G. Krotov [1], V. G. Krotov, L. Leindler [1], L. Leindler [1–9, 12–15], L. Leindler, E. M. Nikišin [1], K. I. Oskol'kov [1], and J. Szabados [1], where the above questions inclusively some generalizations are considered. It turned out that (1) is closely related to smoothness properties of the function f. This is by no means a surprising fact if we have in mind the results of the preceding subsection (cf. in particular Theorem 3.7.2/2). From the point of view of strong summability we dealt with the strong summability of dyadic partial sums and dyadic means. Hence, it seems to be very reasonable to study problems of type (1) within the framework of the spaces $B^s_{p,q}(T_n)$ and $F^s_{p,q}(T_n)$. Similar as in Corollary 3.7.1(ii) we have to investigate the question whether the quasi-norms $\|f \mid F^s_{p,q}(T_n)\|_S$ in (3.7.2/23) can be replaced by terms of type (1). This is possible as we shall see in this subsection. Moreover, we are able to state analogous results where we replace the partial sums $S_\nu f$ by appropriate smooth means $M^\psi_\nu f$ (cf. (3.7.2/2)). Note that in this case $\min(p, q) \leqq 1$ is admitted. We refer to 3.7.4.

Theorem 1. *Let* $1 < p < \infty$ *and let* $0 < s < \infty$.

(i) *If* $0 < q \leqq \infty$, *then*

$$B^s_{p,q}(T_n) = \Bigg\{ f \mid f \in D'(T_n),\ \|f \mid B^s_{p,q}(T_n)\|^+_S = \|S_1 f(x) \mid L_p(T_n)\| + \left(\sum_{\nu=1}^{\infty} \nu^{\left(s - \frac{1}{q}\right)q} \|(f - S_\nu f)(x) \mid L_p(T_n)\|^q \right)^{\frac{1}{q}} < \infty \Bigg\}. \tag{2}$$

Furthermore, $\|f \mid B^s_{p,q}(T_n)\|^+_S$ *is an equivalent quasi-norm (norm if* $q \geqq 1$) *in* $B^s_{p,q}(T_n)$.

(ii) *If* $1 < q < \infty$, *then*

$$F^s_{p,q}(T_n) = \Big\{ f \mid f \in D'(T_n),\ \|f \mid F^s_{p,q}(T_n)\|^+_S = \|S_1 f(x) \mid L_p(T_n)\| + \Big\| \Big(\sum_{\nu=1}^{\infty} \nu^{\left(s-\frac{1}{q}\right)q} |(f - S_\nu f)(x)|^q \Big)^{\frac{1}{q}} \Big| L_p(T_n) \Big\| < \infty \Big\}. \tag{3}$$

Furthermore, $\|f \mid F^s_{p,q}(T_n)\|^+_S$ *is an equivalent norm in* $F^s_{p,q}(T_n)$.

Proof. Step 1. We prove that

$$\|f \mid F^s_{p,q}(T_n)\|^+_S \leqq c \|f \mid F^s_{p,q}(T_n)\|_S. \tag{4}$$

If $g_\nu(x)$, $\nu = 1, 2, \ldots$, has the meaning of (3.7.2/24), then

$$\begin{aligned}
&\Big\| \Big(\sum_{\nu=1}^{\infty} \nu^{sq-1} |(f - S_\nu f)(x)|^q \Big)^{1/q} \Big| L_p(T_n) \Big\| \\
&= \Big\| \Big(\sum_{j=0}^{\infty} \Big(\sum_{\nu=2^j}^{2^{j+1}-1} \nu^{sq-1} |(f - S_{2^{j+1}} f)(x) + (S_{2^{j+1}} f - S_\nu f)(x)|^q \Big) \Big)^{1/q} \Big| L_p(T_n) \Big\| \\
&\leqq c \Big\| \Big(\sum_{j=0}^{\infty} 2^{sjq} |g_{2^{j+1}}(x)|^q \Big)^{1/q} \Big| L_p(T_n) \Big\| \\
&\quad + c \Big\| \Big(\sum_{j=0}^{\infty} \sum_{\nu=2^j}^{2^{j+1}-1} \nu^{sq-1} |(S_{2^{j+1}} f - S_\nu f)(x)|^q \Big)^{1/q} \Big| L_p(T_n) \Big\|
\end{aligned} \tag{5}$$

holds. Let $\chi_{j,\nu} = \chi_{j,\nu}(k)$, $k \in Z_n$, be the characteristic function of the corridor

$$\{k \mid k \in Z_n, |k_1| < 2^{j+1}, \ldots, |k_n| < 2^{j+1}\} \setminus \{k \mid k \in Z_n, |k_1| < \nu, \ldots, |k_n| < \nu\}, \tag{6}$$

where $\nu = 2^j, 2^j + 1, \ldots, 2^{j+1} - 1$ and $j = 0, 1, 2, \ldots$ We have

$$\begin{aligned}
(S_{2^{j+1}} f - S_\nu f)(x) &= \sum_{k \in Z_n} \chi_{j,\nu}(k) \hat{f}(k) e^{ikx} \\
&= \sum_{k \in Z_n} \chi_{j,\nu}(k) \hat{g}_{2^j}(k) e^{ikx}; \quad \nu = 2^j, \ldots, 2^{j+1} - 1.
\end{aligned} \tag{7}$$

Using (7) we estimate the second term on the right-hand side of (5) by means of Theorem 3.4.3/3 (cf. also Remark 3.4.3/3). One obtains

$$\begin{aligned}
&\Big\| \Big(\sum_{j=0}^{\infty} \sum_{\nu=2^j}^{2^{j+1}-1} \nu^{sq-1} |(S_{2^{j+1}} f - S_\nu f)(x)|^q \Big)^{1/q} \Big| L_p(T_n) \Big\| \\
&= \Big\| \Big(\sum_{j=0}^{\infty} \sum_{\nu=2^j}^{2^{j+1}-1} \Big| \sum_{k \in Z_n} \chi_{j,\nu}(k)\, \nu^{s-\frac{1}{q}} \hat{g}_{2^j}(k) e^{ikx} \Big|^q \Big)^{\frac{1}{q}} \Big| L_p(T_n) \Big\| \\
&\leqq c \Big\| \Big(\sum_{j=0}^{\infty} \sum_{\nu=2^j}^{2^{j+1}-1} \nu^{sq-1} |g_{2^j}(x)|^q \Big)^{1/q} \Big| L_p(T_n) \Big\| \\
&\leqq c' \Big\| \Big(\sum_{j=0}^{\infty} 2^{sjq} |g_{2^j}(x)|^q \Big)^{1/q} \Big| L_p(T_n) \Big\|.
\end{aligned} \tag{8}$$

Now, (5) and (8) yield (4).

Step 2. We prove that

$$\|f \mid F^s_{p,q}(T_n)\|^* \leqq c \|f \mid F^s_{p,q}(T_n)\|^+_S. \tag{9}$$

Recall that $\|f \mid F^s_{p,q}(T_n)\|^*$ and K_j, $j = 0, 1, 2, \ldots$, have been defined in (3.5.3/3) and (3.5.3/1), respectively. Let $j = 1, 2, 3, \ldots$ and $1 < q < \infty$. Then we have

$$\sum_{k \in K_j} \hat{f}(k)\, e^{ikx} = \sum_{k \in K_j} \hat{g}_\nu(k)\, e^{ikx}; \quad \nu = 2^{j-1}, \ldots, 2^j - 1,$$

and

$$\Big|\sum_{k \in K_j} \hat{f}(k)\, e^{ikx}\Big|^q = 2^{-(j-1)} \sum_{\nu = 2^{j-1}}^{2^j - 1} \Big|\sum_{k \in K_j} \hat{g}_\nu(k)\, e^{ixk}\Big|^q. \tag{10}$$

Consequently, by (10) and Theorem 3.4.3/3 (cf. also Remark 3.4.3/3)

$$\begin{aligned} &\Big\|\Big(\sum_{j=1}^{\infty} 2^{sjq} \Big|\sum_{k \in K_j} \hat{f}(k)\, e^{ikx}\Big|^q\Big)^{1/q} \,\Big|\, L_p(T_n)\Big\| \\ &= \Big\|\Big(\sum_{j=1}^{\infty} \sum_{\nu=2^{j-1}}^{2^j-1} 2^{sjq-(j-1)} \Big|\sum_{k \in K_j} \hat{g}_\nu(k)\, e^{ikx}\Big|^q\Big)^{1/q} \,\Big|\, L_p(T_n)\Big\| \\ &\leqq c\, \Big\|\Big(\sum_{j=1}^{\infty} \sum_{\nu=2^{j-1}}^{2^j-1} \Big|\sum_{k \in K_j} \nu^{s-\frac{1}{q}} \hat{g}_\nu(k)\, e^{ikx}\Big|^q\Big)^{\frac{1}{q}} \,\Big|\, L_p(T_n)\Big\| \\ &\leqq c' \Big\|\Big(\sum_{\nu=1}^{\infty} \nu^{sq-1} |g_\nu(x)|^q\Big)^{1/q} \,\Big|\, L_p(T_n)\Big\|. \end{aligned} \tag{11}$$

Now, (9) follows immediately from (11). Finally, part (ii) of the theorem is a consequence of (4), (9), Theorem 3.5.3(ii), and Theorem 3.7.2/2(ii).

Step 3. Part (i) of the theorem can be proved in a similar way. In this case the scalar version of Theorem 3.4.3/3 is sufficient.

Remark 1. Part (ii) of the theorem gives a complete solution of the problem of strong summability as described in (1), provided that $1 < p < \infty$, $1 < q < \infty$, and $s + \frac{1}{q} > 0$. Our main tool was the multiplier theorem 3.4.3/3. Thus it becomes clear that the above way cannot be extended to values p, q with $\min(p, q) \leqq 1$ or $\max(p, q) = \infty$. The theorem coincides with H.-J. Schmeisser, W. Sickel [1, Theorem 5].

Remark 2. Using the embedding theorems of Corollary 3.5.5 we can describe the smoothness properties of functions f which satisfy (1) in the language of Hölder-Zygmund spaces as it is customary in the theory of strong summability. In particular, we have the following corollary.

Corollary. (i) *Let* $1 < p < \infty$, $0 < q \leqq \infty$, *and* $s + \frac{1}{q} > \frac{n}{p}$. *If* $f \in L_p(T_n)$, *such that*

$$\sum_{\nu=1}^{\infty} \nu^{sq} \|(f - S_\nu f)(x) \mid L_p(T_n)\|^q < \infty, \quad \textit{then} \quad f \in \mathscr{C}^{s+\frac{1}{q}-\frac{n}{p}}(T_n). \tag{12}$$

(ii) *Let* $1 < p < \infty$, $1 < q < \infty$, *and* $s + \frac{1}{q} > \frac{n}{p}$. *If* $f \in L_p(T_n)$ *such that*

$$\Big\|\Big(\sum_{\nu=1}^{\infty} \nu^{sq} |(f - S_\nu f)(x)|^q\Big)^{\frac{1}{q}} \,\Big|\, L_p(T_n)\Big\| < \infty, \quad \textit{then} \quad f \in \mathscr{C}^{s+\frac{1}{q}-\frac{n}{q}}(T_n). \tag{13}$$

As already mentioned in Remark 1 our methods do not work in the case $p = \infty$. However, just the case $p = \infty$ attracted much attention in the literature (cf. the references at the beginning of this subsection). We can give a partial answer in the sense of the above corollary. For this purpose we need the spaces $F^s_{\infty,q}(T_n)$ introduced in Definition 3.5.2.

Theorem 2. *Let* $1 < q < \infty$ *and* $s + \frac{1}{q} > 0$. *If* $f \in L_\infty(T_n)$ *such that*

$$\left\| \left(\sum_{\nu=1}^{\infty} \nu^{sq} |(f - S_\nu f)(x)|^q \right)^{\frac{1}{q}} \mid L_\infty(T_n) \right\| < \infty, \quad \textit{then}$$
$$f \in F^{s+\frac{1}{q}}_{\infty,q}(T_n) \subset \mathscr{C}^{s+\frac{1}{q}}(T_n). \tag{14}$$

Proof. We prove

$$\left\| f \mid F^{s+\frac{1}{q}}_{\infty,q}(T_n) \right\|$$
$$\leqq c \left(\| f \mid L_\infty(T_n) \| + \left\| \left(\sum_{\nu=1}^{\infty} \nu^{sq} |(f - S_\nu f)(x)|^q \right)^{\frac{1}{q}} \mid L_\infty(T_n) \right\| \right). \tag{15}$$

This is sufficient, cf. the embedding (3.5.5/5). Let $\varphi = \{\varphi_j(x)\}_{j=0}^{\infty} \subset S(R_n)$ be a sequence of functions with the properties

(i) $\operatorname{supp} \varphi_0 \subset \{x \mid |x_1| \leqq 2, \ldots, |x_n| \leqq 2\}$,
$\operatorname{supp} \varphi_j \subset \{x \mid |x_1| < 2^{j+1}, \ldots, |x_n| < 2^{j+1}\} \setminus \{x \mid |x_1| < 2^{j-1}, \ldots, |x_n| < 2^{j-1}\}$;
$j = 1, 2, \ldots,$

(ii) $|D^\alpha \varphi_j(x)| \leqq c_\alpha \cdot 2^{-j|\alpha|}$ for all α, j,

(iii) $\sum_{j=0}^{\infty} \varphi_j(x) = 1$ for all $x \in R_n$.

Systems of this type are similar to the systems belonging to $\Phi(R_n)$. We have dyadic corridors instead of dyadic differences of balls. Note that in Definition 3.5.2, $\varphi \in \Phi(R_n)$ can be replaced by a sequence φ with the above properties (i), (ii), and (iii) (the assumptions concerning $\operatorname{supp} \varphi_j$ cause no problems). Consequently, we have

$$\| f \mid F^s_{\infty,q}(T_n) \| \leqq c \| 2^{sj} f_j(x) \mid L_\infty(T_n, l_q) \| \tag{16}$$

if

$$f = \sum_{j=0}^{\infty} \left(\sum_{k \in Z_n} \varphi_j(k) \hat{f}_j(k) \, e^{ikx} \right) \quad \text{in} \quad D'(T_n). \tag{17}$$

We construct a special sequence $\{f_j(x)\}_{j=0}^{\infty}$ with (16) and (17). Suppose f to be as in the theorem. It holds

$$\sum_{k \in Z_n} \varphi_j(k) \hat{f}(k) \, e^{ikx} = \sum_{k \in Z_n} \varphi_j(k) \, \hat{g}_\nu(k) \, e^{ikx}, \tag{18}$$

where $\nu = 1$ if $j = 1$ and $\nu = 2^{j-2} + 1, \ldots, 2^{j-1}$ if $j = 2, 3, \ldots$ Recall that $g_\nu(x) = (f - S_\nu f)(x)$, $\nu = 1, 2, \ldots$ (18) yields

$$\sum_{k \in Z_n} \varphi_j(k) \hat{f}(k) \, e^{ikx} = 2^{-j+2} \sum_{\nu = 2^{j-2}+1}^{2^{j-1}} \sum_{k \in Z_n} \varphi_j(k) \, \hat{g}_\nu(k) \, e^{ikx}, \tag{19}$$

$j = 2, 3, \ldots$ We put

$$f_0(x) = f(x), \qquad f_1(x) = g_1(x), \qquad f_j(x) = 2^{-j+2} \sum_{\nu = 2^{j-2}+1}^{2^{j-1}} g_\nu(x);$$

$j = 2, 3, \ldots$ Then we obtain from (iii), (18), and (19)

$$f = \sum_{j=0}^{\infty} \left(\sum_{k \in Z_n} \varphi_j(k) \hat{f}_j(k) \, e^{ikx} \right).$$

Moreover, using Hölder's inequality it follows

$$\left\| \left(\sum_{j=2}^{\infty} 2^{sjq} |f_j(x)|^q \right)^{1/q} \mid L_\infty(T_n) \right\|$$

$$\leqq 2 \left\| \left(\sum_{j=0}^{\infty} 2^{sjq} \cdot 2^{-jq} \left| \sum_{\nu=2^j+1}^{2^{j+1}} g_\nu(x) \right|^q \right)^{1/q} \mid L_\infty(T_n) \right\|$$

$$\leqq 2 \left\| \left(\sum_{j=0}^{\infty} 2^{sjq-j} \sum_{\nu=2^j+1}^{2^{j+1}} |g_\nu(x)|^q \right)^{1/q} \mid L_\infty(T_n) \right\|$$

$$\leqq c \left\| \left(\sum_{\nu=2}^{\infty} \nu^{sq-1} |g_\nu(x)|^q \right)^{1/q} \mid L_\infty(T_n) \right\|. \tag{20}$$

We replace s by $s + \frac{1}{q}$. Then (15), and hence the theorem, follow from (16) and (20).

Remark 3. The theorem and also the above proof are due to W. Sickel [1]. For the one-dimensional case we refer also to L. Leindler [10, 13] where classical methods are used. If we compare (14) with the one-dimensional results of the papers cited at the beginning of this subsection then it turns out that we do not cover the case $s + \frac{1}{q} > 0$ and $0 < q \leqq 1$. However, an assertion of type (14) holds true also in this case (cf. L. Leindler [10, 13], H.-J. Schmeisser, W. Sickel [2]). Furthermore, from our point of view of strong summability the papers which we mentioned at the beginning of this subsection deal with the limiting case $p = \infty$. Note, that (14) remains true if one replaces the cubic partial sum $S_\nu f$ by spheric partial sums, i.e. by

$$S_\nu^B f(x) = \sum_{|k| < \nu} \hat{f}(k) \, e^{ikx}; \quad \nu = 1, 2, \ldots \tag{21}$$

Then the same proof works.

What can be said about the case $0 < p < \infty$ and $0 < q \leqq \infty$? If $p \leqq 1$, then we cannot expect an equivalent characterization of type (2) or (3), because we used essentially the multiplier theorem 3.4.3/3 and the Lizorkin representations of $B^s_{p,q}(T_n)$ and $F^s_{p,q}(T_n)$ in the proofs. However a careful examination of the proof of Theorem 1 shows that these tools are necessary only in Step 1. Moreover, a modification of Step 2 of that proof enables us to establish the analogon of the Corollary and Theorem 2 if $0 < p < \infty$, $0 < q \leqq \infty$, and $s + \frac{1}{q} > 0$.

Theorem 3. (i) *Let* $0 < p \leqq \infty$, $0 < q \leqq \infty$, *and* $s + \frac{1}{q} > \frac{n}{p}$. *If* $f \in D'(T_n) \cap L_p(T_n)$ *such that*

$$\sum_{\nu=1}^{\infty} \nu^{sq} \| (f - S_\nu f)(x) \mid L_p(T_n) \|^q < \infty, \quad \textit{then} \quad f \in B^{s+\frac{1}{q}}_{p,q}(T_n) \subset \mathscr{C}^{s+\frac{1}{q}-\frac{n}{p}}(T_n). \tag{22}$$

(ii) *Let* $0 < p < \infty$, $0 < q \leqq \infty$, *and* $s + \frac{1}{q} > \frac{n}{p}$. *If* $f \in D'(T_n) \cap L_p(T_n)$ *such that*

$$\left\| \left(\sum_{\nu=1}^{\infty} \nu^{sq} |(f - S_\nu f)(x)|^q \right)^{\frac{1}{q}} \mid L_p(T_n) \right\| < \infty, \quad \textit{then}$$
$$f \in F^{s+\frac{1}{q}}_{p,q}(T_n) \subset \mathscr{C}^{s+\frac{1}{q}-\frac{n}{p}}(T_n) \tag{23}$$

(*modification if* $q = \infty$).

Proof. We restrict ourselves to the proof of (23). (22) can be obtained in the same way. Let $f \in D'(T_n) \cap L_p(T_n)$. Let $\varphi(x) \in S(R_n)$ and $\varphi_0(x) \in S(R_n)$ such that

$$\operatorname{supp} \varphi \subset \{x \mid |x_i| \leqq 8; i = 1, \dots n\}, \setminus \{x \mid |x_i| < 2;\ i = 1, \dots, n\},$$
$$\operatorname{supp} \varphi_0 \subset \{x \mid |x_i| \leqq 8; i = 1, \dots, n\},$$

and $\varphi_0(x) + \sum_{j=1}^{\infty} \varphi(2^{-j}x) = 1$ for all $x \in R_n$. Then we have

$$\left\| f \mid F_{p,q}^{s+\frac{1}{q}}(T_n) \right\| \leqq c \left(\|f \mid L_p(T_n)\| + \left\| \left(\sum_{j=1}^{\infty} \left| \sum_{k \in Z_n} \varphi(2^{-j}k) \hat{f}(k) \,\mathrm{e}^{ikx} \right|^q \cdot 2^{sjq+j} \right)^{\frac{1}{q}} \mid L_p(T_n) \right\| \right). \tag{24}$$

We put

$$\psi(x) = \begin{cases} 1 & \text{if } |x_i| < 1;\ i = 1, \dots, n \\ 0 & \text{otherwise} \end{cases}$$

and $\eta(x) = \psi\left(\frac{x}{16}\right) - \psi(x)$. If $j = 1, 2, \dots$ and $\nu = 2^{j-1}, \dots, 2^j - 1$, then

$$\sum_{k \in Z_n} \varphi(2^{-j}k) \hat{f}(k) \,\mathrm{e}^{ikx} = \sum_{k \in Z_n} \varphi(2^{-j}k)\, \eta(\nu^{-1}k) \hat{f}(k) \,\mathrm{e}^{ikx}$$

and consequently

$$\begin{aligned} 2^{jsq+j} \left| \sum_{k \in Z_n} \varphi(2^{-j}k) \hat{f}(k) \,\mathrm{e}^{ikx} \right|^q &= 2 \sum_{\nu=2^{j-1}}^{2^j-1} 2^{sjq} \left| \sum_{k \in Z_n} \varphi(2^{-j}k)\, \eta(\nu^{-1}k) \hat{f}(k) \,\mathrm{e}^{ikx} \right|^q \\ &\leqq c \sum_{\nu=2^{j-1}}^{2^j-1} \nu^{sq} \left| \sum_{k \in Z_n} \varphi(2^{-j}k)\, \eta(\nu^{-1}k) \hat{f}(k) \,\mathrm{e}^{ikx} \right|^q \end{aligned} \tag{25}$$

hold. According to Theorem 3.4.1/2 and (25) we have

$$\begin{aligned} &\left\| \left(\sum_{j=1}^{\infty} 2^{\left(s+\frac{1}{q}\right)jq} \left| \sum_{k \in Z_n} \varphi(2^{-j}k) \hat{f}(k) \,\mathrm{e}^{ikx} \right|^q \right)^{\frac{1}{q}} \mid L_p(T_n) \right\| \\ &\leqq c \left\| \left(\sum_{j=1}^{\infty} \sum_{\nu=2^{j-1}}^{2^j-1} \nu^{sq} \left| \sum_{k \in Z_n} \varphi(2^{-j}k)\, \eta(\nu^{-1}k) \hat{f}(k) \,\mathrm{e}^{ikx} \right|^q \right)^{\frac{1}{q}} \mid L_p(T_n) \right\| \\ &\leqq c' \sup_{\frac{1}{2} \leqq \tau \leqq 1} \|\varphi(\tau x) \mid H_2^{\varkappa}\| \left\| \left(\sum_{\nu=1}^{\infty} \nu^{sq} \left| \sum_{k \in Z_n} \eta(\nu^{-1}k) \hat{f}(k) \,\mathrm{e}^{ikx} \right|^q \right)^{\frac{1}{q}} \mid L_p(T_n) \right\| \\ &\leqq c'' \left\| \left(\sum_{\nu=1}^{\infty} \nu^{sq} |(f - S_\nu f)(x)|^q \right)^{\frac{1}{q}} \mid L_p(T_n) \right\|. \end{aligned} \tag{26}$$

Here $\varkappa > n\left(\frac{1}{\min(p, q)} + \frac{1}{2}\right)$. Finally, (24), (26), and the embedding (3.5.5/4) yield (23). This completes the proof.

Remark 4. Note that (22) and (23) remain true if the partial sums $S_\nu f$ are replaced by $S_\nu^B f$ $(\nu = 1, 2, \dots)$ from (21). This can be proved by the same method.

3.7.4. Strong Summability of Smooth Means

Theorem 3.7.2/1 suggests to consider the problem of strong summability raised in (3.7.3/1) with the smooth approximations $M_\nu^\varphi f$ instead of the partial sums $S_\nu f$. Recall that $M_\nu^\varphi f$, $\nu = 1, 2, \dots$, has been defined in 3.7.2/(1), (2). Further, σ_p, $\tilde{\sigma}_p$, and $\tilde{\sigma}_{p,q}$ have the meaning of 3.7.1/(1), (2).

Theorem. *Let ψ be a real continuous function on R_n with (3.7.2/1). Let $0 < \sigma < \infty$ and $0 < \lambda < \infty$ with*

$$\| |x|^{-\sigma} (\psi(x) - \psi(\tau x)) \mid H_2^\lambda \| \leqq c < \infty \tag{1}$$

for all τ, $1 \leqq \tau \leqq 2$, and let d be a natural number with

$$\psi\left(\frac{x}{d}\right) - \psi(x) > 0 \quad \text{if} \quad 2 \leqq |x| \leqq 16. \tag{2}$$

(i) *If $0 < p \leqq \infty$, $0 < q \leqq \infty$, λ natural with $\lambda > \sigma_p$, and $\tilde{\sigma}_p < s < \sigma - \frac{n}{2}$, then*

$$B^s_{p,q}(T_n) = \Big\{ f \mid f \in D'(T_n),\ \|f \mid B^s_{p,q}(T_n)\|^+_{M^\psi} = \|M^\psi_1 f(x) \mid L_p(T_n)\|$$
$$+ \left(\sum_{\nu=1}^{\infty} \nu^{sq-1} \|(f - M^\psi_\nu f)(x) \mid L_p(T_n)\|^q \right)^{1/q} < \infty \Big\}. \tag{3}$$

Furthermore, $\|f \mid B^s_{p,q}(T_n)\|^+_{M^\psi}$ is an equivalent quasi-norm in $B^s_{p,q}(T_n)$.

(ii) *If $0 < p < \infty$, $0 < q < \infty$, λ natural with $\lambda > n\left(\frac{1}{\min(p, q)} + \frac{1}{2}\right)$ and $\tilde{\sigma}_{p,q} < s < \sigma - \frac{n}{2}$ then*

$$F^s_{p,q}(T_n) = \Big\{ f \mid f \in D'(T_n),\ \|f \mid F^s_{p,q}(T_n)\|^+_{M^\psi} = \|M^\psi_1 f(x) \mid L_p(T_n)\|$$
$$+ \left\| \left(\sum_{\nu=1}^{\infty} \nu^{sq-1} |(f - M^\psi_\nu f)(x)|^q \right)^{1/q} \mid L_p(T_n) \right\| < \infty \Big\}. \tag{4}$$

Furthermore, $\|f \mid F^s_{p,q}(T_n)\|^+_{M^\psi}$ is an equivalent quasi-norm in $F^s_{p,q}(T_n)$.

Proof. Step 1. We prove that

$$\|f \mid F^s_{p,q}(T_n)\|^+_{M^\psi} \leqq c \|f \mid F^s_{p,q}(T_n)\|. \tag{5}$$

Similarly as in (3.7.3/5) we have

$$\left\| \left(\sum_{\nu=1}^{\infty} \nu^{sq-1} |(f - M^\psi_\nu f)(x)|^q \right)^{1/q} \mid L_p(T_n) \right\|$$
$$= \left\| \left(\sum_{j=0}^{\infty} \sum_{\nu=2^j}^{2^{j+1}-1} \nu^{sq-1} |(f - M^\psi_\nu f)(x)|^q \right)^{1/q} \mid L_p(T_n) \right\|$$
$$\leqq c \left\| \left(\sum_{j=0}^{\infty} 2^{s(j+1)q} |(f - M^\psi_{2^{j+1}} f)(x)|^q \right)^{1/q} \mid L_p(T_n) \right\|$$
$$+ c \left\| \left(\sum_{j=0}^{\infty} \sum_{\nu=1}^{2^j} 2^{sjq-j} |(M^\psi_{2^{j+1}} f - M^\psi_{2^{j+1}-\nu} f)(x)|^q \right)^{1/q} \mid L_p(T_n) \right\|. \tag{6}$$

The first summand can be estimated from above by $c\|f \mid F^s_{p,q}(T_n)\|$ according to Theorem 3.7.2/1(ii). To deal with the second one we observe

$$(M^\psi_{2^{j+1}} f - M^\psi_{2^{j+1}-\nu} f)(x) = \sum_{k \in Z_n} \varrho_{j,\nu}(2^{-j-1}k) \hat{f}(k)\, e^{ikx}, \tag{7}$$

where

$$\varrho_{j,\nu}(x) = \psi(x) - \psi\left(\left(1 - \frac{\nu}{2^{j+1}}\right)^{-1} x\right)$$

and $j = 0, 1, 2, \ldots$, $\nu = 1, 2, \ldots, 2^j$. Note that $1 < \left(1 - \frac{\nu}{2^{j+1}}\right)^{-1} \leqq 2$ in this case. The functions $\varrho_{j,\nu}(x)$ are the counterparts of the function $\varrho(x)$ from (3.7.2/9). Further we put

$$f_l(x) = \sum_{k \in Z_n} |2^{-l}k|^\sigma \varphi_l(k) \hat{f}(k) \, e^{ikx} \quad \text{if} \quad l = 0, 1, 2, \ldots$$

and $f_l(x) = 0$ if $l = -1, -2, \ldots$ Then we obtain similarly as in 3.7.2/(11), (12) with $r = \min(1, p, q)$

$$\begin{aligned} &\left\| \left(\sum_{j=0}^{\infty} \sum_{\nu=1}^{2^j} 2^{\left(s-\frac{1}{q}\right)jq} \left| \sum_{k \in Z_n} \varrho_{j,\nu}(2^{-j-1}k) \hat{f}(k) \, e^{ikx} \right|^q \right)^{\frac{1}{q}} \mid L_p(T_n) \right\|^r \\ &\leqq c \sum_{l=-\infty}^{0} 2^{\left(\sigma - s + \frac{1}{q}\right) rl} \\ &\times \left\| \left(\sum_{j=0}^{\infty} \sum_{\nu=1}^{2^j} 2^{(sq-1)(l+j)} \left| \sum_{k \in Z_n} |2^{-j}k|^{-\sigma} \varrho_{j,\nu}(2^{-j-1}k) \hat{f}_{l+j+2}(k) e^{ikx} \right|^q \right)^{\frac{1}{q}} \mid L_p(T_n) \right\|^r \end{aligned} \tag{8}$$

(1) yields

$$\| |2^l x|^{-\sigma} \varrho_{j,\nu}(2^l x) \mid H_2^\lambda \| \leqq c \cdot 2^{-\frac{n}{2}l}; \quad l = 0, -1, -2, \ldots$$

for all $j = 0, 1, 2, \ldots$, and $\nu = 1, \ldots, 2^j$. Hence, by Theorem 3.4.1/2 the right-hand side of (8) can be estimated by

$$\begin{aligned} &c \sum_{l=-\infty}^{0} 2^{\left(\sigma - s + \frac{1}{q} - \frac{n}{2}\right) lr} \left\| \left(\sum_{j=0}^{\infty} \sum_{\nu=1}^{2^j} 2^{sjq-j} |f_j(x)|^q \right)^{\frac{1}{q}} \mid L_p(T_n) \right\|^r \\ &\leqq c' \| 2^{sj} f_j(x) \mid L_p(T_n, l_q) \| \leqq c'' \| f \mid F^s_{p,q}(T_n) \| . \end{aligned} \tag{9}$$

Here we used that $s < \sigma - \dfrac{n}{2} + \dfrac{1}{q}$. Now, (6)–(9) give

$$\left\| \left(\sum_{\nu=1}^{\infty} \nu^{sq-1} |(f - M_\nu^\varphi f)(x)|^q \right)^{1/q} \mid L_p(T_n) \right\| \leqq c \| f \mid F^s_{p,q}(T_n) \| . \tag{10}$$

Furthermore,

$$\begin{aligned} &\| (M_1^\varphi f)(x) \mid L_p(T_n) \| \\ &\leqq c(\| f \mid L_p(T_n) \| + \| f \mid F^s_{p,q}(T_n) \|) \leqq c' \| f \mid F^s_{p,q}(T_n) \| \end{aligned} \tag{11}$$

because of (10) and $s > \tilde{\sigma}_{p,q}$. (10) and (11) yield the desired estimate (5).

Step 2. We prove

$$\| f \mid F^s_{p,q}(T_n) \| \leqq c \| f \mid F^s_{p,q}(T_n) \|^{+}_{M^\varphi} . \tag{12}$$

We choose a system $\{\varphi_j\}_{j=0}^\infty$ of type $\Phi(R_n)$ such that

$$\operatorname{supp} \varphi_0 \subset \{x \mid |x| \leqq 8\},$$
$$\operatorname{supp} \varphi_j \subset \{x \mid 2^{j+1} \leqq |x| \leqq 2^{j+3}\};$$

$j = 1, 2, \ldots$ Then

$$\begin{aligned} \| f \mid F^s_{p,q}(T_n) \| \leqq c \Bigg(&\left\| \sum_{k \in Z_n} \varphi_0(k) \hat{f}(k) \, e^{ikx} \mid L_p(T_n) \right\| \\ &+ \left\| \left(\sum_{j=1}^{\infty} 2^{sjq} \left| \sum_{k \in Z_n} \varphi_j(k) \hat{f}(k) \, e^{ikx} \right|^q \right)^{1/q} \mid L_p(T_n) \right\| \Bigg) . \end{aligned} \tag{13}$$

We can find a natural number ν_0 such that $\psi\left(\frac{x}{\nu_0}\right) > 0$ if $|x| \leqq 8$. Therefore, by Theorem 3.3.4

$$\left\|\sum_{k\in Z_n} \varphi_0(k)\hat{f}(k)\,\mathrm{e}^{ikx} \mid L_p(T_n)\right\|$$
$$= \left\|\sum_{k\in Z_n} \frac{\varphi_0(k)}{\psi(\nu_0^{-1}k)}\psi(\nu_0^{-1}k)\hat{f}(k)\,\mathrm{e}^{ikx} \mid L_p(T_n)\right\| \leqq c\|(M^\psi_{\nu_0}f)(x) \mid L_p(T_n)\|$$
$$\leqq c'\left(\|(M^\psi_1 f)(x) \mid L_p(T_n)\| + \left\|\sum_{\nu=1}^{\nu_0-1} |(M^\psi_{\nu+1}f - M^\psi_\nu f)(x)| \mid L_p(T_n)\right\|\right)$$
$$\leqq c''\|f \mid F^s_{p,q}(T_n)\|^+_{M^\psi}. \tag{14}$$

On the other hand, by assumption (2) $\eta(x) = \psi\left(\frac{x}{d}\right) - \psi(x) > 0$ if $2 \leqq |x| \leqq 16$. Consequently, $\eta(\tau x) > 0$ if $2 \leqq |x| \leqq 8$ and $1 \leqq \tau \leqq 2$. Let $j = 1, 2, \ldots$ be fixed and let $\nu = 2^{j-1}, \ldots, 2^j - 1$. If we choose $\tau = \frac{2^j}{\nu}$, then we have $\eta(\nu^{-1}x) > 0$ for $2^{j+1} \leqq |x| \leqq 2^{j+3}$. Hence, we obtain

$$2^{sjq}\left|\sum_{k\in Z_n}\varphi_j(k)\hat{f}(k)\,\mathrm{e}^{ikx}\right|^q$$
$$= 2^{sjq-j+1}\sum_{\nu=2^{j-1}}^{2^j-1}\left|\sum_{k\in Z_n}\frac{\varphi_j(k)}{\eta(\nu^{-1}k)}\eta(\nu^{-1}k)\hat{f}(k)\,\mathrm{e}^{ikx}\right|^q$$
$$\leqq c\sum_{\nu=2^{j-1}}^{2^j-1}\left|\sum_{k\in Z_n}\frac{\varphi_j(k)}{\eta(\nu^{-1}k)}\nu^{s-\frac{1}{q}}\eta(\nu^{-1}k)\hat{f}(k)\,\mathrm{e}^{ikx}\right|^q. \tag{15}$$

Note that

$$\sup_{j=1,2,\ldots}\ \sup_{\nu=2^{j-1},\ldots,2^j-1}\left\|\frac{\varphi_j(\nu x)}{\eta(x)} \mid H^\lambda_2\right\| \leqq c < \infty$$

holds. Therefore, with the help of (15) and Theorem 3.4.1/2 we have

$$\left\|\left(\sum_{j=1}^\infty 2^{sjq}\left|\sum_{k\in Z_n}\varphi_j(k)\,\hat{f}(k)\,\mathrm{e}^{ikx}\right|^q\right)^{\frac{1}{q}} \mid L_p(T_n)\right\|$$
$$\leqq \left\|\left(\sum_{j=0}^\infty\sum_{\nu=2^{j-1}}^{2^j-1}\left|\sum_{k\in Z_n}\frac{\varphi_j(k)}{\eta(\nu^{-1}k)}\nu^{s-\frac{1}{q}}\eta(\nu^{-1}k)\hat{f}(k)\,\mathrm{e}^{ikx}\right|^q\right)^{\frac{1}{q}} \mid L_p(T_n)\right\|$$
$$\leqq c\left\|\left(\sum_{\nu=1}^\infty \nu^{sq-1}\left|\sum_{k\in Z_n}\eta(\nu^{-1}k)\hat{f}(k)\,\mathrm{e}^{ikx}\right|^q\right)^{\frac{1}{q}} \mid L_p(T_n)\right\|$$
$$= c\left\|\left(\sum_{\nu=1}^\infty \nu^{sq-1}|(M^\psi_{d\nu}f - M^\psi_\nu f)(x)|^q\right)^{\frac{1}{q}} \mid L_p(T_n)\right\| \leqq c'\|f \mid F^s_{p,q}(T_n)\|^+_{M^\psi}. \tag{16}$$

Finally, (13), (14), and (16) prove (12) and complete the proof of part (ii) of the theorem.

Step 3. We obtain part (i) on the same line as in Step 1 and Step 2 (with the usual modifications). Note that instead of Theorem 3.4.1/2 we use now Theorem 3.3.4. This proves the theorem.

Remark 1. If we compare Theorem 3.7.2/1 and the just proved theorem then we see that $\lambda > \sigma_{p,q}$ in part (ii) has been replaced by $\lambda > n\left(\frac{1}{\min(p,q)} + \frac{1}{2}\right) > \sigma_{p,q}$. This is a consequence of the fact that we are forced to use Theorem 3.4.1/2 instead of Theorem 3.6.4. Further, we remark that part (ii) of the theorem can be extended to $q = \infty$ with more restrictive conditions concerning s (cf. Remark 3.7.2/3).

As in 3.7.2. the above theorem yields some results for the generalized de la Vallée-Poussin means $V_\nu^\psi f(x)$. Recall that $V_\nu^\psi f(x)$, $f \in D'(T_n)$; $\nu = 1, 2, \ldots$; have been defined in 3.7.2/(17), (18). We have the following corollary.

Corollary. *Let* $\psi(x) \in H_2^\lambda(R_n)$, $0 < \lambda < \infty$, *be a real continuous function on* R_n *which satisfies* (3.7.2/17).

(i) *If* $0 < p \leqq \infty$, $0 < q \leqq \infty$, $\tilde{\sigma}_p < s$, *and* λ *natural with* $\lambda > \sigma_p$, *then*

$$B_{p,q}^s(T_n) = \Big\{ f \mid f \in D'(T_n),\ \|f \mid B_{p,q}^s(T_n)\|_{V^\psi}^+ = \|(V_1^\psi f)(x) \mid L_p(T_n)\| + \Big(\sum_{\nu=1}^{\infty} \nu^{sq-1} \|(f - V_\nu^\psi f)(x) \mid L_p(T_n)\|^q \Big)^{1/q} < \infty \Big\}. \tag{17}$$

Furthermore, $\|f \mid B_{p,q}^s(T_n)\|_{V^\psi}^+$ *is an equivalent quasi-norm in* $B_{p,q}^s(T_n)$.

(ii) *If* $0 < p < \infty$, $0 < q < \infty$, $\tilde{\sigma}_{p,q} < s$, *and* λ *natural with* $\lambda > n\left(\frac{1}{\min(p,q)} + \frac{1}{2}\right)$, *then*

$$F_{p,q}^s(T_n) = \Big\{ f \mid f \in D'(T_n),\ \|f \mid F_{p,q}^s(T_n)\|_{V^\psi}^+ = \|(V_1^\psi f)(x) \mid L_p(T_n)\| + \Big\| \Big(\sum_{\nu=1}^{\infty} \nu^{sq-1} |(f - V_\nu^\psi f)(x)|^q \Big)^{1/q} \mid L_p(T_n) \Big\| < \infty \Big\}. \tag{18}$$

Furthermore, $\|f \mid F_{p,q}^s(T_n)\|_{V^\psi}^+$ *is an equivalent quasi-norm in* $F_{p,q}^s(T_n)$.

Remark 2. Similar as in Corollary 3.7.3 we can derive from (17) and (18) results in the language of the Hölder-Zygmund spaces. Further, the analogon of Theorem 3.7.3/2 holds true (with $M_\nu^\psi f$ or $V_\nu^\psi f$ instead of $S_\nu f$). However, in this case we can also use the method of Theorem 3.7.3/3 which also works for $V_\nu^\psi f$, $\nu = 1, 2, \ldots$ In particular, suppose that $\psi(x)$ is a real continuous function with $\psi(x) = 1$ if $|x| \leqq 1$ and $\psi(x) = 0$ if $|x| > 2$, then $S_\nu f$ in Theorem 3.7.3/3 can be replaced by $V_\nu^\psi f(x)$ ($\nu = 1, 2, \ldots$).

Remark 3. The theorem shows that for smooth approximation processes the problem of strong summability can be studied with the methods developed here also for values p, q with $\min(p, q) \leqq 1$. In particular, we obtained results for the generalized de la Vallée-Poussin means. In this direction we refer also to P. Oswald [2], W. Sickel [1, 2] and H.-J. Schmeisser, W. Sickel [2].

Remark 4. We discussed the problem of strong summability for general smooth means $M_\nu^\psi f$, $\nu = 1,2,\ldots$, with an appropriate function ψ, as a model case. The question arises whether by our methods results as in Theorem 3.7.2/1 and the above theorem can be obtained for further classical approximation procedures such as Cesaro-means, Riesz-means, Abel-Cartwright-means or means of Bessel-potential type. This question has been extensively studied in W. Sickel [1, 2]. There at least partial results can be found. As an example let us consider the Riesz-means

$$R_\nu^\beta f(x) = \sum_{|k| \leqq \nu} \left(1 - \frac{|k|^2}{\nu^2}\right)^\beta \hat{f}(k)\, e^{ikx}, \tag{19}$$

where $f \in D'(T_n)$, $\nu = 1, 2, \ldots$, and $0 < \beta < \infty$. Obviously, in the sense of (3.7.2/2) the means $R_\nu^\beta f$ are generated by the functions

$$\psi(x) = \begin{cases} (1 - |x|^2)^\beta & \text{if } |x| \leqq 1 \\ 0 & \text{if } |x| > 1, \end{cases} \tag{20}$$

$0 < \beta < \infty$. We have

$$\| |x|^{-2}(\psi(x) - \psi(\tau x)) \mid H_2^\lambda \| \leqq c < \infty \tag{21}$$

for all τ, $1 \leqq \tau \leqq 2$, if $\lambda < \beta + \frac{1}{2}$, (cf. H.-J. Schmeisser, W. Sickel [2, Lemma 4]). Now, Theorem 3.7.2./1 and the above theorem can be applied to obtain results on strong summability of the Riesz-means defined in (19). We do not go into the details. We refer to H.-J. Schmeisser, W. Sickel [2], where we proved new assertions about strong summability complementing the results presented in this book.

Remark* 5. The summability of trigonometric series, approximation processes, inequalities for trigonometric polynomials etc. played and play a crucial role in the long history of Fourier analysis. The fundamental books in this field of research are A. Zygmund [1], N. K. Bari [1], R. E. Edwards [1], and P. L. Butzer, R. J. Nessel [1], where the latter book comes nearest to the problems treated here, cf. also P. L. Butzer, H. Berens [1]. A systematic treatment of problems of strong summability has ben given recently in the book by L. Leindler [16].

3.8. Remarks

3.8.1. Periodic Spaces and Weighted Spaces

In 3.2.3. we have shown that the distributions from $D'(T_n)$ can be identified with the periodic distributions from $S'(R_n)$ $(S'_\pi(R_n))$. We used this fact in order to derive multiplier theorems and inequalities for trigonometric polynomials and periodic functions (cf. 3.3. and 3.4.). On this basis we have studied the periodic spaces $B^s_{p,q}(T_n)$ and $F^s_{p,q}(T_n)$. Another approach to these spaces is to consider $B^s_{p,q}(T_n)$ and $F^s_{p,q}(T_n)$ as subspaces of the weighted spaces $B^s_{p,q}(R_n, \varrho(x))$ and $F^s_{p,q}(R_n, \varrho(x))$ (cf. Definition 5.1.1/2) with appropriate weights $\varrho(x)$. Namely, we have the following theorem.

Theorem. *Let* $-\infty < s < \infty$ *and* $0 < q \leqq \infty$.

(i) *If* $0 < p \leqq \infty$ *and if* $\varrho(x) = (1 + |x|)^{-\varkappa}$ *with* $\varkappa p > n$, *then*

$$B^s_{p,q}(T_n) = \{f \mid f \in S'_\pi(R_n),\ \|f \mid B^s_{p,q}(R_n, \varrho(x))\| < \infty\}. \tag{1}$$

Furthermore, $\|f \mid B^s_{p,q}(R_n, \varrho(x))\|$ *is an equivalent quasi-norm in* $B^s_{p,q}(T_n)$.

(ii) *If* $0 < p < \infty$, *and if* $\varrho(x) = (1 + |x|)^{-\varkappa}$ *with* $\varkappa p > n$, *then*

$$F^s_{p,q}(T_n) = \{f \mid f \in S'_\pi(R_n),\ \|f \mid F^s_{p,q}(R_n, \varrho(x))\| < \infty\}. \tag{2}$$

Furthermore, $\|f \mid F^s_{p,q}(R_n, \varrho(x))\|$ *is an equivalent quasi-norm in* $F^s_{p,q}(T_n)$.

Note that in the case of these weight functions we do not need the ultra-distributions used in Chapter 5. The proof is not very difficult. It is a consequence of (3.3.4/2) and the definitions of $B^s_{p,q}(R_n, \varrho(x))$ and $F^s_{p,q}(R_n, \varrho(x))$. See also H. Triebel [17]. Similar relations as in (1) and (2) hold true for the spaces $L^A_p(T_n, l_q)$, cf. (3.4.1/1).

3.8.2. Periodic Spaces and Abstract Spaces

The spaces $B^s_{p,q}(T_n)$, $1 \leqq p \leqq \infty$, $1 \leqq q \leqq \infty$, can be characterized as special abstract Besov spaces. This will be studied in 6.5.2. From this point of view we shall deal with some further problems of approximation theory. In particular, we shall be interested in lacunary series and in the behaviour of Fourier coefficients, cf. also H. Triebel [16].

4. Anisotropic Spaces

4.1. Introduction

This chapter deals with anisotropic Sobolev spaces $W_p^{\bar{s}}$ and anisotropic Besov spaces $B_p^{\bar{s}} = B_{p,p}^{\bar{s}}$ (and modifications of them) on the plane and on the unit circle, where $\bar{s} = (s_1, s_2)$ stands for the anisotropic smoothness and $1 < p < \infty$.

In [T, Chapter 2] we developed the theory of the isotropic spaces $B_{p,q}^s(R_n)$ and $F_{p,q}^s(R_n)$ of Besov-Hardy-Sobolev type on the Euclidean n-space R_n with $-\infty < s < \infty$, $0 < p \leqq \infty$, and $0 < q \leqq \infty$. There is no doubt that this theory has a more or less full counterpart for corresponding anisotropic spaces $B_{p,q}^{\bar{s}}(R_n)$ and $F_{p,q}^{\bar{s}}(R_n)$: definitions, embeddings for different metrics, traces on hyperplanes, Fourier multipliers, interpolation, equivalent quasi-norms etc. It is not our aim to give a systematic theory of spaces of type $B_{p,q}^{\bar{s}}(R_n)$ and $F_{p,q}^{\bar{s}}(R_n)$ more or less parallel to [T, Chapter 2] or to Chapter 2 of this book. Some remarks and several references about such a theory have been given in [T, 10.1] and in 2.2.2. On the other hand there exist genuine problems for anisotropic spaces which have no isotropic counterparts. Typical examples are the domain-problem and the trace-problem. If Ω is a (smooth bounded) domain in R_n then the anisotropic space $A(\Omega)$ can be defined as the restriction of the corresponding space $A(R_n)$ on Ω. The domain-problem is characterized by the question whether there exists a linear and bounded extension operator from $A(\Omega)$ in $A(R_n)$. If S is a (smooth) surface in R_n (e.g. the boundary $\partial\Omega$ of a C^∞-domain Ω in R_n) then one asks for an exact description of traces on S of anisotropic spaces (trace-problem). One of the main aims of this chapter is to study traces on curves of anisotropic spaces on the plane (i.e. $n = 2$). But we are not interested in the study of most general anisotropic spaces of $B_{p,q}^{\bar{s}}(R_2)$ and $F_{p,q}^{\bar{s}}(R_2)$ type, on contrary we restrict ourselves to Sobolev spaces $W_p^{\bar{s}}(R_2)$ and special Besov spaces $B_p^{\bar{s}}(R_2) = B_{p,p}^{\bar{s}}(R_2)$ with $1 < p < \infty$. More precisely, the chapter is organized as follows. Section 4.2. contains the necessary definitions and preliminaries. Here we describe (for the first and the last time in this chapter) the Fourier-analytic origin of the spaces in question, otherwise we work mainly with derivatives and differences. Furthermore we quote some known results for classical anisotropic Sobolev-Besov spaces partly from S. M. Nikol'skij [2] and O. V. Besov, V. P. Il'in, S. M. Nikol'skij [1]. In Section 4.3. we prove an anisotropic Hardy inequality which holds a key position in our theory. The Sections 4.4. and 4.5. deal with the spaces $\dot{W}_p^{\bar{s}}(R_2)$ and $\dot{B}_p^{\bar{s}}(R_2)$ (which are near to $W_p^{\bar{s}}(R_2)$ and $B_p^{\bar{s}}(R_2)$, respectively) and their traces on curves in the plane. In the remaining Sections 4.6.–4.8. we describe further results in this direction: anisotropic spaces on the unit circle, weighted anisotropic spaces of Sobolev type on domains, and boundary value problems for semi-elliptic differential equations on the unit circle. This chapter is mainly based on H. Triebel [12, 13, 15].

4.2. Definitions and Preliminaries

4.2.1. Definitions

This chapter is essentially independent of the other chapters of this book. For this purpose we recall some basic definitions and also few assertions which we mentioned in 2.2.2.

Let R_2 be the two-dimensional Euclidean space: the plane. Let $0 < a_2 \leqq a_1 < \infty$ with $a_1 + a_2 = 2$. The anisotropic distance of $x = (x_1, x_2) \in R_2$ from the origin is given by

$$|x|_{\bar{a}} = (|x_1|^{2/a_1} + |x_2|^{2/a_2})^{1/2}, \tag{1}$$

where $\bar{a} = (a_1, a_2)$ characterizes the anisotropy. As in the previous chapters, $S(R_2)$ stands for the collection of all complex-valued infinitely differentiable rapidly decreasing functions on R_2. The dual space $S'(R_2)$ is the collection of all tempered distributions on R_2. We recall that F and F^{-1} stand for the Fourier transform and its inverse on $S'(R_2)$, respectively.

Let $\Phi^{\bar{a}}(R_2)$ be the collection of all systems $\varphi = \{\varphi_j(x)\}_{j=0}^{\infty} \subset S(R_2)$ with the following properties:

(i) $$\operatorname{supp} \varphi_0 \subset \{x \mid |x|_{\bar{a}} \leqq 2\},$$
$$\operatorname{supp} \varphi_j \subset \{x \mid 2^{j-1} \leqq |x|_{\bar{a}} \leqq 2^{j+1}\} \quad \text{if} \quad j = 1, 2, 3, \ldots$$

(ii) For every multi-index $\alpha = (\alpha_1, \alpha_2)$ there exists a positive number c_α with

$$2^{ja_1\alpha_1 + ja_2\alpha_2} |D^\alpha \varphi_j(x)| \leqq c_\alpha \tag{2}$$

for any $j = 0, 1, 2, \ldots$ and any $x \in R_2$.

(iii) $$\sum_{j=0}^{\infty} \varphi_j(x) = 1 \quad \text{if} \quad x \in R_2. \tag{3}$$

This is a smooth resolution of unity adapted to the given anisotropy $\bar{a} = (a_1, a_2)$. One proves in exactly the same way as in the isotropic case that $\Phi^{\bar{a}}(R_2)$ is not empty, cf. e.g. [T, Remark 2.3.1/1]. In particular it is possible to choose φ_1 in such a way that $\varphi_j(x) = \varphi_1(2^{-(j-1)a_1}x_1, 2^{-(j-1)a_2}x_2)$ if $j = 1, 2, 3, \ldots$ This makes also clear that (2) is the best possible choice. If $0 < p < \infty$ then

$$\|f \mid L_p(R_2)\| = \left(\int_{R_2} |f(x)|^p \, dx\right)^{1/p}, \tag{4}$$

where dx stands for the usual Lebesgue measure on R_2.

Definition. *Let* $\bar{a} = (a_1, a_2)$ *with* $0 < a_2 \leqq a_1 < \infty$ *and* $a_1 + a_2 = 2$. *Let* $\varphi = \{\varphi_j(x)\}_{j=0}^{\infty} \in \Phi^{\bar{a}}(R_2)$. *Let* $-\infty < s < \infty$ *and* $\bar{s} = (s_1, s_2)$ *with* $s_1 = \frac{s}{a_1}$ *and* $s_2 = \frac{s}{a_2}$. *Let* $1 < p < \infty$. *Then*

$$B_p^{\bar{s}}(R_2) = \Big\{ f \mid f \in S'(R_2), \|f \mid B_p^{\bar{s}}(R_2)\|^\varphi = \left(\sum_{j=0}^{\infty} 2^{jsp} \|F^{-1}[\varphi_j Ff](\cdot) \mid L_p(R_2)\|^p\right)^{1/p} < \infty \Big\} \tag{5}$$

and

$$H_p^{\bar{s}}(R_2) = \Big\{ f \mid f \in S'(R_2), \|f \mid H_p^{\bar{s}}(R_2)\|^\varphi = \left\| \left(\sum_{j=0}^{\infty} 2^{2js} |F^{-1}[\varphi_j Ff](\cdot)|^2\right)^{1/2} \mid L_p(R_2) \right\| < \infty \Big\}. \tag{6}$$

If in addition s_1 *and* s_2 *are natural numbers then*

$$W_p^{\bar{s}}(R_2) = H_p^{\bar{s}}(R_2). \tag{7}$$

Remark 1. The above definitions are special cases of the general anisotropic spaces of Besov-Hardy-Sobolev type $B^{\bar{s}}_{p,q}(R_2)$ with $-\infty < s < \infty, 0 < p \leqq \infty, 0 < q \leqq \infty$, and $F^{\bar{s}}_{p,q}(R_2)$ with $-\infty < s < \infty$, $0 < p < \infty$, and $0 < q \leqq \infty$, cf. 2.2.2. We have

$$B^{\bar{s}}_p(R_2) = B^{\bar{s}}_{p,p}(R_2) \quad \text{and} \quad H^{\bar{s}}_p(R_2) = F^{\bar{s}}_{p,2}(R_2) \tag{8}$$

with $1 < p < \infty$ and $-\infty < s < \infty$. Furthermore there is no difficulty to extend the above definitions (and also the just mentioned generalizations) from R_2 to R_n. In this case the anisotropy $\bar{a} = (a_1, \ldots, a_n)$ is given by $0 < a_n \leqq a_{n-1} \leqq \ldots \leqq a_1 < \infty$ with $\sum_{j=1}^{n} a_j = n$.

Remark 2. If $\bar{s} = (s_1, s_2)$ with either $0 < s_1 \leqq s_2 < \infty$ or $s_1 = s_2 = 0$ or $-\infty < s_2 \leqq s_1 < 0$ is given, then the above number s and the anisotropy $\bar{a} = (a_1, a_2)$ are uniquely determined by

$$\frac{1}{s} = \frac{1}{2}\left(\frac{1}{s_1} + \frac{1}{s_2}\right) \tag{9}$$

and $a_1 = \dfrac{s}{s_1}$ and $a_2 = \dfrac{s}{s_2}$. We call s the mean smoothness. This is justified by (9).

Remark 3. In the isotropic case, i.e. if $a_1 = a_2 = 1$, we developed the theory of the above spaces, including the indicated generalizations, in [T, Chapter 2]. One can extend these considerations to the anisotropic case. In particular, the above spaces $B^{\bar{s}}_p(R_2)$ and $H^{\bar{s}}_p(R_2)$ are Banach spaces. They are independent of the chosen system $\varphi \in \Phi^{\bar{a}}(R_2)$ (in the sense of equivalent norms). As we said, as far as our main assertions are concerned we shall not base our further studies on the above definitions. The reason why we began in the above way is twofold. First we wanted to emphasize that the spaces in this chapter fit in our general scheme to define spaces via Fourier decompositions. Secondly in our context it is useful to handle occasionally also spaces $B^{\bar{s}}_p(R_2)$ with $s \leqq 0$. In this case one has no equivalent description of the coresponding spaces via derivatives and differences at hand.

4.2.2. Equivalent Norms

In the isotropic case we have $a_1 = a_2 = 1$ and $\bar{s} = (s, s)$. Then $H^s_p(R_2) = H^{\bar{s}}_p(R_2)$ with $1 < p < \infty$ are the well-known Bessel-potential spaces, which can be normed via

$$\|f \mid H^s_p(R_2)\| = \|F^{-1}[(1 + |\xi|^2)^{s/2}\, Ff] \mid L_p(R_2)\|. \tag{1}$$

In comparison with (4.2.1/6) this is essentially an isotropic Littlewood-Paley theorem. In particular $H^0_p(R_2) = L_p(R_2)$. This result can be extended to the anisotropic case. One has to use anisotropic Fourier multiplier theorems, cf. P. Krée [2, p. 69/70] and W. Littman, C. McCarthy, N. Riviere [1, p. 211]. Then it follows that $\|f \mid H^{\bar{s}}_p(R_2)\|^{\varphi}$ in (4.2.1/6) can be replaced by

$$\|f \mid H^{\bar{s}}_p(R_2)\| = \|F^{-1}[(1 + |\xi_1|^2)^{s_1/2} + (1 + |\xi_2|^2)^{s_2/2}]\, Ff \mid L_p(R_2)\|. \tag{2}$$

In other words, $H^{\bar{s}}_p(R_2)$ from (4.2.1/6) is an anisotropic Bessel-potential space. For details we refer to B. Stöckert, H. Triebel [1, Sect. 4.4.]. In particular if s_1 and s_2 in $\bar{s} = (s_1, s_2)$ are natural numbers then $H^{\bar{s}}_p(R_2) = W^{\bar{s}}_p(R_2)$ are the well-known anisotropic Sobolev spaces which can also be normed via

$$\|f \mid W^{\bar{s}}_p(R_2)\| = \|f \mid L_p(R_2)\| + \left\|\frac{\partial^{s_1} f}{\partial x_1^{s_1}} \mid L_p(R_2)\right\| + \left\|\frac{\partial^{s_2} f}{\partial x_2^{s_2}} \mid L_p(R_2)\right\|. \tag{3}$$

In this context we refer also to S. M. Nikol'skij [2, Sects. 9.1. and 9.4.], cf. also [F, 2.5.2.] and [T, 10.1.].

In 4.2.4. we shall describe some interpolation results for anisotropic function spaces. In particular, by (4.2.4/1) the above Besov spaces $B^{\bar{s}}_p(R_2)$ can be obtained by real interpolation of Bessel-potential spaces, where we may assume that the latter are normed via (2). Similar as in the isotropic case one obtains the equivalent norms on the spaces $B_p^{\bar{s}}(R_2)$ which we describe in the sequel. We use the partial differences

$$(\Delta_{h,1}^1 f)(x) = f(x_1 + h, x_2) - f(x_1, x_2), \qquad \Delta_{h,1}^m = \Delta_{h,1}^1 \Delta_{h,1}^{m-1}, \tag{4}$$

if $m = 2, 3, \ldots$, and their obvious counterparts $\Delta_{h,2}^m$. Let $\bar{a} = (a_1, a_2)$ with $0 < a_2 \leq a_1 < \infty$ and $a_1 + a_2 = 2$ be a given anisotropy. Let $\bar{s} = (s_1, s_2)$ with $s_1 = \frac{s}{a_1}$, $s_2 = \frac{s}{a_2}$ and $s > 0$. Let $1 < p < \infty$. If m_1 and m_2 are natural numbers with $m_1 > s_1$ and $m_2 > s_2$ then

$$\|f \mid B_p^{\bar{s}}(R_2)\| = \|f \mid L_p(R_2)\| + \left(\int_{R_1} [|h|^{-s_1 p} \|(\Delta_{h,1}^{m_1} f)(\cdot) \mid L_p(R_2)\|^p + |h|^{-s_2 p} \|(\Delta_{h,2}^{m_2} f)(\cdot) \mid L_p(R_2)\|^p] \frac{dh}{|h|} \right)^{1/p} \tag{5}$$

is an equivalent norm in $B_p^{\bar{s}}(R_2)$. Of course, the left-hand side of (5) depends on m_1 and m_2. But all these norms are mutually equivalent. By the usual abuse of notations this justifies to omit the indices m_1 and m_2 on the left-hand side of (5). Similar as in the isotropic case one can try to replace the one-dimensional differences $\Delta_{h,1}^{m_1}$ and $\Delta_{h,2}^{m_2}$ in (5) by two-dimensional differences. Although we shall not need such an assertion in the sequel we formulate a result which may be found in Tran Duc Long, H. Triebel [1]. We recall that

$$(\Delta_h f)(x) = f(x + h) - f(x), \quad x \in R_2 \text{ and } h \in R_2 \tag{6}$$

are the usual differences in R_2. If $0 < s_1 \leq s_2 < 1$ and if the other above hypotheses are satisfied then

$$\|f \mid L_p(R_2)\| + \left(\int_{R_2} |h|_{\bar{a}}^{-sp} \|(\Delta_h f)(\cdot) \mid L_p(R_2)\|^p \frac{dh}{|h|_{\bar{a}}^2} \right)^{1/p}$$

is an equivalent norm in $B_p^{\bar{s}}(R_2)$. Further equivalent norms may be found in S. M. Nikol'skij [2; 4.3.4], O. V. Besov, V. P. Il'in, S. M. Nikol'skij [1] and H.-J. Schmeisser [7]. In particular, in the same way as in the isotropic case on can replace some differences in (5) by corresponding derivatives.

4.2.3. Embeddings and Traces

Let the anisotropy $\bar{a} = (a_1, a_2)$ with $0 < a_2 \leq a_1 < \infty$ and $a_1 + a_2 = 2$ be given. Let $1 < p < \infty$ and $\bar{s} = (s_1, s_2)$ with $s_1 = \frac{s}{a_1}$, $s_2 = \frac{s}{a_2}$ and $s > 0$, where s is the mean smoothness. We formulate some well-known embedding theorems and trace theorems for the anisotropic Besov spaces $B_p^{\bar{s}}(R_2)$ and the anisotropic Bessel-potential spaces $H_p^{\bar{s}}(R_2)$. We recall that $W_p^{\bar{s}}(R_2) = H_p^{\bar{s}}(R_2)$ are anisotropic Sobolev spaces if s_1 and s_2 in $\bar{s} = (s_1, s_2)$ are natural numbers.

Embeddings. Let $C(R_2)$ be the collection of all complex-valued uniformly continuous and bounded functions in R_2, normed via

$$\|f \mid C(R_2)\| = \sup_{x \in R_2} |f(x)|. \tag{1}$$

The numbers p, s_1 and s_2 have the above meaning. *If m_1 and m_2 are non-negative integers with*

$$\frac{1}{s_1}\left(m_1 + \frac{1}{p}\right) + \frac{1}{s_2}\left(m_2 + \frac{1}{p}\right) < 1, \tag{2}$$

then there exists a positive number c such that

$$\left\| \frac{\partial^{m_1+m_2} f}{\partial x_1^{m_1} \partial x_2^{m_2}} \mid C(R_2) \right\| \leqq c \| f \mid B_p^{\bar{s}}(R_2) \| \tag{3}$$

and

$$\left\| \frac{\partial^{m_1+m_2} f}{\partial x_1^{m_1} \partial x_2^{m_2}} \mid C(R_2) \right\| \leqq c \| f \mid H_p^{\bar{s}}(R_2) \| \tag{4}$$

hold for all $f \in B_p^{\bar{s}}(R_2)$ and all $f \in H_p^{\bar{s}}(R_2)$, respectively. A proof may be found in S. M. Nikol'skij [2; 6.3., 5.6.3.] as far as the Besov spaces are concerned. (4) follows from (3) and $H_p^{\bar{s}}(R_2) \subset B_p^{\varepsilon\bar{s}}(R_2)$ with $0 < \varepsilon < 1$ and $\varepsilon\bar{s} = (\varepsilon s_1, \varepsilon s_2)$. In particular by the usual interpretation we have under the above hypotheses

$$\frac{\partial^{m_1+m_2} f}{\partial x_1^{m_1} \partial x_2^{m_2}}(x) \in C(R_2). \tag{5}$$

Traces on Lines. Let $\sigma > 0$ and $1 < p < \infty$. Then $B_p^\sigma(R_1) = B_{p,p}^\sigma(R_1)$ stands for the nowadays classical Besov spaces. We are interested in traces of the above spaces $B_p^{\bar{s}}(R_2)$ and $H_p^{\bar{s}}(R_2)$ on lines parallel to the axes of coordinates. Let Γ_1 and Γ_2 be the trace-operators defined by

$$(\Gamma_1 f)(x_1) = f(x_1, 0) \quad \text{and} \quad (\Gamma_2 f)(x_2) = f(0, x_2). \tag{6}$$

Furthermore let

$$\sigma_1 = s_1 \left(1 - \frac{1}{ps_2}\right) \quad \text{and} \quad \sigma_2 = s_2 \left(1 - \frac{1}{ps_1}\right). \tag{7}$$

If $\sigma_1 > 0$ then Γ_1 is a retraction

$$\text{from} \quad B_p^{\bar{s}}(R_2) \quad \text{onto} \quad B_p^{\sigma_1}(R_1) \tag{8}$$

and

$$\text{from} \quad H_p^{\bar{s}}(R_2) \quad \text{onto} \quad B_p^{\sigma_1}(R_1),$$

respectively. We recall what is meant by a retraction. Γ_1 is called a retraction if it is a linear and bounded mapping from $B_p^{\bar{s}}(R_2)$ (or $H_p^{\bar{s}}(R_2)$) onto $B_p^{\sigma_1}(R_1)$ and if there exists a linear and bounded mapping Λ_1 from $B_p^{\sigma_1}(R_1)$ into $B_p^{\bar{s}}(R_2)$ (or $H_p^{\bar{s}}(R_2)$) with

$$\Gamma_1 \Lambda_1 = E \text{ (identity mapping in } B_p^{\sigma_1}(R_1)). \tag{9}$$

In other words, the above assertion includes both direct and inverse embedding theorems. A proof of the above assertion may be found in S. M. Nikol'skij [2; 6.7., 6.8., 9.5.]. Obviously, if $\sigma_2 > 0$ then Γ_2 is a retraction

$$\text{from} \quad B_p^{\bar{s}}(R_2) \quad \text{onto} \quad B_p^{\sigma_2}(R_1) \tag{10}$$

and

$$\text{from} \quad H_p^{\bar{s}}(R_2) \quad \text{onto} \quad B_p^{\sigma_2}(R_1). \tag{11}$$

Derivatives. Let the above hypotheses be satisfied and let l be a natural number with $l < s_2$. Then exists a positive number c such that

$$\left\| \frac{\partial^l f}{\partial x_2^l}(\cdot) \mid B_p^{\left(s_1 - \frac{s_1}{s_2} l,\, s_2 - l\right)}(R_2) \right\| \leqq c \| f \mid B_p^{\bar{s}}(R_2) \| \tag{12}$$

holds for all $f \in B_p^{\bar{s}}(R_2)$, cf. S. M. Nikol'skij [2; 5.6.3.]. If one combines (12) with (7), (8), then it follows that

$$\left\| \frac{\partial^l f}{\partial x_2^l}(\cdot, 0) \mid B_p^{s_1 - \frac{s_1}{s_2}\left(l + \frac{1}{p}\right)}(R_1) \right\| \leqq c \| f \mid B_p^{\bar{s}}(R_2) \| \tag{13}$$

holds, provided that $l < s_2 - \dfrac{1}{p}$. Similarly we have

$$\left\| \frac{\partial^l f}{\partial x_2^l}(\cdot) \mid H_p^{\left(s_1 - \frac{s_1}{s_2} l,\, s_2 - l\right)}(R_2) \right\| \leqq c \| f \mid H_p^{\bar{s}}(R_2) \|, \tag{14}$$

cf. S. M. Nikol'skij [2; 9.2.2.] and

$$\left\| \frac{\partial^l f}{\partial x_2^l}(\cdot, 0) \mid B_p^{s_1 - \frac{s_1}{s_2}\left(l + \frac{1}{p}\right)}(R_1) \right\| \leqq c \| f \mid H_p^{\bar{s}}(R_2) \|, \tag{15}$$

the latter under the additional assumption $l < s_2 - \dfrac{1}{p}$.

4.2.4. Interpolation and Duality

Let the anisotropy $\bar{a} = (a_1, a_2)$ with $0 < a_2 \leqq a_1 < \infty$ and $a_1 + a_2 = 2$ be given. Let $1 < p < \infty$ and $\bar{s} = (s_1, s_2)$ with $s_1 = \dfrac{s}{a_1}$, $s_2 = \dfrac{s}{a_2}$ and $-\infty < s < \infty$. It will be useful to describe some interpolation results and duality formulas.

Interpolation. Let A_0 and A_1 be two Banach spaces (more precisely: an interpolation couple). Then $(A_0, A_1)_{\theta, q}$ with $0 < \theta < 1$ and $1 \leqq q \leqq \infty$ denotes the real interpolation method by Lions-Peetre. Extensive descriptions may be found in [I] or J. Bergh, J. Löfström [1]. $\bar{a}$, $\bar{s}$, and p have the above meaning. If $\lambda \in R_1$ then we put $\lambda\bar{s} = (\lambda s_1, \lambda s_2)$. Let $\lambda_0 \in R_1$, $\lambda_1 \in R_1$ and $\lambda_0 \neq \lambda_1$. Let $0 < \theta < 1$, $s \neq 0$ (i.e. $\bar{s} \neq (0, 0)$) and $\lambda = (1 - \theta)\lambda_0 + \theta\lambda_1$. Then

$$(B_p^{\lambda_0 \bar{s}}(R_2), B_p^{\lambda_1 \bar{s}}(R_2))_{\theta, p} = (H_p^{\lambda_0 \bar{s}}(R_2), H_p^{\lambda_1 \bar{s}}(R_2))_{\theta, p} = B_p^{\lambda \bar{s}}(R_2). \tag{1}$$

The proof can be based on Definition 4.2.1 and is essentially the same as in the isotropic case, cf. [I, 2.4.1.]. We refer also to [I, 2.13.2.] and H.-J. Schmeisser [7, p. 70].

Duality. In [I, 2.6] we determined the dual spaces of the isotropic Besov spaces and the isotropic Bessel-potential spaces in the sense of the dual pairing $(S(R_2), S'(R_2))$. There is no problem to extend this method to the anisotropic spaces from Definition 4.2.1. $\bar{s}$ and p have the above meaning. Let $\dfrac{1}{p} + \dfrac{1}{p'} = 1$ and $-\bar{s} = (-s_1, -s_2)$. Then we have

$$(B_p^{\bar{s}}(R_2))' = B_{p'}^{-\bar{s}}(R_2) \quad \text{and} \quad (H_p^{\bar{s}}(R_2))' = H_{p'}^{-\bar{s}}(R_2). \tag{2}$$

As far as interpretations are concerned we refer to [I, 2.6].

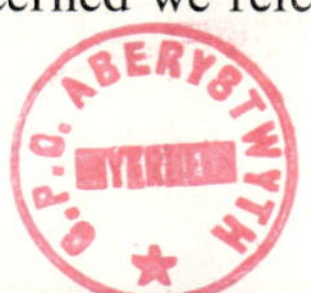

4.3. Anisotropic Inequalities of Hardy Type

4.3.1. Inequalities in One Dimension

First we recall a special case of Hardy's famous inequality, cf. G. H. Hardy, J. E. Littlewood, G. Polya [1, Theorem 330]. *Let $1 < p < \infty$ and let m be a natural number. Then exists a positive number c such that*

$$\int_{R_1} |t|^{-mp} |u(t)|^p \,\mathrm{d}t \leqq c \int_{R_1} \left| \frac{\mathrm{d}^m u(t)}{\mathrm{d}t^m} \right|^p \mathrm{d}t \tag{1}$$

holds for all $u(t) \in S(R_1)$ with $\frac{\mathrm{d}^j u}{\mathrm{d}t^j}(0) = 0$ for $j = 0, \ldots, m-1$. A fractional counterpart of (1) reads as follows. *Let $1 < p < \infty$, $0 < s < 1$ and $s \neq \frac{1}{p}$. Then exists a positive number c such that*

$$\int_{R_1} |t|^{-sp} |u(t)|^p \,\mathrm{d}t \leqq c \int_{R_1 \times R_1} \frac{|u(t) - u(\tau)|^p}{|t-\tau|^{1+sp}} \,\mathrm{d}t \,\mathrm{d}\tau \tag{2}$$

holds for all $u(t) \in S(R_1)$ with $u(0) = 0$ in the case $sp > 1$. A proof of (2) may be found in [I, 3.2.6.]. However the inequality itself is due to J.-L. Lions and P. Grisvard. By standard arguments (based on (2) and an extension of (1)) one obtains the following generalization of (2). *Let $1 < p < \infty$ and let $0 < s = [s] + \{s\}$ where $[s]$ is an integer and $0 < \{s\} < 1$ with $\{s\} \neq \frac{1}{p}$. Then exists a positive number c such that*

$$\int_{R_1} |t|^{-sp} |u(t)|^p \,\mathrm{d}t \leqq c \int_{R_1 \times R_1} \frac{|u^{([s])}(t) - u^{([s])}(\tau)|^p}{|t-\tau|^{1+\{s\}p}} \,\mathrm{d}t \,\mathrm{d}\tau \tag{3}$$

holds for all $u(t) \in S(R_1)$ with $u^{(j)}(0) = 0$ for $j = 0, \ldots, \left[s - \frac{1}{p}\right]$. We need a generalization of (3). $B^s_p(R_1) = B^s_{p,p}(R_1)$ with $1 < p < \infty$ and $s > 0$ are the usual Besov spaces on R_1. We recall that $(\Delta^1_h f)(t) = f(t+h) - f(t)$ if $t \in R_1$ and $h \in R_1$, and $\Delta^m_h = \Delta^{m-1}_h \Delta^1_h$ if $m = 2, 3, \ldots$

Proposition. *Let $1 < p < \infty$. Let $0 < s = [s]^- + \{s\}^+$, where $[s]^-$ is an integer, $0 < \{s\}^+ \leqq 1$ and $\{s\}^+ p \neq 1$. Let k and n be integers with $0 \leqq k \leqq [s]^-$ and $n > s - k$. Then exists a positive number c such that*

$$\int_{R_1} |t|^{-sp} |u(t)|^p \,\mathrm{d}t \leqq c \int_{R_1} |h|^{-(s-k)p} \|(\Delta^n_h u^{(k)})(\cdot) \mid L_p(R_1)\|^p \frac{\mathrm{d}h}{|h|} \tag{4}$$

holds for all $u \in B^s_p(R_1)$ with $u^{(j)}(0) = 0$ for $j = 0, \ldots, \left[s - \frac{1}{p}\right]$.

Proof. We recall that

$$\|u \mid B^s_p(R_1)\| = \|u \mid L_p(R_1)\| + \left(\int_{R_1} |h|^{-(s-k)p} \|(\Delta^n_h u^{(k)})(\cdot) \mid L_p(R_1)\|^p \frac{\mathrm{d}h}{|h|} \right)^{1/p} \tag{5}$$

is an equivalent norm in $B^s_p(R_1)$ (by the usual abuse of notations we omitted k and n

on the left-hand side of (5)). Let us assume that we proved (4) for

$$u(t) \in S(R_1), \quad \operatorname{supp} u \subset [-1, 1] \text{ and } u^{(j)}(0) = 0 \text{ for } j = 0, \ldots, \left[s - \frac{1}{p}\right]. \tag{6}$$

Then one obtains (4) with $u(t) \in C_0^\infty(R_1)$ and $u^{(j)}(0) = 0$ for $j = 0, \ldots, \left[s - \frac{1}{p}\right]$ by a homogeneity argument. Then the assertion of the proposition follows by completion. However under the restriction (6) the first term on the right-hand side of (5) can be omitted, i.e. the $\frac{1}{p}$-power of the right-hand side of (4) is a $B_p^s(R_1)$-norm for those functions. If $s \neq$ integer then (4) (with (6)) follows from (3). If s is a natural number, then one obtains (4) by real interpolation, cf. [I, 2.10.4.] for details.

Remark 1. Inequality (4) can be reformulated in the form

$$\int_{R_1} |t|^{-sp} |u(t)|^p \,\mathrm{d}t \leqq c \|u \mid B_p^s(R_1)\|^p, \tag{7}$$

where again $s > 0$, $s - \frac{1}{p} \neq$ integer, $u(t) \in B_p^s(R_1)$ and $u^{(j)}(0) = 0$ for $j = 0, \ldots, \left[s - \frac{1}{p}\right]$. This follows from (5). An extension of (4) or (7) to $B_p^s(R_1)$ with $0 \leqq s - \frac{1}{p} =$ integer is not possible. The corresponding counterpart of (1) reads as follows,

$$\int_{R_1} |t|^{-sp} |u(t)|^p \,\mathrm{d}t \leqq c \|u \mid W_p^s(R_1)\|^p, \tag{8}$$

where s is a natural number, $1 < p < \infty$, $u(t) \in W_p^s(R_1)$ and $u^{(j)}(0) = 0$ if $j = 0, \ldots, s - 1$. In particular if s is a natural number then we have two inequalities: (1) and (4), or (8) and (7), respectively.

Remark 2. The inequalities (4) and (7) can be extended from $B_p^s(R_1) = B_{p,p}^s(R_1)$ to $B_{p,q}^s(R_1)$, where $B_{p,q}^s(R_1)$ are the classical Besov spaces on the real line with $s > 0$, $1 < p < \infty$ and $1 < q < \infty$: Let $1 < p < \infty$ and $1 < q < \infty$. Let $0 < s = [s]^- + \{s\}^+$ where $[s]^-$ is an integer, $0 < \{s\}^+ \leqq 1$ and $\{s\}^+ p \neq 1$. Let k and n be integers with $0 \leqq k \leqq [s]^-$ and $n > s - k$. Then exists a positive number c such that

$$\int_0^\infty t^{-sq} \left(\int_{-t}^{t} |u(\tau)|^p \,\mathrm{d}\tau \right)^{q/p} \frac{\mathrm{d}t}{t} \leqq c \int_{R_1} |h|^{-(s-k)q} \|(\Delta_h^n u^{(k)})(\cdot) \mid L_p(R_1)\|^q \frac{\mathrm{d}h}{|h|} \tag{9}$$

holds for all $u \in B_{p,q}^s(R_1)$ with $u^{(j)}(0) = 0$ for $j = 0, \ldots, \left[s - \frac{1}{p}\right]$. Furthermore, (9) can be reformulated as

$$\int_0^\infty t^{-sq} \left(\int_{-t}^{t} |u(\tau)|^p \,\mathrm{d}\tau \right)^{q/p} \frac{\mathrm{d}t}{t} \leqq c \|u \mid B_{p,q}^s(R_1)\|^q. \tag{10}$$

(9) and (10) can be obtained from (4) or (7) by real interpolation, cf. also [I, 4.3.2.]. If $p = q$ then (9) and (10) coincide with (4) and (7), respectively.

Remark 3. Finally we mention the following more sophisticated generalization which covers both (7) and (8) (or (4) and (1), respectively). In [T] we studied the spaces $F_{p,q}^s(R_1)$. If $s > 0$, $1 < p < \infty$, $1 < q < \infty$ and $n > s$ then

$$\|u \mid F_{p,q}^s(R_1)\| = \|u \mid L_p(R_1)\| + \left\| \left(\int_0^\infty r^{-sq} \left[\int_{-1}^{1} |(\Delta_{rh}^n u)(\cdot)| \,\mathrm{d}h \right]^q \frac{\mathrm{d}r}{r} \right)^{1/q} \middle| L_p(R_1) \right\| \tag{11}$$

are equivalent norms in $F^s_{p,q}(R_1)$, cf. [T, 2.5.11.]. (Again we omitted n on the left-hand side of (11)). This is the counterpart of (5). There are further equivalent norms, cf. [T, 2.5.11. and 2.5.13.]. We recall that

$$B^s_p(R_1) = B^s_{p,p}(R_1) = F^s_{p,p}(R_1) \quad \text{and} \quad H^s_p(R_1) = F^s_{p,2}(R_1) \tag{12}$$

if $s > 0$ and $1 < p < \infty$. In particular if s is a natural number then $W^s_p(R_1) = F^s_{p,2}(R_1)$ is a Sobolev space. The counterpart of (7) reads as follows. *Let* $1 < p < \infty$, $1 < q < \infty$, $s > 0$ *and* $s - \frac{1}{p} \neq$ *integer. Then exists a positive number c such that*

$$\int_{R_1} |t|^{-sp} |u(t)|^p \, dt \leqq c \|u \mid F^s_{p,q}(R_1)\|^p \tag{13}$$

holds for all $u \in F^s_{p,q}(R_1)$ *with* $u^{(j)}(0) = 0$ *for* $j = 0, \ldots, \left[s - \frac{1}{p}\right]$. One can replace $\|u \mid F^s_{p,q}(R_1)\|^p$ on the right-hand side of (13) by the second term of the right-hand side of (11). One can prove (13) by small modifications of the proofs of Proposition 2.8.6/1 and Step 2 of the proof of Theorem 2.9.2 in [T]. We shall not go into detail. By (12) we have also a modification of (7) with the Bessel-potential spaces $H^s_p(R_1)$ instead of $B^s_p(R_1)$. This generalizes (8). In connection with (13) we refer also to J. Franke [3, 2.2.1].

4.3.2. Inequalities in Two Dimensions

After the preparations in 4.3.1. we come to the main result of Section 4.3. which is the basis for the further considerations in this chapter.

Definition 1. *Let* $\bar{a} = (a_1, a_2)$ *with* $0 < a_2 \leqq a_1 < \infty$ *and* $a_1 + a_2 = 2$ *be the given anisotropy. Let* $1 < p < \infty$ *and* $\bar{s} = (s_1, s_2)$ *with* $s_1 = \frac{s}{a_1}$, $s_2 = \frac{s}{a_2}$ *and* $s > 0$. *Then* $\bar{s}$ *is called critical if there exist non-negative integers* m_1 *and* m_2 *with*

$$\frac{1}{s_1}\left(m_1 + \frac{1}{p}\right) + \frac{1}{s_2}\left(m_2 + \frac{1}{p}\right) = 1. \tag{1}$$

Otherwise $\bar{s}$ *is called non-critical.*

Remark 1. In the isotropic case, i.e. $a_1 = a_2 = 1$, (1) reduces to $0 \leqq s - \frac{2}{p} =$ integer. It is well-known that $0 \leqq s - \frac{2}{p} =$ integer are critical values in the theory of the isotropic Besov spaces and Bessel-potential spaces.

Definition 2. *Let* $1 < p < \infty$ *and let* $\bar{s} = (s_1, s_2)$ *with* $0 < s_1 \leqq s_2 < \infty$ *be non-critical in the sense of Definition* 1. *Then*

$$\dot{B}^{\bar{s}}_p(R_2) = \left\{ f \mid f \in B^{\bar{s}}_p(R_2) \quad \text{and} \quad \frac{\partial^{m_1+m_2} f}{\partial x_1^{m_1} \partial x_2^{m_2}}(0) = 0 \quad \text{if} \quad \frac{1}{s_1}\left(m_1 + \frac{1}{p}\right) + \frac{1}{s_2}\left(m_2 + \frac{1}{p}\right) < 1 \right\} \tag{2}$$

and

$$\dot{H}^{\bar{s}}_p(R_2) = \left\{ f \mid f \in H^{\bar{s}}_p(R_2) \quad \text{and} \quad \frac{\partial^{m_1+m_2} f}{\partial x_1^{m_1} \partial x_2^{m_2}}(0) = 0 \quad \text{if} \quad \frac{1}{s_1}\left(m_1 + \frac{1}{p}\right) + \frac{1}{s_2}\left(m_2 + \frac{1}{p}\right) < 1 \right\}. \tag{3}$$

If in addition s_1 *and* s_2 *are natural numbers then we put* $\dot{W}^{\bar{s}}_p(R_2) = \dot{H}^{\bar{s}}_p(R_2)$.

Remark 2. Of course, m_1 and m_2 in (2) and (3) are non-negative integers. This definition makes sense, cf. (4.2.3/3) and (4.2.3/4). We add a warning. In [T, Chapter 5] we dealt with homogeneous isotropic spaces of Besov-Hardy-Sobolev type which we denoted (following a proposal by J. Peetre) as $\dot{B}^s_{p,q}(R_n)$ and $\dot{F}^s_{p,q}(R_n)$. The above dotted spaces from (2) and (3) have nothing to do with these spaces.

We formulate and prove part (ii) of the following theorem for the Bessel-potential spaces $H^{\bar{s}}_p(R_2)$ where we need (4.3.1/13). However in the sequel only the case $W^{\bar{s}}_p(R_2) = H^{\bar{s}}_p(R_2)$ is of interest, where s_1 and s_2 in $\bar{s} = (s_1, s_2)$ are natural numbers (anisotropic Sobolev spaces), and then (4.3.1/8) is sufficient.

Theorem. *Let $\bar{a} = (a_1, a_2)$ with $0 < a_2 \leqq a_1 < \infty$ and $a_1 + a_2 = 2$ be the given anisotropy. Let $|x|_{\bar{a}}$ be the anisotropic distance from* (4.2.1/1). *Let $1 < p < \infty$. Let $\bar{s} = (s_1, s_2)$ with $s_1 = \dfrac{s}{a_1}$, $s_2 = \dfrac{s}{a_2}$ and $s > 0$ be non-critical.*

(i) *There exists a positive number c such that*

$$\int_{R_2} |x|_{\bar{a}}^{-sp}\, |f(x)|^p \,\mathrm{d}x \leqq c \|f \mid B^{\bar{s}}_p(R_2)\|^p \tag{4}$$

holds for all $f \in \dot{B}^{\bar{s}}_p(R_2)$.

(ii) *There exists a positive number c such that*

$$\int_{R_2} |x|_{\bar{a}}^{-sp}\, |f(x)|^p \,\mathrm{d}x \leqq c \|f \mid H^{\bar{s}}_p(R_2)\|^p \tag{5}$$

holds for all $f \in \dot{H}^{\bar{s}}_p(R_2)$.

Proof. Step 1. Let $0 < s_1 < \dfrac{1}{p}$ and $f \in S(R_2)$. Then (4.3.1/4) yields

$$\int_{R_1} |x_1|^{-s_1 p}\, |f(x_1, x_2)|^p \,\mathrm{d}x_1$$

$$\leqq c \int_{R_1} |h|^{-s_1 p}\, \|(\Delta^1_{h,1} f)(\cdot, x_2) \mid L_p(R_1)\|^p \frac{\mathrm{d}h}{|h|}, \tag{6}$$

where c is independent of $x_2 \in R_1$. We integrate over $x_2 \in R_1$ and observe that $|x|_{\bar{a}}^{-sp} \leqq |x_1|^{-s_1 p}$. Then (4) follows from (4.2.2/5) and a completion argument. By (4.3.1/12) and (4.3.1/13) we have

$$\int_{R_1} |x_1|^{-s_1 p}\, |f(x_1, x_2)|^p \,\mathrm{d}x_1 \leqq c \int_{R_1} |(F_1^{-1}(1 + |\xi_1|^2)^{s_1/2} F_1 f)(x_1, x_2)|^p \,\mathrm{d}x_1, \tag{7}$$

where F_1 and F_1^{-1} are the Fourier transform and its inverse on R_1. Then (5) follows from (7) and (4.2.2/2) in the same way as above.

Step 2. Let $\dfrac{1}{p} \leqq s_1 \leqq s_2$ and $s_2 - \dfrac{1}{p} \neq$ integer. Then we have $m + \dfrac{1}{p} < s_2 < m + 1 + \dfrac{1}{p}$ where m is an appropriate non-negative integer. Let $f \in \dot{B}^{\bar{s}}_p(R_2)$, where we may assume that $f \in S(R_2)$, cf. Remark 3 below. We decompose $f(x)$

by

$$f(x_1, x_2) = \sum_{k=0}^{m} \chi(x_2) \frac{x_2^k}{k!} \frac{\partial^k f}{\partial x_2^k}(x_1, 0) + g(x_1, x_2), \quad x = (x_1, x_2) \in R_2, \tag{8}$$

where $\chi(t) \in S(R_1)$ has a compact support and $\chi(t) = 1$ if $|t| \leqq 1$. We fix x_1 and apply (4.3.1/4) (with modified indices) to $u(t) = g(x_1, t) \in S(R_1)$. We have

$$\begin{aligned} \int_{R_1} |x_2|^{-s_2 p} |g(x_1, x_2)|^p \, dx_2 &\leqq c \int_{R_1} |h|^{-s_2 p} \, \|(\Delta_{h,2}^{m+2} g)(x_1, \cdot) \mid L_p(R_1)\|^p \frac{dh}{|h|} \\ &\leqq c' \int_{R_1} |h|^{-s_2 p} \, \|(\Delta_{h,2}^{m+2} f)(x_1, \cdot) \mid L_p(R_1)\|^p \frac{dh}{|h|} + c' \sum_{k=0}^{m} \left| \frac{\partial^k f}{\partial x_2^k}(x_1, 0) \right|^p, \end{aligned} \tag{9}$$

cf. (8). We integrate (9) over $x_1 \in R_1$. Because $|x_2|^{-s_2 p} \geqq |x|_{\bar{a}}^{-sp}$ we obtain

$$\begin{aligned} \int_{R_2} |x|_{\bar{a}}^{-sp} |g(x)|^p \, dx &\leqq c \|f \mid B_p^{\bar{s}}(R_2)\|^p + c \sum_{k=0}^{m} \left\| \frac{\partial^k f}{\partial x_2^k}(x_1, 0) \mid L_p(R_1) \right\|^p \\ &\leqq c \|f \mid B_p^{\bar{s}}(R_2)\|^p, \end{aligned} \tag{10}$$

cf. (4.2.2/5) and (4.2.3/13). Next we deal with the terms $\chi(x_2)\, x_2^k \dfrac{\partial^k f}{\partial x_2^k}(x_1, 0)$ in (8) where $k = 0, \ldots, m$. We have

$$\begin{aligned} &\int_{R_2} |x|_{\bar{a}}^{-sp} |x_2|^{kp} \left| \frac{\partial^k f}{\partial x_2^k}(x_1, 0) \right|^p dx \\ &= \int_{R_1} \left| \frac{\partial^k f}{\partial x_2^k}(x_1, 0) \right|^p \left[\int_{R_1} |x_2|^{kp} \left(|x_1|^{2/a_1} + |x_2|^{2/a_2} \right)^{-sp/2} dx_2 \right] dx_1 . \end{aligned} \tag{11}$$

Because $s_2 > m + \dfrac{1}{p}$ the integral in the brackets $[\ldots]$ converges if $x_1 \neq 0$ and equals $c|x_1|^{-s_1 p + \frac{s_1}{s_2} + \frac{s_1}{s_2} kp}$ for some $c > 0$. Then we obtain

$$\int_{R_2} |x|_{\bar{a}}^{-sp} |x_2|^{kp} \left| \frac{\partial^k f}{\partial x_2^k}(x_1, 0) \right|^p dx = c \int_{R_1} |x_1|^{-\sigma_k p} \left| \frac{\partial^k f}{\partial x_2^k}(x_1, 0) \right|^p dx_1 \tag{12}$$

with $\sigma_k = s_1 - \dfrac{s_1}{p s_2} - \dfrac{s_1}{s_2} k$. We have

$$\frac{\sigma_k}{s_1} = 1 - \frac{1}{s_2} \left(k + \frac{1}{p} \right) > 0 \tag{13}$$

and

$$\frac{1}{s_1} \left(\sigma_k - \frac{1}{p} + \frac{1}{p} \right) + \frac{1}{s_2} \left(k + \frac{1}{p} \right) = 1 . \tag{14}$$

Because $\bar{s}$ is non-critical it follows $\sigma_k - \dfrac{1}{p} \neq$ integer, cf. (13), (14), and (1). By (14) and (2) we have

$$\frac{\partial^n}{\partial x_1^n} \left(\frac{\partial^k f}{\partial x_2^k} \right)(0, 0) = 0 \quad \text{if} \quad n = 0, \ldots, \left[\sigma_k - \frac{1}{p} \right].$$

We apply (4.3.1/4) (with σ_k instead of s) to $u(t) = \frac{\partial^k f}{\partial x_2^k}(t, 0)$. Then (12) yields

$$\int_{R_2} |x|_{\bar{a}}^{-sp} |x_2|^{kp} \left|\frac{\partial^k f}{\partial x_2^k}(x_1, 0)\right|^p \mathrm{d}x \leqq c \left\| \frac{\partial^k f}{\partial x_2^k}(\cdot, 0) \mid B_p^{\sigma_k}(R_1)\right\|^p. \tag{15}$$

Because $0 \leqq k \leqq m < s_2 - \frac{1}{p}$ we can apply (4.2.3/13) and obtain

$$\int_{R_2} |x|_{\bar{a}}^{-sp} |x_2|^{kp} \left|\frac{\partial^k f}{\partial x_2^k}(x_1, 0)\right|^p \mathrm{d}x \leqq c\|f \mid B_p^{\bar{s}}(R_2)\|^p. \tag{16}$$

Now (4) follows from (8), (10) and (16).

Step 3. Let $\frac{1}{p} \leqq s_1 \leqq s_2$ and $s_2 - \frac{1}{p} =$ integer. Because $\bar{s}$ is non-critical, the admissible numbers m_1 and m_2 in (2) remain unchanged if one replaces $\bar{s}$ by $(1 + \varepsilon)\,\bar{s}$ or $(1 - \varepsilon)\,\bar{s}$ with $\varepsilon > 0$ small. Then it follows from (4.2.4/1) and the interpolation theory for finite co-dimensional subspaces from [I, 1.17.1.] that

$$(\dot{B}_p^{(1+\varepsilon)\bar{s}}(R_2), \dot{B}_p^{(1-\varepsilon)\bar{s}}(R_2))_{1/2, p} = \dot{B}_p^{\bar{s}}(R_2). \tag{17}$$

By Step 2 we know (4) with $(1 + \varepsilon)\, s$, $(1 + \varepsilon)\,\bar{s}$ and $(1 - \varepsilon)\, s$, $(1 - \varepsilon)\,\bar{s}$ instead of s and $\bar{s}$, respectively. Then (4) follows from (17) and a corresponding interpolation formula for weighted L_p-spaces. The proof of (i) is complete.

Step 4. We indicate the necessary modifications in order to prove (ii) if $\frac{1}{p} \leqq s_1 \leqq s_2$. First we again assume $m + \frac{1}{p} < s_2 < m + 1 + \frac{1}{p}$. We use (8) and (4.3.1/13) with $q = 2$ (instead of (4.3.1/4)). Then the counterpart of (10) reads as follows,

$$\int_{R_2} |x|_{\bar{a}}^{-sp} |g(x)|^p \,\mathrm{d}x \leqq c\|f \mid H_p^{\bar{s}}(R_2)\|^p. \tag{18}$$

We take over (11)–(15) and use (4.2.3/15) instead of (4.2.3/13). This proves (16) with $H_p^{\bar{s}}(R_2)$ instead of $B_p^{\bar{s}}(R_2)$ and consequently (5). If $s_2 - \frac{1}{p} =$ integer then we use the complex interpolation formula

$$[\dot{H}_p^{(1+\varepsilon)\bar{s}}(R_2), \dot{H}_p^{(1-\varepsilon)\bar{s}}(R_2)]_{1/2} = \dot{H}_p^{\bar{s}}(R_2) \tag{19}$$

with $\varepsilon > 0$ small.

Remark 3. Let $f_j(x) \in S(R_2)$ be an approximating sequence of a given function $f \in \dot{B}_p^{\bar{s}}(R_2)$, where the parameters satisfy the hypotheses of the theorem. Then $\frac{\partial^{m_1+m_2} f_j}{\partial x_1^{m_1}\,\partial x_2^{m_2}}(0) \to 0$ if $j \to \infty$ and

$$\frac{1}{s_1}\left(m_1 + \frac{1}{p}\right) + \frac{1}{s_2}\left(m_2 + \frac{1}{p}\right) < 1. \tag{20}$$

Obviously there exist functions $h_j(x) \in S(R_2)$ with $h_j \to 0$ in $B_p^{\bar{s}}(R_2)$ and $\frac{\partial^{m_1+m_2}(f_j - h_j)}{\partial x_1^{m_1}\,\partial x_2^{m_2}}(0) = 0$.

Then $f_j - h_j$ approximates f in $\dot{B}_p^{\bar{s}}(R_2)$. Hence $S(R_2) \cap \dot{B}_p^{\bar{s}}(R_2)$ is dense in $\dot{B}_p^{\bar{s}}(R_2)$. This justifies our assumption at the beginning of Step 2 of the above proof.

Remark 4. On the one hand one can generalize (5) to corresponding spaces $\dot{F}_{p,q}^{\bar{s}}(R_2)$ with $1 < p < \infty$, $1 < q < \infty$ and $\bar{s}$ non-critical. This is based on (4.3.1/13). On the other hand we are mainly interested in the case $\dot{H}_p^{\bar{s}}(R_2) = \dot{W}_p^{\bar{s}}(R_2)$, where s_1 and s_2 in $\bar{s} = (s_1, s_2)$ are natural numbers. In this case the proof of (5) is much more simpler: One does not need Step 1 and (19). Furthermore, (4.3.1/1) instead of (4.3.1/13) is sufficient.

4.4. The Spaces $\dot{B}_p^{\bar{s}}(R_2)$ and $\dot{W}_p^{\bar{s}}(R_2)$

4.4.1. The System $\Psi^{\bar{a}}(R_2)$, an Inequality

Let $\bar{a} = (a_1, a_2)$ with $0 < a_2 \leqq a_1 < \infty$ and $a_1 + a_2 = 2$ be the given anisotropy. Let $|x|_{\bar{a}}$ be the anisotropic distance from (4.2.1/1). Let $\dot{R}_2 = R_2 - \{0\}$. Let $\Psi^{\bar{a}}(R_2)$ be the collection of all systems $\psi = \{\psi_j(x)\}_{j=0}^{\infty}$ of infinitely differentiable functions on R_2 with the following properties:

(i) $\operatorname{supp} \psi_0 \subset \{x \mid |x|_{\bar{a}} > \tfrac{1}{2}\}$,

$\operatorname{supp} \psi_j \subset \{x \mid 2^{-j-1} < |x|_{\bar{a}} < 2^{-j+1}\}$ if $j = 1, 2, 3, \ldots$

(ii) For every multi-index $\alpha = (\alpha_1, \alpha_2)$ there exists a positive number c_α with

$$2^{-ja_1\alpha_1 - ja_2\alpha_2} |D^\alpha \psi_j(x)| \leqq c_\alpha \tag{1}$$

for any $j = 0, 1, 2, \ldots$ and $x \in R_2$.

(iii) $$\sum_{j=0}^{\infty} \psi_j(x) = 1 \text{ if } x \in \dot{R}_2. \tag{2}$$

Cf. with the system $\Phi^{\bar{a}}(R_2)$ from 4.2.1.

Proposition. *Let $s > 0$ and $\bar{s} = (s_1, s_2)$ with $s_1 = \dfrac{s}{a_1}$ and $s_2 = \dfrac{s}{a_2}$. Let $\psi \in \Psi^{\bar{a}}(R_2)$ and let $1 < p < \infty$.*

(i) *There exists a positive number c such that*

$$\sum_{j=0}^{\infty} \|\psi_j f \mid B_p^{\bar{s}}(R_2)\|^p \leqq c \| |x|_{\bar{a}}^{-s} f \mid L_p(R_2)\|^p + c\|f \mid B_p^{\bar{s}}(R_2)\|^p \tag{3}$$

holds for any $f \in B_p^{\bar{s}}(R_2)$ for which the right-hand side of (3) *is finite.*

(ii) *Let in addition s_1 and s_2 be two natural numbers. There exists a positive number c such that*

$$\sum_{j=0}^{\infty} \|\psi_j f \mid W_p^{\bar{s}}(R_2)\|^p \leqq c \| |x|_{\bar{a}}^{-s} f \mid L_p(R_2)\|^p + c\|f \mid W_p^{\bar{s}}(R_2)\|^p \tag{4}$$

holds for any $f \in W_p^{\bar{s}}(R_2)$ for which the right-hand side of (4) *is finite.*

Proof. Step 1. We prove (i). Let $m_1 > s_1$, $b > 0$ and $j = 1, 2, \ldots$ Then we have

$$\begin{aligned} &\int_{R_1} |h|^{-s_1 p} \int_{R_2} |(\Delta_{h,1}^{m_1} \psi_j f)(x)|^p \, dx \frac{dh}{|h|} \\ &\leqq c \int_{|h| \geqq b2^{-ja_1}} |h|^{-s_1 p} \|\psi_j f \mid L_p(R_2)\|^p \frac{dh}{|h|} \\ &\quad + \int_{|h| \leqq b2^{-ja_1}} |h|^{-s_1 p} \int_{R_2} |(\Delta_{h,1}^{m_1} \psi_j f)(x)|^p \, dx \frac{dh}{|h|}. \end{aligned} \tag{5}$$

The first term on the right-hand side can be estimated from above by

$$c \cdot 2^{ja_1 s_1 p} \|\psi_j f \mid L_p(R_2)\|^p \leqq c' \| |x|_{\bar{a}}^{-s} \psi_j f \mid L_p(R_2)\|^p, \tag{6}$$

where c and c' depend on b (and m_1) but not on j and f. By mathematical induction we have

$$(\Delta^{m_1}_{h,1}\psi_j f)(x) = \sum_{k=0}^{m_1} c_k(\Delta^k_{h,1} f)(x)(\Delta^{m_1-k}_{h,1}\psi_j)(x_1 + kh, x_2), \tag{7}$$

where the c_k's are appropriate numbers. We choose the above number $b > 0$ small. If $|h| \leqq b \cdot 2^{-ja_1}$ then (7) yields

$$|(\Delta^{m_1}_{h,1}\psi_j f)(x)| \leqq c \sum_{k=0}^{m_1} |(\Delta^k_{h,1} f)(x)| \, 2^{ja_1(m_1-k)} \, |h|^{m_1-k}$$

if $2^{-j-2} < |x|_{\bar{a}} < 2^{-j+2}$ and $(\Delta^{m_1}_{h,1}\psi_j f)(x) = 0$ otherwise. We put these estimates in the second term on the right-hand side of (5) and obtain

$$\begin{aligned}
&\int\limits_{|h| \leqq b2^{-ja_1}} |h|^{-s_1 p} \int\limits_{R_2} |(\Delta^{m_1}_{h,1}\psi_j f)(x)|^p \, dx \frac{dh}{|h|} \\
&\leqq c \sum_{0 \leqq k < m_1 - s_1} 2^{ja_1(m_1-k)p} \int\limits_{|h| \leqq b2^{-ja_1}} |h|^{-s_1 p + (m_1-k)p} \frac{dh}{|h|} \int\limits_{2^{-j-3} < |x|_{\bar{a}} < 2^{-j+3}} |f(x)|^p \, dx \\
&\quad + c \sum_{m_1 \geqq k \geqq m_1 - s_1} 2^{ja_1(m_1-k)p} \int\limits_{|h| \leqq b2^{-ja_1}} |h|^{-s_1 p + (m_1-k)p} \int\limits_{2^{-j-2} < |x|_{\bar{a}} < 2^{-j+2}} |(\Delta^k_{h,1} f)(x)|^p \, dx \frac{dh}{|h|}.
\end{aligned} \tag{8}$$

The first term on the right-hand side of (8) can be estimated in the same way as in (6). The second term on the right-hand side of (8) can be estimated from above by

$$c \sum_{m_1 \geqq k \geqq m_1 - s_1} \int\limits_{|h| \leqq b2^{-ja_1}} |h|^{-s_1 p} \int\limits_{2^{-j-2} < |x|_{\bar{a}} < 2^{-j+2}} |(\Delta^k_{h,1} f)(x)|^p \, dx \frac{dh}{|h|}. \tag{9}$$

We may assume that m_1 is large. Then $k \geqq m_1 - s_1$ in (9) is also large. In particular we assume that $k > s_1$. Hence by (5), (6), (8) and (9) we have

$$\begin{aligned}
&\int\limits_{R_1} |h|^{-s_1 p} \int\limits_{R_1} |(\Delta^{m_1}_{h,1}\psi_j f)(x)|^p \, dx \frac{dh}{|h|} \\
&\leqq c \sum_{r=-3}^{3} \| |x|_{\bar{a}}^{-s} \psi_{j+r} f \mid L_p(R_2) \|^p \\
&\quad + \sum_{m_1 \geqq k > s_1} \int\limits_{R_1} |h|^{-s_1 p} \int\limits_{2^{-j-2} < |x|_{\bar{a}} < 2^{-j+2}} |(\Delta^k_{h,1} f)(x)|^p \, dx \frac{dh}{|h|}
\end{aligned} \tag{10}$$

(with $\psi_{j+r} = 0$ if $j + r < 0$). Of course, one has a similar estimate with respect to the x_2-direction. We use (4.2.2/5). Then summation over j in (10) and its x_2-counterpart yields (3).

Step 2. We prove (ii). We have

$$\left\| \frac{\partial^{s_1}}{\partial x_1^{s_1}} \psi_j f \mid L_p(R_2) \right\|^p \leq c \sum_{k=0}^{s_1} 2^{j(s_1-k)a_1 p} \int\limits_{2^{-j-1}<|x|_{\bar{a}}<2^{-j+1}} \left| \frac{\partial^k f}{\partial x_1^k}(x) \right|^p \mathrm{d}x$$

$$\leq c' \int\limits_{2^{-j-1}<|x|_{\bar{a}}<2^{-j+1}} \left| \frac{\partial^{s_1} f}{\partial x_1^{s_1}}(x) \right|^p \mathrm{d}x$$

$$+ c' \cdot 2^{js_1 a_1 p} \int\limits_{2^{-j-1}<|x|_{\bar{a}}<2^{-j+1}} |f(x)|^p \,\mathrm{d}x, \tag{11}$$

where the latter inequality follows from a corresponding inequality with $j = 1$ and a homogeneity argument. We have a similar estimate with respect to the x_2-direction. Summation over j yields (4).

4.4.2. The Spaces $\dot{B}_p^{\bar{s}}(R_2)$ and $\dot{W}_p^{\bar{s}}(R_2)$ with $s > 0$ and $\bar{s}$ Non-Critical

We study the spaces from Definition 4.3.2/2. We introduce

$$C_0^\infty(\dot{R}_2) = \{f \mid f \in S(R_2), \operatorname{supp} f \text{ is compact}, 0 \notin \operatorname{supp} f\}. \tag{1}$$

We recall that $\dot{R}_2 = R_2 - \{0\}$.

Proposition. *Let $1 < p < \infty$ and let $\bar{s} = (s_1, s_2)$ with $0 < s_1 \leq s_2 < \infty$ be non-critical. Then $C_0^\infty(\dot{R}_2)$ is dense in $\dot{B}_p^{\bar{s}}(R_2)$ and (if in addition s_1 and s_2 are natural numbers) dense in $\dot{W}_p^{\bar{s}}(R_2)$.*

Proof. We prove the assertion for the spaces $\dot{B}_p^{\bar{s}}(R_2)$. By Remark 4.3.2/3 it is sufficient to approximate functions $f \in S(R_2) \cap \dot{B}_p^{\bar{s}}(R_2)$ which have a compact support by functions from $C_0^\infty(\dot{R}_2)$. Let $f(x)$ be such a function. Let

$$f(x) = \sum c_{m_1, m_2} x_1^{m_1} x_2^{m_2} \varkappa(x) + g(x)$$

where the finite sum $\sum$ stands for the beginning of the Taylor series of $f(x)$ at the origin with

$$1 < \frac{1}{s_1}\left(m_1 + \frac{1}{p}\right) + \frac{1}{s_2}\left(m_2 + \frac{1}{p}\right) < N, \tag{2}$$

where N is an arbitrary number. $\varkappa(x) \in S(R_2)$ has a compact support and $\varkappa(x) = 1$ if, say, $|x| \leq 1$. If N is large then $g(x)$ can be approximated in $W_p^M(R_2)$ by functions from $C_0^\infty(\dot{R}_2)$, where M is at our disposal. If M is large then this is also an approximation in $B_p^{\bar{s}}(R_2)$. Hence it is sufficient to approximate $x_1^{m_1} x_2^{m_2} \varkappa(x)$ with (2) in the desired way. Let $\varrho(t) \in S(R_1)$ with $\varrho(t) = 1$ if $|t| \leq 1$ and $\varrho(t) = 0$ if $|t| \geq 2$. Let $\varrho_j(x) = \varrho(2^{ja_1} x_1)\, x_1^{m_1}\, \varrho(2^{ja_2} x_2)\, x_2^{m_2}$ where $\bar{a} = (a_1, a_2)$ is the given anisotropy. We use (4.2.2/5). If $M_1 > s_1$ then we have

$$\int\limits_{R_1} |h|^{-s_1 p} \, \|(\Delta_{h,1}^{M_1} \varrho_j)(\cdot) \mid L_p(R_2)\|^p \frac{\mathrm{d}h}{|h|}$$

$$\leq c \cdot 2^{-ja_2(m_2 p+1)} \, \|\varrho(2^{ja_1} x_1)\, x_1^{m_1} \mid B_p^{s_1}(R_1)\|^p. \tag{3}$$

By straightforward calculations we have

$$\|\varrho(2^{ja_1} x_1)\, x_1^{m_1} \mid W_p^k(R_1)\|^p \leq c \cdot 2^{-ja_1(m_1 p - kp + 1)}$$

if $k = 0, 1, 2, \ldots$ It follows by interpolation or the so-called multiplicative inequalities that the $B_p^{s_1}$-norm on the right-hand side of (3) can be estimated from above by $c \cdot 2^{-ja_1(m_1 p - s_1 p + 1)}$. Then (4.2.2/5), (3) and its x_2-counterpart yield

$$\|\varrho_j(x) \mid B_p^{\bar{s}}(R_2)\|^p \leqq c \cdot 2^{-spj\left[\frac{1}{s_1}\left(m_1+\frac{1}{p}\right)+\frac{1}{s_2}\left(m_2+\frac{1}{p}\right)\right]} 2^{spj}.$$

By (2) we have $\varrho_j(x) \to 0$ in $B_p^{\bar{s}}(R_2)$ if $j \to \infty$. This shows that $\varkappa(x)\, x_1^{m_1} x_2^{m_2}$ can be approximated in $B_p^{\bar{s}}(R_2)$ by functions from $C_0^\infty(\dot{R}_2)$. This proves the proposition for the spaces $\dot{B}_p^{\bar{s}}(R_2)$. The proof for the spaces $\dot{W}_p^{\bar{s}}(R_2)$ is the same.

Theorem. *Let* $1 < p < \infty$ *and let* $\bar{s} = (s_1, s_2)$ *with* $0 < s_1 \leqq s_2 < \infty$ *be non-critical. Let* $\psi \in \Psi^{\bar{a}}(R_2)$.

(i) *Then*

$$\|f \mid \dot{B}_p^{\bar{s}}(R_2)\|^\psi = \left(\sum_{j=0}^{\infty} \|\psi_j f \mid B_p^{\bar{s}}(R_2)\|^p\right)^{1/p} \tag{4}$$

is an equivalent norm in $\dot{B}_p^{\bar{s}}(R_2)$.

(ii) *If in addition* s_1 *and* s_2 *are natural numbers then*

$$\|f \mid \dot{W}_p^{\bar{s}}(R_2)\|^\psi = \left(\sum_{j=0}^{\infty} \|\psi_j f \mid W_p^{\bar{s}}(R_2)\|^p\right)^{1/p} \tag{5}$$

is an equivalent norm in $\dot{W}_p^{\bar{s}}(R_2)$.

Proof. We prove (i). By the above Proposition we may assume $f \in C_0^\infty(\dot{R}_2)$. In particular $f(x) = \sum_{j=0}^{\infty} \psi_j(x)\, f(x)$. Let $m_1 > s_1$. Then exists a natural number N such that for any choice of $x \in R_2$ and $h \in R_1$ at most N terms $(\Delta_{h,1}^{m_1} \psi_j f)(x)$ are different from zero. Hence

$$\int_{R_1} |h|^{-s_1 p} \int_{R_2} |(\Delta_{h,1}^{m_1} f)(x)|^p \, dx \frac{dh}{|h|} \leqq c \sum_{j=0}^{\infty} \int_{R_1} |h|^{-s_1 p} \int_{R_2} |(\Delta_{h,1}^{m_1} \psi_j f)(x)|^p \, dx \frac{dh}{|h|}. \tag{6}$$

By (4.2.2/5) and the x_2-counterpart of (6) we have

$$\|f \mid B_p^{\bar{s}}(R_2)\| \leqq c \|f \mid \dot{B}_p^{\bar{s}}(R_2)\|^\psi. \tag{7}$$

The converse assertion follows from Proposition 4.4.1 and Theorem 4.3.2. The proof of (ii) is the same.

4.4.3. The Spaces $\dot{B}_p^{\bar{s}}(R_2)$ and $\dot{W}_p^{\bar{s}}(R_2)$, the General Case

Let $\bar{a} = (a_1, a_2)$ with $0 < a_2 \leqq a_1 < \infty$ and $a_1 + a_2 = 2$ be the given anisotropy. Let $-\infty < s < \infty$. We recall that we introduced in Definition 4.2.1 the spaces $B_p^{\bar{s}}(R_2)$ for all couples $\bar{s} = (s_1, s_2)$ with $s_1 = \dfrac{s}{a_1}$ and $s_2 = \dfrac{s}{a_2}$. On the other hand, the spaces $\dot{B}_p^{\bar{s}}(R_2)$ are defined for $s > 0$ and $\bar{s}$ non-critical, cf. Definition 4.3.2/2. We wish to extend this definition to arbitrary couples $\bar{s}$. Let $D'(\dot{R}_2)$ be the collection of all complex-valued distributions on $\dot{R}_2 = R_2 - \{0\}$. We recall that $\Psi^{\bar{a}}(R_2)$ has been defined in 4.4.1. Furthermore, $\|f \mid \dot{B}_p^{\bar{s}}(R_2)\|^\psi$ for all couples $\bar{s}$ and $\|f \mid \dot{W}_p^{\bar{s}}(R_2)\|^\psi$ have the meaning of (4.4.2/4) and (4.4.2/5), respectively.

Definition 1. *Let* $1 < p < \infty$, $-\infty < s < \infty$ *and* $\bar{s} = (s_1, s_2)$ *with* $s_1 = \frac{s}{a_1}$ *and* $s_2 = \frac{s}{a_2}$. *Let* $\psi \in \Psi^{\bar{a}}(R_2)$.

(i) *Then*

$$\dot{B}_p^{\bar{s}}(R_2) = \{f \mid f \in D'(\dot{R}_2),\ \|f \mid \dot{B}_p^{\bar{s}}(R_2)\|^{\psi} < \infty\}. \tag{1}$$

(ii) *If in addition* s_1 *and* s_2 *are natural numbers then*

$$\dot{W}_p^{\bar{s}}(R_2) = \{f \mid f \in D'(\dot{R}_2),\ \|f \mid \dot{W}_p^{\bar{s}}(R_2)\|^{\psi} < \infty\}. \tag{2}$$

Remark 1. We prefer the above definition. Another possibility would be to replace $D'(\dot{R}_2)$ by $D'(R_2)$ of $S'(R_2)$ and to handle these spaces modulo combinations of the δ-distribution and its derivatives. These two approaches are essentially the same. Of course, the motivation of the above definition comes from Theorem 4.4.2. In order to justify the above definition of $\dot{B}_p^{\bar{s}}(R_2)$ and $\dot{W}_p^{\bar{s}}(R_2)$ we must prove that these spaces coincide with the spaces $\dot{B}_p^{\bar{s}}(R_2)$ and $\dot{W}_p^{\bar{s}}(R_2)$ from Definition 4.3.2/2 if $s > 0$ and $\bar{s}$ non-critical (appropriate interpretation as far as the different underlying spaces of distributions $D'(\dot{R}_2)$ and $S'(R_2)$ are concerned). Furthermore we must prove that in any case the above defined spaces $\dot{B}_p^{\bar{s}}(R_2)$ and $\dot{W}_p^{\bar{s}}(R_2)$ are independent of the chosen system $\psi \in \Psi^{\bar{a}}(R_2)$. Both assertions justify the notations (1) and (2).

Definition 2. *Let* $1 < p < \infty$, $-\infty < s < \infty$ *and* $\bar{s} = (s_1, s_2)$ *with* $s_1 = \frac{s}{a_1}$ *and* $s_2 = \frac{s}{a_2}$. *Then* $\bar{s}$ *is called critical if*

either $s > 0$ *and* $\bar{s}$ *is critical in the sense of Def.* 4.3.2/1,

or $s < 0$ *and there exist natural numbers* m_1 *and* m_2 *with*

$$\frac{1}{s_1}\left(m_1 - \frac{1}{p}\right) + \frac{1}{s_2}\left(m_2 - \frac{1}{p}\right) = -1. \tag{3}$$

Otherwise, $\bar{s}$ *is called non-critical.*

Remark 2. This is an extension of Definition 4.3.2/1. In particular, $\bar{s} = (0, 0)$ is always non-critical.

In this subsection we clarify the problems sketched in Remark 1. Of course $|x|_{\bar{a}}$ has the meaning of (4.2.1/1). If $1 < p < \infty$, $s > 0$ and $\bar{s} = (s_1, s_2)$ with $s_1 = \frac{s}{a_1}$ and $s_2 = \frac{s}{a_2}$ then we but

$$\|f \mid \dot{B}_p^{\bar{s}}(R_2)\|^* = \|f \mid B_p^{\bar{s}}(R_2)\| + \|\, |x|_{\bar{a}}^{-s} f \mid L_p(R_2)\| \tag{4}$$

and (if in addition s_1 and s_2 are natural numbers)

$$\|f \mid \dot{W}_p^{\bar{s}}(R_2)\|^* = \|f \mid W_p^{\bar{s}}(R_2)\| + \|\, |x|_{\bar{a}}^{-s} f \mid L_p(R_2)\|. \tag{5}$$

Furthermore, if $s > 0$ then the spaces from Definition 1 can be considered as subspaces of $L_p(R_2)$. In particular one has no difficulties to compare these spaces with their counterparts from Definition 4.3.2/2.

Lemma. *Let* $1 < p < \infty$, $s > 0$ *and* $\bar{s} = (s_1, s_2)$ *with* $s_1 = \frac{s}{a_1}$ *and* $s_2 = \frac{s}{a_2}$.

(i) *Let* $f \in B_p^{\bar{s}}(R_2)$ *and* $\|f \mid \dot{B}_p^{\bar{s}}(R_2)\|^* < \infty$. *Then*

$$\frac{\partial^{m_1+m_2} f}{\partial x_1^{m_1} \partial x_2^{m_2}}(0) = 0 \quad \text{if} \quad \frac{1}{s_1}\left(m_1 + \frac{1}{p}\right) + \frac{1}{s_2}\left(m_2 + \frac{1}{p}\right) < 1. \tag{6}$$

(ii) *Let in addition* s_1 *and* s_2 *be natural numbers. Let* $f \in W_p^{\bar{s}}(R_2)$ *and* $\|f \mid \dot{W}_p^{\bar{s}}(R_2)\|^* < \infty$. *Then* (6) *holds.*

Proof. (ii) follows from (i) and well-known embeddings. Similarly we may assume that $\bar{s}$ is non-critical. Furthermore by Remark 4.3.2/3 and Theorem 4.3.2 it is sufficient to deal with $f \in S(R_2)$ such that $|x|_{\bar{a}}^{-s} f(x) \in L_p(R_2)$. Let

$$f(x) = \sum \frac{x_1^{m_1} x_2^{m_2}}{m_1!\, m_2!} \frac{\partial^{m_1+m_2} f}{\partial x_1^{m_1} \partial x_2^{m_2}}(0) + g(x) \quad \text{if} \quad |x| < 1, \tag{7}$$

where the sum in (7) runs over all integers with

$$\frac{1}{s_1}\left(m_1 + \frac{1}{p}\right) + \frac{1}{s_2}\left(m_2 + \frac{1}{p}\right) < 1. \tag{8}$$

Let $x_1 = r^{a_1} \cos\varphi$ and $x_2 = r^{a_2} \sin\varphi$ with $0 \leqq r < \infty$ and $0 \leqq \varphi < 2\pi$ (anisotropic polar coordinates). We have

$$|x|_{\bar{a}}^{-sp}\, |x_1^{m_1} x_2^{m_2}|^p \,\mathrm{d}x \sim r^{-sp+a_1 m_1 p + a_2 m_2 p} |\cos\varphi|^{m_1 p}\, |\sin\varphi|^{m_2 p}\, r \,\mathrm{d}r \,\mathrm{d}\varphi. \tag{9}$$

If (8) holds then it follows

$$-sp + a_1 m_1 p + a_2 m_2 p = sp\left(-1 + \frac{m_1}{s_1} + \frac{m_2}{s_2}\right) < -\frac{sp}{p}\left(\frac{1}{s_1} + \frac{1}{s_2}\right) = -2. \tag{10}$$

Now it is not difficult to see that $|x|_{\bar{a}}^{-s} f(x) \in L_p(R_2)$ and (7)–(10) yield the desired result.

The spaces $\dot{B}_p^{\bar{s}}(R_2)$ and $\dot{W}_p^{\bar{s}}(R_2)$ in the theorem below must be understood in the sense of Definition 1.

Theorem. *Let* $1 < p < \infty$, $s > 0$, $\bar{s} = (s_1, s_2)$ *with* $s_1 = \dfrac{s}{a_1}$ *and* $s_2 = \dfrac{s}{a_2}$.

(i) *Then* $\dot{B}_p^{\bar{s}}(R_2)$ *is a Banach space, it is independent of the chosen system* $\psi \in \Psi^{\bar{a}}(R_2)$. *Furthermore*

$$\dot{B}_p^{\bar{s}}(R_2) = \{f \mid f \in L_p(R_2),\ \|f \mid \dot{B}_p^{\bar{s}}(R_2)\|^* < \infty\}. \tag{11}$$

If $\bar{s}$ *is non-critical, then* $\dot{B}_p^{\bar{s}}(R_2)$ *coincides with the corresponding spaces from Definition 4.3.2/2.*

(ii) *Let in addition* s_1 *and* s_2 *be natural numbers. Then* $\dot{W}_p^{\bar{s}}(R_2)$ *is a Banach space, it is independent of the chosen system* $\psi \in \Psi^{\bar{a}}(R_2)$. *Furthermore,*

$$\dot{W}_p^{\bar{s}}(R_2) = \{f \mid f \in L_p(R_2),\ \|f \mid \dot{W}_p^{\bar{s}}(R_2)\|^* < \infty\}. \tag{12}$$

If $\bar{s}$ *is non-critical, then* $\dot{W}_p^{\bar{s}}(R_2)$ *coincides with the corresponding spaces from Definition 4.3.2/2.*

Proof. Step 1. We prove (i), the proof of (ii) is essentially the same. Of course the space on the right-hand side of (11) is a Banach space. We prove that $C_0^\infty(\dot{R}_2)$ is a dense subset, cf. (4.4.2/1). We use the above Lemma and follow the arguments from the proof of Proposition 4.4.2, where (4.4.2/2) must be replaced by

$$1 \leqq \frac{1}{s_1}\left(m_1 + \frac{1}{p}\right) + \frac{1}{s_2}\left(m_2 + \frac{1}{p}\right) < N. \tag{13}$$

We use the same notations and obtain

$$\|\varrho_j(x) \mid B_p^{\bar{s}}(R_2)\| \leqq c, \tag{14}$$

where c is independent of j. By (4.2.4/2) $B_p^{\bar{s}}(R_2)$ is reflexive (as a matter of fact it is isomorphic to l_p). Hence, for any natural number M the set $\{\varrho_j(x)\}_{j=M}^{\infty}$ is weakly compact and $\varrho_j(x) \to 0$ in $L_p(R_2)$ if $j \to \infty$. Now we can apply Mazur's theorem, cf. K. Yosida [1, Theorem 2 in 5.1.]. This shows that any f which belongs to the right-hand side of (11) can be approximated in $B_p^{\bar{s}}(R_2)$ by $f_j \in C_0^{\infty}(\dot{R}_2)$, where we may assume that also $\| |x|_{\bar{a}}^{-s} (f - f_j) \,|\, L_p(R_2)\| \to 0$ holds if $j \to \infty$. This proves the desired density property.

Step 2. We prove that the spaces $\dot{B}_p^{\bar{s}}(R_2)$ are independent of $\psi \in \Psi^{\bar{a}}(R_2)$. Let $\psi' = \{\psi_j'(x)\}_{j=0}^{\infty} \in \Psi^{\bar{a}}(R_2)$ be a second system. Then (4.4.1/10) yields

$$\|\psi_j f \,|\, B_p^{\bar{s}}(R_2)\| \leqq \sum_{r=-3}^{3} (\| |x|_{\bar{a}}^{-s} \psi_{j+r}' f \,|\, L_p(R_2)\| + \|\psi_{j+r}' f \,|\, B_p^{\bar{s}}(R_2)\|). \tag{15}$$

We have

$$\begin{aligned} &\| |x|_{\bar{a}}^{-s} \psi_j' f \,|\, L_p(R_2)\| \\ &\leqq c\Big(\int_{R_1} [|h|^{-s_1 p} \|(\Delta_{h,1}^{m_1} \psi_j' f)(\cdot) \,|\, L_p(R_2)\|^p \\ &\qquad + |h|^{-s_2 p} \|(\Delta_{h,2}^{m_2} \psi_j' f)(\cdot) \,|\, L_p(R_2)\|^p] \frac{\mathrm{d}h}{|h|}\Big)^{1/p}, \end{aligned} \tag{16}$$

where c is independent of j. If $j = 1$ then this follows by standard arguments (as in the isotropic case). The rest is a homogeneity argument. If we put (16) in (15) then we obtain the desired independence. Now it is easy to see that $C_0^{\infty}(\dot{R}_2)$ is dense in $\dot{B}_p^{\bar{s}}(R_2)$.

Step 3. Let $f \in C_0^{\infty}(\dot{R}_2)$. Then (4.4.2/6) and (16) yield

$$\|f \,|\, \dot{B}_p^{\bar{s}}(R_2)\|^* \leqq c \|f \,|\, \dot{B}_p^{\bar{s}}(R_2)\|^{\psi} \tag{17}$$

with $\psi \in \Psi^{\bar{a}}(R_2)$. The converse assertion follows from Proposition 4.4.1. Hence, on $C_0^{\infty}(\dot{R}_2)$ the two norms are equivalent. (11) follows now from the two preceding steps. If $\bar{s}$ is non-critical, then we have Proposition 4.4.2 and Theorem 4.4.2. This shows that in this case the above space coincides with the space from (4.3.2/2).

4.4.4. Properties of the Spaces $\dot{B}_p^{\bar{s}}(R_2)$

Theorem 4.4.3 gives a satisfactory description of the spaces $\dot{B}_p^{\bar{s}}(R_2)$ with $s > 0$ and of the spaces $\dot{W}_p^{\bar{s}}(R_2)$. This subsection deals mainly with the spaces $\dot{B}_p^{\bar{s}}(R_2)$ with $s \leqq 0$ in the sense of the Definitions 1 and 2 from 4.4.3., including a duality assertion for all spaces $\dot{B}_p^{\bar{s}}(R_2)$ with $-\infty < s < \infty$. The study of the spaces $\dot{B}_p^{\bar{s}}(R_2)$ with $s < 0$ is not only of interest for its own sake, but one needs also some assertions in the framework of an L_2-theory for homogeneous Dirichlet problems for semi-elliptic differential equations which we sketch in Section 4.8.

We recall our general assumptions. $\bar{a} = (a_1, a_2)$ with $0 < a_2 \leqq a_1 < \infty$ and $a_1 + a_2 = 2$ stands for the given anisotropy. Furthermore, the anisotropic smoothness is given by $\bar{s} = (s_1, s_2)$ with $s_1 = \frac{s}{a_1}$ and $s_2 = \frac{s}{a_2}$, where $s \in R_1$ is the mean smoothness. Duality must always be understood in the sense of the dual pairing $(D(\dot{R}_2), D'(\dot{R}_2))$. If $s > 0$ then one has no difficulties to compare $B_p^{\bar{s}}(R_2)$ and $\dot{B}_p^{\bar{s}}(R_2)$, because all these spaces can be interpreted as subspaces of $L_p(R_2)$. If $s \leqq 0$ then the situation s different. Our point of view is the following. Let $s \leqq 0$, $f \in \dot{B}_p^{\bar{s}}(R_2)$ and $\varphi \in S(R_2)$

(or $D(R_2)$). Then we split

$$\varphi(x) = \sum \frac{\partial^{m_1+m_2}\varphi}{\partial x_1^{m_1}\,\partial x_2^{m_2}}(0)\,\varkappa_{m_1,m_2}(x) + \tilde{\varphi}(x), \tag{1}$$

where the finite sum $\sum$ near the origin is the beginning of the Taylor series of $\varphi(x)$ and $\varkappa_{m_1,m_2}(x) \in D(R_2)$. We put

$$F(\varphi) = f(\tilde{\varphi}) + \sum a_{m_1,m_2} \frac{\partial^{m_1+m_2}\varphi}{\partial x_1^{m_1}\,\partial x_2^{m_2}}(0), \tag{2}$$

where a_{m_1,m_2} are complex numbers and the sum in (2) is the same as in (1). One hopes that $f(\tilde{\varphi})$ makes sense and that any $F \in B^{\bar{s}}_p(R_2)$ can be uniquely represented in this way. Formula (4) below must be understood in this way.

Theorem. *Let* $1 < p < \infty$ *and* $\frac{1}{p} + \frac{1}{p'} = 1$. *Let* $\bar{s} = (s_1, s_2)$ *with* $s_1 = \frac{s}{a_1}$, $s_2 = \frac{s}{a_2}$ *and* $-\infty < s < \infty$.

(i) *Then* $\dot{B}^{\bar{s}}_p(R_2)$ *is a Banach space, it is independent of the chosen system* $\psi \in \Psi^{\bar{a}}(R_2)$ *and* $C_0^\infty(\dot{R}_2)$ *is a dense subset. Furthermore*

$$(\dot{B}^{\bar{s}}_p(R_2))' = \dot{B}^{-\bar{s}}_{p'}(R_2) \tag{3}$$

(*in the sense of the above interpretation*).

(ii) *Let in addition* $-\infty < s < 0$ *and* $\bar{s}$ *non-critical in the sense of Definition* 4.4.3/2. *Then*

$$B^{\bar{s}}_p(R_2) = \dot{B}^{\bar{s}}_p(R_2) \oplus \left\{\frac{\partial^{n_1+n_2-2}}{\partial x_1^{n_1-1}\,\partial x_2^{n_2-1}}\delta \text{ with } \frac{1}{s_1}\left(n_1 - \frac{1}{p}\right) + \frac{1}{s_2}\left(n_2 - \frac{1}{p}\right) > -1\right\}, \tag{4}$$

where δ *is the usual* δ*-distribution and* n_1 *und* n_2 *in* (4) *are natural numbers.*

(iii) *If* $\frac{2}{p} - 2 < s < \frac{2}{p}$ *then*

$$B^{\bar{s}}_p(R_2) = \dot{B}^{\bar{s}}_p(R_2). \tag{5}$$

Proof. Step 1. Let $s > 0$ and $\bar{s}$ be non-critical. We prove (3). Let $\psi \in \Psi^{\bar{a}}(R_2)$ and let $f \in \dot{B}^{-\bar{s}}_{p'}(R_2)$ with respect to this system ψ, cf. Definition 4.4.3/1. If $\varphi \in C_0^\infty(\dot{R}_2)$ (and $\psi_{-1} = 0$) then we have

$$|f(\varphi)| = \left|\sum_{j=0}^{\infty} (\psi_j f)\,(\psi_{j-1}\varphi + \psi_j\varphi + \psi_{j+1}\varphi)\right|$$
$$\leqq c \sum_{j=0}^{\infty} \|\psi_j f \mid B^{-\bar{s}}_{p'}(R_2)\|\,\|\psi_{j-1}\varphi + \psi_j\varphi + \psi_{j+1}\varphi \mid B^{\bar{s}}_p(R_2)\|,$$

cf. (4.2.4/2). Then it follows that

$$|f(\varphi)| \leqq c\|f \mid \dot{B}^{-\bar{s}}_{p'}(R_2)\|^{\psi}\,\|\varphi \mid \dot{B}^{\bar{s}}_p(R_2)\|^{\psi}. \tag{6}$$

By Proposition 4.4.2 the set $C_0^\infty(\dot{R}_2)$ is dense in $\dot{B}^{\bar{s}}_p(R_2)$. Then (6) yields $f \in (\dot{B}^{\bar{s}}_p(R_2))'$. In order to prove the converse assertion we assume that $f \in (\dot{B}^{\bar{s}}_p(R_2))'$. If $\psi = \{\psi_j(x)\}_{j=0}^{\infty} \in \Psi^{\bar{a}}(R_2)$ and $\varphi \in C_0^\infty(\dot{R}_2)$ then we have

$$|(\psi_j f)(\varphi)| = |f(\psi_j\varphi)| \leqq \|f \mid (\dot{B}^{\bar{s}}_p(R_2))'\|\,\|\psi_j\varphi \mid \dot{B}^{\bar{s}}_p(R_2)\|$$
$$\leqq c\|f \mid (\dot{B}^{\bar{s}}_p(R_2))'\|\,\|\varphi \mid B^{\bar{s}}_p(R_2)\|, \tag{7}$$

where we used that $\bar{s}$ is non-critical and that

$$\|\psi_j\varphi \mid \dot{B}_p^{\bar{s}}(R_2)\| \leqq c\|\varphi \mid \dot{B}_p^{\bar{s}}(R_2)\| \tag{8}$$

holds for all $\varphi \in \dot{B}_p^{\bar{s}}(R_2)$ and $j = 0, 1, 2, \ldots$ (cf. again (4.4.1/10)). Hence by the Hahn-Banach theorem we have $\psi_j f \in (B_p^{\bar{s}}(R_2))' = B_{p'}^{-\bar{s}}(R_2)$, cf. (4.2.4/2). Furthermore there exist a number $c > 0$ and functions $\varphi_j \in C_0^\infty(\dot{R}_2)$ with

$$\|\varphi_j \mid B_p^{\bar{s}}(R_2)\| = 1, \quad (\psi_j f)(\varphi_j) \geqq c\|\psi_j f \mid B_{p'}^{-\bar{s}}(R_2)\|. \tag{9}$$

Let $\tilde{\varphi}_j = \varphi_j(\psi_{j-1} + \psi_j + \psi_{j+1})$ (with $\psi_{-1} = 0$). By (8) we have

$$\|\tilde{\varphi}_j \mid B_p^{\bar{s}}(R_2)\| \leqq c', \quad (\psi_j f)(\tilde{\varphi}_j) = (\psi_j f)(\varphi_j) \tag{10}$$

where c' is independent of j. Let $\{c_j\}_{j=0}^\infty \in l_p$ be a sequence of complex numbers and let $\varphi = \sum_{j=0}^N c_{6j}\psi_{6j}\tilde{\varphi}_{6j}$, where N is an arbitrary natural number. Then we have

$$\begin{aligned}\left|\sum_{j=0}^N c_{6j}(\psi_{6j}f)(\tilde{\varphi}_{6j})\right| &= |f(\varphi)| \leqq \|f \mid (\dot{B}_p^{\bar{s}}(R_2))'\| \, \|\varphi \mid \dot{B}_p^{\bar{s}}(R_2)\| \\ &\leqq c\|f \mid (\dot{B}_p^{\bar{s}}(R_2))'\| \left(\sum_{j=1}^N |c_{6j}|^p\right)^{1/p},\end{aligned} \tag{11}$$

where we used (8) and (10). Because $\{c_j\}_{j=0}^\infty \in l_p$ is an arbitrary sequence we have $\{(\psi_{6j}f)(\tilde{\varphi}_{6j})\}_{j=0}^\infty \in l_{p'}$. Similar arguments hold if we replace $6j$ be $6j+1, \ldots, 6j+5$. By (9) and (10) it follows that

$$\|f \mid \dot{B}_{p'}^{-\bar{s}}(R_2)\|^\psi \leqq c\| f \mid (\dot{B}_p^{\bar{s}}(R_2))'\| \tag{12}$$

for any $\psi \in \Psi^{\bar{a}}(R_2)$. In particular by (6) and (12) the space $\dot{B}_{p'}^{-\bar{s}}(R_2)$ is independent of $\psi \in \Psi^{\bar{a}}(R_2)$. Furthermore we have (3). By duality (8) yields

$$\|\psi_j\varphi \mid \dot{B}_{p'}^{-\bar{s}}(R_2)\| \leqq c\|\varphi \mid \dot{B}_{p'}^{-\bar{s}}(R_2)\|, \tag{13}$$

where c is independent of j. With the help of this estimate it is easy to see that $C_0^\infty(\dot{R}_2)$ is dense in $\dot{B}_{p'}^{-\bar{s}}(R_2)$.

Step 2. Let $s > 0$ and $0 < \varepsilon < 1$. We claim that

$$(\dot{B}_p^{(1+\varepsilon)\bar{s}}(R_2), \dot{B}_p^{(1-\varepsilon)\bar{s}}(R_2))_{1/2,p} = \dot{B}_p^{\bar{s}}(R_2), \tag{14}$$

where $(\cdot, \cdot)_{1/2,p}$ stands for the real interpolation method. This follows from (4.2.4/1) and a direct calculation of Peetre's K-functional, cf. [I, 1.3.1.] as far as the latter is concerned. We give an outline. We assume that both $(1-\varepsilon)\bar{s}$ and $(1+\varepsilon)\bar{s}$ are non-critical. Let $\psi = \{\psi_j\} \in \Psi^{\bar{a}}(R_2)$. If $f \in \dot{B}_p^{(1-\varepsilon)\bar{s}}(R_2)$ and $t > 0$ then

$$\begin{aligned}K^p(t,f) &\sim \inf_{f=f_0+f_1} (\|f_0 \mid \dot{B}_p^{(1+\varepsilon)\bar{s}}(R_2)\|^p + t^p\|f_1 \mid \dot{B}_p^{(1-\varepsilon)\bar{s}}(R_2)\|^p) \\ &\sim \sum_{j=0}^\infty \inf_{\psi_j f = f_0{}^j + f_1{}^j} (\|\psi_j f_0^j \mid B_p^{(1+\varepsilon)\bar{s}}(R_2)\|^p + t^p\|\psi_j f_1^j \mid B_p^{(1-\varepsilon)\bar{s}}(R_2)\|^p) \\ &\sim \sum_{j=0}^\infty K^p(t, \psi_j f, B_p^{(1+\varepsilon)\bar{s}}(R_2), B_p^{(1-\varepsilon)\bar{s}}(R_2)).\end{aligned} \tag{15}$$

In the middle equivalence we used (8). This equivalence and (4.2.4/1) prove (14) because we know from the very beginning that $C_0^\infty(\dot{R}_2)$ is dense in the two involved spaces. Let $\bar{s}$ be critical. By duality, cf. [I, 1.11.2] we have

$$(\dot{B}_p^{\bar{s}}(R_2))' = ((\dot{B}_p^{(1+\varepsilon)\bar{s}}(R_2))', (\dot{B}_p^{(1-\varepsilon)\bar{s}}(R_2))')_{1/2,p'}$$
$$= (\dot{B}_{p'}^{-(1+\varepsilon)\bar{s}}(R_2), \dot{B}_{p'}^{-(1-\varepsilon)\bar{s}}(R_2))_{1/2,p'}. \qquad (16)$$

By the above method, cf. (13) and (15), if follows that $(\dot{B}_p^{\bar{s}}(R_2))'$ is the completion of $C_0^\infty(\dot{R}_2)$ in $\dot{B}_{p'}^{-\bar{s}}(R_2)$. On the other hand we have (6) in any case. This proves (3) for critical $\bar{s}$, including the fact that $C_0^\infty(\dot{R}_2)$ is dense in $\dot{B}_{p'}^{-\bar{s}}(R_2)$ and that $\dot{B}_{p'}^{-\bar{s}}(R_2)$ is independent of $\psi \in \Psi^{\bar{a}}(R_2)$.

Step 3. Let $s < 0$. Then $(\dot{B}_p^{\bar{s}}(R_2))'$ makes sense (in the above interpretation) because $C_0^\infty(\dot{R}_2)$ is a dense subset. If $\bar{s}$ is non-critical then $\dot{B}_{p'}^{-\bar{s}}(R_2)$ is reflexive. This follows from (4.2.4/2) and (4.3.2/2). Then we have (3) by reflexivity. The remaining cases follow by interpolation similarly as in (16). This proves (i) if $s \neq 0$.

Step 4. We prove (iii) (this covers also (i) with $s = 0$). If $0 < s < \frac{2}{p}$ then $\dot{B}_p^{\bar{s}}(R_2) = B_p^{\bar{s}}(R_2)$, cf. (4.3.2/2). In particular if $\psi = \{\psi_j(x)\}_{j=0}^\infty \in \Psi^{\bar{a}}(R_2)$ then we have

$$\|\psi_j f \mid B_p^{\bar{s}}(R_2)\| \leqq c\|f \mid B_p^{\bar{s}}(R_2)\| \qquad (17)$$

for all $f \in B_p^{\bar{s}}(R_2)$ and all $j = 0, 1, 2, \ldots$ By duality one can extend (17) to $-2 + \frac{2}{p} < s < 0$ and by interpolation to $s = 0$. Now (5) follows by the above technique.

Step 5. We prove (ii). Let $s > 0$ and $\bar{s}$ be non-critical. Then we have (4.3.2/2). Let P be the projection parallel to $B_p^{\bar{s}}(R_2)/\dot{B}_p^{\bar{s}}(R_2)$. If $F \in (B_p^{\bar{s}}(R_2))'$ and $\varphi \in B_p^{\bar{s}}(R_2)$ then we have a decomposition

$$F(\varphi) = F(P\varphi) + F(\varphi - P\varphi)$$

with $FP \in (\dot{B}_p^{\bar{s}}(R_2))'$, while $F[E - P]$ is a linear combination of $\frac{\partial^{m_1+m_2}}{\partial x_1^{m_1}\partial x_2^{m_2}}\delta$ with $\frac{1}{s_1}\left(m_1 + \frac{1}{p}\right) + \frac{1}{s_2}\left(m_2 + \frac{1}{p}\right) < 1$, cf. also (1) and (2). The reformulation of this assertion (with $-\bar{s}$ and p' instead of $\bar{s}$ and p, respectively) and (3) yield (4).

4.5. Traces of $\dot{B}_p^{\bar{s}}(R_2)$ and $\dot{W}_p^{\bar{s}}(R_2)$ on Curves

4.5.1. Introduction

Let $\bar{a} = (a_1, a_2)$ with $0 < a_2 \leqq a_1 < \infty$ and $a_1 + a_2 = 2$ be a given anisotropy. Let $s > 0$ and $\bar{s} = (s_1, s_2)$ with $s_1 = \frac{s}{a_1}$ and $s_2 = \frac{s}{a_2}$. In 4.2.3. we described traces of $B_p^{\bar{s}}(R_2)$ on the axes of the coordinates. If $\sigma_1 = s_1\left(1 - \frac{1}{ps_2}\right) > 0$ then $\Gamma_1 : f(x) \to f(x_1, 0)$ is a retraction from $B_p^{\bar{s}}(R_2)$ onto $B_p^{\sigma_1}(R_1)$, cf. 4.2.3. We recall that this assertion includes both direct and inverse embedding theorems. In the isotropic case, i.e. $a_1 = a_2 = 1$, this assertion is completely sufficient in order to handle traces on smooth curves, e.g. on the boundary of C^∞-domains in R_2. One has to use the above result, smooth resolutions of unity and diffeomorphic maps of the plane onto itself. In the anisotropic case the latter tool is not available. There is an important exception of this negative statement: $B_p^{\bar{s}}(R_2)$ is stable under the fibre-preserving diffeomorphic maps of R_2 onto itself of the type

$$(x_1, x_2) \to (x_1, x_2 - \lambda(x_1)), \tag{1}$$

where $\lambda(t)$ is a C^∞-function with bounded derivatives, cf. Step 1 of the proof of Theorem 4.5.4. Then it is obvious that the trace from $\{(x_1, x_2) \mid x_2 = 0\}$ can be transferred to the curve $\{(x_1, x_2) \mid x_2 = \lambda(x_1) \text{ with } x_1 \in R_1\}$. This fact is well-known and has been observed almost twenty years ago by S. V. Uspenskij, cf. also S. V. Uspenskij [1]. The situation changes drastically if one deals with other curves, e.g. boundaries of bounded C^∞-domains in R_2. These problems attracted some attention, however there seem to be only very few papers in this direction. We refer to S. V. Uspenskij [2], his successors S. B. Sadykova [1] and G. A. Šmirev [1], and also M. D. Ramazanov [1]. A survey has been given in the recent book S. B. Uspenskij, G. V. Demidenko, V. G. Perepelkin [1]. Our approach is completely independent of those in the cited papers and can be described as follows. Via anisotropic resolutions of unity the problem can be reduced to a local problem. We deal with a model case: Traces on the curve C_ϱ which is given near the origin by $\{(x_1, x_2) \mid x_2 = x_1^\varrho, 0 < x_1 < 1\}$. There are three cases: (i) The case $1 \leqq \varrho < \infty$ is essentially covered by the method sketched above, (ii) $\frac{s_1}{s_2} = \frac{a_2}{a_1} \leqq \varrho < 1$ and (iii) $0 < \varrho < \frac{a_2}{a_1}$. Presumably the description of traces of the considered spaces on C_ϱ in the two latter cases is quite different. We deal exclusively with the case $\frac{a_2}{a_1} \leqq \varrho < 1$. Furthermore we are mainly interested in the traces of the spaces $\dot{B}_p^{\bar{s}}(R_2)$ and $\dot{W}_p^{\bar{s}}(R_2)$ from Definition 4.4.3/1. If $\bar{s}$ is non-critical then these spaces are very near to the original spaces $B_p^{\bar{s}}(R_2)$ and $W_p^{\bar{s}}(R_2)$, respectively, cf. Definition 4.3.2/2, cf. also 4.5.6. Our main result (cf. 4.5.4.) has the same final character as the embedding on lines and will again be formulated in the language of retractions. Later on we shall use these results in order to describe traces of anisotropic Besov-Sobolev spaces in the unit circle K on the circumference ∂K, cf. Section 4.6. These results, in turn, are the basis for a study of boundary value problems for semi-elliptic differential equations, cf. Section 4.8.

4.5.2. Weighted Besov Spaces on R_1^+

The traces of the spaces $\dot{B}_p^{\bar{s}}(R_2)$ and $\dot{W}_p^{\bar{s}}(R_2)$ on C_ϱ are described by weighted Besov spaces on C_ϱ. This subsection contains the necessary preparations: Weighted Besov spaces on R_1^+. In the next subsection we carry over these results from R_1^+ to C_ϱ.

First we recall that the classical Besov spaces $B_p^s(R_1) = B_{p,p}^s(R_1)$ with $s > 0$ and $1 < p < \infty$ can be normed via

$$\|f \mid B_p^s(R_1)\| = \|f \mid L_p(R_1)\| + \left(\int_0^\infty t^{-sp} \|(\Delta_t^m f)(\cdot) \mid L_p(R_1)\|^p \frac{dt}{t} \right)^{1/p} \tag{1}$$

where $m > s$ is a natural number, and Δ_t^m are the usual differences,

$$(\Delta_t^1 f)(\tau) = f(t+\tau) - f(\tau), \qquad \Delta_t^k = \Delta_t^1 \Delta_t^{k-1} \quad \text{if} \quad k = 2, 3, \ldots$$

Again, by the usual abuse of notations we omitted m on the left-hand side of (1). Let $R_1^+ = \{\tau \mid \tau > 0\}$ and $\overline{R_1^+} = \{\tau \mid \tau \geqq 0\}$. If $s > 0$ and $1 < p < \infty$ then we introduce

$$\tilde{B}_p^s(R_1^+) = \{f \mid f \in B_p^s(R_1),\ \operatorname{supp} f \subset \overline{R_1^+}\}, \tag{2}$$

cf. [I, 2.10.3]. Let $d(\tau)$ be a positive infinitely differentiable function on R_1^+ with $d(\tau) = \tau$ if $0 < \tau < 1$ and $d(\tau) = 1$ if $\tau > 2$. Then $\tilde{B}_p^s(R_1^+)$ can be normed via

$$\|f \mid \tilde{B}_p^s(R_1^+)\| = \left(\int_0^\infty d^{-sp}(\tau)\,|f(\tau)|^p\,\mathrm{d}\tau\right)^{1/p} + \left(\int_0^\infty t^{-sp}\|(\Delta_t^m f)(\cdot) \mid L_p(R_1)\|^p \frac{\mathrm{d}t}{t}\right)^{1/p}, \tag{3}$$

where m is a natural number with $m > s$, cf. [I, 4.3.2.]. If s is non-critical, i.e. $s - \frac{1}{p} \neq$ integer, then the first summand on the right-hand side of (3) can be replaced by $\|f \mid L_p(R_1^+)\|$, cf. (4.3.1/4). In other words, if $s - \frac{1}{p} \neq$ integer then $\|f \mid B_p^s(R_1^+)\|$ is an equivalent norm in $\tilde{B}_p^s(R_1^+)$ and we have in this case $\tilde{B}_p^s(R_1^+) = \mathring{B}_p^s(R_1^+)$, where the latter space is the completion of $C_0^\infty(R_1^+)$ in $B_p^s(R_1^+)$. We refer again to [I, 4.3.2.].

Next we introduce the one-dimensional counterpart of the system $\Psi^{\bar{a}}(R_2)$ from 4.4.1. Let $\varkappa = \{\varkappa_j(\tau)\}_{j=0}^\infty$ be a system of infinitely differentiable functions on R_1^+ with the following properties:

(i) $$\operatorname{supp} \varkappa_0 \subset \{\tau \mid \tau \geqq \tfrac{1}{2}\},$$
$$\operatorname{supp} \varkappa_j \subset \{\tau \mid 2^{-j-1} < \tau < 2^{-j+1}\} \quad \text{if} \quad j = 1, 2, 3, \ldots \tag{4}$$

(ii) For every $k = 0, 1, 2, \ldots$ there exists a positive number c_k with

$$2^{-kj}|\varkappa_j^{(k)}(\tau)| \leqq c_k \tag{5}$$

for any $j = 0, 1, 2, \ldots$ and $\tau > 0$.

(iii) $$\sum_{j=0}^\infty \varkappa_j(\tau) = 1 \quad \text{if} \quad \tau > 0. \tag{6}$$

Definition. *Let $s > 0$, $1 < p < \infty$ and $\mu \in R_1$. Let $\varkappa$ be the above system. Then*

$$B_p^s(R_1^+, \mu) = \Big\{f \mid f \in L_p^{\mathrm{loc}}(R_1^+),\ \|f \mid B_p^s(R_1^+, \mu)\| = \Big[\sum_{j=0}^\infty (2^{j\mu p}\|\varkappa_j f \mid B_p^s(R_1)\|^p + 2^{j(\mu+s)p}\|\varkappa_j f \mid L_p(R_1)\|^p)\Big]^{1/p} < \infty\Big\}. \tag{7}$$

Remark 1. The spaces $B_p^s(R_1^+, \mu)$ coincide essentially with the spaces $B_{p,p}^s(R_1^+, \varrho^{\mu p}, \varrho^{(\mu+s)p})$ with $\varrho(\tau) = d^{-1}(\tau)$ from [I, 3.2.3.] where $d(\tau)$ has the above meaning. (There is a small difference as far as the behaviour of the above function $\varrho(\tau)$ and the function $\varrho(\tau)$ from [I, 3.2.3.] for large values of τ is concerned. However this is completely uninteresting for our purposes.) Very roughly, our study of the spaces $\dot{B}_p^{\bar{s}}(R_2)$ and $\dot{W}_p^{\bar{s}}(R_2)$ in the preceding sections is the anisotropic counterpart of corresponding considerations in [I, Chapter 3] for respective isotropic spaces.

Proposition. *Let $1 < p < \infty$, $s > 0$ and $-\infty < \mu < \infty$. Let $d(\tau)$ be the above function on R_1^+. Let m be a natural number with $m > s$. Then*

$$\|f \mid B_p^s(R_1^+, \mu)\|^* = \left(\int_0^\infty d^{-sp-\mu p}(\tau)\, |f(\tau)|^p \, d\tau\right)^{1/p}$$

$$+ \left(\int_0^\infty \int_0^\infty t^{-sp} |[\Delta_t^m (d^{-\mu} f)]\,(\tau)|^p \, d\tau \frac{dt}{t}\right)^{1/p} \tag{8}$$

is an equivalent norm in $B_p^s(R_1^+, \mu)$.

Proof. If n is a natural number, then $\tilde{W}_p^n(R_1^+) = \mathring{W}_p^n(R_1^+)$ is the completion of $C_0^\infty(R_1^+)$ in the norm

$$\|f \mid W_p^n(R_1^+)\| = \left[\int_0^\infty (|f(\tau)|^p + |f^{(n)}(\tau)|^p)\, d\tau\right]^{1/p}. \tag{9}$$

Let $\varkappa = \{\varkappa_j(\tau)\}_{j=0}^\infty$ be the above system. Then it follows from [I, Theorem 3.2.6 and Theorem 3.2.4/2] that

$$\|f \mid W_p^n(R_1^+)\|^* = \left(\sum_{j=0}^\infty \|\varkappa_j f \mid W_p^n(R_1)\|^p + 2^{jnp} \|\varkappa_j f \mid L_p(R_1)\|^p\right)^{1/p} \tag{10}$$

is an equivalent norm in $\tilde{W}_p^n(R_1)$. It is easy to see that

$$\|d^{-\mu} f \mid W_p^n(R_1^+)\|^* \sim \left(\sum_{j=0}^\infty 2^{j\mu p} \|\varkappa_j f \mid W_p^n(R_1)\|^p + 2^{j\mu p + jnp} \|\varkappa_j f \mid L_p(R_1)\|^p\right)^{1/p}$$

$$= \|f \mid W_p^n(R_1^+, d^{-\mu p}, d^{-\mu p - np})\| \tag{11}$$

holds (equivalent norms) where we used again the notations from [I, 3.2.3.] (with the above indicated small modifications for large values of τ). In other words, $f \to d^{-\mu} f$ yields an isomorphic mapping from $W_p^n(R_1^+, d^{-\mu p}, d^{-\mu p - np})$ onto $\tilde{W}_p^n(R_1^+)$, where $n = 0, 1, 2, \ldots$ We recall two interpolation results: (i) The real interpolation $(\cdot, \cdot)_{\theta, p}$ of two spaces $\tilde{W}_p^n(R_1^+)$ with different n's yields the above spaces $\tilde{B}_p^s(R_1^+)$, cf. [I, 2.10.4.], (ii) The same real interpolation of two spaces $W_p^n(R_1^+, d^{-\mu p}, d^{-\mu p - np})$ with the same different n's as in (i) yields the above spaces

$$B_{p,p}^s(R_1^+, d^{-\mu p}, d^{-\mu p - sp}) = B_p^s(R_1^+, \mu),$$

cf. [I, 3.4.2.] and the above Remark 1. These interpolation results and (3) yield the proposition.

Remark 2. The proposition yields a quite satisfactory description of the weighted Besov spaces $B_p^s(R_1^+, \mu)$ on R_1^+. In the next subsection we transfer these spaces from R_1^+ to some curves in the plane. It comes out that just these spaces are the exact traces of the anisotropic spaces from the preceding subsections.

4.5.3. Weighted Besov Spaces on Curves

Let $0 < \varrho < 1$ and let $\gamma(\tau)$ be an infinitely differentiable monotonically increasing (i.e. not decreasing) function on R_1^+ with

$$\gamma(\tau) = \tau^\varrho \quad \text{if} \quad 0 < \tau < 1 \quad \text{and} \quad \gamma(\tau) = 2 \quad \text{if} \quad \tau > 2.$$

Let

$$C_\varrho = \{x \mid x = (x_1, \gamma(x_1)) \text{ with } 0 < x_1 < \infty\} \tag{1}$$

be the curve in the plane which is of interest in our context. Let σ be the arc length on C_ϱ measured from the origin (i.e. the origin corresponds to $\sigma = 0$). Let $d(\sigma)$ be a positive infinitely differentiable function on R_1^+ with $d(\sigma) = \sigma$ if $0 < \sigma < 1$ and $d(\sigma) = 1$ if $\sigma > 2$.

Definition. *Let $1 < p < \infty$, $s > 0$ and $-\infty < \mu < \infty$. Let m be a natural number with $m > s$. Then $B_p^s(C_\varrho, \mu)$ is the collection of all functions $g(\sigma)$, defined on C_ϱ, which are locally p-integrable (with respect to the measure $d\sigma$) such that*

$$\|g \mid B_p^s(C_\varrho, \mu)\|^* = \left(\int_0^\infty d^{-sp-\mu p}(\sigma)\, |g(\sigma)|^p \, d\sigma \right)^{1/p} + \left(\int_0^\infty \int_0^\infty \lambda^{-sp} |[\Delta_\lambda^m (d^{-\mu} g)]\,(\sigma)|^p \, d\sigma \frac{d\lambda}{\lambda} \right)^{1/p} \tag{2}$$

is finite. Let $\widetilde{B}_p^s(C_\varrho) = B_p^s(C_\varrho, 0)$.

Remark. Obviously, this definition is the direct counterpart of Proposition 4.5.2 and the underlying Definition 4.5.2. There is an immediate one-to-one relation between $B_p^s(C_\varrho, \mu)$ and $B_p^s(R_1^+, \mu)$. In particular, $B_p^s(C_\varrho, \mu)$ is independent of the choice of m (equivalent norms). By the usual abuse of notations this justifies to omit m on the left-hand side of (2). In the sense of the one-to-one relation between the spaces on R_1^+ and C_ϱ we shall also write $\|h \mid B_p^s(R_1)\|$ if $h = h(\sigma)$ is a function on C_ϱ with compact support and if we need the B_p^s-norm for $h(\sigma)$ extended by zero in an appropriate way. The notation $\widetilde{B}_p^s(C_\varrho)$ comes from (4.5.2/2) and (4.5.2/3).

We recall that $\bar{a} = (a_1, a_2)$ with $0 < a_2 \leqq a_1 < \infty$ and $a_1 + a_2 = 2$ stands for the given anisotropy. Let $\Psi^{\bar{a}}(R_2)$ be the system from 4.4.1.

Proposition. *Let $s > 0$, $1 < p < \infty$ and $-\infty < \mu < \infty$. Let $\psi = \{\psi_j(x)\}_{j=0}^\infty \in \Psi^{\bar{a}}(R_2)$ with respect to the anisotropy $\bar{a} = (a_1, a_2)$. Let C_ϱ be the curve (1) with $\frac{a_2}{a_1} \leqq \varrho < 1$ and $\sigma = \sigma(x_1)$ with $0 < x_1 < \infty$ be the arc length of C_ϱ and $x_1(\sigma)$ its inverse function. Let*

$$\varkappa_j(\sigma) = \psi_j(x_1(\sigma), \gamma(x_1(\sigma))) \quad \text{with} \quad j = 0, 1, 2, \ldots$$

Then

$$\|g \mid B_p^s(C_\varrho, \mu)\| = \left[\sum_{j=0}^\infty (2^{j\mu a_1 \varrho p} \|\varkappa_j g \mid B_p^s(R_1)\|^p + 2^{j(\mu+s)a_1\varrho p} \|\varkappa_j g \mid L_p(R_1)\|^p) \right]^{1/p} \tag{3}$$

is an equivalent norm on $B_p^s(C_\varrho, \mu)$.

Proof. We prove that $\varkappa = \{\varkappa_j(\sigma)\}_{j=0}^\infty$ is a system in the sense of 4.5.2. if one replaces 2^{-j} by $2^{-ja_1\varrho}$ in (4.5.2/5) and similarly in (4.5.2/4). The proof is quite straightforward. Near the origin we have

$$\frac{dx_1}{d\sigma}(\sigma) = \left(\frac{d\sigma}{dx_1} \right)^{-1} (x_1(\sigma)) = (1 + \varrho^2 x_1^{2(\varrho-1)})^{-1/2} \sim x_1^{1-\varrho}(1 + \varrho^{-2} x_1^{2(1-\varrho)})^{-1/2}$$

with $x_1 = x_1(\sigma)$. By iteration it follows that near the origin

$$\frac{d^k x_1}{d\sigma^k}(\sigma) \sim x_1^{1-k\varrho} \sim \sigma^{\frac{1}{\varrho}-k} \tag{4}$$

holds, $k = 1, 2, \ldots$ Because $\frac{a_2}{a_1} \leqq \varrho$ we may assume (essentially without restriction of generality) by simple geometric reasoning that $\varkappa_j(\sigma) = \varkappa(2^{ja_1}x_1(\sigma))$, where $\varkappa \in C_0^\infty(R_1)$ is supported, say, in the interval $(\frac{1}{2}, 2)$ and $j = 1, 2, 3, \ldots$ We recall that $x_1(\sigma) \sim 2^{-ja_1}$ is equivalent to $\sigma \sim 2^{-ja_1\varrho}$. Then we have

$$\left|\frac{d^k\varkappa_j(\sigma)}{d\sigma^k}\right| \leqq c \cdot 2^{ja_1\varrho k}.$$

This proves that (4.5.2/4)–(4.5.2/6) is satisfied with $2^{-ja_1\varrho}$ instead of 2^{-j} and its obvious counterparts where j is replaced by $j-1$ and $j+1$. However this replacement in Definition 4.5.2 is immaterial: One has again Proposition 4.5.2. This proves the above proposition.

4.5.4. Traces of the Spaces $\dot{B}_p^{\bar{s}}(R_2)$ and $\dot{W}_p^{\bar{s}}(R_2)$: The Main Theorem

We recall that $\bar{a} = (a_1, a_2)$ with $0 < a_2 < a_1 < \infty$ and $a_1 + a_2 = 2$ stands for the given anisotropy. Let $\bar{s} = (s_1, s_2)$ with $s_1 = \frac{s}{a_1}$, $s_2 = \frac{s}{a_2}$ and $s > 0$ be the anisotropic smoothness. We denoted s as the mean smoothness, which can also be calculated via $\frac{1}{s} = \frac{1}{2}\left(\frac{1}{s_1} + \frac{1}{s_2}\right)$. The curve C_ϱ is given by (4.5.3/1), the spaces $B_p^\sigma(C_\varrho, \mu)$ have the same meaning as in Definition 4.5.3. The set $C_0^\infty(\dot{R}_2)$ has been introduced in (4.4.2/1). The assertion that the trace operator $R: f \to f \mid C_\varrho$ is a retraction includes both the direct and the inverse embedding theorem, cf. 4.2.3. for explanations.

Theorem. *Let $1 < p < \infty$. Let $\bar{a} = (a_1, a_2)$ with $0 < a_2 < a_1 < \infty$ and $a_1 + a_2 = 2$ be the anisotropy. Let $\frac{a_2}{a_1} \leqq \varrho < 1$. Let $\bar{s} = (s_1, s_2)$ with $s_1 = \frac{s}{a_1}$, $s_2 = \frac{s}{a_2}$ and $s > 0$.*

(i) Let $s_2 \leqq \frac{1}{p}$. Then

$$\{f \mid f \in C_0^\infty(\dot{R}_2),\ \operatorname{supp} f \cap C_\varrho = \emptyset\} \tag{1}$$

is dense in $\dot{B}_p^{\bar{s}}(R_2)$.

(ii) Let $s_2 > \frac{1}{p}$, $\sigma_1 = s_1\left(1 - \frac{1}{ps_2}\right)$ and $\mu = \left(\frac{1}{\varrho} - 1\right)\left(\sigma_1 - \frac{1}{p}\right)$. Then the trace operator $R: f(x) \to f \mid C_\varrho$ (restriction of $f(x)$ on the curve C_ϱ) is a retraction from $\dot{B}_p^{\bar{s}}(R_2)$ onto $B_p^{\sigma_1}(C_\varrho, \mu)$.

(iii) Let s_1 and s_2 be natural numbers. Then the above trace operator R is a retraction from $\dot{W}_p^{\bar{s}}(R_2)$ onto $B_p^{\sigma_1}(C_\varrho, \mu)$ where σ_1 and μ have the same meaning as in (ii).

Proof. Step 1. We begin with some preparations. Let $\lambda(\tau)$ be a real C^∞-function on the real line with

$$|\lambda^{(k)}(\tau)| \leqq c_k \quad \text{if} \quad \tau \in R_1 \quad \text{and} \quad k = 0, 1, 2, \ldots \tag{2}$$

Then

$$y_1 = x_1, \qquad y_2 = x_2 - \lambda(x_1) \tag{3}$$

is a diffeomorphic mapping of R_2 onto itself which maps the curve $\Lambda = \{(x_1, \lambda(x_1)) \mid x_1 \in R_1\}$ onto $\{(y_1, 0) \mid y_1 \in R_1\}$. We claim that (3) yields an one-to-one mapping from $B_p^{\bar{s}}(R_2)$ onto itself, where $\bar{s}$ and p have the same meaning as in the theorem. Let $B_p^{s_0}(R_2) = B_{p,p}^{s_0}(R_2)$ be the usual isotropic Besov space, which can be normed e.g. via (4.2.2/5) with s_0 instead of s_1 and s_2. Then

$$\left(\int_{R_1} |h|^{-s_2 p} \|(\Delta_{h,2}^{m_2} f)(\cdot) \mid L_p(R_2)\|^p \frac{dh}{|h|}\right)^{1/p} + \|f \mid B_p^{s_1}(R_2)\| \tag{4}$$

is an equivalent norm in $B_p^{\bar{s}}(R_2)$ (again we assume that $m_2 > s_2$). The diffeomorphism (3) yields an one-to-one mapping of $B_p^{s_1}(R_2)$ onto itself. This is well known (and also an immediate consequence of a corresponding assertion for isotropic Sobolev spaces and real interpolation). A corresponding assertion for the first summand in (4) follows by direct calculations and the structure of (3). If $\varkappa > 0$ then the spaces $B_p^{\varkappa}(R_1)$ can be normed via (4.5.2/1) (with $\varkappa$ instead of s). In an obvious way one can introduce Besov spaces $B_p^{\varkappa}(\Lambda)$ on the curve Λ where distances are measured via the arc length on Λ. The diffeomorphism (3) yields also an one-to-one mapping from $B_p^{\varkappa}(\Lambda)$ onto $B_p^{\varkappa}(R_1)$. We remark that the distortion coefficients of the isomorphic mappings from $B_p^{\bar{s}}(R_2)$ onto itself and from $B_p^{\varkappa}(\Lambda)$ onto $B_p^{\varkappa}(R_1)$ depend only on a finite number of the c_k's from (2). Let $\bar{s} = (s_1, s_2)$ with $s_2 > \frac{1}{p}$ and let $\sigma_1 = s_1\left(1 - \frac{1}{ps_2}\right)$ be the above number. Then it follows from (4.2.3/8) and the above considerations that the trace operator $f(x) \to f \mid \Lambda$ (restriction on Λ) is a retraction from $B_p^{\bar{s}}(R_2)$ onto $B_p^{\sigma_1}(\Lambda)$, where the (direct and inverse) embedding constants depend only on a finite number of the c_k's in (2). If in addition the support of $g \in B_p^{\sigma_1}(\Lambda)$ is contained in, say, $\Lambda \cap \{x \mid |x| < 1\}$ then we assume that the extension operator yields a function $f(x) \in B_p^{\bar{s}}(R_2)$ with $\operatorname{supp} f \subset \{x \mid |x| < 2\}$.

Step 2. Let $1 < p < \infty$ and $s_2 > \frac{1}{p}$. Let $\psi = \{\psi_j(x)\}_{j=0}^{\infty} \in \Psi^{\bar{a}}(R_2)$ and $\varkappa_j(\sigma) = \psi_j(x_1(\sigma), \gamma(x_1(\sigma)))$ in the sense of Proposition 4.5.3. Let $j = 1, 2, \ldots$ and let

$$y_1 = 2^{ja_1} x_1, \qquad y_2 = 2^{ja_2} x_2.$$

We put $g_j(y) = (\psi_j f)(x)$. Then (4.2.2/5) yields

$$\|\psi_j f \mid B_p^{\bar{s}}(R_2)\| \sim 2^{js - \frac{2j}{p}} \|g_j \mid B_p^{\bar{s}}(R_2)\|, \tag{5}$$

where $\sim$ indicates an equivalence where the corresponding constants are independent of j. The curve C_ϱ is transformed in a curve $C_{\varrho,j}$ which is given by $y_2 = 2^{ja_2 - ja_1\varrho} y_1^\varrho$ if $0 < y_1 \leqq 2^{ja_1}$ (the other parts of the curve are out of interest). Because $\varrho \geqq \frac{a_2}{a_1}$ the curves $C_{\varrho,j}$ satisfy locally in $\frac{1}{2} \leqq y_1 \leqq 2$ the hypotheses of the curve Λ from the first step, where we may assume that locally the counterparts of the c_k's from (2) are independent of j. We apply Step 1 to $g_j(y)$ and obtain that the restriction of $g_j(y)$ to $C_{\varrho,j}$ belongs to $B_p^{\sigma_1}$ where the corresponding embedding constants are independent of j. The dilation of the arc lengths of C_ϱ and $C_{\varrho,j}$ near $y_1 = 1$ is given by $2^{-ja_1} \times \varrho(2^{-ja_1})^{\varrho-1} \sim 2^{-ja_1\varrho}$. Consequently

$$\|\varkappa_j Rf \mid B_p^{\sigma_1}(C_\varrho)\| \sim 2^{ja_1\varrho\left(\sigma_1 - \frac{1}{p}\right)} \|g_j \mid B_p^{\sigma_1}(C_{\varrho,j})\|. \tag{6}$$

By (5) and (6) we have

$$2^{\nu a_1 \varrho j} \|\varkappa_j Rf \mid B_p^{\sigma_1}(C_\varrho)\| \leqq c \|\psi_j f \mid B_p^{\bar{s}}(R_2)\| \tag{7}$$

with

$$\frac{s}{s_1}\varrho\nu = a_1\varrho\nu = -\frac{s}{s_1}\varrho\left(\sigma_1 - \frac{1}{p}\right) + s - \frac{2}{p}. \tag{8}$$

However

$$\sigma_1 - \frac{1}{p} = s_1\left(1 - \frac{1}{ps_2} - \frac{1}{ps_1}\right) = \frac{s_1}{s}\left(s - \frac{2}{p}\right). \tag{9}$$

By (8) and (9) follows $\nu = \mu$, where μ has the meaning of the theorem. By (7) with $\nu = \mu$ and Proposition 4.5.3 we have that R is a continuous mapping from $\dot{B}_p^{\bar{s}}(R_2)$ into $B_p^{\sigma_1}(C_\varrho, \mu)$ (as far as the second summand on the right-hand side of (4.5.3/3) is concerned we use a homogeneity argument). The last remarks at the end of the first step show that R is even a retraction. This proves (ii).

Step 3. Let $0 < s_1 < s_2 \leqq \frac{1}{p}$ and $f \in \dot{B}_p^{\bar{s}}(R_2)$. In order to prove (i) we may assume $f \in C_0^\infty(\dot{R}_2)$. We recall that any $h(x) \in C_0^\infty(R_2)$ can be approximated in the isotropic Besov space $B_p^{s_2}(R_2) = B_{p,p}^{s_2}(R_2)$ by functions from $C_0^\infty(R_2)$ with support off $\{x \mid x = (x_1, 0)\}$, cf. [I, 2.9.3.]. By the above constructions and $B_p^{s_2}(R_2) \subset B_p^{\bar{s}}(R_2)$ follows (i).

Step 4. The proof of (iii) is the same as the proof of (ii) where one uses the corresponding assertions from 4.2.3.

Remark 1. Let Λ_1 be the extension operator which we described in 4.2.3., cf. in particular (4.2.3/9). We may assume that Λ_1 is a common coretraction with respect to the retraction Γ_1 from 4.2.3. for all spaces $B_p^{\bar{s}}(R_2)$ and $W_p^{\bar{s}}(R_2)$ with the same anisotropy $\bar{a} = (a_1, a_2)$ and $s_2 > \frac{1}{p}$. A corresponding assertion for isotropic Besov-Sobolev spaces is known, cf. [S, proof of Theorem 2.4.2] or [T, proo of Theorem 2.7.2]. There is no doubt that this assertion can be extended to anisotropic Besov-Sobolev spaces. By the above procedure this assertion can be extended to our situation: If the anisotropy $\bar{a}$ is fixed then there exists a common coretraction (extension operator) for the trace operator R from the parts (ii) and (iii) of the above theorem. This observation is useful for interpolation purposes. On this way one can extend the above theorem to spaces of type $\dot{B}_{p,q}^{\bar{s}}(R_2)$ and $\dot{H}_p^{\bar{s}}(R_2)$, where it is quite clear what is meant by these notations. We do not go into detail.

Remark 2. Of peculiar interest is the case $s = \frac{2}{p}$, where s is the mean smoothness. Then $\bar{s} = (s_1, s_2)$ is critical because (4.3.2/1) is satisfied with $m_1 = m_2 = 0$. We have $s_2 \geqq \frac{2}{p}$ and $\sigma_1 - \frac{1}{p} = 0$, cf. (9). Then it follows that the trace of this space $\dot{B}_p^{\bar{s}}(R_2)$ (or $\dot{W}_p^{\bar{s}}(R_2)$ if s_1 and s_2 are natural numbers) is the unweighted Besov space $\tilde{B}_p^{1/p}(C_\varrho) = B_p^{1/p}(C_\varrho, 0)$, which, in turn, is a space with a critical index, cf. 4.5.2. In particular in this case, the first term on the right-hand side of (4.5.3/2) with $s = \frac{1}{p}$ and $\mu = 0$ cannot be replaced by $\|g \mid L_p(C_\varrho)\|$, in contrast to the cases with $\mu = 0$ and $s - \frac{1}{p} +$ integer.

Remark 3. Another interesting case of the above theorem is $\varrho = \frac{a_2}{a_1} = \frac{s_1}{s_2}$. Then (9) yields

$$\sigma_1 - \frac{1}{p} + \mu = \frac{s_1}{s}\left(s - \frac{2}{p}\right)\left(1 + \frac{1}{\varrho} - 1\right) = \frac{s_2}{s}\left(s - \frac{2}{p}\right) = \sigma_2 - \frac{1}{p}, \tag{10}$$

cf. (4.2.3/7). In [S, 2.7.3.] we described what is meant by a "differential dimension" of a function space, which reflects the homogeneity behaviour of the spaces under the mapping $f(x) \to f(\lambda x)$. By (10) the differential dimension of the trace space $B_p^{\sigma_1}(C_\varrho, \mu)$ is the same as for the spaces $B_p^{\sigma_2}(R_1)$. This is quite natural because $B_p^{\sigma_2}(R_1)$ is the trace of $B_p^{\bar{s}}(R_2)$ on the line $\{x \mid x = (0, x_2)\}$, cf. (4.2.3/10).

4.5.5. Traces of the Spaces $\dot{B}_p^{\bar{s}}(R_2)$ and $\dot{W}_p^{\bar{s}}(R_2)$: Complements

Let $\bar{a} = (a_1, a_2)$ with $0 < a_2 \leqq a_1 < \infty$ and $a_1 + a_2 = 2$ be the given anisotropy. Let C_ϱ be the curve given by (4.5.3/1). As usual $\bar{s} = (s_1, s_2)$ with $s_1 = \frac{s}{a_1}$, $s_2 = \frac{s}{a_2}$ and $s > 0$ measures the anisotropic smoothness. Theorem 4.5.4 dealt with the traces of the spaces $\dot{B}_p^{\bar{s}}(R_2)$ (and $\dot{W}_p^{\bar{s}}(R_2)$) on C_ϱ if $\frac{a_2}{a_1} \leqq \varrho < 1$. There are two other cases: (i) $1 \leqq \varrho < \infty$ and (ii) $0 < \varrho < \frac{a_2}{a_1}$. The limiting case $\varrho = \frac{a_2}{a_1}$ which we described in Remark 4.5.4/3 gives a faint hint that the trace of $\dot{B}_p^{\bar{s}}(R_2)$ on C_ϱ with $0 < \varrho < \frac{a_2}{a_1}$ could be a mixture of a space $B_p^{\sigma_1}(C_\varrho, \mu)$ and a space $B_p^{\sigma_2}(C_\varrho, 0)$, where σ_1 and σ_2 have the meaning of (4.2.3/7). There are no substantial results at this moment in the framework of our approach. However we refer to the papers of the Russian school which we mentioned in 4.5.1. The other case, i.e. $1 \leqq \varrho < \infty$, can be treated rather easily in our framework. We recall that $\tilde{B}_p^{\sigma_1}(C_\varrho) = B_p^{\sigma_1}(C_\varrho, 0)$ are the unweighted Besov spaces on C_ϱ, cf. Definition 4.5.3.

Theorem 1. *Let $1 < p < \infty$. Let $\bar{a} = (a_1, a_2)$ with $0 < a_2 \leqq a_1 < \infty$ and $a_1 + a_2 = 2$ be the anisotropy. Let $1 \leqq \varrho < \infty$. Let $\bar{s} = (s_1, s_2)$ with $s_1 = \frac{s}{a_1}$, $s_2 = \frac{s}{a_2}$ and $s > 0$.*

(i) *Let $s_2 \leqq \frac{1}{p}$. Then the set (4.5.4/1) is dense in $\dot{B}_p^{\bar{s}}(R_2)$.*

(ii) *Let $s_2 > \frac{1}{p}$ and $\sigma_1 = s_1 \left(1 - \frac{1}{ps_2}\right)$. Then the trace operator R: $f(x) \to f \mid C_\varrho$ (restriction of $f(x)$ on the curve C_ϱ) is a retraction from $\dot{B}_p^{\bar{s}}(R_2)$ onto $\tilde{B}_p^{\sigma_1}(C_\varrho)$.*

(iii) *Let s_1 and s_2 be natural numbers. Then the trace operator R is a retraction from $\dot{W}_p^{\bar{s}}(R_2)$ onto $\tilde{B}_p^{\sigma_1}(C_\varrho)$.*

Proof. (*Hints*). One can follow the proof of Theorem 4.5.4 with some minor modifications as far as Proposition 4.5.3 is concerned. We omit details. Cf. also H. Triebel [12, II, 3.2.].

Remark 1. The theorem is not a surprise, cf. 4.2.3. and the method from Step 1 of the proof of Theorem 4.5.4. Furthermore, one has an immediate counterpart of Remark 4.5.4/1.

One of the main aims of the series H. Triebel [12, 13] is to apply the above theory to boundary value problems for semi-elliptic differential equations in the unit circle. We shall give a brief description of this theory in Section 4.8. As far as the underlying theory in R_2 is concerned we are interested in spaces $\dot{B}_p^{\bar{s}}(R_2)$ and $\dot{W}_p^{\bar{s}}(R_2)$ with $\bar{s} = (s_1, 2s_1)$, i.e. $s_2 = 2s_1$, and in traces of functions and of some of its derivatives on the curve C_ϱ from (4.5.3/1) with $\varrho = \frac{1}{2}$. More precisely: Let $R = (R_1, R_2)$ with

$$R_1 f = f \mid C_\varrho \quad \text{and} \quad R_2 f = \frac{\partial f}{\partial x_2} \mid C_\varrho. \tag{1}$$

We ask under what conditions R is a retraction from $\dot{B}_p^{\bar{s}}(R_2)$ (or $\dot{W}_p^{\bar{s}}(R_2)$) onto some spaces $B_p^{\varkappa_1}(C_\varrho, \mu_1) \times B_p^{\varkappa_2}(C_\varrho, \mu_2)$, where obviously, R_1 and R_2 map in $B_p^{\varkappa_1}(C_\varrho, \mu_1)$ and $B_p^{\varkappa_2}(C_\varrho, \mu_2)$, respectively. We recall what is meant by a retraction in this context. There exists a linear and bounded operator T from $B_p^{\varkappa_1}(C_\varrho, \mu_1) \times B_p^{\varkappa_2}C_\varrho, \mu_2)$ in

$\dot{B}_p^{\bar{s}}(R_2)$ (or $\dot{W}_p^{\bar{s}}(R_2)$) with

$$RT = I \quad \text{(identity in } B_p^{\varkappa_1}(C_\varrho, \mu_1) \times B_p^{\varkappa_2}(C_\varrho, \mu_2)).$$

This formulation includes both direct and inverse embeddings on C_ϱ.

Theorem 2. *Let* $1 < p < \infty$, $s_1 > \frac{1}{2}\left(1 + \frac{1}{p}\right)$, *and* $\bar{s} = (s_1, 2s_1)$. *Let* $\varrho = \frac{1}{2}$.

(i) *R is a retraction from* $\dot{B}_p^{\bar{s}}(R_2)$ *onto*

$$B_p^{s_1 - \frac{1}{2p}}\left(C_\varrho, s_1 - \frac{3}{2p}\right) \times B_p^{s_1 - \frac{1}{2}\left(1 + \frac{1}{p}\right)}\left(C_\varrho, s_1 - \frac{3}{2p} - \frac{1}{2}\right). \tag{2}$$

(ii) *Let* s_1 *be a natural number. Then R is a retraction from* $\dot{W}_p^{\bar{s}}(R_2)$ *onto the space in* (2).

Proof. (*Outline*). We have $\sigma_1 = s_1 - \frac{1}{2p}$ and $\mu = s_1 - \frac{3}{2p}$ in Theorem 4.5.4. This explains the first factor in (2). By (4.2.3/12) we have $\frac{\partial f}{\partial x_2} \in B_p^{(s_1 - \frac{1}{2}, 2s_1 - 1)}(R_2)$. This explains the second factor in (2). Of course this is not a full proof. We refer to H. Triebel [13, 2.4.] for the necessary details.

Remark 2. One can extend this theorem to other cases. The above restrictions come mainly from the application of this theorem to boundary value problems for semi-elliptic differential equations.

4.5.6. Traces of the Spaces $B_p^{\bar{s}}(R_2)$ and $W_p^{\bar{s}}(R_2)$

We are mostly interested in the dotted spaces $\dot{B}_p^{\bar{s}}(R_2)$ and $\dot{W}_p^{\bar{s}}(R_2)$. They are the basis of our considerations on boundary value problems for semi-elliptic differential equations which we describe briefly in 4.8. But of course a description of traces of the undotted spaces $B_p^{\bar{s}}(R_2)$ and $W_p^{\bar{s}}(R_2)$ is of interest for its own sake. Let again $\bar{a} = (a_1, a_2)$ with $0 < a_2 < a_1 < \infty$ and $a_1 + a_2 = 2$ be a given anisotropy. Let $\bar{s} = (s_1, s_2)$ with $s_1 = \frac{s}{a_1}$, $s_2 = \frac{s}{a_2}$ and $s > 0$ be non-critical in the sense of Definition 4.3.2/1. Let $\nu(x) \in C_0^\infty(R_2)$ be a function which is identically 1 near the origin. By (4.3.2/2) and (4.3.2/3) we have

$$B_p^{\bar{s}}(R_2) = \dot{B}_p^{\bar{s}}(R_2) \oplus \left\{\nu(x)\, x_1^{m_1} x_2^{m_2} \text{ with } \frac{1}{s_1}\left(m_1 + \frac{1}{p}\right) + \frac{1}{s_2}\left(m_2 + \frac{1}{p}\right) < 1\right\} \tag{1}$$

and (if in addition s_1 and s_2 are natural numbers)

$$W_p^{\bar{s}}(R_2) = \dot{W}_p^{\bar{s}}(R_2) \oplus \left\{\nu(x)\, x_1^{m_1} x_2^{m_2} \text{ with } \frac{1}{s_1}\left(m_1 + \frac{1}{p}\right) + \frac{1}{s_2}\left(m_2 + \frac{1}{p}\right) < 1\right\}. \tag{2}$$

Here m_1 and m_2 in (1) and (2) are non-negative integers. Let C_ϱ be the curve from (4.5.3/1) where we assume that $\frac{s_1}{s_2} = \frac{a_2}{a_1} \leqq \varrho < 1$, i.e. we restrict our attention to our main case from 4.5.4. Let again σ be the arc length on C_ϱ, cf. the beginning of

4.5.3. Then the curve C_ϱ can be described near the origin by $x_2 = x_2(\sigma)$ and $x_1(\sigma) = x_2^{1/\varrho}(\sigma)$ with $x_2 = \sigma(1 + o(1))$. On C_ϱ we have

$$x_1^{m_1}(\sigma)\, x_2^{m_2}(\sigma) = x_2^{\frac{m_1}{\varrho}+m_2}(\sigma). \tag{3}$$

Then the trace of $B_p^{\bar{s}}(R_2)$ on C_ϱ near the origin is given by the direct product of the trace of $\dot{B}_p^{\bar{s}}(R_2)$ on C_ϱ (near the origin) and the finite-dimensional space spanned by those functions (3) which are not contained in the trace of $\dot{B}_p^{\bar{s}}(R_2)$. We remark that functions from (3) with different exponents $\frac{m_1}{\varrho} + m_2$ are linearly independent. This follows from $x_2(\sigma) = \sigma(1 + o(1))$. Let $\varkappa(\sigma) \in C^\infty[0, \infty)$ with $\varkappa(\sigma) = 1$ if $0 \leqq \sigma < 1$ and $\varkappa(\sigma) = 0$ if $\sigma \geqq 2$.

Theorem. *Let $1 < p < \infty$. Let $\bar{a} = (a_1, a_2)$ with $0 < a_2 < a_1 < \infty$ and $a_1 + a_2 = 2$ be the anisotropy. Let $\frac{a_2}{a_1} \leqq \varrho < 1$. Let $\bar{s} = (s_1, s_2)$ with $s_1 = \frac{s}{a_1}$, $s_2 = \frac{s}{a_2}$ and $s > 0$ be non-critical.*

(i) *Let $s_2 \leqq \frac{1}{p}$. Then the set from (4.5.4/1) is dense in $B_p^{\bar{s}}(R_2)$.*

(ii) *Let $s_2 > \frac{1}{p}$, $\sigma_1 = s_1\left(1 - \frac{1}{ps_2}\right)$ and $\mu = \left(\frac{1}{\varrho} - 1\right)\left(\sigma_1 - \frac{1}{p}\right)$. Then the trace operator R: $f(x) \to f \mid C_\varrho$ (restriction of $f(x)$ on the curve C_ϱ) is a retraction from $B_p^{\bar{s}}(R_2)$ onto*

$$B_p^{\sigma_1}(C_\varrho, \mu) \oplus \operatorname{span}\left\{\varkappa(\sigma)\, x_2^{\frac{m_1}{\varrho}+m_2}(\sigma);\ \frac{m_1}{\varrho} + m_2 \leqq \frac{s_1}{\varrho}\left(1 - \frac{2}{sp}\right)\right\} \tag{4}$$

(m_1 and m_2 are non-negative integers).

(iii) *Let s_1 and s_2 be natural numbers. Then the above trace operator R is a retraction from $W_p^{\bar{s}}(R_2)$ onto the space in (4) where σ_1 and μ have the above meaning.*

Proof. If $s_2 \leqq \frac{1}{p}$ then $\dot{B}_p^{\bar{s}}(R_2) = B_p^{\bar{s}}(R_2)$ and part (i) coincides with part (i) of Theorem 4.5.4. We prove (ii). We claim that $\varkappa(\sigma)\, x_2^{\frac{m_1}{\varrho}+m_2}(\sigma) \in B_p^{\sigma_1}(C_\varrho, \mu)$ if and only if

$$\frac{m_1}{\varrho} + m_2 > \sigma_1 + \mu - \frac{1}{p} = \frac{s_1}{\varrho}\left(1 - \frac{2}{sp}\right), \tag{5}$$

cf. (4.5.4/9). The only-if-part follows from $x_2(\sigma) \sim \sigma$ and Definition 4.5.3. The if-part is a consequence of [T, Proposition 2.8.4]. Hence, (4) is the trace space, cf. also Remark 1 below. The existence of a coretraction (extension operator) can be based on Theorem 4.5.4 and the fact that the second space in (4) is finite-dimensional. The proof of (iii) is the same.

Remark 1. The condition for m_1 and m_2 in (1) can be reformulated as $\frac{m_1}{s_1} + \frac{m_2}{s_2} < 1 - \frac{2}{sp}$ and hence as

$$\frac{m_1}{\varrho} + \frac{s_1}{\varrho s_2} m_2 < \frac{s_1}{\varrho}\left(1 - \frac{2}{sp}\right). \tag{6}$$

Because $\bar{s}$ is non-critical one can replace $<$ in (6) by $\leqq$. Furthermore because $\frac{s_1}{\varrho s_2} \leqq 1$ the condition for m_1 and m_2 in (4) is more restrictive than in (6) (as it must be). If $\varrho = \frac{s_1}{s_2}$ then these two conditions coincide. If $\varrho > \frac{s_1}{s_2}$ then it may happen that some functions $x_2^{\frac{m_1}{\varrho}+m_2}(\sigma)$ where m_1 and m_2 satisfy (6) belong to $B_p^{\sigma_1}(C_\varrho, \mu)$.

If $\frac{1}{\varrho}$ is a natural number then one has a more handsome description of (4). We recall that $[\alpha]$ stands for the largest integer smaller than or equal to α.

Proposition. *Let the hypotheses of the above theorem be satisfied. Let* $s_2 > \frac{1}{p}$ *and let in addition* $\frac{1}{\varrho} = k$ *be a natural number. Then* (4) *can be replaced by*

$$B_p^{\sigma_1}(C_\varrho, \mu) \oplus \operatorname{span}\left\{\varkappa(\sigma)\,\sigma^l \quad \textit{with} \quad l = 0, 1, \ldots, \left[ks_1\left(1 - \frac{2}{sp}\right)\right]\right\}. \tag{7}$$

Proof. Because $\frac{1}{\varrho} = k$, the arc length

$$\sigma = \int_0^{x_2} \sqrt{1 + \frac{1}{\varrho}\lambda^{2\left(\frac{1}{\varrho}-1\right)}}\, d\lambda = x_2(1 + x_2 P(x_2))$$

is an analytic function of x_2 for small values of $|x_2|$. Hence $x_2(\sigma) = \sigma(1 + \ldots)$ is also an analytic function for small values of σ. Then (7) follows from (4).

Remark 2. Let $\mu = 0$. Then we have $\sigma_1 = \frac{1}{p}$ and by (4.5.4/9) $s = \frac{2}{p}$. Then the space in (7) is reduced to $\tilde{B}_p^{1/p}(C_\varrho) \oplus \{\varkappa(\sigma)\}$, cf. Definition 4.5.3. By (4.5.2/3) this space is different from $B_p^{1/p}(C_\varrho)$. This example shows that there is no easy way to incorporate the second space in (7) in the first one.

4.6. Anisotropic Spaces on the Unit Circle

4.6.1. Definitions

In Section 4.6. we deal with anisotropic spaces of Sobolev-Besov type on the unit circle K in the plane,

$$K = \{x \mid x = (x_1, x_2), x_1^2 + x_2^2 < 1\}. \tag{1}$$

In contrast to the corresponding spaces on R_2 we restrict ourselves to a description. Proofs and further details may be found in the relevant parts of the underlying papers H. Triebel [12, 13] and the cited references. We recall that the Besov spaces $B_p^{\bar{s}}(R_2)$ and the Sobolev spaces $W_p^{\bar{s}}(R_2)$ on the plane R_2 have been introduced in Definition 4.2.1. All notations have the previous meaning. In particular, $\bar{a} = (a_1, a_2)$ with $0 < a_2 \leqq a_1 < \infty$ and $a_1 + a_2 = 2$ stands for the anisotropy. We introduce the additional restriction $a_1 \leqq 2a_2$ for reasons which we discuss in the next subsection.

Definition 1. *Let* $\bar{a} = (a_1, a_2)$ *with* $0 < a_2 \leqq a_1 \leqq 2a_2$ *and* $a_1 + a_2 = 2$. *Let* $-\infty < s < \infty$ *and* $\bar{s} = (s_1, s_2)$ *with* $s_1 = \frac{s}{a_1}$ *and* $s_2 = \frac{s}{a_2}$. *Let* $1 < p < \infty$. *Then* $B_p^{\bar{s}}(K)$ *is the restriction of* $B_p^{\bar{s}}(R_2)$ *to* K, *normed by*

$$\|f \mid B_p^{\bar{s}}(K)\| = \inf \|g \mid B_p^{\bar{s}}(R_2)\|, \tag{2}$$

where the infimum is taken over all $g \in B_p^{\bar{s}}(R_2)$ *with* $g(x) = f(x)$ *if* $x \in K$ *(in the sense of distributions on* K*). If in addition* s_1 *and* s_2 *are natural numbers then* $W_p^{\bar{s}}(K)$ *is the restriction of* $W_p^{\bar{s}}(R_2)$ *to* K*, normed by*

$$\|f \mid W_p^{\bar{s}}(K)\| = \inf \|g \mid W_p^{\bar{s}}(R_2)\|, \tag{3}$$

where the infimum is taken over all $g \in W_p^{\bar{s}}(R_2)$ *with* $g(x) = f(x)$ *if* $x \in K$ *(in the sense of distributions on* K*).*

Remark 1. Of course, this definition works for all (bounded or unbounded) domains on R_2. Furthermore, in the same way one can introduce anisotropic Bessel-potential spaces $H_p^{\bar{s}}(K)$. The restriction $a_1 \leqq 2a_2$ ensures that one has intrinsic descriptions of the spaces $B_p^{\bar{s}}(K)$ and $W_p^{\bar{s}}(K)$, cf. 4.6.2.

Next we introduce the counterparts of the system $\Psi^{\bar{a}}(R_2)$ and of the spaces $\dot{B}_p^{\bar{s}}(R_2)$ and $\dot{W}_p^{\bar{s}}(R_2)$ from 4.4.1. and 4.4.3., respectively. Let again $\bar{a} = (a_1, a_2)$ with $0 < a_2 \leqq a_1 < \infty$ and $a_1 + a_2 = 2$ be a given anisotropy. The anisotropic distance $|x|_{\bar{a}}$ has been defined in (4.2.1/1). Let $x^0 = (-1, 0)$ and $x^1 = (1, 0)$. These are the two singular points of the circle K in our theory (the substitutes of the origin in the plane). Then $\Psi^{\bar{a}}(K)$ is the collection of all systems $\psi = \{\psi_j(x)\}_{j=-\infty}^{\infty} \subset S(R_2)$ with the following properties:

(i) $$\operatorname{supp} \psi_j \subset \{x \mid 2^{-j-1} < |x - x^0|_{\bar{a}} < 2^{-j+1}\} \quad \text{if} \quad j = 1, 2, 3, \ldots,$$
$$\operatorname{supp} \psi_j \subset \{x \mid 2^{j-1} < |x - x^1|_{\bar{a}} < 2^{j+1}\} \quad \text{if} \quad j = -1, -2, -3, \ldots,$$
$$\operatorname{supp} \psi_0 \subset \{x \mid |x - x^0|_{\bar{a}} > \tfrac{1}{2}, |x - x^1|_{\bar{a}} > \tfrac{1}{2}\}.$$

(ii) For every multi-index $\alpha = (\alpha_1, \alpha_2)$ there exists a positive number c_α with

$$2^{-|j|a_1\alpha_1 - |j|a_2\alpha_2} |D^\alpha \psi_j(x)| \leqq c_\alpha$$

for any integer j and $x \in R_2$.

(iii) $$\sum_{j=-1}^{\infty} \psi_j(x) = 1 \quad \text{if} \quad x \in K.$$

This is the counterpart of the system $\Psi^{\bar{a}}(R_2)$ from 4.4.1. It is a smooth anisotropic resolution of unity in the unit circle K with respect to the singular points x^0 and x^1. As usual, $D'(K)$ stands for the complex distributions on K.

Definition 2. *Let* $\bar{a} = (a_1, a_2)$ *with* $0 < a_2 \leqq a_1 \leqq 2a_2$ *and* $a_1 + a_2 = 2$. *Let* $-\infty < s < \infty$ *and* $\bar{s} = (s_1, s_2)$ *with* $s_1 = \dfrac{s}{a_1}$ *and* $s_2 = \dfrac{s}{a_2}$. *Let* $1 < p < \infty$ *and let* $\psi \in \Psi^{\bar{a}}(K)$. *Then*

$$\dot{B}_p^{\bar{s}}(K) = \left\{ f \mid f \in D'(K), \|f \mid \dot{B}_p^{\bar{s}}(K)\|^\psi = \left(\sum_{j=-\infty}^{\infty} \|\psi_j f \mid B_p^{\bar{s}}(K)\|^p \right)^{1/p} < \infty \right\}. \tag{4}$$

If in addition s_1 *and* s_2 *are natural numbers then*

$$\dot{W}_p^{\bar{s}}(K) = \left\{ f \mid f \in D'(K), \|f \mid \dot{W}_p^{\bar{s}}(K)\|^\psi = \left(\sum_{j=-\infty}^{\infty} \|\psi_j f \mid W_p^{\bar{s}}(K)\|^p \right)^{1/p} < \infty \right\}. \tag{5}$$

Remark 2. This is the counterpart of Definition 4.4.3/1. The restriction $a_1 \leqq 2a_2$ ensures that we have intrinsic descriptions of these spaces which will be described in the next subsection. In particular the spaces $\dot{B}_p^{\bar{s}}(K)$ and $\dot{W}_p^{\bar{s}}(K)$ are independent of the choice of $\psi \in \Psi^{\bar{a}}(K)$ (in the sense of equivalent norms). Furthermore, they are Banach spaces.

4.6.2. Equivalent Norms

We recall that the anisotropy $\bar{a} = (a_1, a_2)$ with $a_1 + a_2 = 2$ is restricted by $0 < a_2 \leqq a_1 \leqq 2a_2$. Let again $\bar{s} = (s_1, s_2)$ with $s_1 = \frac{s}{a_1}$ and $s_2 = \frac{s}{a_2}$, where $-\infty < s < \infty$. If $s > 0$ then we are interested in descriptions of the spaces $B_p^{\bar{s}}(K)$, $W_p^{\bar{s}}(K)$, and $\dot{B}_p^{\bar{s}}(K)$, $\dot{W}_p^{\bar{s}}(K)$ which are the respective counterparts of (4.2.2/5), (4.2.2/3), and (4.4.3/11), (4.4.3/12). Furthermore, if $s < 0$ we ask for a counterpart of (4.4.4/4). First we introduce the necessary notations. The partial differences $\Delta_{h,1}^m$ and $\Delta_{h,2}^m$ have the same meaning as in 4.2.2. Let

$$[\Delta_{h,1}^m(K) f](x) = \begin{cases} (\Delta_{h,1}^m f)(x) & \text{if} \quad x \in K \quad \text{and} \quad (x_1 + mh, x_2) \in K \\ 0 & \text{otherwise}, \end{cases} \tag{1}$$

where $m = 1, 2, 3, \ldots$ and $h \in R_1$. In other words, $[\Delta_{h,1}^m(K) f](x) = (\Delta_{h,1}^m f)(x)$ if all the needed points $(x_1 + jh, x_2)$ with $j = 0, 1, \ldots, m$ are in K, otherwise we put $[\Delta_{h,1}^m(K) f](x) = 0$. In the same way, $[\Delta_{h,2}^m(K) f](x)$ is defined. $\|f \mid L_p(K)\| = \left(\int_K |f(x)|^p \, dx\right)^{1/p}$ has the usual meaning.

Proposition. *Let $1 < p < \infty$ and let $\bar{s} = (s_1, s_2)$ with $0 < s_1 \leqq s_2 \leqq 2s_1$.*

(i) *Let m_1 and m_2 be two natural numbers with $m_1 > s_1$ and $m_2 > s_2$. Then*

$$\begin{aligned} &\|f \mid B_p^{\bar{s}}(K)\|_{\bar{m}} \\ &= \|f \mid L_p(K)\| + \left(\int_0^1 [h^{-s_1 p} \|[\Delta_{h,1}^{m_1}(K) f](\cdot) \mid L_p(K)\|^p \right. \\ &\qquad \left. + h^{-s_2 p} \|[\Delta_{h,2}^{m_2}(K) f](\cdot) \mid L_p(K)\|^p] \frac{dh}{h}\right)^{1/p} \end{aligned} \tag{2}$$

is an equivalent norm in $B_p^{\bar{s}}(K)$, ($\bar{m} = (m_1, m_2)$).

(ii) *Let in addition s_1 and s_2 be natural numbers. Then*

$$\|f \mid W_p^{\bar{s}}(K)\| = \|f \mid L_p(K)\| + \left\| \frac{\partial^{s_1} f}{\partial x_1^{s_1}} \mid L_p(K) \right\| + \left\| \frac{\partial^{s_2} f}{\partial x_2^{s_2}} \mid L_p(K) \right\| \tag{3}$$

is an equivalent norm in $W_p^{\bar{s}}(K)$.

Remark 1. This is the point where the restriction $a_1 \leqq 2a_2$ (i.e. $s_2 \leqq 2s_1$ in the above proposition) comes in. Under this assumption part (i) and part (ii) can be found in O. V. Besov, V. P. Il'in, S. M. Nikol'skij [1, § 18, in particular 18.5.] and O. V. Besov, V. P. Il'in, S. M. Nikol'skij [1, 9.6. and 9.8.], respectively. However we must add a technical remark. If $s_1 \neq s_2$ then K has not the strong (s_1, s_2)-edge property which is required in O. V. Besov, V. P. Il'in, S. M. Nikol'skij [1, 18.5.]. But one can overcome this difficulty in the same way as in O. V. Besov, V. P. Il'in, S. M. Nikol'skij [1, 9.8.] (we recall that $s_2 \leqq 2s_1$). Cf. also O. V. Besov, V. P. Il'in, S. M. Nikol'skij [1, 8.1. and 16.7.]. On the basis of the just given references one can also prove that the restriction operator from $B_p^{\bar{s}}(R_2)$ onto $B_p^{\bar{s}}(K)$ and from $W_p^{\bar{s}}(R_2)$ onto $W_p^{\bar{s}}(K)$ is a retraction, i.e. there exists a linear and bounded extension operator from $B_p^{\bar{s}}(K)$ into $B_p^{\bar{s}}(R_2)$ and from $W_p^{\bar{s}}(K)$ into $W_p^{\bar{s}}(R_2)$, respectively. In connection with the extension problem we refer also to B. L. Fajn [1].

Next we summarize some properties of the spaces $\dot{B}_p^{\bar{s}}(K)$ and $\dot{W}_p^{\bar{s}}(K)$ from Definition 4.6.1/2. We recall Definition 4.4.3/2 where critical and non-critical couples $\bar{s} = (s_1, s_2)$ have been described. Furthermore we introduce the anisotropic distance

$$|x|_{\bar{a}}(K) = \min(|x - x^0|_{\bar{a}}, |x - x^1|_{\bar{a}}), \quad x \in K, \tag{4}$$

where $x^0 = (-1, 0)$ and $x^1 = (1, 0)$ have the same meaning as in 4.6.1. (anisotropic distance of $x \in K$ to the two critical points x^0 and x^1). $\|f \mid B_p^{\bar{s}}(K)\| = \|f \mid B_p^{\bar{s}}(K)\|_{\bar{m}}$ for some $\bar{m}$ and $\|f \mid W_p^{\bar{s}}(K)\|$ are given by (2) and (3), respectively. We recall that $\bar{s} = (s_1, s_2)$ with $s_1 = \frac{s}{a_1}$ and $s_2 = \frac{s}{a_2}$ if $\bar{a} = (a_1, a_2)$ and $-\infty < s < \infty$ stand for the anisotropy and the mean smoothness, respectively.

Theorem. *Let $\bar{a} = (a_1, a_2)$ with $0 < a_2 \leqq a_1 \leqq 2a_2$ and $a_1 + a_2 = 2$. Let $1 < p < \infty$.*

(i) *Let $-\infty < s < \infty$. Then $\dot{B}_p^{\bar{s}}(K)$ from Definition 4.6.1/2 is a Banach space and all norms $\|f \mid \dot{B}_p^{\bar{s}}(K)\|^{\psi}$ with $\psi \in \Psi^{\bar{a}}(K)$ are mutually equivalent.*

(ii) *Let $s > 0$. Then*

$$\dot{B}_p^{\bar{s}}(K) = \{f \mid f \in D'(K), \|f \mid \dot{B}_p^{\bar{s}}(K)\|^* < \infty\} \tag{5}$$

with

$$\|f \mid \dot{B}_p^{\bar{s}}(K)\|^* = \|f \mid B_p^{\bar{s}}(K)\| + \left(\int_K |x|_{\bar{a}}(K)^{-sp} |f(x)|^p \,dx\right)^{1/p}.$$

If in addition $\bar{s}$ is non-critical then

$$\dot{B}_p^{\bar{s}}(K) = \left\{f \mid f \in B_p^{\bar{s}}(K), \frac{\partial^{m_1+m_2} f}{\partial x_1^{m_1} \partial x_2^{m_2}}(x^0) = \frac{\partial^{m_1+m_2} f}{\partial x_1^{m_1} \partial x_2^{m_2}}(x^1) = 0 \right.$$
$$\left. \text{if } \frac{1}{s_1}\left(m_1 + \frac{1}{p}\right) + \frac{1}{s_2}\left(m_2 + \frac{1}{p}\right) < 1\right\}. \tag{6}$$

(iii) *Let $s \leqq 0$ and let $\bar{s}$ be non-critical. Then*

$$\dot{B}_p^{\bar{s}}(K) = B_p^{\bar{s}}(K). \tag{7}$$

(iv) *Let $\bar{s} = (s_1, s_2)$ where s_1 and s_2 are natural numbers. Then $\dot{W}_p^{\bar{s}}(K)$ from Definition 4.6.1/2 is a Banach space and all norms $\|f \mid \dot{W}_p^{\bar{s}}(K)\|^{\psi}$ with $\psi \in \Psi^{\bar{a}}(K)$ are mutually equivalent. Furthermore*

$$\dot{W}_p^{\bar{s}}(K) = \{f \mid f \in D'(K), \|f \mid \dot{W}_p^{\bar{s}}(K)\|^* < \infty\} \tag{8}$$

with

$$\|f \mid \dot{W}_p^{\bar{s}}(K)\|^* = \|f \mid W_p^{\bar{s}}(K)\| + \left(\int_K |x|_{\bar{a}}(K)^{-sp} |f(x)|^p \,dx\right)^{1/p}.$$

If in addition $\bar{s}$ is non-critical then

$$\dot{W}_p^{\bar{s}}(K) = \left\{f \mid f \in W_p^{\bar{s}}(K), \frac{\partial^{m_1+m_2} f}{\partial x_1^{m_1} \partial x_2^{m_2}}(x^0) = \frac{\partial^{m_1+m_2} f}{\partial x_1^{m_1} \partial x_2^{m_2}}(x^1) = 0 \right.$$
$$\left. \text{if } \frac{1}{s_1}\left(m_1 + \frac{1}{p}\right) + \frac{1}{s_2}\left(m_2 + \frac{1}{p}\right) < 1\right\}. \tag{9}$$

Remark 2. This theorem coincides essentially with Theorem 4 in H. Triebel [12, I]. As for the proof we refer to this paper. At least partly the proof is a technical modification of the arguments of the corresponding assertions for the spaces on R_2. Part (ii) and part (iv) are the counterparts of Theorem 4.4.3. Part (iii) is the counterpart of Theorem 4.4.4 (ii) which looks now simpler.

Remark 3. As a consequence of the above theorem one obtains the following Hardy inequality. Let $1 < p < \infty$ and $s > 0$. Let $\bar{a} = (a_1, a_2)$ with $0 < a_2 \leqq a_1 \leqq 2a_2$ and $a_1 + a_2 = 2$. Let $\bar{s} = (s_1, s_2)$ with $s_1 = \frac{s}{a_1}$ and $s_2 = \frac{s}{a_2}$ be non-ciritical (cf. Definition 4.3.2/1). Then there exists a positive number c such that

$$\int_K |x|_{\bar{a}}(K)^{-sp} \, | f(x)|^p \, dx \leqq c \| f \mid B_p^{\bar{s}}(K)\|^p \tag{10}$$

holds for all f from the right-hand side of (6), where $\|f \mid B_p^{\bar{s}}(K)\| = \|f \mid B_p^{\bar{s}}(K)\|_{\bar{m}}$ is given by (2). If in addition s_1 and s_2 are natural numbers then there exists a positive number c such that

$$\int_K |x_{\bar{a}}|(K)^{-sp} \, |f(x)|^p \, dx \leqq c \| f \mid W_p^{\bar{s}}(K)\|^p \tag{11}$$

holds for all f from the right-hand side of (9) where $\|f \mid W_p^{\bar{s}}(K)\|$ is given by (3).

4.6.3. Traces

In 4.5.4. we studied traces of $\dot{B}_p^{\bar{s}}(R_2)$ and $\dot{W}_p^{\bar{s}}(R_2)$ on the curve C_ϱ from (4.5.3/1). If we replace the Besov-Sobolev spaces on the plane R_2 by the corresponding spaces $\dot{B}_p^{\bar{s}}(K)$ and $\dot{W}_p^{\bar{s}}(K)$ on the unit circle K, then one can ask for traces on the circumference

$$\partial K = \{x \mid x = (x_1, x_2), x_1^2 + x_2^2 = 1\}.$$

The critical points are $x^0 = (-1, 0)$ and $x^1 = (1, 0)$. The boundary ∂K behaves near these points like a curve C_ϱ with $\varrho = \frac{1}{2}$. In order to apply Theorem 4.5.4 we need the additional restriction $\frac{a_2}{a_1} \leqq \varrho = \frac{1}{2}$, i.e. $a_1 \geqq 2a_2$, for the anisotropy $\bar{a} = (a_1, a_2)$. On the other hand in the two preceding subsections we always assumed $0 < a_2 \leqq a_1 \leqq 2a_2$. This justifies to specialize the anisotropy $\bar{a} = (a_1, a_2)$ in this subsection by $\bar{a} = (2a_2, a_2)$ with $a_2 > 0$, or the anisotropic smoothness $\bar{s} = (s_1, s_2)$ by $\bar{s} = (s_1, 2s_1)$. Under this additional restriction it is not complicated to use Theorem 4.5.4 in order to prove a corresponding trace theorem for the spaces $\dot{B}_p^{\bar{s}}(K)$ and $\dot{W}_p^{\bar{s}}(K)$. We introduce the necessary notations. Let λ be the arc length on ∂K such that $\lambda = 0$ corresponds to $x^1 = (1, 0)$. Then $0 \leqq \lambda < 2\pi$ and $\lambda = \pi$ corresponds to $x^0 = (-1, 0)$. Let

$$d(\lambda) = \begin{cases} \min(\lambda, \pi - \lambda) & \text{if} \quad 0 \leqq \lambda \leqq \pi \\ \min(\lambda - \pi, 2\pi - \lambda) & \text{if} \quad \pi < \lambda < 2\pi. \end{cases}$$

In the following definition the arc length on ∂K is extended 2π-periodically in the usual way.

Definition. *Let $1 < p < \infty$, $s > 0$ and $-\infty < \mu < \infty$. Let m be a natural number with $m > s$. Then $B_p^s(\partial K, \mu)$ is the collection of all functions $g(\lambda)$, defined on ∂K, such that*

$$\|g \mid B_p^s(\partial K, \mu)\|^* = \left(\int_0^{2\pi} d^{-sp-\mu p}(\lambda) \, |g(\lambda)|^p \, d\lambda\right)^{1/p} + \left(\int_0^{2\pi}\int_0^{2\pi} \tau^{-sp} |[\Delta_\tau^m(d^{-\mu}g)](\lambda)|^p \, d\lambda \frac{d\tau}{\tau}\right)^{1/p}$$

is finite.

Remark 1. This is the counterpart of Definition 4.5.3.

Theorem. *Let* $1 < p < \infty$ *and* $0 < 2s_1 = s_2$.

(i) *Let* $s_2 \leqq \frac{1}{p}$. *Then* $C_0^\infty(K) = D(K)$ *is dense in* $\dot{B}_p^{\bar{s}}(K)$.

(ii) *Let* $s_2 > \frac{1}{p}$, $\sigma_1 = s_1\left(1 - \frac{1}{ps_2}\right) = s_1 - \frac{1}{2p}$ *and* $\mu = \sigma_1 - \frac{1}{p} = s_1 - \frac{3}{2p}$. *Then the trace operator* $R_0\colon f(x) \to f \mid \partial K$ *(restriction of* $f(x)$ *on* ∂K*) is a retraction from* $\dot{B}_p^{\bar{s}}(K)$ *onto* $B_p^{\sigma_1}(\partial K, \mu)$.

(iii) *If in addition* s_1 *is a natural number then* R_0 *is a retraction from* $\dot{W}_p^{\bar{s}}(K)$ *onto* $B_p^{\sigma_1}(\partial K, \mu)$ *where* σ_1 *and* μ *have the same meaning as in* (ii).

Remark 2. This theorem follows easily from Theorem 4.5.4. We recall that "retraction" covers both direct and inverse embedding assertions, cf. 4.2.3. for explanations.

Remark 3. Similar as in 4.5.6. one can use the above theorem in order to determine the traces of the undotted spaces $B_p^{\bar{s}}(K)$ and $W_p^{\bar{s}}(K)$ from Definition 4.6.1/1 and Proposition 4.6.2. Let $\varkappa_0(\lambda) \in C^\infty(\partial K)$ (resp. $\varkappa_1(\lambda) \in C^\infty(\partial K)$) be 1 in a neighbourhood of x^0 (resp. x^1) and 0 in a neighbourhood of x^1 (resp. x^0). Let $1 < p < \infty$, $0 < 2s_1 = s_2$ and $\bar{s} = (s_1, s_2)$ be non-critical. Then one has the following assertions: (i) If $s_2 \leqq \frac{1}{p}$ then $C_0^\infty(K)$ is dense in $B_p^{\bar{s}}(K)$, (ii) If $s_2 > \frac{1}{p}$, $\sigma_1 = s_1 - \frac{1}{2p}$ and $\mu = s_1 - \frac{3}{2p}$ then the trace operator R_0 is a retraction from $B_p^{\bar{s}}(K)$ onto

$$B_p^{\sigma_1}(\partial K, \mu) \oplus \operatorname{span}\left\{\varkappa_0(\lambda)\, d^l(\lambda),\ \varkappa_1(\lambda)\, d^l(\lambda) \quad \text{with} \quad l = 0, \ldots, \left[s_2 - \frac{3}{p}\right]\right\}. \tag{1}$$

(iii) If (in addition) s_1 is a natural number then R_0 is a retraction from $W_p^{\bar{s}}(K)$ onto the space in (1). These assertions follow from Proposition 4.5.6, where we used that the mean smoothness s is given by $\frac{1}{s} = \frac{1}{2}\left(\frac{1}{s_1} + \frac{1}{s_2}\right) = \frac{3}{4s_1}$.

Remark 4. In connection with semi-elliptic differential operators one needs the trace operator

$$R\colon f(x) \to \left\{f \,\middle|\, \partial K, \frac{\partial f}{\partial x_2} \,\middle|\, \partial K\right\}.$$

Minor modifications of Theorem 4.5.5/2 yield the following result. Let $1 < p < \infty$. If $s_2 > 1 + \frac{1}{p}$ then R is a retraction from $\dot{B}_p^{\bar{s}}(K)$ onto

$$B_p^{s_1 - \frac{1}{2p}}\left(\partial K, s_1 - \frac{3}{2p}\right) \times B_p^{s_1 - \frac{1}{2}\left(1 + \frac{1}{p}\right)}\left(\partial K, s_1 - \frac{3}{2p} - \frac{1}{2}\right). \tag{2}$$

If in addition s_1 is a natural number, then R is a retraction from $\dot{W}_p^{\bar{s}}(K)$ onto the space in (2). For details we refer to H. Triebel [13, 2.4.].

4.7. Weighted Anisotropic Sobolev Spaces on Domains

4.7.1. Definitions and Problems

Let $0 < \varrho < 1$ and let $H(t)$ be a positive C^∞-function on the interval $(-1, 1)$ with

$$\lim_{t \downarrow -1} \frac{H(t)}{(t+1)^\varrho} = b_{-1} > 0 \quad \text{and} \quad \lim_{t \uparrow 1} \frac{H(t)}{(1-t)^\varrho} = b_1 > 0. \tag{1}$$

This section deals with weighted anisotropic Sobolev spaces on the plane R_2 and on the domain

$$\Omega = \{x \mid x = (x_1, x_2) \in R_2, |x_1| < 1, |x_2| < H(x_1)\}. \tag{2}$$

The most prominent example is the unit circle $\Omega = K$ from (4.6.1/1). In some sense Section 4.7. is the continuation of Section 4.6. and the problems treated there. We recall the definition of (unweighted) anisotropic Sobolev spaces on R_2. Let $1 < p < \infty$ and $\bar{l} = (l_1, l_2)$ be a couple of natural numbers with $0 < l_1 \leqq l_2$. Then

$$W_p^{\bar{l}}(R_2) = \Big\{ f \mid f \in L_p(R_2), \|f \mid W_p^{\bar{l}}(R_2)\| = \|f \mid L_p(R_2)\| + \left\| \frac{\partial^{l_1} f}{\partial x_1^{l_1}} \mid L_p(R_2) \right\| + \left\| \frac{\partial^{l_2} f}{\partial x_2^{l_2}} \mid L_p(R_2) \right\| < \infty \Big\}, \tag{3}$$

cf. 4.2.2. Furthermore, let

$$\|f \mid W_p^{\bar{l}}(\Omega)\| = \|f \mid L_p(\Omega)\| + \left\| \frac{\partial^{l_1} f}{\partial x_1^{l_1}} \mid L_p(\Omega) \right\| + \left\| \frac{\partial^{l_2} f}{\partial x_2^{l_2}} \mid L_p(\Omega) \right\|. \tag{4}$$

As in Definition 4.6.1/1 we define $W_p^{\bar{l}}(\Omega)$ as the restriction of $W_p^{\bar{l}}(R_2)$ on Ω normed via the obvious counterpart of (4.6.1/3). The two problems of interest read as follows.

(i) (Intrinsic description and extension). Is $W_p^{\bar{l}}(\Omega)$ the collection of all $f \in L_p(\Omega)$ with $\|f \mid W_p^{\bar{l}}(\Omega)\| < \infty$ (in the sense of the distributions from $D'(\Omega)$) and is the restriction of $W_p^{\bar{l}}(R_2)$ onto $W_p^{\bar{l}}(\Omega)$ a retraction? (The second part of this question coincides with the problem whether there exists a linear and bounded extension operator from $W_p^{\bar{l}}(\Omega)$ into $W_p^{\bar{l}}(R_2)$.)

(ii) (traces) What about the traces of $W_p^{\bar{l}}(R_2)$ or $W_p^{\bar{l}}(\Omega)$ on $\partial\Omega$?

It is known that the first question has an affirmative answer if and only if $\varrho \leqq \frac{l_1}{l_2}$, cf. Remark 4.6.2/1 for references. On the other hand we determined the traces of $\mathring{W}_p^{\bar{l}}(R_2)$ and $W_p^{\bar{l}}(R_2)$ on the curve C_ϱ (and hence via technical modifications also on $\partial\Omega$) under the additional restriction $\frac{l_1}{l_2} \leqq \varrho$, cf. Theorem 4.5.4(iii) and 4.5.6 (and also 4.6.3 as far as $\varrho = \frac{1}{2}$ is concerned). In other words: Only in the "optimal" case $\frac{l_1}{l_2} = \varrho$ we have answers for the two problems as far as unweighted anisotropic Sobolev spaces are concerned.

The aim of this section is to deal with the above problems (i) and (ii) for weighted anisotropic Sobolev spaces $W_p^{\bar{l}}(\Omega, \varkappa)$, where $\varkappa$ stands for a weight and $\frac{l_1}{l_2} < \varrho$ (strong anisotropies). The basic idea reads as follows. Appropriate weights are able to relax the anisotropy quotient $\frac{l_1}{l_2}$ such that $W_p^{\bar{l}}(\Omega, \varkappa)$ can be treated as anisotropic spaces of Sobolev-Besov type with the optimal anisotropy quotient ϱ. This effect is not new. In somewhat other contexts it can be found in V. G. Perepelkin [1], O. V. Besov [2], P. I. Lizorkin [5]. Section 4.7. is based on H. Triebel [15]. We restrict ourselves to a description of the main results and refer for proofs and details to the just cited paper.

We introduce some definitions. Let $\bar{a} = (a_1, a_2)$ with $0 < a_2 \leqq a_1 < \infty$ and $a_1 + a_2 = 2$ be a given anisotropy which fits optimal to the given domain Ω, i.e.

$a_2 = \varrho a_1$. In other words, $a_1 = \dfrac{2}{1+\varrho}$ and $a_2 = \dfrac{2\varrho}{1+\varrho}$. As in 4.6.1. and 4.6.2. the critical points of Ω are given by $x^0 = (-1, 0)$ and $x^1 = (1, 0)$. The anisotropic distance $|x|_{\bar{a}}$ has been defined in (4.2.1/1). Similar as in (4.6.2/4) we introduce

$$|x|_{\bar{a}}(\Omega) = \min(|x - x^0|_{\bar{a}}, |x - x^1|_{\bar{a}}), \quad x \in R_2.$$

Let $\bar{l} = (l_1, l_2)$ be a couple of natural numbers with $0 < \dfrac{l_1}{l_2} < \varrho$ and let

$$\varkappa = \left(l_2 - \frac{l_1}{\varrho}\right) a_2 = 2\frac{\varrho l_2 - l_1}{1+\varrho} > 0. \tag{5}$$

Let $1 < p < \infty$. Then

$$W_p^{\bar{l}}(R_2, \varkappa) = \left\{ f \mid f \in L_p(R_2), \|f \mid W_p^{\bar{l}}(R_2, \varkappa)\| = \|f \mid L_p(R_2)\| \right.$$
$$\left. + \left\| \frac{\partial^{l_1} f}{\partial x_1^{l_1}} \mid L_p(R_2) \right\| + \left\| |x|_{\bar{a}}(\Omega)^{\varkappa} \frac{\partial^{l_2} f}{\partial x_2^{l_2}} \mid L_p(R_2) \right\| < \infty \right\}. \tag{6}$$

Furthermore, let

$$\|f \mid W_p^{\bar{l}}(\Omega, \varkappa)\| = \|f \mid L_p(\Omega)\| + \left\| \frac{\partial^{l_1} f}{\partial x_1^{l_1}} \mid L_p(\Omega) \right\| + \left\| |x|_{\bar{a}}(\Omega)^{\varkappa} \frac{\partial^{l_2} f}{\partial x_2^{l_2}} \mid L_p(\Omega) \right\|. \tag{7}$$

Finally $W_p^{\bar{l}}(\Omega, \varkappa)$ is the restriction of $W_p^{\bar{l}}(R_2, \varkappa)$ on Ω equipped in the usual way with the norm

$$\|f \mid W_p^{\bar{l}}(\Omega, \varkappa)\|^* = \inf \|g \mid W_p^{\bar{l}}(R_2, \varkappa)\|, \tag{8}$$

where the infimum is taken over all $g \in W_p^{\bar{l}}(R_2, \varkappa)$ with $g \mid \Omega = f$ (in the sense of the distributions on Ω).

Let $f \in W_p^{\bar{l}}(R_2, \varkappa)$ and let m_1 and m_2 be two non-negative integers with

$$m_1 + \frac{1}{p} + \varrho\left(m_2 + \frac{1}{p}\right) < l_1. \tag{9}$$

Then it follows from the considerations in H. Triebel [15] that $\dfrac{\partial^{m_1+m_2} f}{\partial x_1^{m_1} \partial x_2^{m_2}}(x)$ is a continuous and bounded function on R_2. In particular,

$$\mathring{W}_p^{\bar{l}}(R_2, \varkappa) = \left\{ f \mid f \in W_p^{\bar{l}}(R_2, \varkappa), \frac{\partial^{m_1+m_2} f}{\partial x_1^{m_1} \partial x_2^{m_2}}(x^0) = \frac{\partial^{m_1+m_2} f}{\partial x_1^{m_1} \partial x_2^{m_2}}(x^1) = 0 \right.$$
$$\left. \text{if (9) is satisfied} \right\} \tag{10}$$

and

$$\mathring{W}_p^{\bar{l}}(\Omega, \varkappa) = \left\{ f \mid f \in W_p^{\bar{l}}(\Omega, \varkappa), \frac{\partial^{m_1+m_2} f}{\partial x_1^{m_1} \partial x_2^{m_2}}(x^0) = \frac{\partial^{m_1+m_2} f}{\partial x_1^{m_1} \partial x_2^{m_2}}(x^1) = 0 \right.$$
$$\left. \text{if (9) is satisfied} \right\} \tag{11}$$

make sense. We restrict the definitions (10) and (11) to non-critical couples $\bar{l} = (l_1, l_2)$. In our context (i.e. anisotropic weighted Sobolev spaces of the above type where $1 < p < \infty$ and $0 < \varrho < 1$ are given) a couple $\bar{l} = (l_1, l_2)$ is called non-critical if there do not exist non-negative integers n_1 and n_2 with

$$n_1 + \frac{1}{p} + \varrho\left(n_2 + \frac{1}{p}\right) = l_1 . \tag{12}$$

In other words, $\left(l_1, \frac{l_1}{\varrho}\right)$ is non-critical in the sense of Definition 4.3.2/1. Of course this is only a condition for l_1 (if p and ϱ are given). But for sake of simplicity we prefer the above formulation.

We deal with the two problems under consideration (extensions and traces) in the language of the dotted spaces $\dot{W}_p^{\bar{l}}(R_2, \varkappa)$ and $\dot{W}_p^{\bar{l}}(\Omega, \varkappa)$. The corresponding problems for the undotted spaces $W_p^{\bar{l}}(R_2, \varkappa)$ and $W_p^{\bar{l}}(\Omega, \varkappa)$ can be treated afterwards in the same way as in 4.5.6. and Remark 4.6.3/3.

4.7.2. Extensions and Traces

We deal with the two problems from 4.7.1. All notations have the same meaning as in 4.7.1. In particular, the couple $\bar{l} = (l_1, l_2)$ of natural numbers is called non-critical (with respect to the given numbers $1 < p < \infty$ and $0 < \varrho < 1$ and the anisotropic weighted Sobolev spaces $\dot{W}_p^{\bar{l}}(\cdot, \varkappa)$) if there do not exist non-negative integers n_1 and n_2 with (4.7.1/12).

Theorem 1. *Let $2 \leqq p < \infty$ and let $0 < \varrho < 1$. Let $\bar{l} = (l_1, l_2)$ be a non-critical couple of natural numbers in the above sense with $\frac{l_1}{l_2} < \varrho$. Let $\varkappa$ be given by* (4.7.1/5). *Then*

$$\dot{W}_p^{\bar{l}}(\Omega, \varkappa) = \{f \mid f \in L_p(\Omega),\ \|f \mid \dot{W}_p^{\bar{l}}(\Omega, \varkappa)\| < \infty\}, \tag{1}$$

where $\|f \mid \dot{W}_p^{\bar{l}}(\Omega, \varkappa)\|$ from (4.7.1/7) *is an equivalent norm in $\dot{W}_p^{\bar{l}}(\Omega, \varkappa)$. Furthermore there exists a linear and bounded extension operator from $\dot{W}_p^{\bar{l}}(\Omega, \varkappa)$ into $\dot{W}_p^{\bar{l}}(R_2, \varkappa)$.*

Remark 1. This theorem coincides essentially with Theorem 1 in H. Triebel [15]. The restriction $2 \leqq p$ is unnatural. It depends on our method (and the used references) in the cited paper. There is hardly any doubt that the theorem remains valid if $1 < p < 2$. If $\varrho = \frac{1}{k}$ with $k = 2, 3, \ldots$, then an extension of the theorem to all parameters $1 < p < \infty$ follows by minor modifications of the method in the cited paper.

In order to formulate the trace theorem we introduce weighted Besov spaces on $\partial\Omega$. Although these spaces are obvious modifications of the corresponding spaces from Definition 4.6.3 we give a detailed description. Let $2L$ be the length of $\partial\Omega$ and let λ be a parameter on $\partial\Omega$ measuring the arc length, $0 \leqq \lambda < 2L$, such that $\lambda = 0$ corresponds to $x^1 = (1, 0)$ and $\lambda = L$ corresponds to $x^0 = (-1, 0)$. Let

$$\mathrm{d}(\lambda) = \begin{cases} \min(\lambda, L - \lambda) & \text{if } 0 \leqq \lambda \leqq L \\ \min(\lambda - L, 2L - \lambda) & \text{if } L < \lambda < 2L . \end{cases}$$

Functions on $\partial\Omega$ are extended $2L$-periodically. Δ_λ^m denotes the usual differences on $\partial\Omega$ measured via the arc length λ. Then the generalization of Definition 4.6.3 reads as follows. Let $1 < p < \infty$, $s > 0$, and $-\infty < \mu < \infty$. Then $B^s_p(\partial\Omega, \mu)$ is the collection of all functions $g(\lambda)$, defined on $\partial\Omega$, such that

$$\|g \mid B^s_p(\partial\Omega, \mu)\|^* = \left(\int_0^{2L} d^{-sp-\mu p}(\lambda)\ |g(\lambda)|^p\ d\lambda\right)^{1/p} + \left(\int_0^{2L}\int_0^{2L} \tau^{-sp}\, |[\Delta^m_\tau(d^{-\mu}g)](\lambda)|^p\ d\lambda \frac{d\tau}{\tau}\right)^{1/p}$$

is finite. Now we are in the position to formulate the trace theorem.

Theorem 2. *Let* $2 \leqq p < \infty$ *and let* $0 < \varrho < 1$. *Let* $\bar{l} = (l_1, l_2)$ *be a non-critical couple of natural numbers in the above sense with* $\frac{l_1}{l_2} < \varrho$. *Let* $\varkappa$ *be given by* (4.7.1/5). *Let*

$$\sigma = l_1\left(1 - \frac{1}{pl_2}\right) \tag{2}$$

and

$$\mu = \left(\frac{1}{\varrho} - 1\right)\left(\sigma - \frac{1}{p}\right) - \left(1 - \frac{l_1}{\varrho l_2}\right)\frac{1}{p}. \tag{3}$$

Then the trace operator $R\colon f(x) \to f \mid \partial\Omega$ *(restriction of* $f(x)$ *on* $\partial\Omega$*) is a retraction from* $\mathring{W}^{\bar{l}}_p(\Omega, \varkappa)$ *onto* $B^\sigma_p(\partial\Omega, \mu)$.

Remark 2. This theorem coincides essentially with Theorem 2 in H. Triebel [15]. We recall that "retraction" covers both direct and inverse embedding assertions, cf. 4.2.3. for explanations. We discuss the numbers σ and μ from (2) and (3), respectively. σ is just the expected smoothness, cf. with the number σ_1 from Theorem 4.5.4(ii). Furthermore, in comparison with the number μ from Theorem 4.5.4(ii) we have now the additional term $\frac{1}{p}\left(1 - \frac{l_1}{\varrho l_2}\right)$ which stands for the deviation of the "optimal" weighted spaces from the "optimal" unweigthed spaces (in the latter case we have $l_1 = \varrho l_2$ and $\varkappa = 0$).

4.8. Boundary Value Problems for Semi-Elliptic Differential Equations

4.8.1. Introduction

Let again K be the unit circle in the plane,

$$K = \{x \mid x = (x_1, x_2), x_1^2 + x_2^2 < 1\}, \tag{1}$$

and let ∂K be its boundary,

$$\partial K = \{x \mid x = (x_1, x_2), x_1^2 + x_2^2 = 1\}.$$

Let $\eta(t) \in C^\infty[-1, 1]$ with $\eta(t) > 0$ if $-1 < t < 1$ and

$$\lim_{t\uparrow 1} \frac{\eta(t)}{1-t} = \lim_{t\downarrow -1} \frac{\eta(t)}{1+t} = 1. \tag{2}$$

We deal mainly with the boundary value problem

$$(\mathfrak{A}_\mu u)(x) = -\frac{\partial^2 u(x)}{\partial x_1^2} + \frac{\partial^4 u(x)}{\partial x_2^4} + \frac{\mu}{\eta^2(x_1)} u(x) = f(x), \quad x \in K, \tag{3}$$

$$u(y) = g_1(y), \qquad \frac{\partial u}{\partial x_2}(y) = g_2(y), \quad y \in \partial K, \tag{4}$$

where $f(x)$ with $x \in K$, and $g_1(y), g_2(y)$ with $y \in \partial K$ are given functions, and $\mu > 0$ is a given number. The unweighted counterpart of the problem (3), (4) reads as follows,

$$(A_\mu u)(x) = -\frac{\partial^2 u(x)}{\partial x_1^2} + \frac{\partial^4 u(x)}{\partial x_2^4} + \mu u(x) = f(x), \quad x \in K, \tag{3'}$$

$$u(y) = g_1(y), \qquad \frac{\partial u}{\partial x_2}(y) = g_2(y), \quad y \in \partial K. \tag{4'}$$

Equations of type (3) and (3′) are called semi-elliptic. We consider these problems in the framework of the anisotropic Besov spaces $\dot{B}_p^{\bar{s}}(K)$ and anisotropic Sobolev spaces $\dot{W}_p^{\bar{s}}(K)$ from Section 4.6. with $\bar{s} = (s_1, 2s_1)$. If $s_1 > 0$ then Theorem 4.6.2 gives a satisfactory description of these spaces. Furthermore by Theorem 4.6.3 and in particular by Remark 4.6.3/4 we have a detailed knowledge about the traces of these spaces. We are mainly interested in problem (3), (4). For this problem we describe a theory which has the same final character as the corresponding theory for boundary value problems for regular elliptic equations (formulated in the language of isotropic Besov-Sobolev spaces). For the problem (3′), (4′) we have only a partial result which we formulate in the final Subsection 4.8.5.

We restrict ourselves to a description. Proofs and further results may be found in the underlying paper H. Triebel [13]. Basically we combine methods from [I, Chapter 5] for regular elliptic boundary value problems in smooth bounded domains with the technique for anisotropic function spaces which we developed in this chapter. The first key assertion is an a priori estimate for the homogeneous problem (3), (4). This will be formulated in 4.8.2. The second main ingredient is an $L_2 - C^\infty$-smoothness theory which we describe in 4.8.3. These two assertions and the trace theorems from 4.6.3. yield our main result by standard arguments, cf. 4.8.4. Finally, in 4.8.5. we add an L_2-result both for (3), (4) and (3′), (4′).

Our method is of qualitative nature. We do not need the special structure of $\mathfrak{A}_\mu$ from (3). What we need is the semi-ellipticity (including the fact that the involved derivatives with respect to the x_2-direction are twice of those ones with respect to the x_1-direction) and a certain behaviour of the weight-factor in front of $u(x)$. In other words we describe a model case, and many other boundary value problems for semi-elliptic equations can be treated by the same technique.

4.8.2. A Priori Estimates

Let $1 < p < \infty$ and $\bar{s} = (s_1, s_2) = (s_1, 2s_1)$. Then $\dot{B}_p^{\bar{s}}(K)$ and $\dot{W}_p^{\bar{s}}(K)$ are the spaces from Section 4.6., cf. in particular Definition 4.6.1/2 (in the latter case s_1 is assumed to be a natural number). Let $\dot{W}_p^{\bar{s}}(K) = L_p(K)$ if $s_1 = 0$. We have the equivalent norms from Theorem 4.6.2 and the traces from Theorem 4.6.3 and Remark 4.6.3/4. If $\bar{s} = (s_1, 2s_1)$ then we put $\overline{s+2} = (s_1 + 2, 2s_1 + 4)$. We recall that $\mathfrak{A}_\mu$ is defined by (4.8.1/3).

Theorem. *Let* $1 < p < \infty$. *There exists a positive number* μ_0 *with the following property:*

(i) *If* $\mu > \mu_0$ *and if* $s_1 = 0, 1, 2, \ldots$, *then exists a positive number c such that*

$$\|f \mid \dot{W}_p^{\overline{s+2}}(K)\| \leqq c\|\mathfrak{A}_\mu f \mid \dot{W}_p^{\bar{s}}(K)\| \tag{1}$$

holds for all $f \in \dot{W}_p^{\overline{s+2}}(K)$ *with* $f\Big|\partial K = \dfrac{\partial f}{\partial x_2}\Big|\partial K = 0$.

(ii) *If* $\mu > \mu_0$ *and if* $s_1 > 0$ *then exists a positive number c such that*

$$\|f \mid \dot{B}_p^{\overline{s+2}}(K)\| \leqq c\|\mathfrak{A}_\mu f \mid \dot{B}_p^{\bar{s}}(K)\| \tag{2}$$

holds for all $f \in \dot{B}_p^{\overline{s+2}}(K)$ *with* $f\Big|\partial K = \dfrac{\partial f}{\partial x_2}\Big|\partial K = 0$.

Remark. This theorem coincides essentially with Theorem 3 in H. Triebel [13]. First one proves a corresponding a priori estimate for the half space $R_2^+ = \{x \mid x = (x_1, x_2), x_2 > 0\}$ instead of K. Afterwards one employs the decomposition technique which we used in Definition 4.6.1/2 in order to introduce the underlying spaces. The above theorem and Remark 4.6.3/4 yield corresponding a priori estimates with non-vanishing boundary data.

4.8.3. $L_2 - C^\infty$-Theory

We recall that $x^0 = (-1, 0)$ and $x^1 = (1, 0)$ are the two critical points in our theory. Let $C^\infty(\bar{K})$ be the collection of all complex-valued infinitely differentiable functions in the closure $\bar{K}$ of the unit circle K. Let

$$\dot{C}^\infty(\bar{K}) = \{g \mid g \in C^\infty(\bar{K}), (D^\alpha g)(x^0) = (D^\alpha g)(x^1) = 0 \text{ for all } \alpha\}.$$

Theorem. *Let* $\mathfrak{A}_\mu$ *be given by* (4.8.1/3). *There exists a number* μ_0 *such that* $\mathfrak{A}_\mu$ *with* $\mu > \mu_0$ *yields an isomorphic mapping from*

$$\left\{f \mid f \in \dot{C}^\infty(\bar{K}), f\Big|\partial K = \frac{\partial f}{\partial x_2}\Big|\partial K = 0\right\}$$

onto $\dot{C}^\infty(\bar{K})$.

Remark. This theorem coincides with Proposition 1 in H. Triebel [13]. The proof is based on an L_2-approach. First one proves by Hilbert space methods (and duality) that $\mathfrak{A}_\mu$ with $\mu > 0$ maps

$$\left\{f \mid f \in \dot{B}_2^{\bar{1}}(K), f\Big|\partial K = \frac{\partial f}{\partial x_2}\Big|\partial K = 0\right\}$$

isomorphically onto $B_2^{\overline{-1}}(K)$ with $\bar{1} = (1,2)$ and $\overline{-1} = (-1, -2)$. (We recall that $\dot{B}_2^{\overline{-1}}(K) = B_2^{\overline{-1}}(K)$, cf. (4.6.2/7).) Afterwards one uses the a priori estimates from Theorem 4.8.2 (with $p = 2$) in connection with well-known bootstrapping procedures and embeddings. In particular the above number μ_0 is the same as in Theorem 4.8.2.

4.8.4. The Main Theorem

The two theorems in the two preceding subsections are sufficient in order to handle the homogeneous boundary value problem (4.8.1/3), (4.8.1/4). The solution of the non-homogeneous boundary value problem (4.8.1/3), (4.8.1/4) is based on the just mentioned two theorems and the trace theorem from Remark 4.6.3/4. Let (as in

Remark 4.6.3/4)

$$R\colon f(x) \to \left\{ f \,\middle|\, \partial K, \frac{\partial f}{\partial x_2} \,\middle|\, \partial K \right\}$$

and

$$(\mathfrak{A}_\mu, R)\colon f(x) \to \left\{ (\mathfrak{A}_\mu f)(x) \quad \text{with} \quad x \in K, f \,\middle|\, \partial K, \frac{\partial f}{\partial x_2} \,\middle|\, \partial K \right\}. \tag{1}$$

We use the same notations as in 4.8.2. In particular $\bar{s} = (s_1, 2s_1)$ and $\overline{s+2} = (s_1 + 2, 2s_1 + 4)$.

Theorem. *Let* $\mathfrak{A}_\mu$ *be given by* (4.8.1/3) *and let* $1 < p < \infty$. *Let* μ_0 *be the number from Theorem* 4.8.2 *and let* $\mu > \mu_0$.

(i) *If* $s_1 = 0, 1, 2, \ldots$ *then* $(\mathfrak{A}_\mu, R)$ *yields an isomorphic mapping from* $\dot{W}_p^{\overline{s+2}}(K)$ *onto*

$$\dot{W}_p^{\bar{s}}(K) \times B_p^{s_1+2-\frac{1}{2p}}\left(\partial K, s_1 + 2 - \frac{3}{2p}\right) \times B_p^{s_1+\frac{3}{2}-\frac{1}{2p}}\left(\partial K, s_1 + \frac{3}{2} - \frac{3}{2p}\right).$$

(ii) *If* $s_1 > 0$ *then* $(\mathfrak{A}_\mu, R)$ *yields an isomorphic mapping from* $\dot{B}_p^{\overline{s+2}}(K)$ *onto*

$$\dot{B}_p^{\bar{s}}(K) \times B_p^{s_1+2-\frac{1}{2p}}\left(\partial K, s_1 + 2 - \frac{3}{2p}\right) \times B_p^{s_1+\frac{3}{2}-\frac{1}{2p}}\left(\partial K, s_1 + \frac{3}{2} - \frac{3}{2p}\right).$$

Remark. As far as the trace spaces are concerned we refer again to Remark 4.6.3/4. The above theorem is a rather final solution of the non-homogeneous boundary value problem (4.8.1/3), (4.8.1/4).

4.8.5. Complements

In Remark 4.8.3 we mentioned a mapping property of $\mathfrak{A}_\mu$ which is not covered by Theorem 4.8.4. We extend this result to non-homogeneous boundary data. This special case has the advantage that it holds not only for the weighted operator $\mathfrak{A}_\mu$ from (4.8.1/3) but also for its unweighted counterpart A_μ from (4.8.1/3′). We use again the above notations. In particular $x^0 = (-1, 0)$ and $x^1 = (1, 0)$. Furthermore let again $\bar{1} = (1, 2)$ and $\overline{-1} = (-1, -2)$. Let $(\mathfrak{A}_\mu, R)$ be given by (4.8.4/1) and let (A_μ, R) be defined similarly.

Proposition. *Let* $\mu > 0$. *Let* $\mathfrak{A}_\mu$ *and* A_μ *be given by* (4.8.1/3) *and* (4.8.1/3′). *Then both* $(\mathfrak{A}_\mu, R)$ *and* (A_μ, R) *yield an isomorphic mapping from*

$$\{f \mid f \in B_2^{\bar{1}}(K), f(x^0) = f(x^1) = 0\}$$

onto

$$B_2^{\overline{-1}}(K) \times B_2^{3/4}(\partial K, \tfrac{1}{4}) \times B_2^{1/4}(\partial K, -\tfrac{1}{4}).$$

Remark 1. We recall that $B_2^{\bar{1}}(K) = W_2^{\bar{1}}(K)$ is a Sobolev space. Furthermore, by (4.6.2/7) we have $B_2^{\overline{-1}}(K) = \dot{B}_2^{\overline{-1}}(K)$.

Remark 2. In connection with boundary value problems for semi-elliptic differential equations with constant coefficients we refer also to G. A. Karapetjan [1].

5. Further Types of Function Spaces

5.1. Weighted Spaces of Besov-Hardy-Sobolev Type

5.1.1. Introduction and Definitions

In [T] we developed the theory of the (unweighted non-homogeneous isotropic) spaces $B^s_{p,q}(R_n)$ and $F^s_{p,q}(R_n)$ of Besov-Hardy-Sobolev type, where $-\infty < s < \infty$, $0 < p \leqq \infty$ and $0 < q \leqq \infty$. These two scales cover many classical function spaces: the Besov-Lipschitz spaces $\Lambda^s_{p,q}(R_n) = B^s_{p,q}(R_n)$ with $s > 0$, $1 < p < \infty$, $1 \leqq q \leqq \infty$; the Hölder-Zygmund spaces $\mathscr{C}^s(R_n) = B^s_{\infty,\infty}(R_n)$ with $s > 0$; the Bessel-potential spaces $H^s_p(R_n) = F^s_{p,2}(R_n)$ with $-\infty < s < \infty$, $1 < p < \infty$ (including the Sobolev spaces $W^m_p(R_n) = H^m_p(R_n)$ with $1 < p < \infty$ and $m = 0, 1, 2, \ldots$ as special cases), the local Hardy spaces $h_p(R_n) = F^0_{p,2}(R_n)$ with $0 < p \leqq 1$, and also a local version $bmo(R_n) = F^0_{\infty,2}(R_n)$ of the spaces $BMO(R_n)$ of functions of bounded mean oscillation. At least in principle the theory of the (unweighted) spaces of entire analytic functions (including its vector-valued version) can serve as a basis in order to study the spaces $B^s_{p,q}(R_n)$ and $F^s_{p,q}(R_n)$. In Chapter 1 of this book we developed extensively the theory of the weighted spaces of entire analytic functions, including some vector-valued assertions. By the above remarks it is now quite natural to ask for a theory of weighted spaces of $B^s_{p,q} - F^s_{p,q}$ type. The basis of such a theory in a somewhat incomplete form has been described in H. Triebel [6]. A modified version of this paper (without proofs) has been given in [T, Chapter 7]. Recently, J. Franke [1] described further properties of these spaces. On the basis of these papers it is now quite clear that one can expect that the theory of the unweighted spaces $B^s_{p,q}(R_n)$ and $F^s_{p,q}(R_n)$ has a more or less full counterpart in the weighted case, at least as far as the underlying weights $\varrho(x)$ have no local singularities. If the weight $\varrho(x)$ has local singularities such as roots or poles then the situation is different. Later on we give some references as far as weights are concerned which obey Muckenhoupt's A_p-condition. However we wish to add a proposal. The theory of the weighted spaces of $B^s_{p,q} - F^s_{p,q}$ type as it will be described in the sequel is based on Theorem 1.9.1, related vector-valued assertions and considerations which avoid local singularities of the involved weights. However for the weighted spaces of $B^s_{p,q}$ type the scalar case of Theorem 1.9.1 or Theorem 1.5.2 would be sufficient. As the latter theorem shows one has then a much more greater flexibility of the involved weights. A careful examination should yield a larger class of admissible weights where some singularities may occur. However nothing has been done in this direction.

We recall that $\omega(x) \in \mathfrak{M}$ and $\varrho(x) \in R(\omega)$ have been introduced in 1.2.1. and 1.4.1., respectively. Furthermore, $L_p(R_n, \varrho(x))$ and $\|f_k \mid L_p(R_n, \varrho(x), l_q)\|$ have been defined in (1.3.1/4) and (1.3.1/8), (1.3.1/9), respectively. Let $\{f_k(x)\}_{k=0}^{\infty}$ be a sequence of functions and let $0 < p \leqq \infty$. We put

$$\|f_k \mid l_q(L_p(R_n, \varrho(x)))\| = \left(\sum_{k=0}^{\infty} \|f_k \mid L_p(R_n, \varrho(x))\|^q\right)^{1/q} \tag{1}$$

if $0 < q < \infty$ and

$$\|f_k \mid l_\infty(L_p(R_n, \varrho(x)))\| = \sup_k \|f_k \mid L_p(R_n, \varrho(x))\|, \tag{2}$$

cf. also the notations in (1.3.1/6) and (1.3.1/7). Finally we specialize $A^{\varphi,\Omega}_{\lambda\omega}$ from (1.9.1/2) by

$$A^{\varphi}_{\lambda\omega} = \sup_k \int_{R_n} |(F\varphi_k(2^k\cdot))(x)| \, e^{\lambda\omega(x)} \, dx, \tag{3}$$

where again $\varphi = \{\varphi_k(x)\}_{k=0}^{\infty} \subset D_\omega$. (Of course the supremum in (3) is taken over $k = 0, 1, 2, \ldots$, and D_ω is the space from Definition 1.2.1/2).

Definition 1. *Let $\omega(x) \in \mathfrak{M}$. Then Φ_ω denotes the collection of all systems $\varphi = \{\varphi_j(x)\}_{j=0}^{\infty} \subset D_\omega$ such that*

$$\begin{cases} \operatorname{supp} \varphi_0 \subset \{x \mid |x| \leqq 2\} \\ \operatorname{supp} \varphi_j \subset \{x \mid 2^{j-1} \leqq |x| \leqq 2^{j+1}\} \quad \text{if} \quad j = 1, 2, 3, \ldots, \end{cases} \tag{4}$$

$$A_{\lambda\omega}^{\varphi} < \infty \quad \textit{for every} \quad \lambda > 0, \tag{5}$$

$$\sum_{j=0}^{\infty} \varphi_j(x) = 1 \quad \textit{for every} \quad x \in R_n. \tag{6}$$

Remark 1. This is a smooth resolution of unity. The class Φ_ω is not empty. This can be proved in exactly the same way as in the classical case which corresponds to $\omega(x) = \log(1 + |x|)$, cf. [T, Remark 2.3.1/1]. In particular we may assume that $\varphi_0 \in D_\omega$, $\varphi_1 \in D_\omega$ and $\varphi_j(x) = \varphi_1(2^{-j+1}x)$ if $j = 2, 3, 4, \ldots$ We refer also to Proposition 1.2.2/1(ii).

Definition 2. *Let $\omega(x) \in \mathfrak{M}$ and $\varrho(x) \in R(\omega)$. Let $-\infty < s < \infty$, $0 < q \leqq \infty$ and $\varphi = \{\varphi_k(x)\}_{k=0}^{\infty} \in \Phi_\omega$.*

(i) *If $0 < p \leqq \infty$ then*

$$B_{p,q}^s(R_n, \varrho(x)) = \{f \mid f \in S'_\omega,\ \|f \mid B_{p,q}^s(R_n, \varrho(x))\|^\varphi = \|2^{js} F^{-1}\varphi_j Ff \mid l_q(L_p(R_n, \varrho(x)))\| < \infty\}. \tag{7}$$

(ii) *If $0 < p < \infty$ then*

$$F_{p,q}^s(R_n, \varrho(x)) = \{f \mid f \in S'_\omega,\ \|f \mid F_{p,q}^s(R_n, \varrho(x))\|^\varphi = \|2^{js} F^{-1}\varphi_j Ff \mid L_p(R_n, \varrho(x), l_q)\| < \infty\}. \tag{8}$$

Remark 2. We recall that S'_ω has been introduced in Definition 1.2.1/3. Furthermore, if $\varrho(x) \equiv 1$, then (7) and (8) coincide with the corresponding unweighted spaces $B_{p,q}^s(R_n)$ and $F_{p,q}^s(R_n)$ from [T, (2.3.1/5) and (2.3.1/6)]. One can also introduce spaces $F_{\infty,q}^s(R_n, \varrho(x))$ with $1 < q \leqq \infty$ in the same way as this has been done in the unweighted case in [T, Definition 2.3.4]. However we restrict our attention to the spaces $F_{p,q}^s(R_n, \varrho(x))$ with $p < \infty$. As we said there is hardly any doubt that the theory of the spaces $B_{p,q}^s(R_n, \varrho(x))$ and $F_{p,q}^s(R_n, \varrho(x))$ can be developed parallelly to [T, Chapter 2] which dealt with the unweighted spaces $B_{p,q}^s(R_n)$ and $F_{p,q}^s(R_n)$. We shall not be concerned with this comprehensive task but restrict ourselves to few significant assertions.

Remark* 3. The above weights $\varrho(x)$ have no local singularities. It should be possible to extend the class of admissible weights $\varrho(x)$ in the same way as it has been described in Remark 1.9.3/2. Weighted spaces of $F_{p,q}^s$-type, where the weights satisfy Muckenhoupt's A_p-condition have been studied by V. M. Kokilašvili [1–6] and Bui Huy Qui [3, 4]. One can expect that weighted spaces of type $F_{p,2}^0$ with $0 < p < \infty$ are near to corresponding Hardy spaces. As far as the latter are concerned we cite few articles where further references can be found: Bui Huy Qui [1, 2], J. Garcia-Cuerva [1], H. P. Heinig, R. Johnson [1], B. Muckenhoupt, R. L. Wheeden [1], J.-O. Strömberg, A. Torchinsky [1], J.-O. Strömberg, R. L. Wheeden [1], R. L. Wheeden [1].

5.1.2. Basic Properties

The first question is whether Definition 5.1.1/2 makes sense, i.e. whether the spaces $B_{p,q}^s(R_n, \varrho(x))$ and $F_{p,q}^s(R_n, \varrho(x))$ are independent of $\varphi \in \Phi_\omega$. Furthermore we explained at the beginning of 1.3.1. what is meant by a quasi-Banach space.

Theorem. *Let $\omega(x) \in \mathfrak{M}$ and $\varrho(x) \in R(\omega)$. Let $-\infty < s < \infty$ and $0 < q \leqq \infty$.*

(i) *Let $0 < p \leqq \infty$. Then $B^s_{p,q}(R_n, \varrho(x))$ is a quasi-Banach space (Banach space if $1 \leqq p \leqq \infty$ and $1 \leqq q \leqq \infty$) and the quasi-norms $\|f \mid B^s_{p,q}(R_n, \varrho(x))\|^\varphi$ with $\varphi \in \Phi_\omega$ and $\|f \mid B^s_{p,q}(R_n, \varrho(x))\|^\psi$ with $\psi \in \Phi_\omega$ are equivalent to each other.*

(ii) *Let $0 < p < \infty$. Then $F^s_{p,q}(R_n, \varrho(x))$ is a quasi-Banach space (Banach space if $1 \leqq p < \infty$ and $1 \leqq q \leqq \infty$) and the quasi-norms $\|f \mid F^s_{p,q}(R_n, \varrho(x))\|^\varphi$ with $\varphi \in \Phi_\omega$ and $\|f \mid F^s_{p,q}(R_n, \varrho(x))\|^\psi$ with $\psi \in \Phi_\omega$ are equivalent to each other.*

Proof. Step 1. Let $\varphi = \{\varphi_k(x)\}_{k=0}^\infty \in \Phi_\omega$ and $\psi = \{\psi_k(x)\}_{k=0}^\infty \in \Phi_\omega$. Let $\psi'_k(x) = \psi_{k-1}(x) + \psi_k(x) + \psi_{k+1}(x)$ with $\psi_{-1}(x) = 0$ where $k = 0, 1, 2, \ldots$ Then we have $\varphi_k(x) = \varphi_k(x)\,\psi'_k(x)$ and consequently

$$(F^{-1}\varphi_k Ff)(x) = \sum_{r=-1}^{1} (F^{-1}\varphi_k FF^{-1}\psi_{k+r}Ff)(x). \tag{1}$$

Let $0 < p < \infty$ and $0 < q \leqq \infty$. Then Definition 5.1.1/1 and (1.9.1/3) yield

$$\|2^{sk}F^{-1}\varphi_k Ff \mid L_p(R_n, \varrho(x), l_q)\| \leqq c\|2^{sk}F^{-1}\psi_k Ff \mid L_p(R_n, \varrho(x), l_q)\|. \tag{2}$$

This proves that $\|f \mid F^s_{p,q}(R_n, \varrho(x))\|^\varphi$ and $\|f \mid F^s_{p,q}(R_n, \varrho(x))\|^\psi$ are equivalent quasi-norms (the properties of a quasi-norm are clear). The corresponding assertion for the quasi-norms $\|f \mid B^s_{p,q}(R_n, \varrho(x))\|^\varphi$ follows from the scalar case of Theorem 1.9.1 which holds also for $p = \infty$, cf. Theorem 1.7.2.

Step 2. The proof of the completeness of $B^s_{p,q}(R_n, \varrho(x))$ and $F^s_{p,q}(R_n, \varrho(x))$ is now essentially the same as in the unweighted case, including some embedding and density assertions which we formulate below. As far as the unweighted case is concerned we refer to [T, 2.3.2., 2.3.3.].

Embedding and Density. Beside minor technical changes many arguments from the unweighted case can be carried over to the weighted case under consideration. The basis is Theorem 1.9.1 and its scalar counterparts (1.5.2/1) and (1.7.2/1). In particular in the same way as in [T, 2.3.3.] one proves the topological embeddings

$$S_\omega \subset B^s_{p,q}(R_n, \varrho(x)) \subset S'_\omega \tag{3}$$

if $-\infty < s < \infty$, $0 < p \leqq \infty$, $0 < q \leqq \infty$ and

$$S_\omega \subset F^s_{p,q}(R_n, \varrho(x)) \subset S'_\omega \tag{4}$$

if $-\infty < s < \infty$, $0 < p < \infty$, $0 < q \leqq \infty$. Furthermore, if $-\infty < s < \infty$, $0 < p < \infty$ and $0 < q < \infty$ then S_ω is dense both in $B^s_{p,q}(R_n, \varrho(x))$ and $F^s_{p,q}(R_n, \varrho(x))$.

Duality. By (3), (4) and the density of S_ω in $B^s_{p,q}(R_n, \varrho(x))$ and $F^s_{p,q}(R_n, \varrho(x))$ if $p < \infty$ and $q < \infty$ we can interprete linear and continuous functionals on these spaces in the usual way as elements of S'_ω. More precisely, $g \in S'_\omega$ belongs to the dual space $(B^s_{p,q}(R_n, \varrho(x)))'$ of $B^s_{p,q}(R_n, \varrho(x))$ with $-\infty < s < \infty$, $0 < p < \infty$ and $0 < q < \infty$ if and only if there exists a positive number c such that

$$|g(h)| \leqq c\|h \mid B^s_{p,q}(R_n, \varrho(x))\| \quad \text{for all} \quad h \in S_\omega. \tag{5}$$

Similarly for $F^s_{p,q}(R_n, \varrho(x))$ with $-\infty < s < \infty$, $0 < p < \infty$, $0 < q < \infty$. In (5) we wrote $\|\cdot\|$ instead of $\|\cdot\|^\varphi$ with $\varphi \in \Phi_\omega$. This is justified by the fact that all quasi-norms $\|\cdot\|^\varphi$ with $\varphi \in \Phi_\omega$ are mutually equivalent. Now one can carry over the consideration from [T, 2.11.] from the unweighted case to the weighted case under consideration. We formulate a result which will be useful later on:

(i) Let $-\infty < s < \infty$, $1 \leqq p < \infty$, $1 \leqq q < \infty$ and $\frac{1}{p} + \frac{1}{p'} = \frac{1}{q} + \frac{1}{q'} = 1$.

Then
$$(B^s_{p,q}(R_n, \varrho(x)))' = B^{-s}_{p',q'}(R_n, \varrho^{-1}(x)). \tag{6}$$

(ii) Let $-\infty < s < \infty$, $1 < p < \infty$, $1 \leqq q < \infty$ and $\frac{1}{p} + \frac{1}{p'} = \frac{1}{q} + \frac{1}{q'} = 1$.

Then
$$(F^s_{p,q}(R_n, \varrho(x)))' = F^{-s}_{p',q'}(R_n, \varrho^{-1}(x)). \tag{7}$$

Of course $\varrho(x) \in R(\omega)$ with $\omega(x) \in \mathfrak{M}$. Then we have also $\varrho^{-1}(x) \in R(\omega)$.

Complex Interpolation. One can try to extend the real and complex interpolation methods from [T, 2.4.] for the unweighted spaces $B^s_{p,q}(R_n)$ and $F^s_{p,q}(R_n)$ to the weighted spaces $B^s_{p,q}(R_n, \varrho(x))$ and $F^s_{p,q}(R_n, \varrho(x))$. We restrict ourselves to complex interpolation formulas which will be useful later on. Let $\omega(x) \in \mathfrak{M}$. Furthermore, let $A = \{z \mid 0 < \operatorname{Re} z < 1\}$ be a strip in the complex plane. Its closure $\{z \mid 0 \leqq \operatorname{Re} z \leqq 1\}$ is denoted by $\bar{A}$. We say that $f(z)$ is an S'_ω-analytic function in A if the following properties are satisfied:

(i) For every fixed $z \in \bar{A}$ we have $f(z) \in S'_\omega$,

(ii) $(F^{-1}\varphi Ff)(x, z)$ is a uniformly continuous and bounded function in $R_n \times \bar{A}$ for every $\varphi \in D_\omega$,

(iii) $(F^{-1}\varphi Ff)(x, z)$ is an analytic function in A for every $\varphi \in D_\omega$ and every fixed $x \in R_n$.

We recall that $(F^{-1}\varphi Ff)(x, z)$ is an entire analytic function in R_n with respect to x if $z \in \bar{A}$ is fixed.

Let $\varrho(x) \in R(\omega)$. We describe the weighted counterparts of the two definitions from [T, 2.4.] as far as spaces of type $F^s_{p,q}$ are concerned. Let $-\infty < s_0 < \infty$, $-\infty < s_1 < \infty$, $0 < q_0 \leqq \infty$, $0 < q_1 \leqq \infty$, $0 < p_0 < \infty$ and $0 < p_1 < \infty$. Then

$$\begin{aligned} &F(F^{s_0}_{p_0,q_0}(R_n, \varrho(x)), F^{s_1}_{p_1,q_1}(R_n, \varrho(x))) \\ &= \Big\{f(z) \mid f(z) \text{ is an } S'_\omega\text{-analytic function in } A, \\ &\qquad f(it) \in F^{s_0}_{p_0,q_0}(R_n, \varrho(x)),\ f(1+it) \in F^{s_1}_{p_1,q_1}(R_n, \varrho(x)) \text{ for every } t \in R_1, \\ &\qquad \|f(z) \mid F(F^{s_0}_{p_0,q_0}(R_n, \varrho(x)), F^{s_1}_{p_1,q_1}(R_n, \varrho(x)))\| \\ &\qquad = \max_{l=0,1} \sup_{t \in R_1} \|f(l+it) \mid F^{s_l}_{p_l,q_l}(R_n, \varrho(x))\| < \infty\Big\}. \end{aligned} \tag{8}$$

If $0 < \theta < 1$ then

$$\begin{aligned} &(F^{s_0}_{p_0,q_0}(R_n, \varrho(x)), F^{s_1}_{p_1,q_1}(R_n, \varrho(x)))_\theta \\ &= \{g \mid \exists f(z) \in F(F^{s_0}_{p_0,q_0}(R_n, \varrho(x)), F^{s_1}_{p_1,q_1}(R_n, \varrho(x))) \text{ with } g = f(\theta)\} \end{aligned} \tag{9}$$

and

$$\begin{aligned} &\|g \mid (F^{s_0}_{p_0,q_0}(R_n, \varrho(x)), F^{s_1}_{p_1,q_1}(R_n, \varrho(x)))_\theta\| \\ &= \inf \|f(z) \mid F(F^{s_0}_{p_0,q_0}(R_n, \varrho(x)), F^{s_1}_{p_1,q_1}(R_n, \varrho(x)))\| \end{aligned} \tag{10}$$

where the infimum is taken over all admissible functions $f(z)$ in the sense of (9).

Of course, one has immediate counterparts of (8)–(10) with $B^{s_0}_{p_0,q_0}(R_n, \varrho(x))$ and $B^{s_1}_{p_1,q_1}(R_n, \varrho(x))$ instead of $F^{s_0}_{p_0,q_0}(R_n, \varrho(x))$ and $F^{s_1}_{p_1,q_1}(R_n, \varrho(x))$, respectively, where $p_0 = \infty$ and $p_1 = \infty$ are admissible values. Now one can follow the arguments from [T, 2.4.6. and 2.4.7.]. Then one obtains the following result: Let $-\infty < s_0 < \infty$, $-\infty < s_1 < \infty$, $0 < q_0 \leqq \infty$, $0 < q_1 \leqq \infty$ and $0 < \theta < 1$.

(i) If $0 < p_0 \leqq \infty$, $0 < p_1 \leqq \infty$ and

$$s = (1-\theta)\, s_0 + \theta s_1, \qquad \frac{1}{p} = \frac{1-\theta}{p_0} + \frac{\theta}{p_1}, \qquad \frac{1}{q} = \frac{1-\theta}{q_0} + \frac{\theta}{q_1} \tag{11}$$

then

$$(B^{s_0}_{p_0,q_0}(R_n, \varrho(x)), B^{s_1}_{p_1,q_1}(R_n, \varrho(x)))_\theta = B^s_{p,q}(R_n, \varrho(x)). \tag{12}$$

(ii) If $0 < p_0 < \infty$ and $0 < p_1 < \infty$ and if (11) is satisfied then

$$(F^{s_0}_{p_0,q_0}(R_n, \varrho(x)), F^{s_1}_{p_1,q_1}(R_n, \varrho(x)))_\theta = F^s_{p,q}(R_n, \varrho(x)). \tag{13}$$

The proof of (12) and (13) is the same as in the unweighted case, cf. [T, 2.4.7.]. One has to use the following maximal inequality: Let $-\infty < s < \infty$, $0 < p < \infty$, $0 < q < \infty$ and $0 < r < \min(p, q)$. Let $\varphi = \{\varphi_k(x)\}_{k=0}^\infty \in \Phi_\omega$. Then exists a constant c such that

$$\left\| 2^{ks} \sup_{y \in R_n} \varrho(y) \frac{|(F^{-1}\varphi_k Ff)(y)|}{1 + |2^k(x-y)|^{n/r}} \mid L_p(R_n, \varrho(x), l_q) \right\| \leqq c \|f \mid F^s_{p,q}(R_n, \varrho(x))\| \tag{14}$$

holds for all $f \in F^s_{p,q}(R_n, \varrho(x))$. However this is just a special case of (1.9.1/3) if one uses (1).

Remark 1. One can consider (6), (7), and (12), (13) as examples how assertions for the unweighted spaces $B^s_{p,q}(R_n)$ and $F^s_{p,q}(R_n)$ can be extended to their weighted counterparts $B^s_{p,q}(R_n, \varrho(x))$ and $F^s_{p,q}(R_n, \varrho(x))$. One can try to prove further properties for the weighted spaces of $B^s_{p,q} - F^s_{p,q}$ type parallelly to Chapter 2 of [T] which deals with the corresponding unweighted spaces. However we describe in the following subsection another possibility. With the help of (6), (7) and (12), (13) we link directly the weighted and the unweighted spaces by an isomorphic mapping. This sheds new light on the weighted spaces. This possibility has been discovered by J. Franke [1].

Remark* 2. The above complex interpolation method $(\cdot, \cdot)_\theta$ is different from the classical complex interpolation method $[\cdot, \cdot]_\theta$ by A. P. Calderón, J.-L. Lions, and S. G. Krejn where the latter one is restricted to Banach spaces. The idea to use S'-analytic functions goes back to A. P. Calderón, A. Torchinsky [1, II, p. 135]. Cf. also L. Päivärinta [1], H. Triebel [10, 11] and [T, 2.4.4–2.4.7, 2.4.9].

5.1.3. Mapping Properties

We recall that $\mathfrak{M}$ and $R(\omega)$ have been introduced in Definition 1.2.1/1 and Definition 1.4.1(i), respectively. We complement the latter definition in the following way.

Definition. *Let $\omega(x) \in \mathfrak{M}$. Then $r(\omega)$ denotes the collection of all infinitely differentiable real functions $\varrho(x)$ on R_n with the following two properties:*

(i) *There exists a positive constant c such that*

$$0 < \varrho(x) \leqq c\varrho(y)\, e^{\omega(x-y)} \tag{1}$$

holds for every $x \in R_n$ and every $y \in R_n$.

(ii) *For all multi-indices* α *there exists a constant* c_α *such that*

$$|D^\alpha \varrho(x)| \leqq c_\alpha \varrho(x) \tag{2}$$

holds for every $x \in R_n$.

Remark 1. Of course, $r(\omega)$ is the smooth version of $R(\omega)$ from Definition 1.4.1(i).

Lemma. *Let* $\omega(x) \in \mathfrak{M}$ *and* $\varrho(x) \in R(\omega)$. *There exist a function* $\tilde{\varrho}(x) \in r(\omega)$ *and two positive numbers* c_1 *and* c_2 *such that*

$$c_1 \varrho(x) \leqq \tilde{\varrho}(x) \leqq c_2 \varrho(x) \tag{3}$$

holds for every $x \in R_n$.

Proof. Let $\varphi(x) \in D_\omega$ with $\varphi(x) \geqq 0$ for all $x \in R_n$ and $\varphi(0) > 0$. Cf. Definition 1.2.1/2(i) and Remark 1.2.2/1. In particular we have (1.2.1/8) for every $\lambda > 0$. Then

$$\tilde{\varrho}(x) = \int_{R_n} \varphi(x-y)\, \varrho(y)\, \mathrm{d}y, \quad x \in R_n, \tag{4}$$

is well-defined, satisfies (3) and consequently also (1) with $\tilde{\varrho}$ instead of ϱ. By (1.2.1/8) and (1.4.1/2) this function has also the desired property (2) with $\tilde{\varrho}$ instead of ϱ.

Remark 2. This simple but very effective observation is due to J. Franke [1].

We are interested in mapping properties of $f \to \varrho f$ with $\varrho \in r(\omega)$ between the above spaces of type $B^s_{p,q}$ and $F^s_{p,q}$. If $f \in S_\omega$ then ϱf must be understood in the sense of pointwise multiplication. Bounded mapping properties for $f \to \varrho f$ between spaces of type $B^s_{p,q}$ and $F^s_{p,q}$ with $p < \infty$, $q < \infty$ are first derived for $f \in S_\omega$ and extended by continuity. Furthermore, $f \to \varrho f$ for spaces of type $B^s_{p,\infty}$ and $F^s_{p,\infty}$ with $p < \infty$ is defined by restriction, provided one knows what $f \to \varrho f$ means in some spaces of type $B^{s-\varepsilon}_{p,q}$ or $F^{s-\varepsilon}_{p,q}$ with $\varepsilon > 0$. Finally, $f \to \varrho f$ in spaces of type $B^s_{\infty,q}$ is defined by duality. We shall not stress this point in the sequel, however all our calculations must be understood in this way.

Proposition. *Let* $\omega(x) \in \mathfrak{M}$ *and* $\varrho_j(x) \in r(\omega)$ *with* $j = 1, 2, 3, 4$. *Let* $\varrho(x) = \varrho_1(x)\, \varrho_2(x) = \varrho_3(x)\, \varrho_4(x)$ *for every* $x \in R_n$. *Let* $0 < q \leqq \infty$.

(i) *Let* $0 < p \leqq \infty$ *and* $s > \frac{n}{p}$. *Then exists a constant c such that*

$$\|\varrho_2 f \mid B^s_{p,q}(R_n, \varrho_1(x))\| \leqq c \|\varrho_3 f \mid B^s_{p,q}(R_n, \varrho_4(x))\| \tag{5}$$

holds for all $f \in B^s_{p,q}(R_n, \varrho(x))$.

(ii) *Let* $0 < p < \infty$ *and* $s > \frac{n}{p}$. *Then exists a constant c such that*

$$\|\varrho_2 f \mid F^s_{p,q}(R_n, \varrho_1(x))\| \leqq c \|\varrho_3 f \mid F^s_{p,q}(R_n, \varrho_4(x))\| \tag{6}$$

holds for all $f \in F^s_{p,q}(R_n, \varrho(x))$.

Proof. Step 1. We beginn with some preparations. Let $\varphi = \{\varphi_k(x)\}_{k=0}^\infty \in \Phi_\omega$ and let $f \in S'_\omega$ be a regular distribution. If $1 \leqq p \leqq \infty$ then we have

$$\|\varrho f \mid L_p\| \leqq \sum_{k=0}^\infty \|\varrho_4 F^{-1} \varphi_k F \varrho_3 f \mid L_p\| = \|\varrho_3 f \mid B^0_{p,1}(R_n, \varrho_4(x))\| \tag{7}$$

where the right-hand side (or both sides) may be infinite. If $0 < p < 1$ then we have

$$\left\| \sup_{y \in R_n} \frac{|\varrho(x-y) f(x-y)|}{1+|y|^a} \mid L_p \right\|^p \leqq \sum_{k=0}^{\infty} \left\| \sup_{y \in R_n} \varrho_4(x-y) \frac{|(F^{-1}\varphi_k F \varrho_3 f)(x-y)|}{1+|y|^a} \mid L_p \right\|^p \leqq \sum_{k=0}^{\infty} 2^{akp} \left\| \sup_{y \in R_n} \varrho_4(x-y) \frac{|(F^{-1}\varphi_k F \varrho_3 f)(x-y)|}{1+|2^k y|^a} \mid L_p \right\|^p . \tag{8}$$

Let $a > \frac{n}{p}$. We use (1.9.1/3) with, say, $f_k = F^{-1}\psi'_k F \varrho_3 f$, cf. the beginning of the proof of Theorem 5.1.2. Then we have

$$\left\| \sup_{y \in R_n} \frac{|\varrho(x-y) f(x-y)|}{1+|y|^a} \mid L_p \right\|^p \leqq c \|\varrho_3 f \mid B^a_{p,p}(R_n, \varrho_4(x))\|^p \tag{9}$$

where the right-hand side (or both sides) may be infinite.

Step 2. Let $0 < p < \infty$, $0 < q \leqq \infty$ and $s > \frac{n}{p}$. Let $\varphi = \{\varphi_k(x)\}_{k=0}^{\infty} \in \Phi_\omega$ where we may assume $\varphi_k(x) = \varphi_1(2^{-k+1}x)$ if $k = 1, 2, 3, \ldots$ Let $f \in F^s_{p,q}(R_n, \varrho(x))$. Then we claim that

$$\varrho_1(x)\,(F^{-1}\varphi_j F \varrho_2 f)(x) = \varrho_1(x) \int_{R_n} (F^{-1}\varphi_j)(y)\, \varrho_2(x-y) f(x-y)\, \mathrm{d}y \tag{10}$$

makes sense in any case. If $1 \leqq p < \infty$ then the desired assertion follows from

$$\varrho_1(x)\,\varrho_2(x-y) \leqq c\, \mathrm{e}^{\omega(y)}\, \varrho(x-y), \tag{11}$$

(7), and the elementary embedding

$$F^s_{p,q}(R_n, \varrho(x)) \subset B^0_{p,1}(R_n, \varrho(x)). \tag{12}$$

Cf. Remark 3 below as fas as (12) is concerned. If $0 < p < 1$ then the desired assertion follows from (11), (7) (with $p = 1$) and the embedding

$$F^s_{p,q}(R_n, \varrho(x)) \subset B^0_{1,1}(R_n, \varrho(x)), \tag{13}$$

cf. again Remark 3 below. Similarly (10) makes sense if $0 < p \leqq \infty$, $0 < q \leqq \infty$, $s > \frac{n}{p}$ and $f \in B^s_{p,q}(R_n, \varrho(x))$.

Step 3. We prove (ii). The proof of (i) is essentially the same. Let $\varrho_5(x) = \frac{\varrho_2(x)}{\varrho_3(x)}$. We have $\varrho_5(x) \in r(2\omega)$. We use

$$\varrho_5(x-y) = \sum_{|\alpha| \leqq N} c_\alpha (D^\alpha \varrho_5)(x)\, y_1^{\alpha_1} \ldots y_n^{\alpha_n} + \sum_{|\alpha| = N+1} c_\alpha (D^\alpha \varrho_5)(x + \vartheta y)\, y_1^{\alpha_1} \ldots y_n^{\alpha_n} \tag{14}$$

with $0 \leqq |\vartheta| = |\vartheta(x,y)| \leqq 1$. By (10), $\varrho_2 = \varrho_3 \varrho_5$, $\varrho_1 \varrho_5 = \varrho_4$, (14) and (2) we have

$$\begin{aligned} &\varrho_1(x)\, |(F^{-1}\varphi_j F \varrho_2 f)(x)| \\ &\leqq \varrho_1(x) \left| \int_{R_n} (F^{-1}\varphi_j)(y)\, \varrho_5(x-y)\, \varrho_3(x-y) f(x-y)\, \mathrm{d}y \right| \\ &\leqq c \varrho_4(x) \sum_{|\alpha| \leqq N} \left| \int_{R_n} (F^{-1} D^\alpha \varphi_j)(y)\, \varrho_3(x-y) f(x-y)\, \mathrm{d}y \right| \\ &\quad + c \varrho_4(x) \sum_{|\alpha| = N+1} \int_{R_n} |(F^{-1}\varphi_j)(y)|\, |y|^{N+1} \mathrm{e}^{2\omega(y)}\, \varrho_3(x-y)\, |f(x-y)|\, \mathrm{d}y. \end{aligned} \tag{15}$$

We denote the term with $\sum_{|\alpha|=N+1}$ in (15) by $R_j(x)$. Let $j = 1, 2, 3, \dots$ By $(F^{-1}\varphi_j)(y) = 2^{(j-1)n}(F^{-1}\varphi_1)(2^{j-1}y)$ we obtain

$$R_j(x) \leqq c \cdot 2^{-jN} \int_{R_n} |(F^{-1}\varphi_1)(y)|\,|y|^{N+1}\,\mathrm{e}^{3\omega(y)}\,\varrho(x - 2^{-j+1}y)\,|f(x - 2^{-j+1}y)|\,\mathrm{d}y. \tag{16}$$

We used (1) with ϱ_4 instead of ϱ. If $1 \leqq p < \infty$ then we obtain

$$\|R_j \mid L_p\| \leqq c \cdot 2^{-jN}\|\varrho f \mid L_p\|, \quad 1 \leqq p < \infty. \tag{17}$$

(Of course this estimate is also valid if $p = \infty$. This is needed in order to prove (5).) If $0 < p \leqq 1$ and $a > 0$ then we obtain

$$\|R_j \mid L_p\| \leqq c \cdot 2^{-jN}\left\| \sup_{y \in R_n} \frac{|\varrho(x-y)\,f(x-y)|}{1+|y|^a} \mid L_p \right\|, \quad 0 < p \leqq 1. \tag{18}$$

In the terms with $\sum_{|\alpha| \leqq N}$ in (15) one can replace $(\varrho_3 f)(x-y)$ by $(F^{-1}\psi_j' F\varrho_3 f)(x-y)$, where ψ_j' has the same meaning as at the beginning of the proof of Theorem 5.1.2. Let $1 \leqq p < \infty$, $0 < q \leqq \infty$ and $0 < s < N$. Then (15) (with the just indicated replacement), (17) and the Fourier multiplier assertion of (1.9.1/3) yield

$$\|\varrho_2 f \mid F^s_{p,q}(R_n, \varrho_1(x))\| \leqq c\|\varrho_3 f \mid F^s_{p,q}(R_n, \varrho_4(x))\| + c\|\varrho f \mid L_p\|. \tag{19}$$

By (7) and (12) we obtain (6). Let $0 < p < 1$, $0 < q \leqq \infty$ and $\frac{n}{p} < a < s < N$. By (15), (18) and again (1.9.1/3) we have

$$\begin{aligned} &\|\varrho_2 f \mid F^s_{p,q}(R_n, \varrho_1(x))\| \\ &\leqq c\|\varrho_3 f \mid F^s_{p,q}(R_n, \varrho_4(x))\| + c\left\| \sup_{y \in R_n} \frac{|\varrho(x-y)\,f(x-y)|}{1+|y|^a} \mid L_p \right\|. \end{aligned} \tag{20}$$

We apply (9) and an embedding of type (12) with $B^a_{p,p}$ instead of $B^0_{p,1}$. This proves (6).

Remark 3. Embeddings of type (12) with the same p's on both sides have nothing to do with the weights involved. They follow from Definition 5.1.1/2, cf. also [T, Proposition 2.3.2/2]. We prove embeddings of type (13). Let $0 < p_1 < p_2 \leqq \infty$ and $\varphi = \{\varphi_k(x)\}_{k=0}^\infty \in \Phi_\omega$. Let ψ_j' be the same functions as at the beginning of the proof of Theorem 5.1.2. Then it follows from the proof of Proposition 1.4.3 that there exists a constant c such that

$$\|\varrho(2^{-j}\cdot)\,F^{-1}\varphi_j(2^j\cdot)\,Ff \mid L_{p_2}\| \leqq c\|\varrho(2^{-j}\cdot)\,F^{-1}\psi_j'(2^j\cdot)\,Ff \mid L_{p_1}\| \tag{21}$$

holds for all $j = 0, 1, 2, \dots$ We replace $f(x)$ by $f(2^{-j}x)$. Then we have

$$\|\varrho F^{-1}\varphi_j Ff \mid L_{p_2}\| \leqq c \cdot 2^{jn\left(\frac{1}{p_1} - \frac{1}{p_2}\right)}\|\varrho F^{-1}\psi_j' Ff \mid L_{p_1}\|. \tag{22}$$

However this and (12) are the basis for (13) and other embeddings with different metrics, cf. [T, 2.7.1.].

Remark 4. The above proof shows that (6) holds if $1 \leqq p < \infty$, $0 < q \leqq \infty$ and $s > 0$. Similarly (5).

If $\varrho_0(x) \equiv 1$ then $B^s_{p,q}(R_n) = B^s_{p,q}(R_n, \varrho_0(x))$ and $F^s_{p,q}(R_n) = F^s_{p,q}(R_n, \varrho_0(x))$ are the unweighted spaces of Besov-Hardy-Sobolev type.

Theorem. *Let $\omega(x) \in \mathfrak{M}$ and $\varrho(x) \in r(\omega)$. Let $0 < q \leqq \infty$ and $-\infty < s < \infty$.*

(i) Let $0 < p \leqq \infty$. Then $f \to \varrho f$ yields an isomorphic mapping from $B^s_{p,q}(R_n, \varrho(x))$ onto $B^s_{p,q}(R_n)$.

(ii) *Let* $0 < p < \infty$. *Then* $f \to \varrho f$ *yields an isomorphic mapping from* $F^s_{p,q}(R_n, \varrho(x))$ *onto* $F^s_{p,q}(R_n)$.

Proof. Step 1. We prove (ii). Let $0 < p < \infty$, $0 < q \leqq \infty$ and $s > \frac{n}{p}$. Then (6) proves that $f \to \varrho f$ is an isomorphic mapping from $F^s_{p,q}(R_n, \varrho(x))$ onto $F^s_{p,q}(R_n)$ and from $F^s_{p,q}(R_n)$ onto $F^s_{p,q}(R_n, \varrho^{-1}(x))$. We recall $\varrho^{-1}(x) \in r(\omega)$. Furthermore, $f \to \varrho f$ is a formally self-adjoint operator with respect to the dual pairing (S_ω, S'_ω), cf. the explanations in front of the above proposition. Let $1 < p < \infty$, $1 \leqq q < \infty$ and $s > \frac{n}{p}$. Then it follows from (5.1.2/7) that $f \to \varrho f$ is also an isomorphic mapping from $F^{-s}_{p',q'}(R_n)$ onto $F^{-s}_{p',q'}(R_n, \varrho^{-1}(x))$ and from $F^{-s}_{p',q'}(R_n, \varrho(x))$ onto $F^{-s}_{p',q'}(R_n)$ with $\frac{1}{p} + \frac{1}{p'} = \frac{1}{q} + \frac{1}{q'} = 1$. Application of the complex interpolation formula (5.1.2/13) yields (ii), cf. Remark 5 below.

Step 2. The proof of (i) is the same. It is based on (5), (5.1.2/6), and (5.1.2/12), where the limiting case $p = \infty$ is included now.

Remark 5. With respect to (5.1.2/13) and (5.1.2/12) we used that the mapping $f \to \varrho f$ has the interpolation property. In contrast to the real interpolation method and the classical complex interpolation method $[\cdot, \cdot]_\theta$ this is not automatically ensured for the above complex method $(\cdot, \cdot)_\theta$. However for multiplications with smooth functions one has no problems, cf. the proof of Corollary 2.8.2 in [T].

Remark 6. The above theorem is due to J. Franke [1]. His proof is different. Instead of the rather explicit calculations from the proposition, duality and complex interpolation he uses the splitting technique form [T, 2.8.2.] (which is now a fashionable tool in para-differential equations).

5.1.4. Equivalent Quasi-Norms

Theorem 5.1.3 gives the possibility to reduce problems for weighted spaces to unweighted spaces. We describe few examples in this subsection. Let $\omega(x) \in \mathfrak{M}$ and $\varrho(x) \in R(\omega)$, cf. Definition 1.2.1/1 and Definition 1.4.1(i). Let $m = 0, 1, 2, \ldots$ and $1 < p < \infty$. Then

$$W^m_p(R_n, \varrho(x)) = \left\{ f \mid f \in S'_\omega, \|f \mid W^m_p(R_n, \varrho(x))\| = \sum_{|\alpha| \leqq m} \|\varrho D^\alpha f \mid L_p\| < \infty \right\} \quad (1)$$

are weighted Sobolev spaces where $\|\cdot \mid L_p\|$ has the same meaning as in 1.3.1. Let $B_n = \{y \mid |y| \leqq 1\}$ be the unit ball in R_n. Let

$$\tilde{\sigma}_p = n\left(\frac{1}{\min(p, 1)} - 1\right) \quad \text{and} \quad \tilde{\sigma}_{p,q} = n\left(\frac{1}{\min(p, q, 1)} - 1\right) \quad \text{if}$$
$$0 < p \leqq \infty, \quad 0 < q \leqq \infty, \quad (2)$$

cf. [T, (2.5.3/8) and (2.5.11/2)]. We recall that

$$(\Delta^1_h f)(x) = f(x + h) - f(x), \qquad \Delta^m_h = \Delta^{m-1}_h \Delta^1_h \quad \text{with} \quad m = 2, 3, \ldots$$

have the usual meaning.

Theorem. *Let* $\omega(x) \in \mathfrak{M}$ *and* $\varrho(x) \in R(\omega)$.

(i) *Let* $1 < p < \infty$ *and* $m = 0, 1, 2, \ldots$ *Then*

$$W^m_p(R_n, \varrho(x)) = F^m_{p,2}(R_n, \varrho(x)). \quad (3)$$

(ii) *Let* $0 < p < \infty$, $0 < q < \infty$ *and* $s > \tilde{\sigma}_{p,q}$. *If* M *is an integer with* $M > s$ *then*

$$\|f \mid F^s_{p,q}(R_n, \varrho(x))\|_M = \|\varrho f \mid L_p\| + \left\| \varrho(\cdot) \left(\int_0^1 r^{-sq} \left(\int_{B_n} |(\Delta^M_{rh} f)(\cdot)| \, \mathrm{d}h \right)^q \frac{\mathrm{d}r}{r} \right)^{1/q} \Bigg| L_p \right\| \tag{4}$$

is an equivalent quasi-norm in $F^s_{p,q}(R_n, \varrho(x))$.

(iii) *Let* $0 < p \leqq \infty$, $0 < q \leqq \infty$ *and* $s > \tilde{\sigma}_p$. *If* M *is an integer with* $M > s$ *then*

$$\|f \mid B^s_{p,q}(R_n, \varrho(x))\|_M = \|\varrho f \mid L_p\| + \left(\int_{|h| \leqq 1} |h|^{-sp} \, \|\varrho(\cdot)\,(\Delta^M_h f)(\cdot) \mid L_p\|^q \frac{\mathrm{d}h}{|h|^n} \right)^{1/q} \tag{5}$$

is an equivalent quasi-norm in $B^s_{p,q}(R_n, \varrho(x))$.

Proof. Step 1. We prove (i). Without restriction of generality we may assume $\varrho \in r(\omega)$, cf. Definition 5.1.3 and Lemma 5.1.3. Let $1 < p < \infty$ and $m = 0, 1, 2, \ldots$ We recall that $F^m_{p,2}(R_n) = W^m_p(R_n)$ are the usual Sobolev spaces, cf. [T, Theorem 2.5.6.]. This is essentially a Littlewood-Paley theorem. Hence by Theorem 5.1.3 we have to prove that $\|f \mid W^m_p(R_n, \varrho(x))\|$ and

$$\|\varrho f \mid W^m_p(R_n)\| = \sum_{|\alpha| \leqq m} \|D^\alpha \varrho f \mid L_p\| \tag{6}$$

are equivalent norms. By (5.1.3/2), $\|\varrho f \mid W^m_p(R_n)\|$ can be estimated from above by $c\|f \mid W^m_p(R_n, \varrho(x))\|$. Conversely, we have

$$\|f \mid W^m_p(R_n, \varrho(x))\| \leqq \|\varrho f \mid W^m_p(R_n)\| + c \sum_{|\alpha| < m} \|\varrho D^\alpha f \mid L_p\| \,. \tag{7}$$

We decompose R_n in congruent cubes. By (5.1.3/1), $\varrho(x)$ can be replaced in these cubes by appropriate constants. Then it follows from well-known properties of the usual Sobolev spaces that the remainder term in (7) can be estimated from above by

$$\varepsilon \|f \mid W^m_p(R_n, \varrho(x))\| + c_\varepsilon \|\varrho f \mid L_p\| \,, \tag{8}$$

where $\varepsilon > 0$ is at our disposal. Then (7) yields the desired assertion.

Step 2. We prove (iii). We again assume $\varrho \in r(\omega)$. First we remark that (5) with 1 instead of $\varrho(x)$ is an equivalent quasi-norm in the unweighted space $B^s_{p,q}(R_n)$, cf. [T, Theorem 2.5.12]. Hence by Theorem 5.1.3 we have to prove that $\|f \mid B^s_{p,q}(R_n, \varrho(x))\|_M$ and $\|\varrho f \mid B^s_{p,q}(R_n)\|_M$ are equivalent to each other. By mathematical induction we find

$$(\Delta^M_h \varrho f)(x) = \sum_{k=0}^{M} c_k (\Delta^k_h f)(x)\,(\Delta^{M-k}_h \varrho)(x + kh) \,. \tag{9}$$

Consequently, by (5.1.3/2) we have

$$|(\Delta^M_h \varrho f)(x)| \leqq c \sum_{k=0}^{M} |h|^{M-k} \varrho(x)\, |(\Delta^k_h f)(x)| \tag{10}$$

and

$$\|\varrho f \mid B^s_{p,q}(R_n)\|_M \leqq c\|\varrho f \mid L_p\| + c \sum_{k \geqq M-s} \left(\int_{|h| \leqq 1} |h|^{(-s+M-k)q} \, \|\varrho(\cdot)\,(\Delta^k_h f)(\cdot) \mid L_p\|^q \frac{\mathrm{d}h}{|h|^n} \right)^{1/q} . \tag{11}$$

We estimated in (11) the terms in the sum $\sum_{k=0}^{M}$ with $k < M - s$, i.e. $0 < -s + M-k$, from above by $\|\varrho f \mid L_p\|$. Let us assume that M is large, in particular $M - s > N > s$ for a given natural number N. Then we have

$$\left(\sum_{M>k\geqq M-s} \int_{|h|\leqq 1} |h|^{(-s+M-k)q} \|\varrho(\cdot)(\Delta_h^k f)(\cdot) \mid L_p\|^q \frac{\mathrm{d}h}{|h|^n}\right)^{1/q}$$

$$\leqq \varepsilon\|f \mid B_{p,q}^s(R_n, \varrho(x))\|_N + c_\varepsilon\|\varrho f \mid L_p\|, \tag{12}$$

where $\varepsilon > 0$ is at our disposal. All quasi-norms $\|\cdot \mid B_{p,q}^s(R_n)\|_N$ with $N > s$ are mutually equialent. If $N > s$ is given we choose $M > N + s$, apply (11) and (12) and obtain

$$\|\varrho f \mid B_{p,q}^s(R_n)\|_N \leqq c\|f \mid B_{p,q}^s(R_n, \varrho(x))\|_N. \tag{13}$$

In order to prove the converse estimate we choose again $N > s$ and $M > N + s$. Then (9), (10), and (12) yield

$$\|f \mid B_{p,q}^s(R_n, \varrho(x))\|_M$$
$$\leqq c\|\varrho f \mid B_{p,q}^s(R_n)\|_M + \varepsilon\|f \mid B_{p,q}^s(R_n, \varrho(x))\|_N + c_\varepsilon\|\varrho f \mid L_p\|. \tag{14}$$

As in the unweighted case, $\|f \mid B_{p,q}^s(R_n, \varrho(x))\|_N$ can be estimated from above by $c\|f \mid B_{p,q}^s(R_n, \varrho(x))\|_M$, where c is independent of f. This follows in the same way as in the unweighted case, cf. in particular [T, (2.5.9/45)]. Hence we can replace M on both sides of (14) by N. This proves the converse of (13).

Step 3. We prove (ii). We again assume $\varrho \in r(\omega)$. We remark that (4) with 1 instead of $\varrho(x)$ is an equivalent quasi-norm in the unweighted space $F_{p,q}^s(R_n)$, cf. [T, Corollary 2.5.11]. We have

$$\|f \mid F_{p,q}^s(R_n, \varrho(x))\|_M \leqq c\|f \mid F_{p,q}^s(R_n, \varrho(x))\|^\varphi \tag{15}$$

with $\varphi = \{\varphi_k(x)\}_{k=0}^\infty \in \Phi_\omega$, cf. (4) and (5.1.1/8). In order to prove (15) one can follow the arguments in Step 1 of the proof of Theorem 2.5.11 in [T] line for line. Because the integration over r in (4) is restricted by $\int_0^1 \ldots \mathrm{d}r$, we have no problems to incorporate $\varrho(x)$ in this calculation. We use $c_1\varrho(y) \leqq \varrho(x) \leqq c_2\varrho(y)$ if $|x - y| \leqq 1$, cf. (5.1.3/1), where c_1 and c_2 are two positive numbers which are independent of x and y. Then the problem can be reduced to the maximal inequality from Theorem 1.3.3/2 with ϱf_k instead of f_k and to the maximal inequality from Theorem 1.9.1. Then one obtains (15). We prove the converse. By Theorem 5.1.3 and the above remark it is sufficient to estimate $\|\varrho f \mid F_{p,q}^s(R_n)\|_M$ from above by $c\|f \mid F_{p,q}^s(R_n, \varrho(x))\|_M$. As in the second step we may assume $M - s > N > s$, where N and M are natural numbers. Then $\|\varrho f \mid F_{p,q}^s(R_n)\|_N$ can be estimated from above by $c\|\varrho f \mid F_{p,q}^s(R_n)\|_M$. We use (9) and (10). We have an immediate counterpart of the terms with $\sum_{k\geqq M-s}$ in (11). By (15) and an appropriate modification of (12) it follows that these terms can be estimated from above by

$$c\|f \mid F_{p,q}^s(R_n, \varrho(x))\|_M + \varepsilon\|f \mid F_{p,q}^s(R_n, \varrho(x))\|^\varphi + c_\varepsilon\|\varrho f \mid L_p\|, \tag{16}$$

where $\varepsilon > 0$ is at our disposal. The terms with $\sum_{0\leqq k<M-s}$ can be estimated as in [T, (2.5.10/8), (2.5.11/10), (2.5.11/11)]. They can be incorporated in the second and in the third term in (16). Hence

$$\|f \mid F_{p,q}^s(R_n, \varrho(x))\|^\varphi \leqq c\|f \mid F_{p,q}^s(R_n, \varrho(x))\|_M \tag{17}$$

with $M - s > N > s$. If f has a compact support then M in (17) can be replaced by N. This follows from the unweighted case and (5.1.3/1). The general case is a consequence of $f = \sum_{j \in Z_n} \psi_j f$, where $\{\psi_j\}$ is a smooth resolution of unity with supports in an appropriate sequence of congruent balls (Z_n is the lattice of all points from R_n with integer-valued coordinates). This shows that M in (17) can always be replaced by N. This completes the proof.

Remark. The proof shows the interplay of Theorem 5.1.3 and direct methods for weighted spaces parallel to the corresponding assertions from [T, Chapter 2] for unweighted spaces. In particular if one has properties, e.g. equivalent quasi-norms, which exhibit the local nature of the considered spaces, then one has a good chance to deal with the above weighted spaces in the same way as with unweighted spaces. $\|f \mid W_p^m(R_n, \varrho(x))\|$ and the quasi-norms in (4) and (5) are examples, in contrast to the original quasi-norms from Definition 5.1.1/2.

5.1.5. Complements

Multiplier Theorems. Theorem 5.1.3 and the sketched possibilities from Remark 5.1.4 can be used to transfer properties of the spaces $B^s_{p,q}(R_n)$ and $F^s_{p,q}(R_n)$ to their weighted counterparts. One can also prove vector-valued Fourier multiplier assertions in the sense of Theorem 1.9.2 and Theorem 1.9.3 with some restrictions for the supports of the system of functions $\{\varphi_k(x)\}_{k=0}^{\infty}$. We refer to J. Franke [1] for details. One obtains Theorem 7.1.2 and Theorem 7.1.3 from [T] for all weights $\varrho(x) \in R(\omega)$, where again $\omega(x) \in \mathfrak{M}$.

Equivalent Quasi-Norms. One can try to derive further equivalent quasi-norms for the spaces $B^s_{p,q}(R_n, \varrho(x))$ and $F^s_{p,q}(R_n, \varrho(x))$ via Theorem 5.1.3 (and direct methods) from corresponding quasi-norms for the spaces $B^s_{p,q}(R_n)$ and $F^s_{p,q}(R_n)$. For example, in [T, 2.12.2] we characterized the spaces $B^s_{p,q}(R_n)$ and $F^s_{p,q}(R_n)$ on the basis of Gauss-Weierstrass and Cauchy-Poisson semi-groups. The problem is to have descriptions for corresponding quasi-norms in the weighted spaces where the weight $\varrho(x)$ stands in front of the other expressions similar as in $\|f \mid W_p^m(R_n, \varrho(x))\|$ from (5.1.4/1), or in (5.1.4/4), (5.1.4/5). As the Gauss-Weierstrass semi-group is concerned we obtained in (1.9.4/9) a rather special result of the desired type.

A priori Estimates. Let $\omega(x) \in \mathfrak{M}$ and $\varrho(x) \in R(\omega)$. By Theorem 1.9.1 one obtains immediately Fourier multiplier assertions for $B^s_{p,q}(R_n, \varrho(x))$ and $F^s_{p,q}(R_n, \varrho(x))$ of a similar type as for their unweighted counterparts. As a consequence we obtain

$$\|-\Delta f + f \mid B^s_{p,q}(R_n, \varrho(x))\| \sim \|f \mid B^{s+2}_{p,q}(R_n, \varrho(x))\| \tag{1}$$

(equivalent quasi-norms), $-\infty < s < \infty$, $0 < p \leqq \infty$, $0 < q \leqq \infty$, and

$$\|-\Delta f + f \mid F^s_{p,}\ (R_n, \varrho(x))\| \sim \|f \mid F^{s+2}_{p,q}(R_n, \varrho(x))\| \tag{2}$$

(equivalent quasi-norms), $-\infty < s < \infty$, $0 < p < \infty$, $0 < q \leqq \infty$. Of course Δ stands for the Laplacian. We recall that $\varrho(x) = e^{|x|^\beta}$ and $\varrho(x) = e^{-|x|^\beta}$ with $0 < \beta < 1$ are admissible weights. On the other hand relations of type (1) and (2) are not valid if, for example, $\varrho(x) = e^{-|x|^\gamma}$ with $\gamma > 1$. Let us assume that

$$\|f \mid W_p^2(R_n, \varrho(x))\| \leqq c\|-\Delta f + f \mid L_p(R_n, \varrho(x))\|, \qquad \varrho(x) = e^{-|x|^\gamma} \tag{3}$$

with $\gamma > 1$ holds, where c is independent of f. Let $1 < p < \infty$ and $n = 2$. We choose $f_k(x) = (x_1 + \mathrm{i}x_2)^k$. We have $(\Delta f_k)(x) \equiv 0$. Then (3) yields

$$\left\| \frac{\partial^2 f_k}{\partial x_1^2} \mathrm{e}^{-|x|^\gamma} \mid L_p \right\| \leqq c \| f_k \, \mathrm{e}^{-|x|^\gamma} \mid L_p \| . \tag{4}$$

However in [F, p. 33] we proved that (4) contradicts well-known properties of the Γ-function.

5.2. Modulation Spaces

5.2.1. Introduction

In 2.2.1. and 2.2.2. we have given a general introduction in the theory of function spaces based on decomposition methods. We recall the dyadic decomposition $\Phi(R_n)$ from Definition 2.2.1/1 and the related spaces $B^s_{p,q}(R_n)$ and $F^s_{p,q}(R_n)$ of Besov-Hardy Sobolev type from 2.2.2. (there is no problem to replace the dimension 2 by n). The anisotropic counterparts have also been mentioned in 2.2.2. (and some aspects have been studied in Chapter 4). The spaces from Definition 2.2.1/2 with dominating mixed smoothness properties are also based on dyadic decompositions. In 2.2.2. we mentioned few other more general possibilities of decompositions. It is the aim of Section 5.2. to deal with a special aspect of one of these possibilities. Instead of dyadic balls from Definition 2.2.1/1 we use congruent cubes (or balls). Furthermore the numbers 2^{sj} from (2.2.2/2) and (2.2.2/3) which stand for the smoothness are replaced by positive numbers a_j. The resulting spaces $B^{\mathfrak{a}}_{p,q}(R_n)$ and $F^{\mathfrak{a}}_{p,q}(R_n)$ with $\mathfrak{a} = \{a_j\}$ are called modulation spaces. This notation has been suggested by H. G. Feichtinger. The aim of Section 5.2. is to give a brief description of the trace problem: What can be said about the trace operator R,

$$R\colon f(x) \to f(x', 0) \quad \text{where} \quad x = (x', x_n),$$

as a mapping from $B^{\mathfrak{a}}_{p,q}(R_n)$ or $F^{\mathfrak{a}}_{p,q}(R_n)$ onto corresponding spaces on R_{n-1}? We omit proofs. We follow H. Triebel [14], where further details and proofs may be found.

The interest in the above spaces comes from two quite different sources. H. G. Feichtinger [3] introduced spaces of Wiener type on locally compact abelian groups, cf. also H. G. Feichtinger [2, 5]. It comes out that the Fourier image $FB^{\mathfrak{a}}_{p,q}(R_n)$ of $B^{\mathfrak{a}}_{p,q}(R_n)$ with $1 \leqq p \leqq \infty$, $1 \leqq q \leqq \infty$, is a space of Wiener type in the sense of Feichtinger. This suggests to deal with spaces of type $B^{\mathfrak{a}}_{p,q}$ with $1 \leqq p \leqq \infty$ and $1 \leqq q \leqq \infty$ on locally compact abelian groups in the framework of the technique used there, cf. H. G. Feichtinger [4]. Our approach is restricted to R_n, but it includes the spaces $F^{\mathfrak{a}}_{p,q}(R_n)$ and extends the range of p and q to $0 < p \leqq \infty$, $0 < q \leqq \infty$ (with $p < \infty$ for $F^{\mathfrak{a}}_{p,q}(R_n)$). The other source is the study of function spaces on R_n based on decomposition methods, cf. also 2.2.2. The first question is for what decompositions (or coverings) of R_n (or locally compact abelian groups etc.) the corresponding spaces of type $B^s_{p,q} - F^s_{p,q}$ make sense. Considerations what coverings are admissible can be found in [F, Chapter 2], V. I. Burenkov [1] and H. G. Feichtinger, P. Gröbner [1], the latter one in locally compact spaces. The second question is whether one cancharacterize ele mentsof corresponding spaces of $B^s_{p,q} - F^s_{p,q}$ type in other terms, e.g. via differences and derivatives of functions, approximation procedures or as traces of harmonic functions or temperatures (just as in the case of the spaces $B^s_{p,q}(R_n)$ and $F^s_{p,q}(R_n)$, their anisotropic or weighted counterparts, or as in the case of spaces with dominating mixed smoothness properties, cf. [T], or Chapter 4, Section 5.1. and Chapter 2 of this book, respectively). Some work in this direction has been done. Beside [F, Chapter 2] we refer to the papers by M. L. Gol'dman [3, 4], G. A. Kaljabin [3, 4] and S. Janson [1]. The feeling is that some regularity assumptions for the admissible coverings of R_n are necessary in order to obtain substantial results (cf. the cited papers by Kaljabin, Gol'dman and Janson). Probably the congruent covering is a limiting case for that purpose and of peculiar interest may be coverings "between" the congruent and the dyadic covering or coverings where the cubes, or rectangles, or balls grow even more rapid than in the dyadic case. In this sense Section 5.2. could be understood as a contribution in order to study the limiting case "congruent covering", also in comparison with the case "dyadic covering".

Finally we wish to mention that the spaces $B^{\mathfrak{a}}_{p,q}(R_n)$ with $1 \leqq p \leqq \infty$ (and $0 < q \leqq \infty$) which we define in the next subsection are also covered by M. L. Gol'dman [4], cf. in particular Theorem 7, Remark 1 and Subsection 4.1. in the just cited paper.

5.2.2. Definitions

We use the same notations as in 1.2.4. and 1.3.1. In particular, R_n stands for the Euclidean n-space, $S(R_n)$ is the Schwartz space of all complex-valued infinitely differentiable rapidly decreasing functions on R_n, and $\|f \mid L_p(R_n)\|$ with $0 < p \leqq \infty$ has been defined in 1.3.1. As usual F and F^{-1} denote the Fourier transform and its inverse on $S(R_n)$, respectively. Let

$$Z_n = \{k \mid k \in R_n,\, k = (k_1, \ldots, k_n),\, k_j \text{ integer}\}.$$

Let $S^c(R_n)$ be the collection of all $f \in S(R_n)$ such hat Ff has a compact support. The counterpart of Definition 2.2.1/1 reads now as follows.

Definition 1. *Let $\Phi_c(R_n)$ be the collection of all systems $\varphi = \{\varphi_k(x)\}_{k \in Z_n} \subset S(R_n)$ of non-negative functions with the following properties:*

(i) $$\operatorname{supp} \varphi_k \subset \{y \mid y = (y_1, \ldots, y_n) \in R_n,\, |y_j - k_j| \leqq 1\}, \tag{1}$$
where $k = (k_1, \ldots, k_n) \in Z_n$.

(ii) *for every multi-index α there exists a number c_α such that $|D^\alpha \varphi_k(x)| \leqq c_\alpha$ for all $k \in Z_n$ and all $x \in R_n$,*

(iii) $$\sum_{k \in Z_n} \varphi_k(x) = 1 \quad \textit{for all} \quad x \in R_n. \tag{2}$$

Remark 1. This is a smooth resolution of unity where the underlying domains are congruent cubes.

Definition 2. *Let $\mathfrak{a} = \{a_k\}_{k \in Z_n}$ be a sequence of positive numbers with the property that there exist two positive numbers c_1 and c_2 such that*

$$0 < c_1 \leqq \frac{a_k}{a_{\tilde{k}}} \leqq c_2 < \infty \quad \textit{for all} \quad k \in Z_n \quad \textit{and} \quad \tilde{k} \in Z_n \quad \textit{with} \quad |k - \tilde{k}| = 1 \tag{3}$$

holds. Let $\varphi \in \Phi_c(R_n)$.

(i) *Let $0 < p \leqq \infty$ and $0 < q \leqq \infty$. Then $S^c(R_n)$ equipped with the quasi-norm*

$$\|f \mid B^{\mathfrak{a}}_{p,q}(R_n)\|^\varphi = \Big(\sum_{k \in Z_n} a_k^q \|F^{-1} \varphi_k Ff \mid L_p(R_n)\|^q\Big)^{1/q} \tag{4}$$

is denoted as $B^{\mathfrak{a}}_{p,q}(R_n)$ (usual modification if $q = \infty$).

(ii) *Let $0 < p < \infty$ and $0 < q \leqq \infty$. Then $S^c(R_n)$ equipped with the quasi-norm*

$$\|f \mid F^{\mathfrak{a}}_{p,q}(R_n)\|^{\varphi} = \left\| \left(\sum_{k \in Z_n} a_k^q \mid (F^{-1}\varphi_k Ff)(\cdot)|^q \right)^{1/q} \mid L_p(R_n) \right\| \tag{5}$$

is denoted as $F^{\mathfrak{a}}_{p,q}(R_n)$ (usual modification if $q = \infty$).

Remark 2. We recall that $F^{-1}\varphi_k Ff$ stands for $(F^{-1}[\varphi_k Ff])(x)$. By the Paley-Wiener-Schwartz theorem $F^{-1}\varphi_k Ff$ is an analytic function, in particular (4) and (5) make sense. Of course, (4) and (5) are quasi-norms (norms if $p \geqq 1$ and $q \geqq 1$). Furthermore, if $\varphi \in \Phi_c(R_n)$ and $\psi \in \Phi_c(R_n)$ then the corresponding quasi-norms $\|f \mid B^{\mathfrak{a}}_{p,q}(R_n)\|^{\varphi}$ and $\|f \mid B^{\mathfrak{a}}_{p,q}(R_n)\|^{\psi}$ are equivalent to each other. One has a similar assertion for the quasi-norms $\|f \mid F^{\mathfrak{a}}_{p,q}(R_n)\|^{\varphi}$ and $\|f \mid F^{\mathfrak{a}}_{p,q}(R_n)\|^{\psi}$. This follows immediately from the unweighted case of Theorem 1.9.1 and its scalar counterpart from 1.7.2. (where the latter holds also for $p = \infty$). This shows that (4) and (5) are reasonable constructions. Furthermore it justifies to write simply $\|f \mid B^{\mathfrak{a}}_{p,q}(R_n)\|$ and $\|f \mid F^{\mathfrak{a}}_{p,q}(R_n)\|$ instead of $\|f \mid B^{\mathfrak{a}}_{p,q}(R_n)\|^{\varphi}$ and $\|f \mid F^{\mathfrak{a}}_{p,q}(R_n)\|^{\varphi}$, respectively. Some parts of the theory of the spaces $B^s_{p,q}(R_n)$ and $F^s_{p,q}(R_n)$ of Besov-Hardy-Sobolev type can be extended immediately to the spaces $B^{\mathfrak{a}}_{p,q}(R_n)$ and $F^{\mathfrak{a}}_{p,q}(R_n)$. For example, maximal inequalities and Fourier multiplier assertions can be proved on the basis of Theorem 1.9.1 and its scalar counterpart. Other assertions for $B^{\mathfrak{a}}_{p,q}(R_n)$ and $F^{\mathfrak{a}}_{p,q}(R_n)$ are different in comparison with $B^s_{p,q}(R_n)$ and $F_{p,q}(R_n)$, respectively. A typical example will be described in the next subsection: Traces on hyper-planes.

Remark 3. In contrast to the spaces $B^s_{p,q}(R_n)$ and $F^s_{p,q}(R_n)$ from [T], the spaces with dominating mixed smoothness properties from Chapter 2, the anisotropic spaces from Chapter 4, or the weighted spaces from Section 5.1., the above spaces $B^{\mathfrak{a}}_{p,q}(R_n)$ and $F^{\mathfrak{a}}_{p,q}(R_n)$ are not complete. Of course one could try to extend the definitions of $B^{\mathfrak{a}}_{p,q}(R_n)$ and $F^{\mathfrak{a}}_{p,q}(R_n)$ to suitable distributions from $S'(R_n)$ or from other appropriate spaces of distributions. From that point of view it would be better to denote $S^c(R_n)$ equipped with the quasi-norm from (4) by $\mathring{B}^{\mathfrak{a}}_{p,q}(R_n)$, and similarly $\mathring{F}^{\mathfrak{a}}_{p,q}(R_n)$. But we prefer the above notations. However to find suitable distribution spaces for an extended definition of the (complete) spaces $B^{\mathfrak{a}}_{p,q}(R_n)$ and $F^{\mathfrak{a}}_{p,q}(R_n)$ is a somewhat delicate question. A detailed discussion may be found in M. L. Gol'dman [4], cf. also [F, 2.2.3.]. However in any case it is quite clear that the assertions of the following two subsections can be extended via limiting processes to extended spaces of type $B^{\mathfrak{a}}_{p,q}(R_n)$ and $F^{\mathfrak{a}}_{p,q}(R_n)$.

5.2.3. Traces

The trace operator R is given by

$$Rf = f(x', 0) \quad \text{where} \quad f(x) \in S^c(R_n) \quad \text{and} \quad x = (x', x_n), \quad x' \in R_{n-1},$$

and $n \geqq 2$. Our aim is to find spaces $B^{\mathfrak{a}'}_{p,q}(R_{n-1})$ and $F^{\mathfrak{a}'}_{p,q}(R_{n-1})$ with suitable sequences $\mathfrak{a}' = \{a'_k\}_{k \in Z_{n-1}}$ of positive numbers such that R is a linear and bounded operator from a given space $B^{\mathfrak{a}}_{p,q}(R_n)$ onto $B^{\mathfrak{a}'}_{p,q}(R_{n-1})$ and from a given space $F^{\mathfrak{a}}_{p,q}(R_n)$ onto $F^{\mathfrak{a}'}_{p,q}(R_{n-1})$. Furthermore, R is called a retraction if there exists a linear and bounded operator T from $B^{\mathfrak{a}'}_{p,q}(R_{n-1})$ into $B^{\mathfrak{a}}_{p,q}(R_n)$, respectively from $F^{\mathfrak{a}'}_{p,q}(R_{n-1})$ into $F^{\mathfrak{a}}_{p,q}(R_n)$ such that

$$RT = I \text{ (identity in } B^{\mathfrak{a}'}_{p,q}(R_{n-1}), \text{ resp. in } F^{\mathfrak{a}'}_{p,q}(R_{n-1})). \tag{1}$$

In this case it is clear that R maps "onto". Obviously, T is an extension operator.

Theorem 1. *Let $0 < p \leqq \infty$, $\bar{p} = \min(1, p)$ and $0 < q \leqq \infty$. Let $\mathfrak{a} = \{a_k\}_{k \in Z_n}$ be a sequence of positive numbers with (5.2.2/3). If $k \in Z_n$ then we put $k = (k', k_n)$ with $k' \in Z_{n-1}$ and $k_n \in Z_1$. Let $k' \in Z_{n-1}$,*

$$a'_{k'} = \inf_{l \in Z_1} a_{(k',l)} > 0 \quad \text{if} \quad 0 < q \leqq \bar{p} \tag{2}$$

and

$$a_{k'}^{\prime\,-\sigma} = \sum_{l=-\infty}^{\infty} a_{(k',l)}^{-\sigma} < \infty \quad \text{with} \quad \frac{1}{\sigma} = \frac{1}{\bar{p}} - \frac{1}{q} \quad \text{if} \quad \bar{p} < q \leqq \infty . \tag{3}$$

Let $\mathfrak{a}' = \{a_{(k',0)}\}_{k' \in Z_{n-1}}$. Then $B^{\mathfrak{a}'}_{p,q}(R_{n-1})$ is a space in the sense of Definition 5.2.2/2(i). If there exists a positive number A such that $a'_{k'} \geqq A a_{(k',0)}$ for $k' \in Z_{n-1}$ then R is a retraction from $B^{\mathfrak{a}}_{p,q}(R_n)$ onto $B^{\mathfrak{a}'}_{p,q}(R_{n-1})$.

Remark 1. As far as proofs are concerned we refer to H. Triebel [14]. Under the hypotheses of the theorem the coretraction T from (1) is independent of $\mathfrak{a}$, p, and q. This is a useful observation in connection with applications of interpolation methods to the spaces in question. What about the additional restriction $a'_{k'} \geqq A a_{(k',0)}$ for some positive A in the theorem? At least in the cases $0 < q \leqq \bar{p}$ and $1 \leqq p < q \leqq \infty$ this condition can be omitted, cf. H. Triebel [14, Corollary]. However the corresponding coretraction T in the sense of (1) depends on $\mathfrak{a}$.

Theorem 2. *Let $0 < p < \infty$ and $0 < q \leqq \infty$. Let $\mathfrak{a} = \{a_k\}_{k \in Z_n}$ be a sequence of positive numbers with (5.2.2/3). If $k \in Z_n$ then we put $k = (k', k_n)$ with $k' \in Z_{n-1}$ and $k_n \in Z_1$. By assumption there exists a positive number c such that*

$$a_{(k',m)} \geqq c|m|^{\varkappa} a_{(k',0)} \quad \text{with} \quad \varkappa > \max\left(1, \frac{1}{p}, \frac{1}{q}\right) \tag{4}$$

holds for all $k' \in Z_{n-1}$ and all $m \in Z_1$. Let $\mathfrak{a}' = \{a_{(k',0)}\}_{k' \in Z_{n-1}}$. Then $F^{\mathfrak{a}'}_{p,q}(R_{n-1})$ is a space in the sense of Definition 5.2.2/2(ii) and R is a retraction from $F^{\mathfrak{a}}_{p,q}(R_n)$ onto $F^{\mathfrak{a}'}_{p,q}(R_{n-1})$.

Remark 2. As far as proofs are concerned we refer again to the above cited paper. Unter the condition of the theorem the coretraction T from (1) is independent of $\mathfrak{a}$, p, and q. What about the restriction (4)? We recall $B^{\mathfrak{a}}_{p,p}(R_n) = F^{\mathfrak{a}}_{p,p}(R_n)$. Then one can interpolate the milder restrictions for $\mathfrak{a}$ from Theorem 1 with the more severe restrictions for $\mathfrak{a}$ from Theorem 2. In this way the restriction (4) can be relaxed, cf. Section 4.2. of the above cited paper.

5.2.4. Further Properties

Maximal Inequalities. As has been said some results for the spaces $B^s_{p,q}(R_n)$ and $F^s_{p,q}(R_n)$ of Besov-Hardy-Sobolev type have immediate counterparts for the spaces $B^{\mathfrak{a}}_{p,q}(R_n)$ and $F^{\mathfrak{a}}_{p,q}(R_n)$. As an example we describe characterizations of the latter spaces via maximal functions. Let $\varphi \in \Phi_c(R_n)$ and $b > 0$. Then we introduce the maximal function

$$(\varphi_k f)^*(x) = \sup_{y \in R_n} \frac{|(F^{-1}\varphi_k Ff)(x-y)|}{1+|y|^b}, \quad x \in R_n, \quad k \in Z_n .$$

If $0 < p \leqq \infty$, $0 < q \leqq \infty$ and $b > \frac{n}{p}$ then

$$\Big(\sum_{k \in Z_n} a_k^q \|(\varphi_k f)^* \mid L_p(R_n)\|^q\Big)^{1/q} \tag{1}$$

is an equivalent quasi-norm in $B^{\mathfrak{a}}_{p,q}(R_n)$ (modification if $q = \infty$). If $0 < p < \infty$, $0 < q \leqq \infty$ and $b > \dfrac{n}{\min(p,q)}$ then

$$\Big\|\Big(\sum_{k \in Z_n} a_k^q |(\varphi_k f)^*(\cdot)|^q\Big)^{1/q} \mid L_p(R_n)\Big\| \tag{2}$$

is an equivalent quasi-norm in $F^{\mathfrak{a}}_{p,q}(R_n)$ (modification if $q = \infty$). These assertions follow from Theorem 1.9.1 and its scalar counterparts.

Continuous Version. As we mentioned in the introduction the interest in these spaces comes at least partly from Feichtinger's Wiener spaces on locally compact abelian groups. In this theory one prefers continuous versions of the counterparts of (5.2.2/4). This is also possible in our context. We describe some results in this direction. Let $\varphi \in S(R_n)$ be a non-negative function with a compact support in R_n and, say, $\varphi(x) = 1$ if $|x| \leqq 1$. Furthermore, let $a(y) > 0$ be a continuous function on R_n with

$$a(y) \leqq ca(z) \quad \text{if} \quad y \in R_n, z \in R_n \quad \text{and} \quad |y - z| \leqq 1,$$

for some $c > 0$ which is independent of $y \in R_n$ and $z \in R_n$. Then $\mathfrak{a} = \{a(k)\}_{k \in Z_n}$ satisfies (5.2.2/3) with $a_k = a(k)$. If $0 < p \leqq \infty$ and $0 < q \leqq \infty$ then

$$\left(\int_{R_n} a^q(y) \| [F^{-1}\varphi(\cdot - y)\, Ff](\cdot) \mid L_p(R_n) \|^q \, dy \right)^{1/q} \tag{3}$$

is an equivalent quasi-norm in $B^{\mathfrak{a}}_{p,q}(R_n)$ (modification if $q = \infty$). If $0 < p < \infty$ and $0 < q \leqq \infty$ then

$$\left\| \left(\int_{R_n} a^q(y) |[F^{-1}\varphi(\cdot - y)\, Ff](\cdot)|^q \, dy \right)^{1/q} \mid L_p(R_n) \right\| \tag{4}$$

is an equivalent quasi-norm in $F^{\mathfrak{a}}_{p,q}(R_n)$ (modification if $q = \infty$). It is almost obvious how to understand the expressions in (3) and (4): If $f \in S^c(R_n)$ is given then one applies F^{-1} to $\varphi(\xi - y)\,(Ff)\,(\xi)$ where $\xi \in R_n$ is variable and $y \in R_n$ is fixed. The result is a function $(F^{-1}\varphi(\cdot - y)\, Ff)\,(x)$ of $x \in R_n$ and $y \in R_n$. The L_p-quasi-norm is always connected with the variable x. One has also continuous versions of (1) and (2).

6. Abstract Spaces

6.1. Introduction

The considerations in the Chapters 2 (Spaces with dominating mixed smoothness properties) and 3 (Periodic spaces) are based on decomposition methods. A given function or distribution is decomposed in an infinite set of smooth (entire analytic) functions. Afterwards these smooth pieces are measured with the help of appropriate norms or quasi-norms. This principle is not restricted to these two distinguished cases: The study of the isotropic spaces of Besov-Hardy-Sobolev type in [T, Chapter 2] was also based on these ideas. The anisotropic generalization has been mentioned in 2.2.2. and 4.2.1. Furthermore the investigation of a class of weighted spaces in 5.1. and of the modulation spaces in 5.2. also used decomposition principles. This chapter tries to give partial answers to the following two questions: (i) What is the abstract background of this successful decomposition principle? (ii) What other types of spaces can be considered in this way? However it is not our aim to give an exhaustive treatment of these topics. On the contrary, this chapter is an updated version of [F, Chapter 3].

The abstract method which will be described in this chapter had been developed independently by J. Peetre [4, Chapter 10] and in [F, Chapter 3], and in the underlying papers. However we wish to mention that priority is due to J. Peetre, cf. the Notes on p. 223 of the just-cited book by J. Peetre. As has been said we follow here rather closely some parts of [F, Chapter 3]. As far as references are concerned we restrict ourselves to those ones from [F, Chapter 3] which are really indispensible or which are of more recent time (later than 1977/78). In other words, the interested reader is asked to consult J. Peetre [4, Chapter 11] and [F, Chapter 3] for further informations and references.

The plan of the chapter is as follows. The Sections 6.2. and 6.3. deal with the abstract versions of the Besov spaces and Bessel-potential spaces, respectively. Fourier series in an abstract setting are treated in Section 6.4. This section may be considered as an abstract addendum to the periodic spaces from Chapter 3. The final Section 6.5. deals with examples and comments: spaces on R_n, periodic spaces, a generalization of the abstract method, classical orthogonal expansions, the Bessel transform. With exception of the last section we give detailed proofs.

6.2. Abstract Besov Spaces

6.2.1. Multipliers for Banach Spaces

Let H be a complex Hilbert space, and let Λ be a self-adjoint positive-definite operator acting in H with $D(\Lambda)$ as its domain of definition. Then

$$\Lambda a = \int_d^\infty t \, \mathrm{d}E(t)\, a, \quad a \in D(\Lambda), \tag{1}$$

stands for the spectral decomposition. $E(\Delta)$ with the Borel set Δ in R_1 is the corresponding spectral measure, $E(t) = E((-\infty, t))$ if $t \in R_1$ and $E(d) = 0$ for some $d > 0$. Let $\lambda(t)$ be a continuous function on R_1 or a step function. Then

$$\lambda(\Lambda)\, a = \int_d^\infty \lambda(t)\, \mathrm{d}E(t)\, a \tag{2}$$

has the usual meaning. Of peculiar interest are the Riesz means $R_{\alpha,\varrho}$ given by

$$R_{\alpha,\varrho}a = \int_d^\infty \left(1 - \frac{t}{\varrho}\right)_+^\alpha \mathrm{d}E(t)\, a, \tag{3}$$

where α is a complex number with $\operatorname{Re}\alpha \geqq 0$, $\varrho > 0$ and

$$\left(1 - \frac{t}{\varrho}\right)_+ = \max\left(0, 1 - \frac{t}{\varrho}\right). \tag{4}$$

Let A be a complex Banach space. We assume that both the above Hilbert space H and A are continuously embedded in a common complex linear Hausdorff space. Furthermore, let $A \cap H$ be dense both in H and A. Let $\lambda(t)$ be a bounded continuous function on R_1 or a bounded step function on R_1. Then $\lambda(t)$ is said to be a multiplier for A (with respect to H and Λ) if there exists a positive number c such that

$$\|\lambda(\Lambda)\, a \mid A\| \leqq c\|a \mid A\| \tag{5}$$

holds for all $a \in H \cap A$. Afterwards by continuity, $\lambda(\Lambda)$ can be extended to a linear and bounded operator in A. Finally, A-measurable and A-integrable functions have the usual meaning, cf. e.g. N. Dunford, J. T. Schwartz [1].

Proposition. *Let H and A be the above spaces and let Λ be the above positive-definite self-adjoint operator in H. Let l be a non-negative integer and let $R_{l,\varrho}\, a$ be an A-Lebesgue-measurable function with respect to ϱ for any $a \in H \cap A$. Let*

$$\sup_{\varrho>0} \|R_{l,\varrho}a \mid A\| \leqq c\|a \mid A\| \quad \textit{for all} \quad a \in H \cap A, \tag{6}$$

where $c \geqq 0$ is an appropriate number. Then there exists a positive number C such that for all $(l + 1)$-times continuously differentiable complex-valued functions $\lambda(t)$ on $[0, \infty)$ with

$$\lambda(t) \to 0 \quad \textit{when} \quad t \to \infty \tag{7}$$

and for all $a \in H \cap A$

$$\|\lambda(\Lambda)\, a \mid A\| \leqq C\left(\int_d^\infty t^{l+1}|\lambda^{(l+1)}(t)| \frac{\mathrm{d}t}{t}\right)\|a \mid A\| \tag{8}$$

holds.

Proof. Let the integral on the right-hand side of (8) be finite. We claim that

$$\lambda(s) = \frac{(-1)^{l+1}}{l!} \int_s^\infty (t - s)^l\, \lambda^{(l+1)}(t)\, \mathrm{d}t \tag{9}$$

holds for any $s > 0$. We denote the right-hand side of (9) by $\mu(s)$. Differentiation yields $\mu^{(l+1)}(s) = \lambda^{(l+1)}(s)$. Hence, $\mu(s) - \lambda(s)$ is a polynomial. Because $\mu(s) - \lambda(s)$

$\to 0$ if $s \to \infty$ we have (9). Let $a \in H \cap A$. Then (2) and (9) yield

$$\lambda(\Lambda)\, a = \int_0^\infty \frac{(-1)^{l+1}}{l!} \left(\int_s^\infty (t-s)^l \lambda^{(l+1)}(t)\, \mathrm{d}t \right) \mathrm{d}E(s)\, a$$

$$= \frac{(-1)^{l+1}}{l!} \int_0^\infty \lambda^{(l+1)}(t) \left(\int_0^t (t-s)^l\, \mathrm{d}E(s)\, a \right) \mathrm{d}t$$

$$= \frac{(-1)^{l+1}}{l!} \int_0^\infty t^l \lambda^{(l+1)}(t) \left(\int_0^\infty \left(1 - \frac{s}{t}\right)_+^l \mathrm{d}E(s)\, a \right) \mathrm{d}t$$

$$= \frac{(-1)^{l+1}}{l!} \int_0^\infty t^l \lambda^{(l+1)}(t)\, R_{l,t} a\, \mathrm{d}t. \tag{10}$$

Now, (8) follows from (6) and (10).

Remark* 1. Both the proposition and its proof are due to P. L. Butzer, R. J. Nessel, W. Trebels [1, p. 341], cf. also R. J. Nessel, G. Wilmes [4] for a more recent version. The above result is sufficient for our purposes. There exist sharper assertions of this type, in particular for fractional l's. Cf. the references in [F, p. 118], the Lecture Notes by W. Trebels [1] and the book by J. Peetre [4].

Remark 2. Of course, (8) is a multiplier assertion. The abstract potential spaces from 6.3. have the same multiplier properties. The abstract Besov spaces from 6.2.2. can be obtained via interpolation from the potential spaces, cf. 6.3.2. Hence, the abstract Besov spaces have also multiplier properties of type (8). But in 6.3.3. we shall prove a much stronger multiplier theorem of Michlin-Hörmander type for the abstract Besov spaces. This is a well-known effect for (concrete) Besov spaces. They have better multiplier properties than the corresponding (Bessel-) potential spaces.

6.2.2. Definitions

First we formalize the hypotheses of Proposition 6.2.1.

Definition 1. *Let H be a complex Hilbert space and let Λ be a self-adjoint positive-definite operator acting in H. Let A be a complex Banach space. Let l be a non-negative integer. Then $\mathfrak{R}^l$ is the collection of all triplets $[H, \Lambda, A]$ with the following properties:*

(i) *There exists a linear Hausdorff space $\mathscr{H}$ such that $H \subset \mathscr{H}$ and $A \subset \mathscr{H}$ (linear continuous embedding). Furthermore $H \cap A$ is dense both in H and in A.*

(ii) *For any $a \in H \cap A$ the Riesz mean $R_{l,\varrho}a$ is an A-measurable function with respect to $\varrho \in [0, \infty)$. Furthermore, there exists a positive number c with*

$$\sup_{\varrho > 0} \|R_{l,\varrho}a \mid A\| \leqq c\|a \mid A\| \quad \text{for all} \quad a \in H \cap A. \tag{1}$$

(iii) *If β is a complex number with $\operatorname{Re} \beta > l + 1$ and if $a \in H \cap A$ then*

$$R_{\beta,\varrho}a \to a \quad \text{when} \quad \varrho \to \infty \ (\text{convergence in } A). \tag{2}$$

Remark 1. The assumptions (i) and (ii) coincide with the hypotheses of Proposition 6.2.1. Hence (6.2.1/8) holds. We choose $\lambda(t) = \left(1 - \frac{t}{\varrho}\right)_+^\beta$ with $\operatorname{Re}\beta > l + 1$ and obtain

$$\sup_{\varrho > 0} \|R_{\beta,\varrho} a \mid A\| \leqq c\|a \mid A\|, \quad a \in H \cap A, \tag{3}$$

where c is an appropriate positive constant. Since $H \cap A$ is dense in A, formula (3) can be extended to A. Hence $\{R_{\beta,\varrho}\}_{\varrho > 0}$ is a bounded sequence of operators in A. Assumption (2) must be understood in the sense of these extended operators. In [F, Remark 3.2.2/2] we proved that (iii) is independent of (i, ii): There exist triplets $[H, \Lambda, A]$ which satisfy the assumptions (i) and (ii) of the above proposition but not (2) for β with $\operatorname{Re}\beta > l + 1$.

We need the following obvious modification of the system $\Phi(R_1)$ from 2.2.1. Let Φ be the collection of all systems $\varphi = \{\varphi_j(t)\}_{j=0}^\infty$ of real-valued infinitely differentiable functions on R_1 with the following properties:

(i) $\operatorname{supp}\varphi_0 \subset (-2, 2), \qquad \operatorname{supp}\varphi_j \subset (2^{j-1}, 2^{j+1})$ if $j = 1, 2, 3, \ldots$

(ii) For any non-negative integer k there exists a positive number c_k such that

$$|\varphi_j^{(k)}(t)| \leqq c_k \cdot 2^{-jk} \tag{4}$$

holds for $t \in [0, \infty)$ and $j = 0, 1, 2, \ldots,$

(iii) $$\sum_{j=0}^\infty \varphi_j(t) = 1 \quad \text{if} \quad t \in [0, \infty). \tag{5}$$

As has been mentioned in 2.2.1. there exist systems φ with the desired properties. Similar as in 1.3.1. we put

$$\|a_j \mid l_q(A)\| = \left(\sum_{j=0}^\infty \|a_j \mid A\|^q\right)^{1/q} \quad \text{if} \quad 1 \leqq q < \infty,$$

$$\|a_j \mid l_\infty(A)\| = \sup_j \|a_j \mid A\|.$$

Of course $\{a_j\}_{j=0}^\infty \subset A$.

Definition 2. *Let l be a non-negative integer and let $[H, \Lambda, A] \in \mathfrak{R}^l$ in the sense of Definition 1. Let $s > 0$, $1 \leqq q \leqq \infty$ and $\varphi = \{\varphi_j(t)\}_{j=0}^\infty \in \Phi$. Then*

$$B_q^s = \{a \mid a \in A, \|a \mid B_q^s\|^\varphi = \|2^{sj}\varphi_j(\Lambda)\, a \mid l_q(A)\| < \infty\}. \tag{6}$$

Remark 2. The definition makes sense because $\varphi_j(t)$ satisfies the hypotheses of Proposition 6.2.1 and consequently $\varphi_j(\Lambda)$ may be considered as a linear bounded operator acting in A. Of course, B_q^s depends on the triplet $[H, \Lambda, A]$. But we assume that this triplet is fixed once and for all in this chapter (if not stated otherwise explicitly). Furthermore we shall see that B_q^s is independent of the chosen system $\varphi \in \Phi$ (in the sense of equivalent norms). This justifies our notation.

Remark 3. There is no problem (neither in the above definition nor in the following considerations) to extend (6) to $0 < q < 1$. Then one obtains quasi-Banach spaces. However in this chapter we restrict ourselves to Banach spaces. Another question is the extension of the definition of B_q^s to values $s \leqq 0$. This is strongly suggested by the concrete realizations of type $B_{p,q}^s$ which we considered in the Chapters 2 and 3 and in [T, Chapter 2]. However an immediate definition in the sense of (6) with $s \leqq 0$ does not make sense. One must modify the general abstract background, cf. 6.5.3. In order to define abstract counterparts of spaces of type $F_{p,q}^s$ from the Chapters 2 and 3 and [T, Chapter 2] the above hypotheses are not sufficient. Maybe one can do something in this direction if the underlying spaces H and A are Banach lattices. Few remarks are given in H. Triebel [20]. Further references connected with the above definition may be found in [F, p. 121].

6.2.3. Basic Theorem

All notations have the same meaning as in the preceding subsection.

Lemma. *Let* $[H, \Lambda, A] \in \mathfrak{R}^l$ *for some* l, *let* $\varphi = \{\varphi_j(t)\}_{j=0}^{\infty} \in \Phi$ *and let* $a \in A$. *Then*

$$a = \sum_{j=0}^{\infty} \varphi_j(\Lambda)\, a \quad (\textit{convergence in } A). \tag{1}$$

Proof. Let β be a fixed complex number with $\operatorname{Re}\beta > l + 1$. Let N be a natural number. Proposition 6.2.1 yields

$$\left\| \sum_{j=0}^{N} \varphi_j(\Lambda)\, a - \sum_{j=0}^{N} \varphi_j(\Lambda)\, R_{\beta,\varrho} a \mid A \right\| \leqq c \|a - R_{\beta,\varrho} a \mid A\|, \tag{2}$$

where c is independent of N, ϱ (and a). Hence by the triangle inequality and (6.2.2/2) we have for any natural number N,

$$\left\| \sum_{j=0}^{N} \varphi_j(\Lambda)\, a - a \mid A \right\| \leqq \varepsilon + \left\| \sum_{j=0}^{N} \varphi_j(\Lambda)\, R_{\beta,\varrho} a - R_{\beta,\varrho} a \mid A \right\|, \tag{3}$$

where $\varepsilon > 0$ is given and $\varrho = \varrho(\varepsilon)$ is chosen sufficiently large. However if N is large, $N \geqq N(\varrho)$, then the second term of the right-hand side of (3) is zero. This completes the proof.

Theorem. *Let* $[H, \Lambda, A] \in \mathfrak{R}^l$ *for some non-negative integer* l. *Let* $s > 0$ *and* $1 \leqq q \leqq \infty$. *Then* B_q^s *from Definition* 6.2.2/2 *is a Banach space. It is independent of the choice of* $\varphi \in \Phi$ (*in the sense of equivalent norms*).

Proof. Step 1. Let $\varphi = \{\varphi_j(t)\}_{j=0}^{\infty} \in \Phi$ and $\psi = \{\psi_j(t)\}_{j=0}^{\infty} \in \Phi$. We have

$$\psi_j(t) = \psi_j(t)\,(\varphi_{j-1}(t) + \varphi_j(t) + \varphi_{j+1}(t)) \tag{4}$$

with $\varphi_{-1}(t) = 0$. By Proposition 6.2.1 we obtain

$$\|\psi_j(\Lambda)\, a \mid A\| \leqq c(\|\varphi_{j-1}(\Lambda)\, a \mid A\| + \|\varphi_j(\Lambda)\, a \mid A\| + \|\varphi_{j+1}(\Lambda)\, a \mid A\|),$$

where c is independent of j. Consequently

$$\|a \mid B_q^s\|^{\psi} \leqq c \|a \mid B_q^s\|^{\varphi}. \tag{5}$$

Step 2. Let $\|a \mid B_q^s\|^{\varphi} = 0$. Then we have $\varphi_j(\Lambda)\, a = 0$ and by the above lemma $a = 0$. Then it follows that $\|a \mid B_q^s\|^{\varphi}$ is a norm. By (5) the space B_q^s is independent of $\varphi \in \Phi$ and different φ's yield equivalent norms. Finally, let $\{a_k\}_{k=1}^{\infty}$ be a fundamental sequence in B_q^s. Because $s > 0$ the above lemma yields

$$\|a_m - a_k \mid A\| \leqq \sum_{j=0}^{\infty} \|\varphi_j(\Lambda)\, a_m - \varphi_j(\Lambda)\, a_k \mid A\| \leqq c \|a_m - a_k \mid B_q^s\|^{\varphi}.$$

Hence $\{a_k\}_{k=1}^{\infty}$ is also a fundamental sequence in A. Let $a_k \to a$ in A. Then it follows by standard arguments $a_k \to a$ in B_q^s. The proof is complete.

Remark 1. In the sequel we write $\|a \mid B_q^s\|$ instead of $\|a \mid B_q^s\|^{\varphi}$ with $\varphi \in \Phi$.

Remark 2. Let $\chi_j(t)$ be the characteristic function of $[2^{j-1}, 2^j]$ if $j = 1, 2, 3, \ldots$, and let $\chi_0(t)$ be the characteristic function of $[0, 1]$. If $A = H$ then

$$\|a \mid B_q^s\| \sim \|2^{sj} \chi_j(\Lambda)\, a \mid l_q(H)\| \tag{6}$$

(equivalent norms). This is a simple consequence of spectral theory.

6.3. Abstract Potential Spaces

6.3.1. The Operators Λ^α

We always assume $[H, \Lambda, A] \in \mathfrak{R}^l$ for some non-negative integer l. If α is an arbitrary complex number then the fractional powers of Λ in H are well-defined by (6.2.1/2) with $\lambda(t) = t^\alpha$. We denote these operators by Λ_H^α with the domain of definition $D(\Lambda_H^\alpha)$. Let

$$D(\mathring{\Lambda}_A^\alpha) = \{a \mid a \in A \cap D(\Lambda_H^\alpha), \Lambda_H^\alpha a \in A\} \tag{1}$$

and

$$\mathring{\Lambda}_A^\alpha a = \Lambda_H^\alpha a \quad \text{if} \quad a \in D(\mathring{\Lambda}_A^\alpha). \tag{2}$$

Of course, $\mathring{\Lambda}_A^\alpha$ is a linear operator in A. We shall prove that $\mathring{\Lambda}_A^\alpha$ is a closable operator in A. Then we put $\Lambda_A^\alpha = \overline{\mathring{\Lambda}_A^\alpha}$ (the closure of $\mathring{\Lambda}_A^\alpha$) and denote the domain of definition of Λ_A^α by $D(\Lambda_A^\alpha)$. We recall that $D(\Lambda_A^\alpha)$ equipped with the norm $\|a \mid A\| + \|\Lambda_A^\alpha a \mid A\|$ becomes a Banach space.

Proposition. (i) *$D(\mathring{\Lambda}_A^\alpha)$ is dense in A and $\mathring{\Lambda}_A^\alpha$ is a closable operator.*

(ii) *Let β be a complex number with* $\operatorname{Re}\beta > l + 1$. *Then*

$$N_\beta = \{a \mid \exists b \in A \cap H \text{ and } \exists \varrho > 0 \text{ with } a = R_{\beta,\varrho} b\} \tag{3}$$

is dense in $D(\Lambda_A^\alpha)$.

Proof. Step 1. We prove that $\mathring{\Lambda}_A^\alpha$ is a closable operator in A. Let $\{a_k\}_{k=1}^\infty \subset D(\mathring{\Lambda}_A^\alpha)$ with

$$a_k \to 0 \quad \text{and} \quad \Lambda_H^\alpha a_k \to b \quad \text{if} \quad k \to \infty \quad (\text{in } A). \tag{4}$$

Let $\varrho > 0$ be fixed. Then Proposition 6.2.1 yields

$$R_{l,\varrho} a_k \to 0 \quad \text{and} \quad R_{2l+2,\varrho} a_k \to 0 \quad \text{if} \quad k \to \infty \ (\text{in } A). \tag{5}$$

On the other hand we have

$$(\Lambda_H^\alpha R_{l+2,\varrho})(R_{l,\varrho} a_k) = R_{2l+2,\varrho} \Lambda_H^\alpha a_k \to R_{2l+2,\varrho} b \tag{6}$$

if $k \to \infty$ (in A). However, $\Lambda_H^\alpha R_{l+2,\varrho}$ can be extended to a bounded operator in A. Hence, by (6) and the first assertion in (5) we have

$$R_{2l+2,\varrho} b = 0 \quad \text{for all} \quad \varrho > 0.$$

Now, (6.2.2/2) shows that $b = 0$. This proves that $\mathring{\Lambda}_A^\alpha$ is a closable operator.

Step 2. We prove that $D(\mathring{\Lambda}_A^\alpha)$ is dense in A. We recall that $A \cap H$ is dense in A. Then (6.2.2/2) and its obvious counterpart with H instead of A prove that N_β is dense in A. If $b \in N_\beta$ then $b \in D(\mathring{\Lambda}_A^\alpha)$. This follows from (1) and Proposition 6.2.1. Hence $N_\beta \subset D(\mathring{\Lambda}_A^\alpha)$. This proves that $D(\mathring{\Lambda}_A^\alpha)$ is dense in A. The proof of (i) is complete.

Step 3. We prove (ii). By definition it is sufficient to prove that N_β is dense in $D(\mathring{\Lambda}_A^\alpha)$. Let $a \in D(\mathring{\Lambda}_A^\alpha)$. Then we have

$$N_\beta \ni R_{\beta,\varrho} a \to a \quad \text{if} \quad \varrho \to \infty \quad (\text{in } A \cap H)$$

and

$$\Lambda_H^\alpha R_{\beta,\varrho} a = R_{\beta,\varrho} \Lambda_H^\alpha a \to \Lambda_H^\alpha a \quad \text{if} \quad \varrho \to \infty \quad (\text{in } A).$$

This yields the desired assertion.

We recall the definition of a positive operator in a complex Banach space. A closed operator Λ with the dense domain of definition $D(\Lambda)$, acting in a Banach space A, is said to be a positive operator if $(-\infty, 0]$ belongs to the resolvent set of Λ and if there exists a positive number C such that for all $a \in D(\Lambda)$ and all $\tau \in (-\infty, 0]$

$$\|(\Lambda - \tau E)^{-1} \mid A \to A\| \leqq \frac{C}{1 + |\tau|} \tag{7}$$

holds (E stands for the identity). For positive operators and their domains of definitions (considered as Banach spaces) there exist an elaborated interpolation theory and a theory of fractional powers, cf. e.g. [I, 1.14. and 1.15.]. We shall use some of these results in the sequel.

As usual we write $\Lambda_A = \Lambda_A^1$, where Λ_A^α has the above meaning. We shall see that Λ_A is a positive operator. Temporarily $(\Lambda_A)^\alpha$ stands for the fractional power of Λ_A, where α is a complex number.

Theorem. *Let $[H, \Lambda, A] \in \mathfrak{R}^l$ for some non-negative integer l.*

(i) *Λ_A is a positive operator.*

(ii) *It holds $(\Lambda_A)^\alpha = \Lambda_A^\alpha$ for any complex number α.*

(iii) *$\Lambda^{i\varkappa}$ is a linear bounded operator for any real number $\varkappa$. Furthermore, there exists a positive number c such that*

$$\|\Lambda^{i\varkappa} \mid A \to A\| \leqq c(1 + |\varkappa|^{l+1}) \tag{8}$$

holds for all real numbers $\varkappa$.

Proof. Step 1. We prove (i). We apply Proposition 6.2.1 to $\lambda(t) = (t - \tau)^{-1}$ with $\tau \in R_1$ and $\tau \leqq 0$ (modification near the origin if $\tau = 0$). Let F_τ be the corresponding operator. We have

$$\|F_\tau \mid A \to A\| \leqq \frac{c}{1 + |\tau|}, \quad \tau \leqq 0, \tag{9}$$

where c is an appropriate positive number. Let $a \in N_\beta$, cf. (3), with $\operatorname{Re} \beta > l + 1$. Then we have

$$(\Lambda_A - \tau E) F_\tau a = F_\tau(\Lambda - \tau E) a = a. \tag{10}$$

However by the above proposition N_β is dense both in A and in $D(\Lambda_A)$ and Λ_A is a closed operator. Then (10) yields

$$(\Lambda_A - \tau E) F_\tau = E \quad \text{(identity in } A\text{)},$$
$$F_\tau(\Lambda_A - \tau E) a = a \quad \text{if} \quad a \in D(\Lambda_A).$$

Consequently, $F_\tau = (\Lambda_A - \tau E)^{-1}$. This proves (i).

Step 2. We prove (ii). Recall that N_β has been defined in (3). If m is a natural number then we have

$$(\Lambda_A)^m N_\beta = \Lambda_H^m N_\beta = N_\beta. \tag{11}$$

Furthermore, $(\Lambda_A)^m$ and Λ_A^m coincide on N_β. By (11) and the mapping properties of the closed operators $(\Lambda_A)^m$ and Λ_A^m we have $\Lambda_A^m = (\Lambda_A)^m$. We recall the definition of the fractional powers of the positive-definite operator Λ_A. Let α be a complex number and let m and n be natural numbers with $-n < \operatorname{Re} \alpha < m - n$. If $a \in D(\Lambda_A^m)$

then

$$(\Lambda_A)^\alpha a = \frac{\Gamma(m)}{\Gamma(\alpha+n)\,\Gamma(m-n-\alpha)} \int_0^\infty t^{\alpha+n-1} \Lambda_A^{m-n} (\Lambda_A + tE)^{-m}\, a\, \mathrm{d}t, \qquad (12)$$

cf. [I, Lemma 1.15.1]. Again $\Lambda_A^{m-n}(\Lambda_A + tE)^{-m}$ maps N_β onto itself and coincides on N_β with $\Lambda_H^{m-n}(\Lambda_H + tE)^{-m}$, where again $\Lambda_H = \Lambda$ is the given self-adjoint operator in H. Consequently,

$$(\Lambda_A)^\alpha a = \Lambda_H^\alpha a = \Lambda_A^\alpha a \quad \text{if} \quad a \in N_\beta. \qquad (13)$$

Recall that N_β is dense in $D(\Lambda_A^m)$. Furthermore, $D(\Lambda_A^m) = D((\Lambda_A)^m)$ is dense in $D((\Lambda_A)^\alpha)$, cf. [I, 1.15.1.]. Hence, N_β is dense in $D((\Lambda_A)^\alpha)$. On the other hand, by the above proposition N_β is also dense in $D(\Lambda_A^\alpha)$. Then (13) yields (ii).

Step 3. We prove (iii). Let $a \in N_\beta \subset D(\Lambda_A^{i\varkappa})$. By (6.2.2/2) and Proposition 6.2.1 with $\lambda(t) = \left(1 - \frac{t}{\varrho}\right)_+^{l+2} t^{i\varkappa}$ (modification near the origin) we have

$$\|\Lambda_A^{i\varkappa} a \mid A\| = \lim_{\varrho \to \infty} \|R_{l+2,\varrho} \Lambda_A^{i\varkappa} a \mid A\| \leqq c(1 + |\varkappa|^{l+1}) \|a \mid A\|, \qquad (14)$$

where c is independent of ϱ and $\varkappa$. Because N_β is dense in $D(\Lambda_A^{i\varkappa})$, (14) yields (8).

Remark. If $\operatorname{Re} \alpha < 0$ then the fractional power Λ_A^α of the positive operator Λ_A is a bounded operator. This is a well-known fact for general positive operators in Banach spaces, cf. e.g. [I, 1.15.]. On the other hand, in the general case the pure imaginary powers $\Lambda^{i\varkappa}$ of a positive operator Λ in a Banach space are not necessarily bounded, cf. H. Komatsu [2, p. 342] for a counter-example. In our case we have (8). Inequalities of this type have far-reaching consequences in interpolation theory, cf. [I, 1.15.3.]. We return to this point in the next subsection.

6.3.2. Interpolation

Let $[H, \Lambda, A] \in \mathfrak{R}^l$ for some non-negative integer l. In particular we have $H \subset \mathscr{H}$ and $A \subset \mathscr{H}$ for some linear Hausdorff space $\mathscr{H}$. Furthermore let

$$[H, \Lambda, A_0] \in \mathfrak{R}^l \quad \text{and} \quad [H, \Lambda, A_1] \in \mathfrak{R}^l \qquad (1)$$

with respect to the same linear Hausdorff space $\mathscr{H}$, i.e. $A_0 \subset \mathscr{H}$ and $A_1 \subset \mathscr{H}$. In particular, $\{A_0, A_1\}$ is an interpolation couple. The classical complex interpolation method and the real interpolation method will be denoted by $[\cdot, \cdot]_\theta$ and $(\cdot, \cdot)_{\theta,p}$ respectively, where $0 < \theta < 1$ and $1 \leqq p \leqq \infty$. As far as interpolation theory is concerned we refer to [I] or J. Bergh, J. Löfström [1]. If the underlying Banach space is A, then B_q^s has the meaning of Definition 6.2.2/2. In this case Λ_A^α and the corresponding domains of definition $D(\Lambda_A^\alpha)$ have the meaning of the preceding subsection. If other Banach spaces are involved, $A_0, A_1, (A_0, A_1)_{\theta,p}$ or $[A_0, A_1]_\theta$ then we write $B_q^s(A_0)$, $B_q^s(A_1)$, $B_q^s((A_0, A_1)_{\theta,p})$ or $B_q^s([A_0, A_1]_\theta)$, respectively.

Theorem. *Let $[H, \Lambda, A] \in \mathfrak{R}^l$, $[H, \Lambda, A_0] \in \mathfrak{R}^l$ and $[H, \Lambda, A_1] \in \mathfrak{R}^l$ for some non-negative integer l, where all spaces A, H, A_0 and A_1 are continuously embedded in a common linear Hausdorff space.*

(i) *Let $0 < s_0 < \infty$, $0 < s_1 < \infty$, $0 < \theta < 1$, $s = (1-\theta)s_0 + \theta s_1$, $1 \leqq q_0 < \infty$, $1 \leqq q_1 < \infty$ and $\frac{1}{q} = \frac{1-\theta}{q_0} + \frac{\theta}{q_1}$. Then*

$$(B_{q_0}^{s_0}(A_0), B_{q_1}^{s_1}(A_1))_{\theta,q} = B_q^s((A_0, A_1)_{\theta,q}) \qquad (2)$$

and

$$[B^{s_0}_{q_0}(A_0), B^{s_1}_{q_1}(A_1)]_\theta = B^s_q([A_0, A_1]_\theta). \tag{3}$$

(ii) *Let* $0 < s_0 < s_1 < \infty$, $0 < \theta < 1$, $s = (1-\theta) s_0 + \theta s_1$, $1 \leqq q_0 \leqq \infty$, $1 \leqq q_1 \leqq \infty$ *and* $1 \leqq q \leqq \infty$. *Then*

$$(B^{s_0}_{q_0}, B^{s_1}_{q_1})_{\theta,q} = B^s_q. \tag{4}$$

(iii) *Let* $\alpha > 0$ *and* $0 < \theta < 1$. *Then*

$$(A, D(\Lambda^\alpha_A))_{\theta,q} = B^{\alpha\theta}_q \tag{5}$$

and

$$[A, D(\Lambda^\alpha_A)]_\theta = D(\Lambda^{\alpha\theta}_A). \tag{6}$$

Proof. Step 1. Let $\varphi = \{\varphi_j(t)\}_{j=0}^\infty \in \Phi$, cf. 6.2.2. Let $\psi = \{\psi_j(t)\}_{j=0}^\infty$ be a system of real-valued infinitely differentiable functions on R_1 which satisfies the conditions (i) and (ii) (with ψ instead of φ) for the functions of the system φ from 6.2.2., and

$$\psi_j(t) = 1 \quad \text{if} \quad t \in \operatorname{supp} \varphi_j, \quad j = 0, 1, 2, \ldots$$

It is easy to see that there exist systems φ and ψ with these properties. If $s > 0$ and $1 \leqq q \leqq \infty$ then we put

$$l^s_q(A) = \Big\{c \mid c = \{c_j\}_{j=0}^\infty \subset A,$$
$$\|c_j \mid l^s_q(A)\| = \left(\sum_{j=0}^\infty 2^{jsq} \|c_j \mid A\|^q\right)^{1/q} < \infty\Big\} \tag{7}$$

(with the usual modification if $q = \infty$). The operator S, given by

$$Sa = \{\varphi_j(\Lambda)\, a\}_{j=0}^\infty, \tag{8}$$

maps B^s_q continuously into $l^s_q(A)$. On the other hand, the operator R, given by

$$R\{a_j\} = \sum_{j=0}^\infty \psi_j(\Lambda)\, a_j, \tag{9}$$

maps $l^s_q(A)$ continuously in B^s_q: This follows from Proposition 6.2.1. If $a \in A \cap H$ then we have $\psi_j(\Lambda)\,\varphi_j(\Lambda)\, a = \varphi_j(\Lambda)\, a$. Completion yields $\psi_j(\Lambda)\,\varphi_j(\Lambda) = \varphi_j(\Lambda)$. Then Lemma 6.2.3 yields

$$RSa = \sum_{j=0}^\infty \varphi_j(\Lambda)\, a = a. \tag{10}$$

Consequently,

$$RS = E \text{ (identity in } B^s_q). \tag{11}$$

In other words, R is a retraction and S is a corresponding coretraction. This is a standard situation in interpolation theory. Recall that $P = SR$ is a projection in $l^s_q(A)$. Let $Pl^s_q(A)$ be its range (a closed subspace of $l^s_q(A)$). Then S yields an one-to-one mapping from B^s_q onto $Pl^s_q(A)$. This assertion holds for different values of s and q, and also if one replaces A by A_0 and A_1. Then it follows from [I, 1.17.1.] that S yields also an one-to-one mapping from

$$[B^{s_0}_{q_0}(A_0), B^{s_1}_{q_1}(A_1)]_\theta \quad \text{onto} \quad P[l^{s_0}_{q_0}(A_0), l^{s_1}_{q_1}(A_1)]_\theta. \tag{12}$$

One has a similar assertion for the real method $(A_0, A_1)_{\theta,q}$. Consequently, (2) and (3) follow from

$$(l_{q_0}^{s_0}(A_0), l_{q_1}^{s_1}(A_1))_{\theta,q} = l_q^s((A_0, A_1)_{\theta,q}) \tag{13}$$

and

$$[l_{q_0}^{s_0}(A_0), l_{q_1}^{s_1}(A_1)]_\theta = l_q^s([A_0, A_1]_\theta), \tag{14}$$

respectively. Under the hypotheses of the above theorem, (13) and (14) may be found in [I, Theorem 1.18.1]. Furthermore by the same method one proves (4). One has to use [I, Theorem 1.18.2].

Step 2. We prove (iii). First we claim

$$B_1^\alpha \subset D(\Lambda_A^\alpha) \subset B_\infty^\alpha, \quad \alpha > 0. \tag{15}$$

If $a \in N_\beta$, cf. (6.3.1/3), then (6.2.3/1) proves

$$\|\Lambda_A^\alpha a \mid A\| \leqq \sum_{j=0}^{\infty} \|\Lambda_H^\alpha \varphi_j(\Lambda)\, a \mid A\|. \tag{16}$$

The functions $\varphi_j(t)$ and $\psi_j(t)$ have the same meaning as in the first step. Then we have

$$t^\alpha \varphi_j(t) = 2^{j\alpha} \psi_j(t)\, \lambda_j(t) \quad \text{with} \quad \lambda_j(t) = \left(\frac{t}{2^j}\right)^\alpha \varphi_j(t), \tag{17}$$

$j = 1, 2, 3, \ldots$ We apply Proposition 6.2.1 with $\lambda_j(t)$ instead of $\lambda(t)$ and obtain

$$\|\Lambda_A^\alpha a \mid A\| \leqq c \sum_{j=0}^{\infty} 2^{j\alpha} \|\psi_j(\Lambda)\, a \mid A\| \leqq c' \|a \mid B_1^\alpha\|. \tag{18}$$

Because N_β is dense in B_1^α, the left-hand side of (15) follows from (18) and a completion argument. In order to prove the right-hand side of (15) we replace (17) by

$$2^{j\alpha} \varphi_j(t) = t^\alpha \tilde{\lambda}_j(t) \quad \text{with} \quad \tilde{\lambda}_j(t) = \left(\frac{2^j}{t}\right)^\alpha \varphi_j(t), \tag{19}$$

$j = 1, 2, 3, \ldots$ If $a \in N_\beta$ then Proposition 6.2.1 with $\tilde{\lambda}_j(t)$ instead of $\lambda(t)$ yields

$$\|a \mid B_\infty^\alpha\| = \sup_j 2^{j\alpha} \|\varphi_j(\Lambda)\, a \mid A\| \leqq c \|\Lambda_A^\alpha a \mid A\|. \tag{20}$$

Completion yields the right-hand side of (15). Let $0 < s_0 < s_1 < \infty$. Then (4), (15) and the reiteration theorem of interpolation theory, cf. [I, 1.10.], yield

$$(D(\Lambda_A^{s_0}), D(\Lambda_A^{s_1}))_{\theta,q} = B_q^s, \qquad s = (1-\theta)\, s_0 + \theta s_1. \tag{21}$$

We use the fact that $\Lambda_A^{-s_0}$ maps $D(\Lambda_A^{s_0})$ onto A and $D(\Lambda_A^{s_1})$ onto $D(\Lambda_A^{s_1-s_0})$. By the above considerations, cf. (17) and (19), $\Lambda_A^{-s_0}$ maps also B_q^s onto $B_q^{s-s_0}$. If we apply $\Lambda_A^{-s_0}$ to (21) then we obtain (5). Finally, (6) is a consequence of (6.3.1/8) and the interpolation theory for fractional powers of positive-definite operators, cf. [I, 1.15.3.].

Remark 1. The above theorem is the abstract counterpart of well-known interpolation properties for Besov-Sobolev spaces on R_n or on T_n. In this case one can take $H = L_2(R_n)$ (resp. $L_2(T_n)$), $A = L_p(R_n)$ (resp. $L_p(T_n)$), and $\Lambda = E - \Delta$, where Δ stands for the Laplacian. We refer to 6.5.1. and 6.5.2. where we describe this possibility in some detail. Cf. also [I, 2.4.1., 2.4.2.] as far as the R_n-case is concerned and 3.6.1. for its periodic counterpart.

Remark 2. In [I, 1.14.] we developed an interpolation theory for positive operators. If one applies this theory to the above theorem then one obtains a lot of equivalent norms for the spaces B_q^s. If Λ is the infinitesimal generator of a strongly countinuous semi-group in A, then one has more equivalent norms for the spaces B_q^s, cf. [I, 1.13.]. Some references can be found in [F, p. 129].

6.3.3. Multipliers

Let again $[H, \Lambda, A] \in \mathfrak{R}^l$ for some non-negative integer l. If β is a complex number with $\operatorname{Re} \beta > l + 1$, then N_β is dense in A and dense in $D(\Lambda_A^\alpha)$ for all $\alpha > 0$, cf. Proposition 6.3.1. In other words, the definition of a multiplier for A from (6.2.1/5) can be modified as follows: Let $\lambda(t)$ be a bounded continuous function on R_1 or a bounded step function on R_1. Then $\lambda(t)$ is said to be a multiplier for A (with respect to H and Λ) if there exists a positive number c such that

$$\|\lambda(\Lambda)\, a \mid A\| \leqq c\|a \mid A\| \quad \text{for} \quad a \in N_\beta \tag{1}$$

holds. If one replaces A in (1) by $D(\Lambda_A^\alpha)$ with $\alpha > 0$ or B_q^s with $s > 0$ and $1 \leqq q < \infty$ then one obtains the definition of a multiplier for $D(\Lambda_A^\alpha)$ or B_q^s, respectively. This definition is reasonable because N_β is dense in $D(\Lambda_A^\alpha)$ and consequently also dense in B_q^s if $q < \infty$. This follows from (6.3.2/5) and well-known properties of interpolation spaces. One can restrict the above definition of multipliers for the spaces B_q^s to $q < \infty$ and extends this definition afterwards to $q = \infty$ via interpolation, cf. (6.3.2/4). By (6.3.1/11) we have $\Lambda_A^\alpha N_\beta = N_\beta$ for all $\alpha \geqq 0$, and consequently for all $\alpha \in R_1$. Now it follows easily that $\lambda(t)$ is a multiplier for $D(\Lambda_A^\alpha)$, $\alpha > 0$, if and only if $\lambda(t)$ is a multiplier for A. Furthermore, if $0 < \alpha < s < \infty$ then $\Lambda_A^\alpha B_q^s = B_q^{s-\alpha}$. Consequently, $\lambda(t)$ is a multiplier for B_q^s if and only if it is a multiplier for $B_q^{s-\alpha}$. By interpolation, cf. (6.3.2/4), it follows that $\lambda(t)$ is a multiplier for $B_{q_0}^{s_0}$ with $s_0 > 0$ and $1 \leqq q_0 < \infty$ if and only if it is a multiplier for $B_{q_1}^{s_1}$ with $s_1 > 0$ and $1 \leqq q_1 < \infty$. The extension to B_∞^s is, by definition, a matter of interpolation. Hence we have two classes of multipliers, the class M_0 of multipliers for $D(\Lambda_A^\alpha)$ with $\alpha \geqq 0$, and the class M_1 of multipliers for B_q^s with $s > 0$ and $1 \leqq q \leqq \infty$. By (6.3.2/5) it follows $M_0 \subset M_1$. In other words, any multiplier for $D(\Lambda_A^\alpha)$ is also a multiplier for B_q^s. The converse assertion is not true in general, cf. the Example below. Proposition 6.2.1 describes a subclass of M_0 which is sufficient for our purposes. On this basis one can select a corresponding subclass of M_1 which includes multipliers of Michlin-Hörmander type for B_q^s. In general, such a multiplier assertion cannot be expected for the spaces $D(\Lambda_A^\alpha)$.

Proposition. *Let $[H, \Lambda, A] \in \mathfrak{R}^l$ for some non-negative integer l. Let $\lambda(t)$ be a $(l + 1)$-times continuously differentiable complex-valued function $\lambda(t)$ on $[0, \infty)$ with*

$$\max_{j=0,\dots,l+1} \sup_{t \in [0,\infty)} t^j \, |\lambda^{(j)}(t)| < \infty . \tag{2}$$

Then $\lambda(t) \in M_1$.

Proof. Let $\varphi = \{\varphi_j(t)\}_{j=0}^\infty \in \Phi$ and $\psi = \{\psi_j(t)\}_{j=0}^\infty$ be the two systems from Step 1 of the proof of Theorem 6.3.2. By (2) and the Propositions 6.3.1 and 6.2.1 we have

$$\|\varphi_j(\Lambda)\, \lambda(\Lambda)\, a \mid A\| = \|\lambda(\Lambda)\, \psi_j(\Lambda)\, \varphi_j(\Lambda)\, a \mid A\| \leqq c\|\varphi_j(\Lambda)\, a \mid A\|,$$

where c is independent of j. Then the desired assertion is an immediate consequence of the definition of B_q^s.

Example. As had been said a multiplier assertion of type (2) cannot be expected for the spaces A and $D(\Lambda_A^\alpha)$ in general. We describe a counter-example. Let $\tilde{C}^m$ be the completion of the Schwartz space $S = S(R_1)$ in the norm

$$\|f \mid \tilde{C}^m\| = \sum_{j=0}^{m} \sup_{t \in R_1} |f^{(j)}(t)| .$$

Of course $m = 0, 1, 2, \ldots$ Similar as in Proposition 6.5.1 one can prove $[L_2(R_1), \Lambda, \tilde{C}(R_1)] \in \Re^2$ with $A = \tilde{C}(R_1) = \tilde{C}^0(R_1)$ and $\Lambda f = f - f''$. Then

$$(A, D(\Lambda_A^m))_{\theta,\infty} = (\tilde{C}(R_1), \tilde{C}^{2m}(R_1))_{\theta,\infty} = \tilde{\mathscr{C}}^{2m\theta}(R_1),$$

$0 < \theta < 1$, where $\tilde{\mathscr{C}}^{2m\theta}(R_1)$ is a space of Hölder-Zygmund type. The hypotheses of the above proposition are satisfied. In particular, $\tilde{\mathscr{C}}^s(R_1)$, $s > 0$, has multipliers of type (2). On the other hand, it is a known fact that there exist smooth functions with (2) which are not multipliers for $\tilde{C}(R_1)$, cf. e.g. [S, p. 21]. One has a similar situation if one deals with $A = L_1(R_1)$ as the basic space. If $1 < p < \infty$ and $A = L_p(R_1)$, then $L_p(R_1)$ has multipliers of type (2). But even in this case the multiplier class M_1 for the Besov spaces $B^s_{p,q}(R_1)$ is larger than the multiplier class M_0 for the Bessel-potential spaces $H^s_p(R_1)$ (appropriate definitions for M_0 and M_1), cf. [T, 2.6.3., 2.6.4.].

6.4. Fourier Series

6.4.1. Introduction

The abstract theory which we developed in the Sections 6.2. and 6.3. can be applied to many concrete situations. In 6.5.1. we consider (isotropic non-homogeneous) Besov-Sobolev spaces on R_n. The periodic counterpart is treated in 6.5.2.: Periodic spaces of Besov-Sobolev type with the n-torus T_n as the underlying domain. In particular it becomes clear that at least some classical periodic spaces of Besov-Sobolev type from Chapter 3 can also be considered as special cases of our abstract approach. In both cases (6.5.1. and 6.5.2.) we use $\Lambda = E - \Delta$ as the underlying operator, where Δ stands for the Laplacian. In the periodic case, Λ is an operator with pure point spectrum, in contrast to its R_n-counterpart. In the periodic case, the previous decompositions, cf. (6.2.3/1), yield the well-known trigonometric series, which attracted so much attention in the history of mathematics. One can ask the question what parts of this elaborated theory fit in our abstract theory and what is a suitable abstract setting for such a theory. Our approach is clear: We begin with $[H, \Lambda, A] \in \Re^l$ for some non-negative integer l in the sense of Definition 6.2.2/1, where now Λ is a self-adjoint positive-definite operator in H with a pure point spectrum. As will be clear just now, (6.2.3/1) becomes the Fourier series of a with respect to the eigenelements of Λ. In the concrete case of periodic spaces, cf. 6.5.2., we have the usual trigonometric series. It is not a surprise that Fourier series in an abstract setting have also been considered extensively. We refer to W. Trebels [1]. The difference is the following. In our approach we work with the Riesz means, cf. Definition 6.2.2/1, in particular (6.2.2/1). In the just cited Lecture Notes by W. Trebels and in the underlying papers the Riesz means in (6.2.2/1) are replaced by Cesàro means. If the eigenvalues of Λ have a sufficiently regular distribution then there is essentially no difference between these two approaches, at least as far as our theory is concerned. A sufficiently regular distribution of eigenvalues means that there exists a function $\Phi(t) = at^{\varkappa_1}(t + b)^{\varkappa_2}$ on $[0, \infty)$ with $a > 0$, $b > 0$, $\varkappa_1 > 0$, $\varkappa_2 \geqq 0$ such that the different eigenvalues σ_k can be calculated by $\sigma_k = \sigma_0 + \Phi(k)$ if $k = 1, 2, \ldots$ This hypothesis is satisfied for many classical orthogonal expansions (trigonometric series, Legendre-Hermite-Jacobi expansions, spherical harmonics etc.). Under this hypothesis we compared in [F, 3.4.1.] the two different approaches (via Riesz means or via Cesàro means) in detail.

We formalize our basic assumptions. Let $[H, \Lambda, A] \in \mathfrak{R}^l$ for some non-negative integer l. We assume that Λ is a self-adjoint positive-definite operator with pure point spectrum, i.e. the spectrum of Λ consists of eigenvalues of finite multiplicity. Let $\{\lambda_k\}_{k=0}^{\infty}$ be the eigenvalues, ordered in the usual way,

$$0 < \lambda_0 \leqq \lambda_1 \leqq \lambda_2 \leqq \ldots \leqq \lambda_n \leqq \ldots, \tag{1}$$

including their multiplicities. Of course $\lambda_n \to \infty$ if $n \to \infty$. Let $\{a_k\}_{k=0}^{\infty} \subset D(\Lambda) \subset H$ be the corresponding orthonormed eigenelements, i.e. $\Lambda a_k = \lambda_k a_k$ and $(a_k, a_l) = \delta_{k,l}$, where $(\cdot, \cdot)$ stands for the scalar product in H. If $a \in H$, then we have the Fourier expansion

$$a = \sum_{j=0}^{\infty} \alpha_j a_j \tag{2}$$

with $\alpha_j = (a, a_j)$ (convergence in H). We wish to extend (2) in a formal sense to A. Let λ be a fixed eigenvalue of Λ and let P be the corresponding projection onto the finite-dimensional eigenspace. If one chooses an appropriate function $\lambda(t)$ then Proposition 6.2.1 can be applied to $P = \lambda(\Lambda)$. In particular,

$$a \to \alpha_k = (a, a_k), \quad k = 0, 1, 2, \ldots,$$

can be extended by continuity from $H \cap A$ to A. As a by-product we obtain $a_k \in A$ and consequently $a_k \in A \cap H$. The complex numbers $\alpha_k = \alpha_k(a)$, $a \in A$, are called the Fourier coefficients of a with respect to $\{a_j\}_{j=0}^{\infty}$. Hence in a formal sense we have

$$a \sim \sum_{k=0}^{\infty} \alpha_k a_k, \quad a \in A. \tag{3}$$

Recall that (6.2.3/1) is an A-convergence. This shows that "$\sim$" in (3) can be replaced by "$=$" in the sense of an A-block-convergence, where the blocks are given by $A_j = \sum_{k=0}^{\infty} \varphi_j(\lambda_k)\, \alpha_k a_k$ with $j = 0, 1, 2, \ldots$ The situation is especially simple and satisfactory if one knows that there exists a natural number N such that every block A_j contains at most N non-vanishing elements $\alpha_k a_k$. This situation is satisfied for lacunary series.

In the Subsections 6.4.2. and 6.4.3. we deal with two special problems: (i) What can be said about the just mentioned lacunary series (Subs. 6.4.2.) and (ii) What can be said about the behaviour of the Fourier coefficients α_k in dependence on smoothness properties of $a \in A$ (Subs. 6.4.3.)?

6.4.2. Lacunary Series

Lacunary series of trigonometric systems play an important role in analysis. They have been studied extensively. The relations between the Fourier coefficients of lacunary trigonometric series and the smoothness of the represented functions are well-known, cf. e.g. N. K. Bari [1]. At the end of the last subsection we mentioned that (abstract) lacunary series seem to fit especially well in our theory. First we give a precise definition.

Definition. *Let $[H, \Lambda, A] \in \mathfrak{R}^l$ for some non-negative integer l, cf. Definition 6.2.2/1. Let Λ be a (self-adjoint positive-definite) operator with pure point spectrum with the eigenvalues λ_k and the (H-orthonormed) eigenelements a_k, cf. (6.4.1/2). Let $\alpha_k = \alpha_k(a)$ be the Fourier coefficients of an element $a \in A$ in the sense of (6.4.1/3). Let $n_0 < n_1 < n_2 < \ldots$ for a given $a \in A$ be the (finite or infinite) set of all non-negative integers with*

$\alpha_{n_k} \neq 0$. *Then (6.4.1/3) is said to be a lacunary series (in A) if there exist a number $d > 1$ and a natural number N with*

$$\lambda_{n_{k+1}} \geqq d\lambda_{n_k} \quad \text{when} \quad k \geqq N. \tag{1}$$

Remark 1. Beside the above definition of a lacunary series we introduced in [F, 3.4.2.] so-called l-lacunary series, where l is a natural number. (6.4.1/3) is called l-lacunary if for any interval $I = [0, 1]$ or $I = [2^j, 2^{j+1}]$ with $j = 0, 1, 2, \ldots$ there exist at most l numbers α_k with $\lambda_k \in I$, which are different from zero. Of course a lacunary series is also an l-lacunary series if l is chosen large enough. The converse assertion is not true, in general. We restrict ourselves here to lacunary series in the sense of the above definition. For assertions connected with the more general concept of l-lacunary series we refer to [F].

Remark 2. In concrete situations one has often the additional information

$$\lambda_k = ck^\sigma(1 + o(1)), \qquad o(1) \to 0 \quad \text{if} \quad k \to \infty, \tag{2}$$

where c and σ are appropriate positive numbers. Then (1) can be reformulated as follows: There exist a positive number $d' > 1$ and a natural number N' with

$$n_{k+1} \geqq d'n_k \quad \text{when} \quad k \geqq N'. \tag{3}$$

(3) is the usual definition for lacunary triconometric systems.

Remark 3. It is quite natural to restrict the conditions (1) and (3) to large values of k: The beginning of the Fourier series (6.4.1/3) has no influence in the behaviour of the Fourier coefficients α_k or in smoothness assertions for the represented element $a \in A$.

Recall that B^s_∞ has been introduced in Definition 6.2.2/2.

Theorem. *Let $[H, \Lambda, A] \in \mathfrak{R}^l$ for some non-negative integer l. Let Λ be an operator with pure point spectrum (with the eigenvalues λ_k and the eigenelements a_k where $k = 0, 1, 2, \ldots$). Let $a \in A$ be represented by a lacunary Fourier series. Let $s > 0$. Then the following two assertions are equivalent to each other:*

(i) $\quad \|a_k \mid A\| \, \alpha_k = O(\lambda_k^{-s}), \quad k = 0, 1, 2, \ldots,$

(ii) $\quad a \in B^s_\infty$.

Proof. In the definition of the system Φ in 6.2.2. one can replace the number 2 by any number $b > 1$. Let $\varphi = \{\varphi_j(t)\}_{j=0}^\infty$ such a modified system. Then

$$B^s_\infty = \{a \mid a \in A, \|a \mid B^s_\infty\|^\varphi = \|b^{sj}\varphi_j(\Lambda)\, a \mid l_\infty(A)\| < \infty\}, \tag{4}$$

cf. (6.2.2/6). This is an easy consequence of Proposition 6.2.1 or Proposition 6.3.3: Let (1) be satisfied. Then one can choose $b > 1$ in such a way that every interval (b^{j-1}, b^{j+1}) with $j \geqq N$ contains at most one eigenvalue $\lambda_{k(j)}$ with $\alpha_{k(j)}(a) \neq 0$. Furthermore we may assume that we have

$$b^{js}\varphi_j(\Lambda)\, a = b^{sj}\alpha_{k(j)}a_{k(j)}$$

for these intervals. Because $b^j \sim \lambda_{k(j)}$ we obtain

$$b^{js}\varphi_j(\Lambda)\, a \sim \lambda^s_{k(j)}\alpha_{k(j)}a_{k(j)}. \tag{5}$$

But (5) and (4) complete the proof.

Remark 4. We assume that there exist two positive numbers c_1 and c_2 with

$$c_1 \leqq \|a_k \mid A\| \leqq c_2 \quad \text{if} \quad k = 0, 1, 2, \ldots$$

Furthermore we assume that (2) is satisfied. If $a \in A$ is represented by a lacunary Fourier series then

$$\text{(i)} \quad \alpha_k = O(k^{-\sigma s}) \quad \text{and} \quad \text{(ii)} \quad a \in B^s_\infty \tag{6}$$

are equivalent assertions, $s > 0$. This is a familiar formulation. We describe a well-known example.

Example. Let $T_1 = [-\pi, \pi]$ be the 1-torus, $H = L_2(T_1)$, $A = C(T_1)$ (the space of all continuous functions on T_1), $\Lambda f = f - f''$. We have $[H, \Lambda, A] \in \mathfrak{R}^1$, cf. 6.5.2. Furthermore, Λ is an operator with pure point spectrum, its eigenvalues are $\lambda_k = 1 + k^2$ and its normalized eigenelements are $\frac{1}{\sqrt{2\pi}}\, e^{ikx}$. In this case lacunary Fourier series are the usual one-dimensional lacunary trigonometric series. As we shall see in 6.5.2. we have $B^s_\infty = \mathscr{C}^{2s}(T_1)$, where $\mathscr{C}^\sigma(T_1)$ with $\sigma > 0$ are the periodic Hölder-Zygmund spaces. Let $f \in C(T_1)$ be represented by a lacunary trigonometric series with the Fourier coefficients α_k. Then

$$\text{(i)} \quad \alpha_k = O(k^{-s}) \quad \text{and} \quad \text{(ii)} \quad f \in \mathscr{C}^s(T_1) \tag{7}$$

are equivalent assertions, $s > 0$. This follows immediately from (6). The equivalence relation (7) is well-known, cf. e.g. N. K. Bari [1].

6.4.3. Fourier Coefficients

Let a 2π-periodic function on the real line be represented by its trigonometric series. The behaviour of the corresponding Fourier coefficients in dependence on the smoothness of the given function has been studied for a long time. It is known that best results can be expected in the framework of a Hilbert space theory. It is our aim to deal with the abstract counterpart in the sense of the preceding theory.

We use the notations of the two preceding subsections. In particular $[H, \Lambda, A] \in \mathfrak{R}^l$ for some non-negative integer l. In this subsection we specialize $A = H$. Furthermore, as in 6.4.1. we assume that Λ is a (self-adjoint positive-definite) operator with pure point spectrum. Again, the eigenvalues (including their multiplicities) are denoted by λ_k, cf. (6.4.1/1), and the corresponding H-orthonormed eigenelements by a_k. We have

$$a = \sum_{j=0}^{\infty} \alpha_j a_j \quad \text{(convergence in } H\text{)}. \tag{1}$$

Furthermore we assume that we have the following additional information about the eigenvalues λ_k: There exist two positive numbers c and ϱ with

$$\lambda_k \geqq c k^{1/\varrho} \quad \text{if} \quad k = 1, 2, 3, \ldots \tag{2}$$

It is easy to see that (2) is equivalent to the familiar assertion

$$N(\lambda) = \sum_{\lambda_k \leqq \lambda} 1 \leqq c' \lambda^\varrho, \quad \lambda > 0, \tag{3}$$

where ϱ has the same meaning as in (2) and c' is an appropriate positive number. Because $A = H$ the spaces B^s_q with $s > 0$ and $1 \leqq q \leqq \infty$ from Definition 6.2.2/2 are especially simple. Recall that in our situation, i.e. $A = H$, Remark 6.2.3/2 can be used. Let $I_0 = [0, 1)$ and $I_j = [2^{j-1}, 2^j)$ if $j = 1, 2, 3, \ldots$ Then (6.2.3/6) yields

$$\|a \mid B^s_q\| = \left(\sum_{j=0}^{\infty} 2^{sjq} \left(\sum_{\lambda_k \in I_j} |\alpha_k|^2 \right)^{q/2} \right)^{1/q}, \tag{4}$$

where $a \in H$ is given by (1), $s > 0$ and $1 \leqq q \leqq \infty$ (obvious modification if $q = \infty$).

Theorem. *Let the above hypotheses be satisfied. In particular, λ_k are the eigenvalues of A, the number $\varrho > 0$ has the same meaning as in (3) (or (2)) and α_k are the Fourier coefficients of a, cf. (1). Let $1 \leqq r \leqq 2$, $w \in R_1$ and $w + \varrho\left(1 - \frac{r}{2}\right) > 0$. Then there exists a positive number c such that*

$$\sum_{k=0}^{\infty} \lambda_k^w |\alpha_k|^r \leqq c \left\| a \mid B_r^{\frac{w}{r} + \varrho\left(\frac{1}{r} - \frac{1}{2}\right)} \right\|^r \tag{5}$$

holds for all $a \in B_r^{\frac{w}{r} + \varrho\left(\frac{1}{r} - \frac{1}{2}\right)}$.

Proof. We have

$$\sum_{k=0}^{\infty} \lambda_k^w |\alpha_k|^r \leqq c \sum_{j=0}^{\infty} 2^{jw} \sum_{\lambda_k \in I_j} |\alpha_k|^r. \tag{6}$$

The number of eigenvalues λ_k with $\lambda_k \in I_j$ can be estimated from above by $N(2^j) \leqq c' \cdot 2^{j\varrho}$, cf. (3). By Hölder's inequality we obtain

$$\sum_{\lambda_k \in I_j} |\alpha_k|^r \leqq c \left(\sum_{\lambda_k \in I_j} |\alpha_k|^2 \right)^{\frac{r}{2}} 2^{j\varrho\left(1 - \frac{r}{2}\right)}. \tag{7}$$

Now (6) and (7) yield

$$\sum_{k=0}^{\infty} \lambda_k^w |\alpha_k|^r \leqq c \sum_{j=0}^{\infty} 2^{jw + j\varrho\left(1 - \frac{r}{2}\right)} \left(\sum_{\lambda_k \in I_j} |\alpha_k|^2 \right)^{\frac{r}{2}}. \tag{8}$$

However (4) and (8) prove (5).

Remark 1. Let $\|a \mid B_q^s\|$ for all $s \in R_1$ and all $0 < q \leqq 2$ be defined by (4). Then the above proof shows that (5) remains valid for all $0 < r \leqq 2$ and all $w \in R_1$. The extension of B_q^s with $s > 0$ from $1 \leqq q \leqq \infty$ to $0 < q \leqq \infty$ causes no problems, even not in the general case from Definition 6.2.2/2, cf. Remark 6.2.2/3. The extension of the above special spaces $B_q^s = B_q^s(H)$ or of the more general spaces $B_q^s = B_q^s(A)$ from Definition 6.2.2/2 to negative values of s is not so clear. The problem is to interprete the corresponding (complete!) spaces B_q^s if $s \leqq 0$. In concrete situations, such as Besov-Sobolev spaces on R_n or on the n-torus T_n, one has satisfactory answers for these questions. All these spaces can be considered as subspaces of appropriate spaces of distributions, such as $S'(R_n)$ or $D'(T_n)$, cf. Chapter 3 as far as the spaces on T_n are concerned. In the abstract situation we descirbe in 6.5.3. a procedure which may be useful in our context here.

Remark 2. The most interesting case is $w = 0$ and $r = 1$ (absolute convergence of the Fourier coefficients). Then (5) yields

$$\sum_{k=0}^{\infty} |\alpha_k| \leqq c \|a \mid B_1^{\varrho/2}\|. \tag{9}$$

Remark 3. In many concrete situations (2) can be strengthened by (6.4.2/2), i.e.

$$\lambda_k = ck^{1/\varrho}(1 + o(1)), \qquad o(1) \to 0 \quad \text{if} \quad k \to \infty. \tag{10}$$

Then (5) can be reformulated by

$$\sum_{k=1}^{\infty} k^{\frac{w}{\varrho}} |\alpha_k|^r \leqq c \left\| a \mid B_r^{\frac{w}{r} + \varrho\left(\frac{1}{r} - \frac{1}{2}\right)} \right\|^r. \tag{11}$$

Example. As in the preceding subsection we illustrate the above results by the one-dimensional trigonometric series. Let again $T_1 = [-\pi, \pi]$ be the 1-torus, $H = L_2(T_1)$

and $Af = f - f''$. As we mentioned in Example 6.4.2 our abstract theory is applicable, cf. also 6.5.2 where we deal with the n-dimensional periodic case in greater detail. We have (10) with $\varrho = \frac{1}{2}$. Furthermore, $B^s_q = B^s_q(L_2(T_1)) = B^{2s}_{2,q}(T_1)$, where $s > 0$ and $1 \leqq q \leqq \infty$, cf. 6.5.2. Recall that $B^\sigma_{2,q}(T_1)$ with $\sigma > 0$ and $1 \leqq q \leqq \infty$ are the usual periodic Besov spaces. Cf. 3.5.4. for well-known equivalent norms. In this special situation (5) and (9) read as follows,

$$\sum_{k=1}^{\infty} k^w |\alpha_k|^r \leqq c \left\| f \mid B_{2,r}^{\frac{w}{r}+\frac{1}{r}-\frac{1}{2}}(T_1) \right\|^r \tag{12}$$

and

$$\sum_{k=0}^{\infty} |\alpha_k| \leqq c \| f \mid B^{1/2}_{2,1}(T_1) \| . \tag{13}$$

At the first moment r and w in (12) are restricted by $1 \leqq r \leqq 2$ and $w + 1 - \frac{r}{2} > 0$. However as had been said in Remark 1 in this concrete situation (12) can be extended immediately to $0 < r \leqq 2$ and $w \in R_1$. The needed spaces $B^\sigma_{2,r}(T_1)$ with $0 < r \leqq \infty$ and $\sigma \in R_1$ have been treated extensively in Chapter 3. We mention a simple consequence of (12) and (13). Recall

$$C^j(T_1) \subset W^j_2(T_1), \quad j = 0, 1, 2, \ldots, \tag{14}$$

where $W^j_2(T_1)$ are the usual periodic Sobolev spaces and $C^j(T_1)$ stands for all periodic functions having continuous derivatives up to the order j. Interpolation yields

$$\mathscr{C}^\sigma(T_1) \subset B^\sigma_{2,\infty}(T_1), \quad \sigma > 0, \tag{15}$$

where $\mathscr{C}^\sigma(T_1) = B^\sigma_{\infty,\infty}(T_1)$ with $\sigma > 0$ are the periodic Hölder-Zygmund spaces, cf. also 3.5.4. Then we have

$$\mathscr{C}^\sigma(T_1) \subset B^{\sigma-\varepsilon}_{2,r}(T_1) \tag{16}$$

for any $\sigma > 0$, $\varepsilon > 0$ and $0 < r \leqq \infty$, cf. 3.5.1. Then (12) and (13) yield

$$\sum_{k=1}^{\infty} k^w |\alpha_k|^r \leqq c \| f \mid \mathscr{C}^s(T_1) \|^r \quad \text{if} \quad s > 0 \quad \text{and} \quad s > \frac{w}{r} + \frac{1}{r} - \frac{1}{2} \tag{17}$$

with $0 < r \leqq 2$, $w \in R_1$ and

$$\sum_{k=1}^{\infty} |\alpha_k| \leqq c \| f \mid \mathscr{C}^s(T_1) \| \quad \text{if} \quad s > \tfrac{1}{2}. \tag{18}$$

Remark* 4. (12), (13), and (17), (18) are more or less known. (18) is the famous inequality of S. N. Bernštejn [1]. (17) with $w = 0$ and $0 < s < 1$ is due to O. Szász [1], (13) has been proved essentially by O. Szász [2]. We refer also to S. B. Stečkin [1] and the survey by J. R. McLaughlin [1]. In [F, p. 142] few further references have been given. More recent papers dealing with problems of this type are G. E. Karadžov [1], P. Oswald [3] and A. Pietsch [4]. As far as the connection to entropy numbers is concerned we refer to B. Carl [1].

Remark 5. The question arises whether the parameters in (12), (13), and (17), (18) are sharp. The space $B^{1/2}_{2,1}(T_1)$ in (13) cannot be replaced by $B^{1/2}_{2,q}(T_1)$ with $q > 1$. This follows simply from the fact that there exist essentially unbounded functions which belong to $B^{1/2}_{2,q}(T_1)$ with $q > 1$, what contradicts (13) with $B^{1/2}_{2,q}(T_1)$ instead of $B^{1/2}_{2,1}(T_1)$. Furthermore (17) with $w = 0$ and $1 \leqq r \leqq 2$ and $s = \frac{1}{r} - \frac{1}{2}$ is not valid, cf. A. Zygmund [1, § 6.33].

6.5. Examples and Comments

6.5.1. Spaces on R_n

The theory which we developed in this chapter can be applied to many concrete situations: Isotropic and anisotropic spaces of Besov-Sobolev type on R_n and on the n-torus T_n, weighted L_p-spaces, classical orthogonal expansions etc. In this Section 6.5. we describe few of these applications, however only in the Subsections 6.5.1. and 6.5.2. we give proofs, otherwise we restrict ourselves to rough descriptions. We refer to [F, 3.5.] for further informations.

This subsection deals with isotropic non-homogeneous spaces of Besov-Sobolev type on R_n which have been treated extensively in [T, Chapter 2]. However it is quite clear that the specialization of our abstract approach to spaces on R_n cannot cover the full scales $B^s_{p,q}(R_n)$ and $F^s_{p,q}(R_n)$ of the Besov-Hardy-Sobolev spaces from [T, Chapter 2] with $-\infty < s < \infty$, $0 < p \leqq \infty$, $0 < q \leqq \infty$. One can expect to obtain spaces $B^s_{p,q}(R_n)$ with $s > 0$, $1 \leqq p \leqq \infty$, $1 \leqq q \leqq \infty$, and the spaces $H^s_p(R_n) = F^s_{p,2}(R_n)$ with $s > 0$ and $1 < p < \infty$.

We use the notations from 1.2.4.: As usual R_n stands for the real Euclidean n-space. The Schwartz space of all complex-valued infinitely differentiable rapidly decreasing functions on R_n and the complex-valued tempered distributions are denoted by $S(R_n)$ and $S'(R_n)$, respectively. F and F^{-1} stand for the Fourier transform on $S'(R_n)$ and its inverse, respectively. Let $W^m_p(R_n)$ with $1 < p < \infty$ and $m = 0, 1, 2, \ldots$ be the usual Sobolev spaces (with respect to the Lebesgue measure). Of course $W^0_p(R_n) = L_p(R_n)$. We are interested in triplets $[H, \Lambda, A]$ where $H = L_2(R_n)$, $A = L_p(R_n)$ with $1 < p < \infty$, and

$$(\Lambda f)(x) = f(x) - (\Delta f)(x), \qquad f \in D(\Lambda) = W^2_2(R_n), \tag{1}$$

where Δ stands for the Laplacian. It is well-known that Λ is a positive-definite self-adjoint operator acting in H. Furthermore

$$\Lambda f = F^{-1}(1 + |x|^2)\, Ff, \quad f \in D(\Lambda). \tag{2}$$

If $\operatorname{Re}\beta \geqq 0$ then the Riesz means from (6.2.1/3) are given by

$$(R_{\beta,\varrho} f)(x) = F^{-1}\left(1 - \frac{1 + |x|^2}{\varrho}\right)^{\beta}_{+} Ff, \quad f \in L_2(R_n), \quad \varrho > 1. \tag{3}$$

This can easily be calculated. Cf. Remark 2 below and also [F, 3.5.1.].

Proposition. *Let* $H = L_2(R_n)$, Λ *be given by* (1), *and* $A = L_p(R_n)$ *with* $1 < p < \infty$. *Then* $[H, \Lambda, A] \in \mathfrak{R}^{n+1}$.

Proof. Part (i) of Definition 6.2.2/1 is satisfied with, say, $\mathscr{H} = S'(R_n)$. Next we prove that the Riesz means $R_{\alpha,\varrho}$ are uniformly bounded in $L_p(R_n)$ with respect to ϱ if $\alpha > \frac{n-1}{2}$. It suffices to show that

$$\sup_{\varrho > 1} \left\| F^{-1}\left(1 - \frac{1 + |x|^2}{\varrho}\right)^{\alpha}_{+} \mid L_1(R_n)\right\| < \infty \tag{4}$$

holds. This is a consequence of

$$\left\| F^{-1} \left(1 - \frac{1 + |x|^2}{\varrho} \right)_+^{\alpha} \mid L_1(R_n) \right\|$$
$$= \left(1 - \frac{1}{\varrho} \right)^{\alpha} \left\| F^{-1} \left(1 - \frac{|x|^2}{\varrho - 1} \right)_+^{\alpha} \mid L_1(R_n) \right\|$$
$$\leqq \| (\varrho - 1)^{n/2} [F^{-1}(1 - |y|^2)_+^{\alpha}] ((\varrho^2 - 1)^{1/2} \cdot) \mid L_1(R_n) \|$$
$$= \| F^{-1}(1 - |y|^2)_+^{\alpha} \mid L_1(R_n) \| < \infty$$

if $\alpha > \frac{n-1}{2}$, cf. E. M. Stein, G. Weiss [1, pp. 253, 255]. Hence (6.2.2/1) with $A = L_p(R_n)$ and $l > \frac{n-1}{2}$ is satisfied. We prove (6.2.2/2) with $A = L_p(R_n)$ and $\operatorname{Re} \beta > n + 2$. Because $S(R_n)$ is dense in $L_p(R_n) \cap L_2(R_n)$ we may assume $f \in S(R_n)$. Because $\operatorname{Re} \beta > n + 1$ we have

$$\left(1 - \frac{1 + |x|^2}{\varrho} \right)_+^{\beta} Ff \to Ff \quad \text{when} \quad \varrho \to \infty \quad \text{in} \quad W_1^{n+1}(R_n). \tag{5}$$

If $n + 1$ us an even number then (3) and (5) yield

$$R_{\beta, \varrho} f \to f \quad \text{when} \quad \varrho \to \infty \quad \text{in} \quad L_{\infty}(R_n, 1 + |x|^{n+1}).$$

If $n + 1$ is an odd number then we replace $n + 1$ in the last calculation by $n + 2$. This proves (6.2.2/2).

Remark 1. The above calculations can be extended to the limiting cases $p = 1$ and $p = \infty$, cf. [F, 3.5.2.].

Remark 2. Recall the well-known fact that F and F^{-1} yield unitary mappings from $L_2(R_n)$ onto itself. Hence by (2) the operator Λ can be reduced on this way to the multiplication operator

$$(Mg)(x) = (1 + |x|^2) g(x), \quad x \in R_n, \tag{6}$$

acting in $L_2(R_n)$. If $\varphi(t)$ is a continuous function or a step function on $[0, \infty)$ then

$$(\varphi(M) g)(x) = \varphi(1 + |x|^2) g(x), \quad x \in R_n. \tag{7}$$

By (2) we have

$$\varphi(\Lambda) f = F^{-1} \varphi(1 + |x|^2) Ff. \tag{8}$$

Then one obtains (3) as a special case. We have $[L_2(R_n), M, L_p(R_n)] \in \mathfrak{R}^0$ with $1 \leqq p < \infty$, as can be checked easily. In [F, 3.5.1.] we dealt with more general multiplication operators, where $(1 + |x|^2)$ in (6) is replaced by a weight $h(x)$. Again the above abstract theory can be applied. If one uses the interpolation formulas from Theorem 6.3.2, then one obtains in a rather natural way the so-called Herz spaces and generalizations of them, cf. [F, 3.5.1.] for the details and A. Kufner, O. John, S. Fučik [1, p. 422] as far as Herz spaces are concerned.

Remark 3. By the above proposition the abstract theory can be applied. Recall that

$$D(\Lambda^m) = W_p^{2m}(R_n) \text{ with } m = 0, 1, 2, \ldots \text{ and } 1 < p < \infty, \tag{9}$$

where Λ given by (2) is now interpreted as an unbounded operator acting in $L_p(R_n)$ with $1 < p < \infty$. By Theorem 6.3.2 we have

$$D(\Lambda^s) = [A, D(\Lambda^m)]_{\theta} = [L_p(R_n), W_p^{2m}(R_n)]_{\theta} = H_p^{2s}(R_n), \tag{10}$$

where $0 < \theta < 1$ and $s = \theta m$. Here $H_p^{\sigma}(R_n)$ with $\sigma > 0$ and $1 < p < \infty$ are the Bessel-potential spaces. If $\sigma = m = 0, 1, 2, \ldots$ then $H_p^{\sigma}(R_n) = W_p^m(R_n)$ are the Sobolev spaces. Furthermore, by

(6.3.2/5) we have

$$B_q^s = (A, D(\Lambda^m))_{\theta,q} = (L_p(R_n), W_p^{2m}(R_n))_{\theta,q} = B_{p,q}^{2s}(R_n) \tag{11}$$

with $0 < \theta < 1$, $s = \theta m$, $1 < p < \infty$ and $1 \leqq q \leqq \infty$. Of course, $B_{p,q}^s(R_n)$ are the well-known classical Besov spaces, cf. also 2.2.2. As far as the right-hand sides of (10) and (11) are concerned we refer also to [I, 2.4.2.]. Let $\varphi = \{\varphi_j(t)\}_{j=0}^\infty \in \Phi$, cf. 6.2.2. Then we have by (8), (11) and Definition 6.2.2/2

$$B_{p,q}^s(R_n) = \{f \mid f \in L_p(R_n), \left\| 2^{\frac{s}{2}j} F^{-1}\varphi_j(1 + |x|^2) Ff \mid l_q(L_p(R_n)) \right\| < \infty\}. \tag{12}$$

However this is a representation of $B_{p,q}^s(R_n)$ of type (2.2.2/2) with $\sqrt{2}$ instead of 2. This modification is immaterial, cf. (6.4.2/4) for a similar modification in the abstract case.

Remark 4. As we mentioned the limiting cases $p = \infty$ and $p = 1$ can be included in the above consideration (few technical modifications are necessary), cf. [F, 3.5.2.]. One obtains the Besov spaces $B_{1,q}^s(R_n)$ with $1 \leqq q \leqq \infty$ and spaces of Hölder-Zygmund type.

6.5.2. Periodic Spaces, Multiple Trigonometric Series

This subsection should be considered as an addendum to Chapter 3. We wish to make clear that at least the classical periodic spaces of Besov-Sobolev type can also be considered as special cases of the abstract theory developed in this chapter. Furthermore we wish to specialize the results about lacunary Fourier series and Fourier coefficients from 6.4.2. and 6.4.3.

Let again T_n be the n-torus represented as usual by $T_n = \{x \mid x = (x_1, \ldots, x_n), |x_j| \leqq \pi \text{ if } j = 1, \ldots, n\}$. The spaces $L_p(T_n)$ with $1 \leqq p < \infty$ have the usual meaning (Lebesgue measure). Let $C(T_n)$ be the collection of all complex-valued continuous functions on T_n, equipped with the norm $\|f \mid C(T_n)\| = \sup_{x \in T_n} |f(x)|$. As in Chapter 3 we introduce

$$C^m(T_n) = \{f \mid D^\alpha f \in C(T_n) \text{ if } 0 \leqq |\alpha| \leqq m\} \tag{1}$$

equipped with the norm $\|f \mid C^m(T_n)\| = \sum_{|\alpha| \leqq m} \|D^\alpha f \mid C(T_n)\|$, and the Sobolev spaces

$$W_p^m(T_n) = \{f \mid f \in L_p(T_n), \|f \mid W_p^m(T_n)\| = \sum_{|\alpha| \leqq m} \|D^\alpha f \mid L_p(T_n)\| < \infty\}, \tag{2}$$

where $m = 0, 1, 2, \ldots$ and $1 < p < \infty$. Of course, $C^0(T_n) = C(T_n)$ and $W_p^0(T_n) = L_p(T_n)$. As far as the classical periodic Besov spaces $B_{p,q}^s(T_n)$ with $s > 0$, $1 < p < \infty$, $1 \leqq q \leqq \infty$, and the periodic Hölder-Zygmund spaces $\mathscr{C}^s(T_n)$ with $s > 0$ are concerned we refer to Chapter 3. In 3.5.4. we described these spaces by norms where only derivatives and differences of the given functions are involved. These are the classical characterizations. Recall the interpolation formulas

$$\mathscr{C}^s(T_n) = (C(T_n), C^m(T_n))_{\theta,\infty} \quad \text{if} \quad 0 < \theta < 1, \quad \text{and} \quad s = \theta m, \tag{3}$$

and

$$B_{p,q}^s(T_n) = (L_p(T_n), W_p^m(T_n))_{\theta,q}, \quad 0 < \theta < 1, \quad s = \theta m, \quad 1 \leqq q \leqq \infty, \tag{4}$$

cf. 3.6.1.

In the sense of the abstract theory developed in this chapter we deal with the triplet $[H, \Lambda, A]$ where $H = L_2(T_n)$,

$$(\Lambda f)(x) = f(x) - (\Delta f)(x), \qquad f \in D(\Lambda) = W_2^2(T_n), \tag{5}$$

and $A = L_p(T_n)$ with $1 < p < \infty$ or $A = C(T_n)$. It is well-known that Λ is a positive-definite self-adjoint operator with pure point spectrum. We have

$$N(\lambda) = c\lambda^{n/2}(1 + o(1)), \qquad o(1) \to 0 \quad \text{if} \quad \lambda \to \infty, \tag{6}$$

where $N(\lambda)$ is the eigenvalue distribution function from (6.4.3/3) and c is an appropriate positive number. Recall that any periodic distribution can be represented as

$$f = \sum_{k \in Z_n} \alpha_k (2\pi)^{-n/2} e^{ikx}, \tag{7}$$

where Z_n is the lattice of all points in R_n with integer-valued components and kx stands for the scalar product of $k \in Z_n$ and $x \in R_n$, cf. 3.2.3. Obviously, $(2\pi)^{-n/2} e^{ikx}$ are the orthonormed eigenelements of A and $\alpha_k = \alpha_k(f)$ are the corresponding Fourier coefficients. Then (7) and (5) yield

$$(Af)(x) = \sum_{k \in Z_n} (1 + |k|^2) \alpha_k (2\pi)^{-n/2} e^{ikx}, \quad f \in W_2^2(T_n). \tag{8}$$

Then it follows easily that the Riesz means $R_{\beta,\varrho}$ of A are given by

$$(R_{\beta,\varrho} f)(x) = \sum_{k \in Z_n} \left(1 - \frac{1 + |k|^2}{\varrho}\right)_+^{\beta} \alpha_k (2\pi)^{-n/2} e^{ikx}, \quad f \in L_2(T_n), \tag{9}$$

where $\varrho > 1$ and $\operatorname{Re} \beta > 0$.

Proposition. *Let $H = L_2(T_n)$, A be given by* (5) *and either $A = L_p(T_n)$ with $1 < p < \infty$ or $A = C(T_n)$. Then $[H, A, A] \in \mathfrak{R}^{1+[(n-1)/2]}$.*

Proof. Part (i) of Definition 6.2.2/1 is obvious. Furthermore, (6.2.2/1) is known if $l > \frac{n-1}{2}$, cf. E. M. Stein, G. Weiss [1, pp. 253–255]. Finally, (6.2.2/2) is valid if a belongs to the linear hull of the functions e^{ikx}. The rest is a matter of completion. The proof is complete.

Remark 1. By the above proposition the abstract theory can be applied. If $A = L_p(T_n)$ then

$$D(A^m) = W_p^{2m}(T_n) \quad \text{with} \quad m = 0, 1, 2, \dots \quad \text{and} \quad 1 < p < \infty, \tag{10}$$

where A is interpreted as an unbounded operator acting in $L_p(T_n)$ with $1 < p < \infty$. Precisely in the same way as in Remark 6.5.1/3 we obtain in this case

$$D(A^s) = H_p^{2s}(T_n), \quad s > 0,\ 1 < p < \infty \tag{11}$$

(periodic Bessel-potential spaces) and

$$B_q^s = B_{p,q}^{2s}(T_n), \quad s > 0,\ 1 < p < \infty,\ 1 \leqq q \leqq \infty. \tag{12}$$

If $A = C(T_n)$ then (12) with $q = \infty$ must be replaced by

$$B_\infty^s = \mathscr{C}^{2s}(T_n), \quad s > 0. \tag{13}$$

The proof of (13) is a matter of interpolation theory. We omit the details.

Finally we wish to apply the assertions for (abstract) lacunary series from 6.4.2. and for the behaviour of Fourier coefficients from 6.4.3. to multiple trigonometric series. The eigenvalues of the operator A from (5) are given by $\lambda_k = 1 + |k|^2$ with $k \in Z_n$. Any periodic distribution $f \in D'(T_n)$ can be represented by (7) (convergence in $D'(T_n)$). Then (7) is called a lacunary multiple trigonometric series if one can find a number $b > 1$ and a natural number N with the following property: For every natural number j with $j \geqq N$ there exists at most one eigenvalue λ_k with $b^{j-1} \leqq \lambda_k < b^j$ such that the corresponding Fourier coefficient α_k in (7) is different from zero. Of course this is simply the concretization of Definition 6.4.2.

Theorem. (i) *Let $f \in C(T_n)$ be represented by a lacunary trigonometric series* (7). *Let $s > 0$. Then*

$$\alpha_k = O(|k|^{-s}), \quad k \in Z_n, \quad \textit{if and only if} \quad f \in \mathscr{C}^s(T_n). \tag{14}$$

(ii) *Let $1 \leqq r \leqq 2$, $w \in R_1$ and $w + \frac{n}{2}\left(1 - \frac{r}{2}\right) > 0$. Then there exists a positive number c such that*

$$\sum_{Z_n \ni k \neq 0} |k|^w \, |\alpha_k|^r \leqq c \left\| f \mid B_{2,r}^{\frac{w}{r}+n\left(\frac{1}{r}-\frac{1}{2}\right)}(T_n) \right\|^r \tag{15}$$

holds for any $f \in B_{2,r}^{\frac{w}{r}+n\left(\frac{1}{r}-\frac{1}{2}\right)}(T_n)$ (where $\alpha_k = \alpha_k(f)$ are the Fourier coefficients of f in the sense of (7)).

Proof. Step 1. We prove (i). By Theorem 6.4.2 and (13) we have

$$\alpha_k = O(|k|^{-2s}), \quad k \in Z_n, \quad \text{if and only if} \quad f \in \mathscr{C}^{2s}(T_n).$$

This proves (14).

Step 2. Part (ii) follows from (6.4.3/5) with $\lambda_k = 1 + |k|^2$, $\varrho = \frac{n}{2}$, cf. (6), and (12) with $p = 2$.

Remark 2. For sake of illustration we considered in 6.4.2. and 6.4.3. the one-dimensional case, which is now completely justified. (6.4.2/7) is a special case of (14) and (6.4.3/12) is a special case of (15). The counterpart of (6.4.3/13) (absolute convergence of multiple trigonometric series) is given by

$$\sum_{k \in Z_n} |\alpha_k| \leqq c \|f \mid B_{2,1}^{n/2}(T_n)\|. \tag{16}$$

This is a special case of (15) ($w = 0$ and $r = 1$). As in the one-dimensional case, there are no problems to extend (15) to $0 < r \leqq 2$ and $w \in R_1$. The spaces $B_{2,r}^{\sigma}(T_n)$ with $\sigma \in R_1$ and $0 < r \leqq 2$ have the same meaning as in Chapter 3. In other words, (15) holds for all r with $0 < r \leqq 2$ and all $w \in R_1$. Finally we mention that few references can be found in [F, p. 153], cf. also Remark 6.4.3/4.

6.5.3. Generalization of the Abstract Approach

If one compares the theory of the periodic Besov-Hardy-Sobolev spaces $B_{p,q}^s(T_n)$, $F_{p,q}^s(T_n)$ from Chapter 3 or of their R_n-counterparts $B_{p,q}^s(R_n)$, $F_{p,q}^s(R_n)$ from [T, Chapter 2] with the abstract theory developed in this chapter, then it is clear that one has no chance to cover the full scales of these concrete spaces in the above abstract approach. It is hard to imagine how to define spaces of $F_{p,q}^s$ type, say, with $s > 0$, $1 < p < \infty$, $1 < q < \infty$, in an abstract theory without additional assumptions for the underlying Banach spaces (a possible proposal in this direction would be to work from the very beginning with Banach lattices or function spaces). Even for the spaces $B_{p,q}^s$ we restricted the considerations to Banach spaces (i.e. $1 \leqq p \leqq \infty$, $1 \leqq q \leqq \infty$), in contrast to the theory of the spaces $B_{p,q}^s(T_n)$ and $B_{p,q}^s(R_n)$, and to $s > 0$. From a point of view of applications the last restriction, i.e. $s > 0$, is especially annoying. For example the restriction $w + \varrho\left(1 - \frac{r}{2}\right) > 0$ in Theorem 6.4.3 is rather irrelevant. It has nothing to do with the proof of this theorem, it comes simply from the restriction $s > 0$ in Definition 6.2.2/2. In concrete situations this restriction can be omitted, cf. Remark 6.5.2/2 and Example 6.4.3. Another example why the restriction $s > 0$ in Definition 6.2.2/2 should be removed comes from the theory of integral operators. We sketched this theory in 2.5.1. in the framework of the theory of function spaces with dominating mixed smoothness properties. From this point of view a duality theory of the spaces B_q^s from Definition 6.2.2/2 is desirable, something like $(B_q^s)' = B_{q'}^{-s}$ with $s > 0$, $1 \leqq q \leqq \infty$, $\frac{1}{q} + \frac{1}{q'} = 1$ (for this purpose one would identify the dual of the Hilbert space H with H itself). In H. Triebel [20] we developed an abstract theory in the sense of this chapter which allows to define abstract spaces B_q^s with $s \leqq 0$ (and $1 \leqq q \leqq \infty$) and which includes also a duality theory for these spaces. Furthermore we applied this theory to abstract integral operators. We do not give a detailed description of this theory and refer the interested reader to the just cited paper. However we formulate in somewhat rough terms few ideas.

The Basic Triplet. The triplet $[H, A, A]$ from Definition 6.2.2/1 is now replaced by the triplet $[\mathfrak{H}, A, A]$ where $\mathfrak{H} = \{\mathscr{H}\}$ is a family of Hilbert spaces with the distinguished Hilbert space $H \in \mathfrak{H}$. It is assumed that there exists a locally convex space N with the dense and continuous embedding $N \subset \mathscr{H}$ for $\mathscr{H} \in \mathfrak{H}$ and

$$N \subset H = H' \subset N' \tag{1}$$

for the distinguished Hilbert space H (dual pairing) and $\mathscr{H}' \in \mathfrak{H}$. Furthermore, by assumption A is a Banach space and there exist two Hilbert spaces $\mathscr{H}_0 \in \mathfrak{H}$ and $\mathscr{H}_1 \in \mathfrak{H}$ with the dense embedding $\mathscr{H}_0 \subset A \subset \mathscr{H}_1$. Finally A is supposed to be a positive-definite self-adjoint operator in the distinguished Hilbert space H and we have formally $A^l \mathfrak{H} = \mathfrak{H}$ for all integers l. This is the substitute of the triplet $[H, A, A]$ from 6.2.2.

The Spaces B_q^s. Let $\varphi = \{\varphi_j(t)\}_{j=0}^{\infty} \in \Phi$, cf. 6.2.2. Let $s \in R_1$ and $0 < q \leqq \infty$. Then the desired extension of (6.2.2/6) reads as follows,

$$B_q^s = \{a \mid a \in N', \|a \mid B_q^s\|^{\varphi} = \|2^{sj}\varphi_j(A)\, a \mid l_q(A)\| < \infty\,. \tag{2}$$

If $s > 0$ then we can replace N' in (2) by A and we have (6.2.2/6) (with $0 < q \leqq \infty$). In order to prove that B_q^s is independent of φ one needs again a multiplier assertion of the type as in Proposition 6.2.1. In the above cited paper we modified this multiplier hypothesis somewhat. But in any case under suitable assumptions similar as in Proposition 6.2.1 one obtains that B_q^s is a quasi-Banach space which is independent of φ (equivalent quasi-norms).

Examples. Let $[H, A, A]$ be the triplet from Proposition 6.5.2. Then $[\mathfrak{H}, A, A]$ with $\mathfrak{H} = \{A^l L_2(T_n)\}_{l=-\infty}^{\infty}$ has the desired properties. Now (2) yields the periodic Besov spaces $B_{p,q}^s(T_n)$ with $s \in R_1$, $1 < p \leqq \infty$ and $0 < q \leqq \infty$. In order to describe a second example we assume that $[H, A, A]$ is the triplet from Proposition 6.5.1. Then $[\mathfrak{H}, A, A]$ with

$$\mathfrak{H} = \{A^l L_2(R_n, (1 + |x|)^{\varkappa})\}_{\substack{\varkappa \in R_1 \\ l \text{ integer}}}$$

has the desired properties. Then one obtains the Besov spaces $B_{p,q}^s(R_n)$ with $s \in R_1$, $1 < p < \infty$, $0 < q \leqq \infty$. One can replace $A = L_p(R_n)$ by $A = L_p(R_n, (1 + |x|^2)^{\lambda})$. Then one obtains corresponding weighted Besov spaces in the sense of 5.1.1. with weights of type $(1 + |x|^2)^{\lambda}$, $\lambda \in R_1$. One can modify the assumptions for the system Φ. Then the above approach yields the spaces $B_{p,q}^s(R_n, \varrho(x))$ from 5.1.1. with $s \in R_1$, $1 < p < \infty$, $0 < q \leqq \infty$.

Integral Operators. In 2.5.1. we described how to deal with integral operators in the framework of function spaces with dominating mixed smoothness properties on the basis of decomposition methods. The above considerations allow us to transfer this method to the abstract case. One has to replace pointwise multiplications by tensor products. Then one can introduce abstract Besov spaces with "dominating mixed smoothness properties". The smoothness properties of the kernels of corresponding abstract integral operators are measured with the help of these spaces. We omit any details and refer to the above cited paper.

6.5.4. Classical Orthogonal Expansions

In 6.5.1. and 6.5.2. we applied the abstract theory to spaces on R_n and on the n-torus T_n. The latter yields the orthogonal expansion of a function in a multiple trigonometric series. This subsection deals with further orthogonal expansions which fit in the abstract theory developed in this chapter.

Hermite Series. Let $H = L_2(R_1)$ (Lebesgue measure) and let $\mathring{A}$ be given by

$$(\mathring{A}f)(x) = -f''(x) + x^2 f(x), \tag{1}$$

defined on the set of all infinitely differentiable functions with compact support on R_1. Then $\mathring{A}$ is an essentially self-adjoint positive-definite operator in H. Let A be its closure. Then A is a self-adjoint positive-definite operator in H with pure point spectrum. Its eigenfunctions and eigenvalues are given by

$$H_k(x) = e^{x^2/2} \frac{d^k}{dx^k} (e^{-x^2}) \quad \text{and} \quad \lambda_k = 2k + 1, \tag{2}$$

respectively, $k = 0, 1, 2, \ldots$ $H_k(x)$ are the well-known Hermite functions. Let $A = L_p(R_n)$ with $1 < p < \infty$. Then $[H, A, A] \in \mathfrak{R}^l$ for some non-negative integer l in the sense of Definition 6.2.2/1. Cf. [F, p. 154] for references.

Laguerre Series. Let $H = L_2(R_1^+)$ (Lebesgue measure) with $R_1^+ = (0, \infty)$ and let $\mathring{A}_\alpha$ be given by

$$(\mathring{A}_\alpha f)(x) = -4xf''(x) - 4f'(x) + \left(x + \frac{\alpha^2}{x}\right) f(x), \quad \alpha > -1, \tag{3}$$

defined on the set of all functions $f(x) = x^{\alpha/2} g(x)$, where $g(x)$ is an infinitely differentiable function on $\overline{R_1^+} = [0, \infty)$ with $g(x) = 0$ for large values of x. Then $\mathring{A}_\alpha$ is an essentially self-adjoint positive-definite operator in H. Let A_α be its closure. Then A_α is a self-adjoint positive-definite operator in H with pure point spectrum. Its eigenfunctions and eigenvalues are given by

$$L_k^\alpha(x) = e^{x/2} x^{-\alpha/2} \frac{d^k}{dx^k} (e^{-x} x^{k+\alpha}) \quad \text{and} \quad \lambda_k = 2(2k + \alpha + 1), \tag{4}$$

respectively, $k = 0, 1, 2, \ldots$ $L_k^\alpha(x)$ are the well-known Laguerre functions. Let $A = L_p(R_1^+)$. If either $\alpha > 0$, $1 \leqq p < \infty$ or $-1 < \alpha \leqq 0$, $\frac{2}{2+\alpha} < p < -\frac{2}{\alpha}$, then $[H, A_\alpha, A] \in \mathfrak{R}^l$ for some non-negative integer l. Cf. [F, pp. 154/155] for references.

Jacobi Series. Let $I = (-1, 1)$ and

$$L_{p,\alpha,\beta}(I) = \left\{ f \,\middle|\, \int_{-1}^{1} |f(x)|^p (1 - x)^\alpha (1 + x)^\beta \, dx < \infty \right\}, \tag{5}$$

normed in the usual way, $1 \leqq p < \infty$, $\alpha \geqq -\frac{1}{2}$, $\beta \geqq -\frac{1}{2}$. Let $H = L_{2,\alpha,\beta}(I)$ and let $\mathring{A}_{\alpha,\beta}$ be given by

$$(\mathring{A}_{\alpha,\beta} f)(x) = -((1-x^2) f')' + ((\alpha+\beta) x + \alpha - \beta) f' + f,$$
$$\alpha \geqq -\tfrac{1}{2}, \quad \beta \geqq -\tfrac{1}{2}, \tag{6}$$

defined on the set of all polynomials in I. Then $\mathring{A}_{\alpha,\beta}$ is an essentially self-adjoint positive-definite operator in H. Let $A_{\alpha,\beta}$ be its closure. Then $A_{\alpha,\beta}$ is a self-adjoint positive-definite operator in H with pure point spectrum. Its eigenfunctions and eigenvalues are given by

$$J_k^{\alpha,\beta}(x) = (1-x)^{-\alpha} (1+x)^{-\beta} \frac{\mathrm{d}^k}{\mathrm{d}x^k} ((1-x)^{k+\alpha} (1+x)^{k+\beta}) \tag{7}$$

and

$$\lambda_k = k(k+\alpha+\beta+1) + 1,$$

respectively, $k = 0, 1, 2, \ldots$ $J_k^{\alpha,\beta}(x)$ are the well-known Jacobi polynomials. Let either $A = L_{p,\alpha,\beta}(I)$ with $1 \leqq p < \infty$ or $A = C(\bar{I})$, the space of all complex-valued continuous functions on $\bar{I} = [-1, 1]$. Then $[H, A_{\alpha,\beta}, A] \in \mathfrak{R}^l$ for some non-negative integer l. Cf. [F, p. 155] for references.

Spherical Harmonics. Let ω_n be the unit sphere in R_n and let $H = L_2(\omega_n)$. Let $\mathring{A}$ be given by

$$(\mathring{A} f)(x) = -(\Delta f)(x) + f(x), \quad x \in \omega_n, \tag{8}$$

where Δ is now the Laplace-Beltrami operator on ω_n, defined on the set of all infinitely differentiable functions on ω_n. Then $\mathring{A}$ is an essentially self-adjoint positive-definite operator in H. Let A be its closure. Then A is a self-adjoint positive-definite operator in H with pure point spectrum. The different (!) eigenvalues are given by $k(k+n-2) + 1$ with $k = 0, 1, 2, \ldots$ For fixed k the corresponding eigenfunctions coincide with the spherical harmonics of degree k. Let A be either $L_p(\omega_n)$ with $1 \leqq p < \infty$ or the space of all complex-valued continuous functions on ω_n. Then $[H, A, A] \in \mathfrak{R}^l$ for some non-negative integer l. Cf. [F, p. 156] for references.

Besov Spaces. In the above four examples (Hermite, Laguerre, Jacobi series and spherical harmonics) the hypotheses of Definition 6.2.2/2 are satisfied. Hence one can introduce corresponding Besov spaces B_q^s, $s > 0$, $1 \leqq q \leqq \infty$. Furthermore the unterlying operators have a pure point spectrum. Consequently the theory from Section 6.4. about Fourier series can be applied. There arise several questions. First one would ask for a better and more explicit description of the corresponding Besov spaces B_q^s, maybe similar as in the periodic case which has been treated extensively in Chapter 3. One can ask whether the translation (semi-) group which played a crucial role in the theory of the spaces on R_n and T_n has appropriate counterparts (the above A-operators should be the corresponding infinitesimal generators). Something can be done in this direction. We refer to P. L. Butzer, R. L. Stens [1, 2], P. L. Butzer, R. L. Stens, M. Wehrens [1, 2], E. Görlich, C. Markett [1], C. Markett [1], T. Runst [1], T. Runst, W. Sickel [1], R. L. Stens, M. Wehrens [1] and the papers cited there. The dissertation by T. Runst [1] gives a rather systematic study of Besov-Sobolev and potential spaces with respect to Jacobi expansions on the basis of multiplier assertions in the sense of W. C. Connett, A. L. Schwartz [1] and G. Gasper, W. Trebels [2, 3].

Strong Summability. In 3.7. we studied the problem of strong summability for trigonometric series. Of course this problem makes also sense for the above orthogonal expansions. As far as the Jacobi series are concerned we refer to T. Runst, W. Sickel [1].

Multipliers. The multiplier problem has been treated in 6.3.3. As for the Besov spaces one has Proposition 6.3.3, which is much more better than the multiplier assertions from Proposition 6.2.1 for the basic space A, and hence for the corresponding potential spaces, cf. the beginning of 6.3.3. One can ask whether the multiplier assertions for the basic space A (and hence for the potential spaces) can be strengthened in the sense of Proposition 6.3.3. Furthermore vector-valued multiplier assertions would be of interest and (as a special case) a Littlewood-Paley theory would be desirable. Something can be done in this direction. Beside the literature mentioned in [F, p. 156] we refer to W. Trebels [1], W. C. Connett, A. L. Schwartz [1, 2], G. Gasper, W. Trebels [2, 3], T. Runst [1] and T. Runst, W. Sickel [1].

6.5.5. Bessel-Transform

In the preceding subsection we dealt with classical orthogonal Fourier expansions, where the corresponding Λ-operators are generated by differential operators of second order. In the case of the spaces on R_n we prefered the representation (6.5.1/2) for the underlying Λ-operator. An expansion via eigenelements in this case is simply not available because Λ is not an operator with pure point spectrum. The question arises whether other cases of interest can be treated in a similar way. In other words, is it possible to replace the Fourier transform and its inverse in (6.5.1/2) by other distinguished (integral) transforms? First candidates are integral transforms of Bessel-Hankel type. In this case the necessary multiplier assertions are known, cf. P. L. Butzer, R. J. Nessel, W. Trebels [1]. The corresponding theory has been developed by G. Altenburg [1, 2]. We give a somewhat rough description of his results.

Let $R_1^+ = (0, \infty)$, $1 < p < \infty$ and $\mu \geqq -\frac{1}{2}$. Then $L_{p,\mu}(R_1^+)$ is the collection of all complex-valued functions on R_1^+ with

$$\|f \mid L_{p,\mu}(R_1^+)\| = \left(\int_0^\infty |f(x)|^p\, x^{2\mu+1}\, dx\right)^{1/p} < \infty .$$

Equipped in the usual way with a scalar product, $L_{2,\mu}(R_1^+)$ becomes a Hilbert space. Let $J_\mu(x)$ with $x > 0$ and $\mu \geqq -\frac{1}{2}$ be the usual Bessel function. Then

$$(B_\mu f)(y) = \int_0^\infty \frac{J_\mu(xy)}{(xy)^\mu} f(x)\, x^{2\mu+1}\, dx, \quad y > 0,$$

denotes the Bessel transform (modified Hankel transform). Then B_μ is an isometric mapping from $L_{2,\mu}(R_1^+)$ onto itself and we have $B_\mu = B_\mu^{-1}$. Let $\mu \geqq -\frac{1}{2}$, $H = L_{2,\mu}(R_1^+)$,

$$\Lambda = B_\mu(1 + x^2)\, B_\mu, \tag{1}$$

and $A = L_{p,\mu}(R_1^+)$ with $1 < p < \infty$. Then $[H, \Lambda, A] \in \mathfrak{R}^l$ for sufficiently large natural numbers l in the sense of Definition 6.2.2/1. On the basis of this result, G. Altenburg [1, 2] studied Besov spaces in the sense of Definition 6.2.2/2 and potential spaces.

Remark 1. Because $B_\mu^{-1} = B_\mu$ the construction (1) is rather natural, cf. (6.5.1/2).

Remark 2. Proposition 6.3.3 describes multipliers for the spaces B_q^s of Michlin-Hörmander type. Again the question arises whether the spaces $L_{p,\mu}(R_1^+)$ and corresponding potential spaces have similar multiplier properties, cf. A. L. Brodskij [1], G. Gasper, W. Trebels [4] and the papers cited there. Furthermore there is a close connection between Hankel multipliers, multipliers for the other orthogonal expansions and radial Fourier multipliers.

References

Achieser, N. I.

1. On weighted approximation of continuous functions by polynomials on the whole real line. (Russian) Uspechi Mat. Nauk 11, 4 (1956), 3–43.

Agalarov, S. I.

1. A Littlewood-Paley theorem and Fourier multipliers in weighted spaces with mixed norms. (Russian) Trudy Mat. Inst. Steklov 140 (1976), 3–13.

Alexits, G.

1. Über die Approximation im starken Sinne. In: Proc. Conf. "Approximation Theory", Oberwolfach 1963, Basel: Birkhäuser Verlag 1964, 89–93.

Altenburg, G.

1. Bessel-Transformationen in Räumen von Grundfunktionen über dem Intervall $\Omega = (0, \infty)$ und deren Dualräume. Math. Nachr. 108 (1982), 197–218.
2. Eine Realisierung der Theorie der abstrakten Besov-Räume $B_q^s(A)$ ($s > 0, 1 \leqq q \leqq \infty$) und der Lebesgue-Räume $H_{p,\mu}^s$ auf der Grundlage Besselscher Differentialoperatoren. Z. Anal. Anwendungen 3 (1984), 43–63.

Amanov, T. I.

1. Representation and embedding theorems for the function spaces $S_{p,\theta}^{(r)}B(R_n)$ and $S_{p^*,\theta}^{(r)}B\,(0 \leqq x_j \leqq 2\pi, j = 1, \dots, n)$. (Russian) Trudy Mat. Inst. Steklov 77 (1965), 5–34.
2. The classes SB with mixed norm. Representation theorems and embeddings. (Russian) In: Matematika i Mechanika I, part. 1, Alma-Ata: 1966.
3. Spaces of Differentiable Functions with Dominating Mixed Derivatives. (Russian) Alma-Ata: Nauka Kaz. SSR 1976.

Andersen, K. F.; John, R. T.

1. Weighted inequalities for vector-valued maximal functions and singular integrals. Studia Math. 69 (1980), 19–31.

Babenko, K. I.

1. On the approximation of periodic functions of several variables by trigonometric polynomials. (Russian) Dokl. Akad. Nauk SSSR 132 (1960), 247–250.
2. Approximation by trigonometric polynomials in some classes of periodic functions of several variables. Dokl. Akad. Nauk SSSR 132 (1960), 982–985.

Bachvalov, N. S.

1. Embedding theorems for classes of functions with some bounded derivatives. (Russian) Vestnik Moskov. Univ. Ser. I Mat. Mech. 3 (1963), 7–16.

Bagby, R. J.

1. An extended inequality for the maximal function. Proc. Amer. Math. Soc. 48 (1975), 419–422.

Bari, N. K.

1. A Treatise on Trigonometric Series, I, II. Pergamon Press 1964 (Translation from Russian).

Benedek, A.; Calderón, A. P.; Panzone, R.

1. Convolution operators on Banach space valued functions. Proc. Nat. Acad. Sci. U.S.A. 48 (1962), 356–365.

Benedek, A.; Panzone, R.

1. The spaces L^p, with mixed norm. Duke Math. J. 28 (1961), 301–324.

Bergh, J.; Löfström, J.

1. Interpolation Spaces. An Introduction. Berlin, Heidelberg, New York: Springer-Verlag 1976.

Bernštejn, S. N.

1. On the absolute convergence of trigonometric series. (Russian) Coll. papers, Vol. I, 217–223.

Bertolo, J. I.; Fernandez, D. L.

1. On the connection between the real and the complex interpolation method for several Banach spaces. Rend. Sem. Mat. Univ. Padova 66 (1982), 193–209.

Besov, O. V.

1. Classes of functions with a generalized mixed Hölder condition. (Russian) Trudy Mat. Inst. Steklov 105 (1969), 21–29.
2. Weighted estimates of mixed derivatives in a domain. (Russian) Trudy Mat. Inst. Steklov 156 (1980), 16–21.

Besov, O. V.; Džabrailov, A. D.
1. Interpolation theorems for some spaces of differentiable functions. (Russian) Trudy Mat. Inst. Steklov 105 (1969), 15–20.

Besov, O. V.; Il'in, V. P.; Kudrjavzev, L. D.; Lizorkin, P. I.; Nikol'skij, S. M.
1. Embedding theorems for classes of differentiable functions of several variables. (Russian) In "Partial Differential Equations", Moskva: Nauka 1970, 38–63.

Besov, O. V.; Il'in, V. P.; Nikol'skij, S. M.
1. Integral Representations of Functions and Embedding Theorems. (Russian) Moskva: Nauka 1975 [English translation: Scripta Series in Math., Washington: Halsted Press; New York, Toronto, London: V. H. Winston & Sons 1978/79].

Beurling, A.
1. Quasi-analyticity and general distributions. Lectures 4 and 5, Stanford: A. M. S. Summer Institute 1961.

Björck, G.
1. Linear partial differential operators and generalized distributions. Ark. Mat. 6 (1966), 351–407.

Boas, R. P.
1. Entire Functions. New York: Academic Press, Inc. Publishers 1954.

Brodskij, A. L.
1. Smooth multipliers of the Fourier-Bessel transform. (Russian) Differencial'nye Uravnenija 16 (1980), 2181–2185.

Bryčkov, Ju. A.
1. On the theory of spaces with a dominating mixed derivative. (Russian) Mat. Zametki 31 (1982), 679–694.

Buchvalov, A. V.
1. The complex interpolation method in spaces of vector-valued functions and generalized Besov spaces. (Russian) Dokl. Akad. Nauk SSSR 260 (1981), 265–269.

Bugrov, Ja. S.
1. Approximation of a class of functions with a dominating mixed derivative. (Russian) Mat. Sb. 106 (1964), 410–418.
2. A representation theorem of a class of functions. (Russian) Dokl. Akad. Nauk SSSR 163 (1965), 799–800.
3. Embedding theorems for some classes of functions. (Russian) Trudy Mat. Inst. Steklov 77 (1965), 49–72.
4. A representation theorem of a class of functions. (Russian) Sibirsk. Mat. Ž. 7 (1966), 242–251.
5. Constructive characterization of classes of functions with a dominating mixed derivative. (Russian) Trudy Mat. Inst. Steklov 131 (1974), 25–32.

Bui Huy Qui
1. Some aspects of weighted and non-weighted Hardy spaces. Kôkyûroku Res. Inst. Math. Sci. 383 (1980), 38–56.
2. Weighted Hardy spaces. Math. Nachr. 103 (1981), 45–62.
3. Weighted Besov and Triebel spaces: Interpolation by the real method. Hiroshima Math. J. 12, 3 (1982), 581–605.
4. Characterizations of weighted Besov and Triebel-Lizorkin spaces via temperatures. J. Functional Analysis 55 (1984), 39–62.

Burenkov, V. I.
1. On resolutions of unity. (Russian) Trudy Mat. Inst. Steklov 150 (1979), 24–30.

Burenkov, V. I.; Gol'dman, M. L.
1. On norming of periodic analoga of normed function spaces. (Russian) Dokl. Akad. Nauk SSSR 264 (1982), 271–274.
2. On estimates of norms of operators in spaces of periodic and non-periodic functions. (Russian) Dokl. Akad. Nauk SSSR 267 (1982), 1289–1293.

Butzer, P. L.
1. The Shannon sampling theorem and some of its generalizations. An overview. In "Constructive Function Theory '81", Publishing House Bulgarian Acad. Sci., Sofia, 1983, 258–274.

Butzer, P. L.; Berens, H.
1. Semi-Groups of Operators and Approximation. Berlin, Heidelberg, New York: Springer-Verlag 1967.

Butzer, P. L.; Nessel, R. J.

1. Fourier Analysis and Approximation. New York, London: Academic Press 1971.

Butzer, P. L.; Nessel, R. J.; Trebels, W.

1. Multipliers with respect to spectral measures in Banach spaces and approximation. I: Radial multipliers in connection with Riesz-bounded spectral measures. J. Approximation Theory 8 (1973), 335–356.

Butzer, P. L.; Stens, R. L.

1. The operational properties of the Chebyshev transform. I. General properties. Funct. Approximatio Comment. Math. 5 (1977), 129–160. II. Fractional derivatives. In "Theory of Approximation of Functions", Proc. Intern. Conf. in Kaluga "Nauka", Moskva 1977, 49–61.
2. Chebyshev transform methods in the solution of the fundamental theorem of best algebraic approximation in the fractional case. In "Fourier Analysis and Approximation Theory", Coll. Math. Soc. János Bolyai 19, Amsterdam: North-Holland 1978, 191–212.

Butzer, P. L.; Stens, R. L.; Wehrens, M.

1. Approximation by algebraic convolution integrals. In "Approximation Theory and Functional Analysis". Amsterdam: North-Holland 1979, 71–120.
2. Higher order moduli of continuity based on the Jacobi translation operator and best approximation. C. R. Math. Rep. Acad. Sci. Canada II (1980), 2, 83–88.

Calderón, A. P.

1. Inequalities for the maximal function relative to a metric. Studia Math. 57 (1976), 297–306.

Calderón, A. P.; Torchinsky, A.

1. Parabolic maximal functions associated with a distribution. I, II. Advances in Math. 16 (1975), 1–64 and 24 (1977), 101–171.

Carl, B.

1. Entropy numbers, s-numbers, and eigenvalue problems. J. Functional Analysis 41 (1981), 290–306.

Carl, B.; Kühn, T.

1. Entropy and eigenvalues of certain integral operators. Preprint, Jena: 1982.

Christ, M.; Fefferman, R.

1. A note on weighted norm inequalities for the Hardy-Littlewood maximal operator. Proc. Amer. Math. Soc. 87 (1983), 447–448.

Coifman, R. R.; Fefferman, C.

1. Weighted norm inequalities for maximal functions and singular integrals. Studia Math. 51 (1974), 241–250.

Colzani, L.

1. Fourier transform of distributions in Hardy spaces. Boll. Un. Mat. Ital. (6) 1–A (1982), 403 – 410.

Connett, W. C.; Schwartz, A. L.

1. The theory of ultraspherical multipliers. Mem. Amer. Math. Soc. 9, 183 (1977).
2. The Littlewood-Paley theory for Jacobi expansions. Trans. Amer. Math. Soc. 251 (1979), 219–234.

Cwikel, M.

1. On $(L^{p_0}(A_0), L^{p_1}(A_1))_{\theta,q}$. Proc. Amer. Math. Soc. 44 (1974), 286–292.

Dunford, N.; Schwartz, J. T.

1. Linear Operators, I. New York, London: Interscience Publishers 1958.

Džabrailov, A. D.

1. On some function spaces—direct and inverse embedding theorems. (Russian) Dokl. Akad. Nauk SSSR 159 (1964), 254–257.
2. On the theory of "embedding theorems". (Russian) Trudy Mat. Inst. Steklov 89 (1967), 80–118.
3. Embedding theorems for spaces of functions whose derivatives satisfy a multiple integral Hölder condition. (Russian) Trudy Mat. Inst. Steklov 117 (1972), 113–138.
4. Families of spaces of functions whose mixed derivatives satisfy a multiple integral Hölder condition. (Russian) Trudy Mat. Inst. Steklov 117 (1972), 139–158.
5. Theorems on the extension of functions from the spaces $S_p^r W$, $S_{p,\theta}^r B$ beyond the boundary of the domain Ω. (Russian) Trudy Mat. Inst. Steklov 131 (1974), 81–93

Edwards, R. E.

1. Fourier Series. A Modern Introduction, I, II. sec. ed. New York, Heidelberg, Berlin: Springer-Verlag 1979, 1982.

2. Functional Analysis. New York, Chicago, San Francisco, Toronto, London: Rinehardt and Winston 1965.

Edwards, R. E.; Gaudry, G. I.

1. Littlewood-Paley and Multiplier Theory. Berlin, Heidelberg, New York: Springer-Verlag 1977.

Elstner, D.

1. Eigenwertverteilungen von Integraloperatoren. Dissertation A, Jena: 1984.

Fajn, B. L.

1. On the extension of functions of anisotropic Sobolev spaces. (Russian) Trudy Mat. Inst. Steklov 170 (1984), 248–272.

Favini, A.

1. Su una extensione del methodo d'interpolazione complesso. Rend. Sem. Mat. Univ. Padova 47 (1972), 243–298.

Fefferman, C.

1. The multiplier problem for the ball. Ann. of. Math. 94 (1971), 330–336.

Fefferman, C.; Stein, E. M.

1. Some maximal inequalities. Amer. J. Math. 93 (1971), 107–115.
2. H^p spaces of several variables. Acta Math. 129 (1972), 137–193.

Feichtinger, H. G.

1. On a new Segal algebra. Monatsh. Math. 92 (1981), 269–289.
2. Banach spaces of distributions of Wiener's type and interpolation. In: Proc. Intern. Conf. "Functional Analysis and Approximation", Oberwolfach, 1980. Basel, Boston, Stuttgart: ISNM 60 Birkhäuser Verlag 1981, 153–165.
3. Banach convolution algebras of Wiener type. In: Proc. Intern. Conf. "Functions, Series, Operators", Budapest 1980. Amsterdam, Oxford, New York: North Holland 1983, 509–524.
4. Modulation spaces on locally compact abelian groups. Preprint. Univ. Wien: 1983.
5. Strong almost periodicity and Wiener type spaces. In: Proc. Conf. "Constructive Function Theory", Varna 1981. Sofia: 1983, 321–327.

Feichtinger, H. G.; Gröbner, P.

1. Banach spaces of distributions defined by decomposition methods, I. Math. Nachr. 123 (1985), 97–120.

Fernandez, D. L.

1. Interpolation of 2^n Banach spaces. Studia Math. 65 (1979), 175–201.
2. Trace spaces in several variables. An. Acad. Brasil Ci. 53 (1981), 223–232.
3. An extension of the complex method of interpolation. Boll. Un. Mat. Ital. (5) 18-B (1981), 721–732.
4. On the duality of interpolation spaces of several Banach spaces. Acta Sci. Math. (Szeged) 44 (1982), 43–51.
5. Interpolation of Sobolev-Nikol'skij spaces. Sem. Brasil. Análise 18 (1983), 185–225.
6. Interpolation of 2^d Banach spaces and multiparametric approximation. J. Approximation Theory 38 (1983), 240–257.

Foias, C.; Lions, J. L.

1. Sur certains théoremes d'interpolation. Acta Sci. Math. (Szeged) 22 (1961), 269–282.

Franke, J.

1. Fourier-Multiplikatoren, Littlewood-Paley-Theoreme und Approximation durch ganze analytische Funktionen in gewichteten Funktionenräumen. Forschungsergebnisse Univ. Jena N/86/8, 1986.
2. On the spaces F^s_{pq} of Triebel-Lizorkin type: Pointwise multipliers and spaces on domains. Math. Nachr. 125 (1986), 29–68.
3. Elliptische Randwertprobleme in Besov-Triebel-Lizorkin-Räumen. Dissertation, Jena 1985.

Freud, G.

1. Orthogonale Polynome. Berlin: VEB Deutscher Verlag der Wissenschaften 1963.
2. Über die Sättigungsklasse der starken Approximation durch Teilsummen der Fourierschen Reihe. Acta Math. Acad. Sci. Hungar. 20 (1969), 275–281.

Garcia-Cuerva, J.

1. Weighted H^p spaces. Dissertationes Math. 162 (1979).

Gasper, G.; Trebels, W.

1. Multiplier criteria of Hörmander type for Fourier series and applications to Jacobi series and Hankel transforms. Math. Ann. 242 (1979), 225–240.

2. Multiplier criteria of Marcinkiewicz type for Jacobi expansions. Trans. Amer. Math. Soc. 231 (1977), 117–132.
3. A characterization of localized Bessel potential spaces and applications to Jacobi and Hankel multipliers. Studia Math. 65 (1979), 225–240.
4. Hankel multipliers and extensions to radial and quasi-radial Fourier multipliers. In "Recent Trends in Mathematics", Conf. Reinhardsbrunn 1982. Teubner-Texte Math. 50 Leipzig: Teubner 1983, 133–142.

Gel'fand, I. M.; Šilov (Schilow), G. E.
1. Verallgemeinerte Funktionen (Distributionen), II. Berlin: VEB Deutscher Verlag der Wissenschaften 1962. (Übersetzung aus dem Russischen.)

Görlich, E.; Markett, C.
1. Estimates for the norm of the Laguerre translation operator. Numer. Funct. Anal. and Optimiz. 1 (1979), 203–222.

Gohberg, I. Z.; Krejn, M. G.
1. Introduction to the Theory of Linear Non-Self-Adjoint Operators in Hilbert Space. (Russian) Moskva: Nauka 1965.

Gol'dman, M. L.
1. On the traces of functions of a generalized Liouville class. (Russian) Sibirsk. Mat. Ž. 17 (1976), 1236–1255.
2. Description of the traces of an anisotropic generalized Liouville class. (Russian) Dokl. Akad. Nauk SSSR 233 (1977), 513–520.
3. Description of the traces of some functional spaces. (Russian) Trudy Mat. Inst. Steklov 150 (1979), 99–127.
4. The covering method for the description of general spaces of Besov type. (Russian) Trudy Mat. Inst. Steklov 156 (1980), 47–81.

Golovkin, K. K.
1. On equivalent norms in fractional spaces. (Russian) Trudy Mat. Inst. Steklov 66 (1962), 364–383.
2. Geometric characterizations of the smoothness of functions of two variables. (Russian) Trudy Mat. Inst. Steklov 73 (1964), 139–156.
3. Parameter-normed spaces and normed classes. (Russian) Trudy Mat. Inst. Steklov 106 (1969), 3–134.

Grisvard, P.
1. Commutativité de deux foncteurs d'interpolations. J. Math. Pures Appl. 45 (1966), 143–290.

Guzmán, M. de
1. Real Variable Methods in Fourier Analysis. Amsterdam, New York, Oxford: North-Holland Publ. Comp. 1981.

Hardy, G. H.; Littlewood, J. E.; Pólya, G.
1. Inequalities. Cambridge: Univ. Press (reprinted sec. ed.) 1964.

Heinig, H. P.
1. Weighted maximal inequalities for l^r-valued functions. Canad. Math. Bull. 19 (1976), 445–453.

Heinig, H. P.; Johnson, R.
1. Weighted norm inequalities for L^r-valued integral operators and applications. Math. Nachr. 107 (1982), 161–174.

Hörmander, L.
1. Estimates for translation invariant operators in L_p spaces. Acta Math. 104 (1960), 93–139.
2. The Analysis of Linear Partial Differential Operators, I, II. Berlin, Heidelberg, New York, Tokyo: Springer-Verlag 1983.

Hunt, R.; Muckenhoupt, B.; Wheeden, R.
1. Weighted norm inequalities for the conjugate function and Hilbert transform. Trans. Amer. Math. Soc. 176 (1973), 227–251.

Janson, S.
1. Generalizations of Lipschitz spaces and an application to Hardy spaces and bounded mean oscillation. Duke Math. J. 47 (1980), 959–982.

Jawerth, B.
1. Some observations on Besov and Lizorkin-Triebel spaces. Math. Scand. 40 (1977), 94–104.
2. Weighted norm inequalities for functions of exponential type. Ark. Mat. 15 (1977), 223–228.

Kaljabin, G. A.
1. Embedding theorems for generalized Besov and Liouville spaces. (Russian) Dokl. Akad. Nauk SSSR 232 (1977), 1245–1248.
2. Characterizations of spaces of Besov-Lizorkin-Triebel type. (Russian) Dokl. Akad. Nauk SSSR 236 (1977), 1056–1059.
3. Description of the traces of anisotropic spaces of Triebel-Lizorkin type. (Russian) Trudy Mat. Inst. Steklov 150 (1979), 160–173.
4. The description of functions of classes of Besov-Lizorkin-Triebel type. (Russian) Trudy Mat. Inst. Steklov 156 (1980), 82–109.

Karadžov, G. E.
1. The trigonometric multiplier problem. In "Constructive Function Theory '81", Sofia: 1983, 82–86.

Karapetian, G. A.
1. Solution of semi-elliptic equations in the half plane. (Russian) Trudy Mat. Inst. Steklov 170 (1984), 119–138.

Kerman, R. A.; Torchinsky, A.
1. Integral inequalities with weights for the Hardy maximal function. Studia Math. 71, 3 (1981/82), 277–284.

Kokilašvili, V. M.
1. On Fourier multipliers. (Russian) Dokl. Akad. Nauk SSSR 220 (1975), 19–22.
2. On singular integrals and Fourier multipliers. (Russian) Akad. Nauk Gruzin. SSR, Trudy Tbiliss. Mat. Inst. 53 (1976), 38–61.
3. Maximal inequalities and multipliers in weighted Lizorkin-Triebel spaces. (Russian) Dokl. Akad. Nauk SSSR 239 (1978), 42–45.
4. Anisotropic Bessel potentials. Embedding theorems for limiting indices. (Russian) Akad. Nauk Gruzin. SSR, Trudy Tbiliss. Mat. Inst. 58 (1978), 134–149.
5. On multiple Hilbert transforms and multipliers in weighted spaces. (Russian) Soobšč. Akad. Nauk Gruzin. SSR 98 (1980), 285–288.
6. On traces for weighted anisotropic Lizorkin-Triebel spaces. Proc. Intern. Conf. "Approximation and Function Spaces", Gdańsk, 1979. Amsterdam: North-Holland 1981, 338–344.

Kokilašvili, V. M.; Rákosnik, J.
1. Weighted inequalities for vector-valued anisotropic maximal functions. Z. Analysis Anwendungen 4 (1985), 503–511.

Komatsu, H.
1. Ultradistributions: I. Structure theorems and a characterization. II: The kernel theorem and ultradistributions with a support on a manifold. III: Vectorvalued ultradistributions and the theory of kernels. J. Fac. Sci. Univ. Tokyo Sect. IA Math. 20 (1973), 25–105; 24 (1977), 607–628; 29 (1982), 653–718.
2. Fractional powers of operators. Pacific J. Math. 19 (1966), 285–346.

Krée, P.
1. Propriétés de continuité dans L^p de certains noyaux. Boll. Un. Mat. Ital. (3) 22 (1967), 330–344.
2. Sur les multiplicateurs dans FL^p. Ann. Inst. Fourier (Grenoble) 16, 2 (1966), 31–89.

Krotov, V. G.
1. Strong approximation by Fourier series and differentiability properties of functions. Anal. Math. 4 (1978), 199–214.

Krotov, V. G.; Leindler, L.
1. On the strong summability of Fourier series and the classes H^ω. Acta Sci. Math. (Szeged) 40 (1978), 93–98.

Kufner, A.; John, O.; Fučik, S.
1. Function Spaces. Prague: Academia 1977.

Kurtz, D.
1. Littlewood-Paley and multiplier theorems on weighted L^p spaces. Trans. Amer. Math. Soc. 259 (1980), 235–254.

Kurtz, D.; Wheeden, R. L.
1. Results on weighted norm inequalities for multipliers. Trans. Amer. Math. Soc. 255 (1979), 343–362.

2. Approximation of Functions of Several Variables and Imbedding Theorems. Sec. ed. (Russian), Moskva: Nauka 1977. (English translation of the first edition: Berlin, Heidelberg, New York: Springer-Verlag 1975.)
3. On boundary properties of differentiable functions of several variables. (Russian) Dokl. Akad. Nauk SSSR 146 (1962), 542–545.
4. On stable boundary values of differentiable functions of several variables. (Russian) Mat. Sb. 61 (1963), 224–252.
5. Functions with a dominating mixed derivative satisfying a multiple Hölder condition. (Russian) Sibirsk. Mat. Ž. 6 (1963), 1342–1364.

Orlovskij, D. G.
1. On multipliers in the spaces $B^r_{p,\theta}$. Anal. Math. 5 (1979), 207–218.

Oskolkov, K. I.
1. On strong summability of Fourier series and differentiability of functions. Anal. Math. 2 (1976), 41–47.

Oswald, P.
1. Some inequalities for trigonometric polynomials in the metric L_p, $0 < p < 1$. (Russian) Izv. Vysš. Učebn. Zaved. Matematika 7 (1976), 65–75.
2. On Besov-Hardy-Sobolev spaces of analytic functions in the unit disc. Czechoslovak Math. J. 33 (108) (1983), 408–426.
3. On coefficient properties of power series. In: Proc. Intern. Conf. "Constructive Function Theory '81", Sofia: 1983, 468–474.

Päivärinta, L.
1. On the spaces $L^A_p(l_q)$: Maximal inequalities and complex interpolation. Ann. Acad. Sci. Fenn. Ser. AI Math. Dissertationes 25, Helsinki: 1980.

Peetre, J.
1. Applications de la théorie des espaces d'interpolation dans l'analyse harmonique. Ricerche Mat. 15 (1966), 1–34.
2. On the structure of Sobolev and Besov spaces. Unpublished manuscript, Lund: 1970.
3. On spaces of Triebel-Lizorkin type. Ark. Mat. 13 (1975), 123–130.
4. New Thoughts on Besov Spaces. Duke Univ. Math. Series I. Durham: Duke Univ. 1976.
5. Sur la transformation de Fourier des fonctions à valeurs vectorielles. Rend. Sem. Mat. Univ. Padova 42 (1969), 15–26.

Perepelkin, V. G.
1. Integral representations of functions of weighted Sobolev classes in domains and some applications. I, II. (Russian) Sibirsk. Mat. Ž. 17 (1976), 119–140, 318–330.

Pietsch, A.
1. Operator Ideals. Berlin: VEB Deutscher Verlag der Wissenschaften, 1978.
2. Eigenvalues of integral operators. (Russian) Dokl. Akad. Nauk SSSR 247 (1979), 1324–1327.
3. Eigenvalues of integral operators. I, II. Math. Ann. 247 (1980), 169–178 and 262 (1983), 343–376.
4. Approximation spaces. J. Approximation Theory 32 (1981), 115–134.

Plancherel, M.; Polya, G.
1. Fonctions entières et intégrales de Fourier multiples. Math. Helv. 9 (1937), 224–248.

Potapov, M. K.
1. On "angular" approximation. (Russian) In: Proc. Intern. Conf. "Constructive Theory of Functions", Budapest 1969. Budapest: Publishing House Acad. Sci. Hungar. 1971, 371–399.
2. Angular approximation and embedding theorems. (Russian) Math. Balkanica 2 (1972), 183–198.
3. Investigation of some classes of functions by means of "angular" approximation. (Russian) Trudy Mat. Inst. Steklov 117 (1972), 256–300.
4. Embedding of classes of functions with a dominating mixed modulus of smoothness. (Russian) Trudy Mat. Inst. Steklov 131 (1974), 199–210.
5. Embedding theorems in the mixed metric. (Russian) Trudy Mat. Inst. Steklov 156 (1980), 143–156.

Ramazanov, M. D.
1. Theorems on traces and extensions of functions on surfaces. (Russian) Dokl. Akad. Nauk SSSR 190, 4 (1970), 784–787.

Roumieu, C.
1. Sur quelques extensions de la notion de distribution. Ann. Sci. École Norm. Sup., Sér. 3, 77 (1960), 41–121.
2. Ultra-distributions définies sur R^n et sur certaines classes de variétés differentiables. J. Analyse Math. 10 (1962/63), 153–192.

Rubio de Francia, J. L.
1. Boundedness of maximal functions and singular integrals in weighted L^p spaces. Proc. Amer. Math. Soc. 83 (1981), 673–679.
2. Weighted norm inequalities and vector valued inequalities. In: "Harmonic Analysis", Proc. Conf. in Minneapolis, Lecture Notes Math. 908. Berlin, Heidelberg, New York: Springer-Verlag 1982, 86–101.

Runst, T.
1. Räume vom Typ $B^s_{p,q,\alpha,\beta}$, $H^s_{p,\alpha,\beta}$ und $W^s_{p,\alpha,\beta}$ bezüglich Jacobi-Fourier-Entwicklungen. Dissertation A, Jena: 1980.

Runst, T.; Sickel, W.
1. On strong summability of Jacobi-Fourier expansions and smoothness properties of functions. Math. Nachr. 99 (1980), 77–85.

Sadykova, S. B.
1. On traces of functions of the Besov class $B_p^{l_1,\ldots,l_n}$ on smooth surfaces. (Russian) In: "Embedding Theorems and their Applications". Alma-Ata: Nauka 1976, 131–135.

Sawyer, E. T.
1. Two weight norm inequalities for certain maximal and integral operators. In: "Harmonic Analysis", Proc. Conf. in Minneapolis, Lecture Notes Math. 908. Berlin, Heidelberg, New York: Springer-Verlag 1982, 102–127.

Schmeisser, H.-J.
1. On spaces of functions and distributions with mixed smoothness properties of Besov-Triebel-Lizorkin type. I (Basic properties), II (Fourier multipliers and approximation representations). Math. Nachr. 98 (1980), 233–250 and 106 (1982), 187–200.
2. Über Räume von Funktionen und Distributionen mit dominierenden gemischten Glattheitseigenschaften vom Besov-Triebel-Lizorkin-Typ. Dissertation B, Jena: 1981.
3. Imbedding theorems for spaces of functions with dominating mixed smoothness properties of Besov-Triebel-Lizorkin type. Wiss. Z. Friedrich-Schiller-Univ. Jena, Math.-Naturw. Reihe 31 (1982), 635–645.
4. Remarks on spaces of functions with dominating mixed smoothness properties in the sense of S. M. Nikol'skij and T. I. Amanov. Czechoslovak Math. J.
5. Maximal inequalities and Fourier multipliers for spaces with mixed quasinorms. Applications. Z. Analysis Anwendungen 3 (1984), 153–166.
6. On spaces of functions and distributions with dominating mixed smoothness properties of Besov-Triebel-Lizorkin type. In: "Constructive Function Theory '81", Publishing House Bulgarian Acad. Sci., Sofia, 1983, 507–512.
7. Anisotropic spaces. II (Equivalent norms for abstract spaces, function spaces with weights of Sobolev-Besov type). Math. Nachr. 79 (1977), 55–73.

Schmeisser, H.-J.; Sickel, W.
1. On strong summability of multiple Fourier series and smoothness properties of functions. Anal. Math. 8 (1982), 57–70.
2. On strong summability of multiple Fourier series and approximation of periodic functions. Math. Nachr.

Schmeisser, H.-J.; Triebel, H.
1. Anisotropic spaces. I (Interpolation of abstract spaces and function spaces). Math. Nachr. 73 (1976), 107–123.

Schwartz, J. T.
1. A remark on inequalities of Calderón-Zygmund type. Comm. Pure Appl. Math. 14 (1961), 758–799.

Schwartz, L.
1. Théorie des distributions. Paris: Hermann 1973.
2. Espaces de fonctions différentiables à valeurs vectorielles. J. Analyse Math. 4 (1954–1955), 88–148.

3. Theorie des distributions à valeurs vectorielles. Ann. Inst. Fourier (Grenoble) 7 (1957), 1–141 and 8 (1958), 1–209.

Shannon, C. E.

1. Communication in the presence of noise. Proc. IRE 37 (1949), 10–21.

Sickel, W.

1. Periodic spaces and relations to strong summability of multiple Fourier series. Math. Nachr. 124 (1985), 15–44.
2. Periodische Räume: Maximalungleichungen, Fouriersche Multiplikatoren und Beziehungen zur starken Summierbarkeit mehrdimensionaler trigonometrischer Reihen. Dissertation A, Jena: 1984.

Šmirev, G. A.

1. On traces of functions of the class $B^l_{p,\theta}$ on smooth surfaces. (Russian) In: "Theory of cubature formulas and applications of functional analysis on problems of mathematical physics", Trudy Sem. Sobolev 1 (1978), 136–164.

Sparr, G.

1. Interpolation of several Banach spaces. Ann. Mat. Pura Appl. 99 (1974), 247–316.

Stečkin, S. B.

1. On the absolute convergence of orthogonal series. (Russian) Mat. Sb. 29 (1951), 225–232.

Stein, E. M.

1. Singular Integrals and Differentiability Properties of Functions. Princeton: Princeton Univ. Press 1970.

Stein, E. M.; Weiss, G.

1. Introduction to Fourier Analysis on Euclidean Spaces. Princeton: Princeton Univ. Press 1971.

Stens, R. L.; Wehrens, M.

1. Legendre transform methods and best algebraic approximation. Comment. Math. Prace Mat. 21 (1979), 351–380.

Stöckert, B.

1. Ungleichungen vom Plancherel-Polya-Nikol'skij-Typ in gewichteten L^Ω_p-Räumen mit gemischten Normen. Math. Nachr. 86 (1978), 19–32.
2. Multiplikatoren in gewichteten anisotropen L^Ω_p-Räumen ganzer analytischer Funktionen exponentiellen Typs. Math. Nachr. 86 (1978), 33–40.

Stöckert, B.; Triebel, H.

1. Decomposition methods for function spaces of $B^s_{p,q}$ type and $F^s_{p,q}$ type. Math. Nachr. 89 (1979), 247–267.

Storoženko, E. A.

1. Approximation of functions of class H^p, $0 < p < 1$. (Russian) Mat. Sb. 105 (1978), 601–621.
2. On theorems of Jackson type for H^p, $0 < p < 1$. (Russian) Izv. Akad. Nauk SSSR Ser. Mat. 44 (1980), 948–962.

Storoženko, E. A.; Krotov, V. G.; Oswald, P.

1. Direct and converse theorems of Jackson type in L^p spaces, $0 < p < 1$. (Russian) Mat. Sb. 98 (140) (1975), 395–415.

Storoženko, E. A.; Oswald, P.

1. Moduli of smoothness and best approximation in the spaces L^p, $0 < p < 1$. Anal. Math. 3 (1977), 141–150.
2. Jackson theorems in the spaces $L^p(R^k)$, $0 < p < 1$. Sibirsk. Mat. Ž. 19 (1978), 888–901.

Strömberg, J.-O.; Torchinsky, A.

1. Weighted Hardy spaces. Preprint

Strömberg, J.-O.; Wheeden, R. L.

1. Relations between H^p_u and L^p_u with polynomial weights. Trans. Amer. Math. Soc. 270 (1982), 439–467.

Szabados, J.

1. On a problem of Leindler concerning strong approximation by Fourier series. Anal. Math. 2 (1976), 155–161.

Szász, O.

1. Über die Konvergenzexponenten der Fourierschen Reihen. Münch. Sitzungsber. (1922), 135–150.

2. On the absolute convergence of trigonometric series. Ann. of Math. 47 (1946), 213–220.

Tel'jakovskij, S. A.

1. On estimates of derivatives of trigonometric polynomials of several variables. (Russian) Sibirsk. Mat. Ž. 4 (1963), 1404–1411.
2. Some estimates for trigonometric series with quasi-convex coefficients. (Russian) Mat. Sb. 63 (105) (1964), 426–444.

Temljakov, V. N.

1. Approximation of periodic functions of several variables with a bounded mixed derivative. (Russian) Trudy Mat. Inst. Steklov 156 (1980), 233–260.

Tran Duc Long; Triebel, H.

1. Equivalent norms and Schauder bases in anisotropic Besov spaces. Proc. Roy. Soc. Edinburgh Sect. A 84 (1979), 177–183.

Trebels, W.

1. Multipliers for (C, α)-Bounded Fourier Expansions in Banach Spaces and Approximation Theory. Lecture Notes in Math. 329. Berlin, Heidelberg, New York: Springer-Verlag 1973.

Triebel, H.

F. Fourier Analysis and Function Spaces. Teubner Texte Math. 7. Leipzig: Teubner 1977.

I. Interpolation Theory, Function Spaces, Differential Operators. Amsterdam, New York, Oxford: North-Holland Publ. Comp. 1978.

S. Spaces of Besov-Hardy-Sobolev Type. Teubner Texte Math. 15. Leipzig: Teubner 1978.

T. Theory of Function Spaces. Leipzig: Akad. Verlagsgesellschaft Geest & Portig K.-G., and Basel, Boston, Stuttgart: Birkhäuser Verlag 1983.

1. Über die Existenz von Schauderbasen in Sobolev-Besov-Räumen. Isomorphiebeziehungen. Studia Math. 46 (1973), 83–100.
2. Multipliers in Besov spaces and in L_p^Ω-spaces. (The cases $0 < p \leqq 1$ and $p = \infty$.) Math. Nachr. 75 (1976), 229–245.
3. Spaces of distributions with weights. Multipliers in L_p-spaces with weights. Math. Nachr. 78 (1977), 339–355.
4. General function spaces. II. J. Approximation Theory 19 (1977), 154–175, III. Anal. Math. 3 (1977), 221–249, IV. Anal. Math. 3 (1977), 299–315, V. Math. Nachr. 87 (1979), 129–152.
5. A note on quasi-normed convolution algebras of entire analytic functions of exponential type. J. Approximation Theory 22 (1978), 368–373.
6. Maximal inequalities, Fourier multipliers, Littlewood-Paley theorems, and approximation by entire analytic functions in weighted spaces. Anal. Math. 6 (1980), 87–103.
7. Theorems of Littlewood-Paley type for *BMO* and for anisotropic Hardy spaces. In: Proc. Intern. Conf. "Constructive Function Theory 77", Sofia: 1980, 525–532.
8. A remark on integral operators from the standpoint of Fourier analysis. In: Proc. Intern. Conf. "Approximation and Function Spaces", Gdańsk 1979. Amsterdam: North-Holland 1981, 821–829.
9. Characterizations of Besov-Hardy-Sobolev spaces via harmonic functions, temperatures and related means. J. Approximation Theory 35 (1982), 275–297.
10. On $L_p(l_q)$-spaces of entire analytic functions of exponential type: Complex interpolation and Fourier multipliers. The case $0 < p < \infty$, $0 < q < \infty$, J. Approximation Theory 28 (1980), 317–328.
11. Complex interpolation and Fourier multipliers for the spaces $B^s_{p,q}$ and $F^s_{p,q}$ of Besov-Hardy-Sobolev type: The case $0 < p \leqq \infty$, $0 < q \leqq \infty$. Math. Z. 176 (1981), 495–510.
12. Anisotropic function spaces. I (Hardy's inequality, decompositions), II (Traces). Anal. Math. 10 (1984), 53–77, 79–96.
13. A priori estimates and boundary value problems for semi-elliptic differential equations: A model case. Comm. Partial Differential Equations 8 (1983), 1621–1664.
14. Modulation spaces on the Euclidean n-space. Z. Anal. Anwendungen 2 (1983), 443–457.
15. Weighted anisotropic Sobolev spaces on domains: Extensions and traces. Math. Nachr. 119 (1984), 309–319.
16. Remarks on multiple Fourier series. Anal. Math. 2 (1976), 305–317.
17. Periodic spaces of Besov-Hardy-Sobolev type and related maximal inequalities for trigonometrical polynomials. In: "Functions, Series, Operators, Budapest 1980". Colloquia Math. Soc. Janos Bolyai 35, II. Amsterdam, New York: North-Holland 1983, 1201–1209.